Gerhard Pfaff (Ed.)

Encyclopedia of Color, Dyes, Pigments

Also of Interest

Encyclopedia of Color, Dyes, Pigments.
Volume 1: Antraquinonoid Pigments - Color Fundamentals
Gerhard Pfaff (Ed.), 2021
ISBN 978-3-11-058588-9, e-ISBN (PDF) 978-3-11-058807-1,
e-ISBN (EPUB) 978-3-11-058663-3

Encyclopedia of Color, Dyes, Pigments.
Volume 3: Mixed Metal Oxide Pigments - Zinc Sulfide Pigments
Gerhard Pfaff (Ed.), 2021
ISBN 978-3-11-058686-2, e-ISBN (PDF) 978-3-11-058712-8,
e-ISBN (EPUB) 978-3-11-058688-6

Inorganic Pigments
Gerhard Pfaff, to be published in 2024
ISBN 978-3-11-074391-3, e-ISBN (PDF) 978-3-11-074392-0,
e-ISBN (EPUB) 978-3-11-074399-9

Applied Inorganic Chemistry.
Volume 1: From Construction to Photofunctional Materials
Rainer Pöttgen, Thomas Jüstel, Cristian A. Strassert (Eds.),
to be published in 2022
ISBN 978-3-11-073814-8, e-ISBN (PDF) 978-3-11-073314-3,
e-ISBN (EPUB) 978-3-11-073332-7

Applied Inorganic Chemistry.
Volume 2: From Magnetic to Bioactive Materials
Rainer Pöttgen, Thomas Jüstel, Cristian A. Strassert (Eds.),
to be published in 2022
ISBN 978-3-11-073837-7, e-ISBN (PDF) 978-3-11-073347-1,
e-ISBN (EPUB) 978-3-11-073357-0

Encyclopedia of Color, Dyes, Pigments

Volume 2: Color Measurement – Metal Effect Pigments

Edited by
Gerhard Pfaff

DE GRUYTER

Editor
Prof. Dr. Gerhard Pfaff
Allee der Kosmonauten 69
12681 Berlin
Germany
pfaff.pigmente@gmx.de

ISBN 978-3-11-058684-8
e-ISBN (PDF) 978-3-11-058710-4
e-ISBN (EPUB) 978-3-11-058694-7

Library of Congress Number: 2021947998

Bibliographic information published by the Deutsche Nationalbibliothek
The Deutsche Nationalbibliothek lists this publication in the Deutsche Nationalbibliografie; detailed bibliographic data are available on the Internet at http://dnb.dnb.de.

Cover image: Gettyimages/Sagittarius Pro
Typesetting: Integra Software Services Pvt. Ltd.
Printing and binding: CPI books GmbH, Leck

www.degruyter.com

FSC www.fsc.org MIX Papier aus verantwortungsvollen Quellen FSC® C083411

Preface

The three-volume "Encyclopedia of Color, Dyes, Pigments" represents this attempt to summarize the current expertise in the fields stated in the title of the book. The main objective is to present the information that is available today encompassed by these three terms in a scientifically and technically correct, high-level and up-to-date manner. All aspects from theory to practical application are covered.

As the title suggests, all major classes of dyes and pigments are covered in detail. Emphasis is given so that the reader obtains an overview of the basic principles, the synthesis possibilities, the production, the chemical and physical properties as well as the technical application of these colorants. Separate chapters are provided for the main areas of application of dyes and pigments, for color fundamentals and color measurement as well as for historical pigments, dyes and binders. The encyclopedia is addressed to color specialists in industry and academia as well as to dye and pigment users in the applications areas of coatings, paints, cosmetics, plastics, printing inks, ceramics, and building materials. In order to make orientation as easy as possible for the interested reader, the topics covered in the book are arranged in alphabetic order.

Color has always played an important role in the lives of humans and animals. Dyes and pigments were therefore important substances early on in the development of mankind for expressing oneself through color, white or black and for shaping life. Colors and their functions have always been fascinating, playing a major role in the human psyche, and are of great importance in the design of a wide variety of surfaces. While natural colorants were initially used by people for thousands of years, the development of modern natural sciences, especially chemistry, led to the introduction of synthetically produced dyes and pigments in the lives of people. These play a predominant role among today's colorants.

The treatment of the individual topics differs in many cases from that in other reference books. New ways of presentation are chosen for different classes of dyes and pigments, but also for the application of colorants in various systems. The encyclopedia is thus up to date, especially since the latest findings from the field have been included. The authors involved are very familiar with the contents of their chapters, most of them having researched and worked on them for many years.

The objective of the Encyclopedia of Color, Dyes and Pigments is to provide a comprehensive overview of the state of knowledge about dyes and pigments, while at the same time identifying the connections to the relevant coloristic and application fundamentals. Special attention was paid to developing a clear structure of the approximately 70 chapters to provide access to the desired information quickly and without a long search. Figures and tables inserted illustrate fundamental aspects, dye and pigment structures, manufacturing details as well as application examples. At the end of each chapter, references to further reading are given. These include a wide range of journal articles, reference books and patents.

https://doi.org/10.1515/9783110587104-202

The chapters to follow are written by some of the most knowledgeable authors in the subject areas covered in the encyclopedia. Most of them were and are involved in innovative developments and practical applications of colorants. They are authors of a variety of publications, patents, presentations and lectures in the respective areas. Their insights will certainly prove to be valuable to the reader and their contributions to the Encyclopedia of Color, Dyes, Pigments are greatly appreciated.

At this point, I would like to take the opportunity to express my sincere thanks to all the authors involved in this book project. Without their support and cooperation, the completion of the encyclopedia would not have been possible. Finally, a big thank you goes to Karin Sora from de Gruyter, who has supported and accompanied the project from the early beginning and to Vivien Schubert and Esther Markus from the same publisher, who have done an extraordinarily good job in the completion of the encyclopedia.

Berlin, December 2021 Gerhard Pfaff

Contents

Volume 1

Volume 2

Volume 3

List of contributors

Adrian Abel
DCC Europe
Rossendale
Lancashire
United Kingdom
adrianfixby@gmail.com

Michael R. Bartelt
Business Unit Effect Pigments
Schlenk Metallic Pigments GmbH
Barnsdorfer
Hauptstr. 5, Roth
91154 Germany
michael.bartelt@schlenk.de

Robert Christie
School of Textiles & Design
Heriot-Watt University
Scottish Borders Campus
Galashiels TD1 3HF, United Kingdom
rmchristie66@gmail.com

Magdalene Gärtner
Maltechnik
Institut für Maltechnik
Muenstergasse
3, Schwaebisch Gmuend
73525 Germany
magdalene.gaertner@yahoo.de

Uwe Hempelmann
Technical Operations, Lanxess Deutschland GmbH
Rheinuferstraße 7-9, Krefeld
Nordrhein-Westfalen
47812 Germany
uwe.hempelmann@-lanxess.com

Heinz Langhals
Department of Chemistry
Ludwig-Maximilians-Universität
Butenandtstr. 13,
Munich D-81377
Germany
langhals@lrz.uni-muenchen.de

Ghita Lanzendörfer-Yu
Owner
DejaYu Kosmetikblog
Fuchsgrube 36
Mülheim
45478 Germany
info@dejayu.de

Frank Maile
Business Unit Effect Pigments
Schlenk Metallic Pigments GmbH
Barnsdorfer
Hauptstr. 5, Roth
91154 Germany
frank.maile@schlenk.de

Heinz Mustroph
Former FEW Chemicals GmbH
Technikumstraße 1,
Bitterfeld-Wolfen
06756 Germany
mustroph@few.de

Gerhard Pfaff
Former Merck KGaA
Allee der Kosmonauten 69, Berlin
12681 Germany
pfaff.pigmente@gmx.de

https://doi.org/10.1515/9783110587104-204

Thomas Rathschlag
Weilburger Graphics GmbH
Am Rosenbühl 5,
Gerhardshofen 91466
Germany
t.rathschlag@weilburger-graphics.de

Werner Rudolf Cramer
Optical research
Cramer Lackdesign
Hafenweg
22, Muenster 48155
Germany
info@wrcramer.de

Andrew Town
Lambson Ltd., Clifford House
York Road, Wetherby
United Kingdom of Great Britain and Northern
Ireland
andrew.towns@lambson.com

Werner Rudolf Cramer

23 Color measurement

Abstract: Our color perception is characterized by subjective influences. Color measurement enables an objective description of colors. In this process, white light is sent onto a color sample and the reflected rays are measured as a percentage compared to a white standard. The physiological standard color values are first calculated from the physical measured values by standardized conversions. From these, the L*a*b* values can be determined as they are given by the CIE (Commission Internationale de l'Éclairage). While one geometry is sufficient for color pigments to describe them, several defined geometries are required for aluminum and interference pigments. These geometries (illumination and observation angles) are specified for the different measuring instruments.

Keywords: color, three-color theory, four-color theory (opponent colors), color measurement, measurement geometries, calibration, color pigments, aluminum pigments, interference pigments

23.1 Introduction

Colors had and have something fascinating about them. They have always impressed people, although they are natural to us when we open our eyes. It is also natural to talk about colors. However, there is a catch in this matter: colors are sensations, which is why color names are also characterized by sensations. For example, the same blue would be called "steel blue" by some people and "intense blue" by others. For this reason, an objective sensation is hardly possible. Especially in today's world with many automotive and industrial applications of colors and pigments a clear color language is required.

The qualitative and quantitative description of colors interested many chemists and physicists as well as artists. The scientific debate led to the present-day representation and description of colors and their measurements. Two results in particular played a major role: Firstly, Johannes von Kries developed the zone theory, which combined the three-color theory with the four-color theory. On the other hand, the CIE (Commission Internationale de l'Éclairage) – formerly also IKB (International Commission on Illumination) – described the linking of physical measurements with physiological color sensations [1].

This article has previously been published in the journal Physical Sciences Reviews. Please cite as: W. R. Cramer, Color Measurement *Physical Sciences Reviews* [Online] 2021, 6. DOI: 10.1515/psr-2020-0165

https://doi.org/10.1515/9783110587104-023

Some important developmental steps and their originators are shown below. They are intended to show that, parallel to modern natural sciences, many experiments were necessary until today's knowledge about color:

Francesco Maria Grimaldi (1618–1663) laid the foundation for the modern approach to light. He was able to create different colors with the help of a prism and described this process as diffraction. In his publication "De lumine" he presented light as a wave. He also investigated the process of interference.

Isaac Newton (1643–1727) used a prism to split white sunlight into individual colors. By combining the two final colors, he had created a color circle. He was able to reassemble the spectral colors into white light. He published this theory in 1704 in his book *Opticks or a Treatise of the Reflections, Refractions, Inflections and Colors of Light*. Isaac Newton held the view that light consists of smallest particles or corpuscles (corpuscle theory). He assumed that light rays are refracted by gravity.

Christiaan Huygens (1629–1695) worked on the wave theory of light. This theory defines light as an electromagnetic wave. Huygens recognized that optical phenomena such as interference and diffraction could not be explained by Newton's corpuscle theory. For him, light propagated as waves similar to water waves. Monochromatic light has only one defined wavelength. With white light many waves with different wavelengths overlapped.

Joseph von Fraunhofer (1787–1826) experimented with optical gratings to observe the diffraction of light. These experiments extended the wave theory proposed by Huygens.

Claude Boutet presented his color circle in 1708, in which he arranged the primary and secondary colors in a circle (Figure 23.1).

Thomas Young (1773–1829) laid the foundation for the three-color theory. This was further developed by Hermann von Helmholtz (1821–1894). The theory is based on the assumption that any color can be mixed from three primary colors (light). Helmholtz assumed that the eye also has three receptors that are sensitive to different spectral ranges. Young was also interested in measuring the wavelengths of light. With the help of a double-slit experiment he showed an interference pattern which was created by diffraction at the double-slit.

Hermann von Helmholtz (1821–1894) describes in his "Handbuch der physiologischen Optik" the difference between the spectral colors observed by Newton and the pigment colors. The former mix additively, the latter subtractively. His three-color theory allowed him to construct various color triangles, but these lacked the physiological components.

Michel Eugène Chevreul (1786–1889) presented his color theory in his work "De la Loi du Contraste Simultané des Couleurs", in which he also dealt extensively with simultaneous contrasts. In addition to the basic colors yellow, red and blue, he created 23 mixed colors each and thus obtained a 72-part color circle.

Hermann Günther Graßmann (1809–1877) dealt with the mixing of these three primary colors. His Graßmann laws also assumed that the individual primary colors

Figure 23.1: Claude Boutet's color circle with primary and secondary colors from 1708.

could not be created by mixing two others. He also defined color as a combination of hue (spectral color), intensity (chroma) and brightness (white intensity).

James Clerk Maxwell (1831–1879) dealt intensively with color vision. He used color spinning tops for his work on additive color mixing, which he also demonstrated with a (first) color photograph in 1861.

Augustin Jean Fresnel (1788–1827) dealt mathematically with the wave theory of light. He also carried out experiments on interference and polarization.

Wilhelm von Bezold (1837–1907) presented in his book "Farbenlehre in Hinblick auf Kunst und Kunstgewerbe" a cone of color which had spectral colors at the edges and pigment colors inside. For him it was clear that even "the outermost pigments are still far from perfect purity". He also postulated a color wheel based on a color triangle with the colors red, green and blue-violet in the corners as shown in Figure 23.2.

Johann Wolfgang Goethe (1749–1832) compiled his theory of colors based on mixtures. His color wheel consists of the three primary colors yellow, red and blue and the mixed colors between them. Goethe saw his results in contrast to the results of Newton. His experiments were incompatible with his own. Goethe published his results in 1810 in his book "Zur Farbenlehre".

Tobias Mayer (1723–1762) used a color triangle to combine the three primary colors yellow, red and blue as corner colors and their mixtures. As a mathematician, he assumed twelve mixtures of two primary colors each, which can be distinguished by

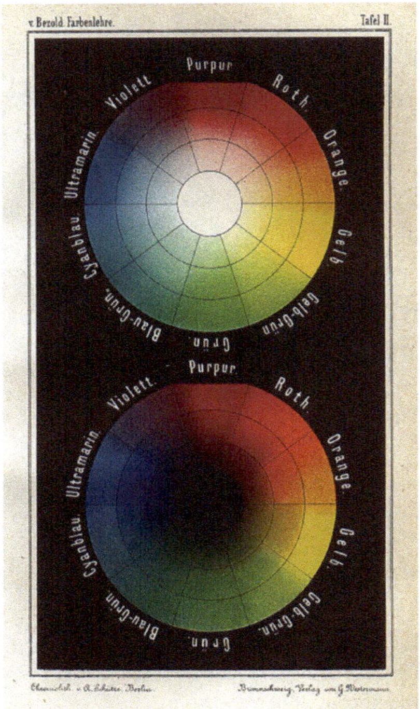

Figure 23.2: For Wilhelm von Bezold, the three colors, violet, red and green, stood in a multi-part color circle that ran inward to white or black.

the eye. That is why he also referred to the colors with units such as g12 for yellow or r12 for red. An orange mixture would then be r6g6. In the three-dimensional representation with triangles in layers, he could also capture mixtures with white and black (Figure 23.3).

Figure 23.3: Tobias Mayer placed the three colors, red, yellow and blue, in a color triangle with the corresponding mixtures between them.

Josef Albers (1888–1976) taught at the Bauhaus before he moved to the USA with his wife. The focus of his work was on optical perceptions and illusions. The combination of colors and surfaces creates unusual impressions and sensory experiences. Like, for example, Victor Vaserely and Richard Anuszkiewicz, he laid the foundations for the art of surgery.

In 1878, Ewald Hering (1834–1918) presented his "doctrine of the sense of light", which was based on counter colors. In Helmholtz's color scheme with the three colors red, green and blue, yellow results from the mixture of red and green. However, the perception of yellow is elementary and not caused by mixing red and green. Hering suggested six basic sensations: yellow–blue, red–green and black–white.

Philipp Otto Runge (1777–1810) corresponded with Johann Wolfgang von Goethe about his ideas of color. He used colors for the first time in a three-dimensional representation of a sphere.

Wilhelm Ostwald (1853–1932) first arranged colors in a 100-part circle. He determined the colors and their positions with the help of spinning tops. He reduced this color circle to 24 colors and also arranged black and white mixtures into a double cone with triangles of the same color. He gave each solid color a number and two letters for the black and white components (Figure 23.4).

Albert Henry Munsell (1858–1918) based his tree of colors on "sensory equality". Thus, a yellow is perceived as brighter and more color-intensive than a blue. For the composition of his system he also used spinning tops to check whether opposite colors mixed to grey. He described the colors by their hue, brightness and intensity (chroma).

Johannes Itten (1888–1967) published his idea in the book "Kunst der Farbe". He was inspired by Adolf Hölzel.

Arthur König (1856–1901), together with Conrad Dieterici (1958–1929), investigated color perception and determined the spectral sensitivity of the light-sensitive cells of the eye (rods and cones). They were thus able to confirm the three-color theory of von Helmholtz and Maxwell.

Johannes von Kries (1853–1928) managed to combine the three-color theory of Hermann von Helmholtz with the four-color theory (theory of counter-colors) of Ewald Hering. In this Kries zone theory, the sensitivities of the three receptors to red–green and yellow–blue reactions as well as an additional black-white reaction are combined as shown in Figure 23.5.

On the one hand, this compilation shows the physical attempts to explain optical phenomena. On the other hand, the discrepancy between the mixing experiments and perception, which lead to different descriptions of the color. Furthermore, the three-color theory, which is based on the visual processes in the eye, and the four-color theory, which reflects our perception, are opposed to each other.

Also of interest are the considerations to classify colors and their mixtures mathematically. This has been successfully achieved by means of the physiological description of colors. And from here the path leads directly to the CIE (Commission

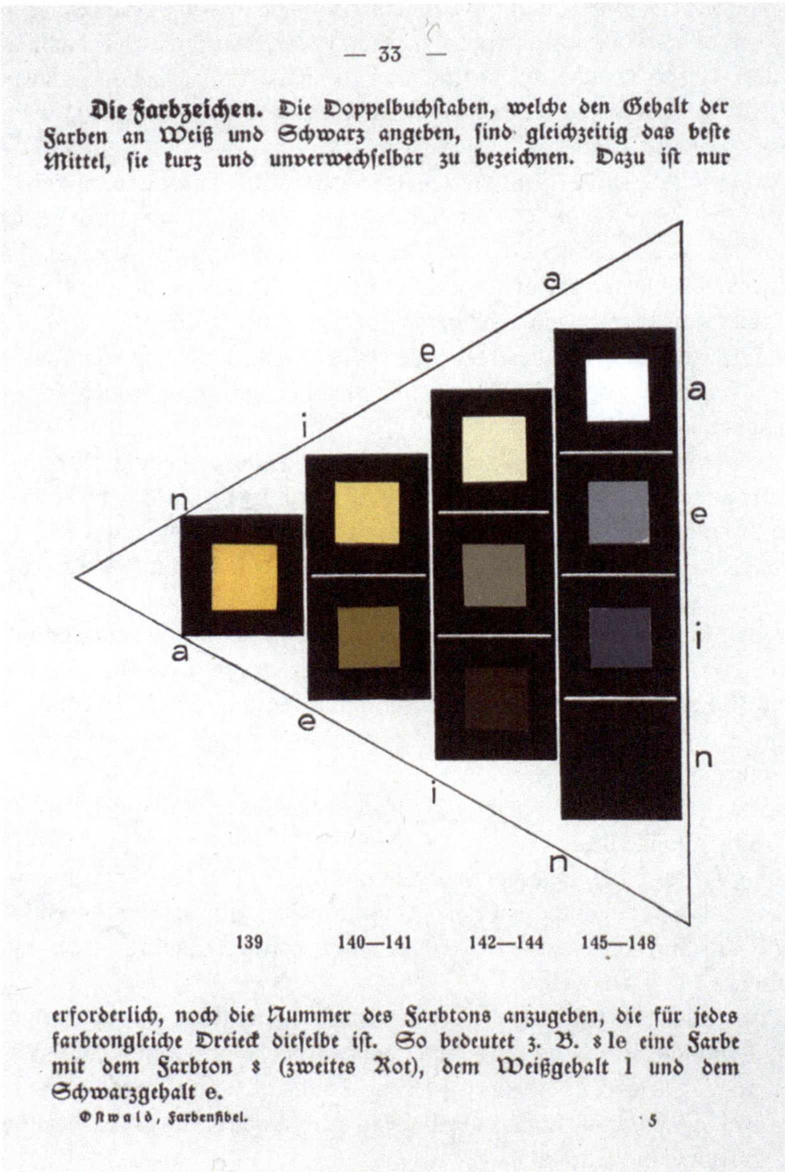

— 33 —

Die Farbzeichen. Die Doppelbuchstaben, welche den Gehalt der Farben an Weiß und Schwarz angeben, sind gleichzeitig das beste Mittel, sie kurz und unverwechselbar zu bezeichnen. Dazu ist nur

erforderlich, noch die Nummer des Farbtons anzugeben, die für jedes farbtongleiche Dreieck dieselbe ist. So bedeutet z. B. 8 le eine Farbe mit dem Farbton 8 (zweites Rot), dem Weißgehalt l und dem Schwarzgehalt e.

Ostwald, Farbenfibel. 5

Figure 23.4: Wilhelm Ostwald mixed the spot colors with white and black in a triangle. Each color was clearly identified by marking it with the number of the color tone as well as double letters for the proportions of white and black.

Internationale de l'Éclairage), which has defined a color system as a combination of physical values with human perception. With the help of a mathematical link, a physical measured value is "translated" into a physiological color perception.

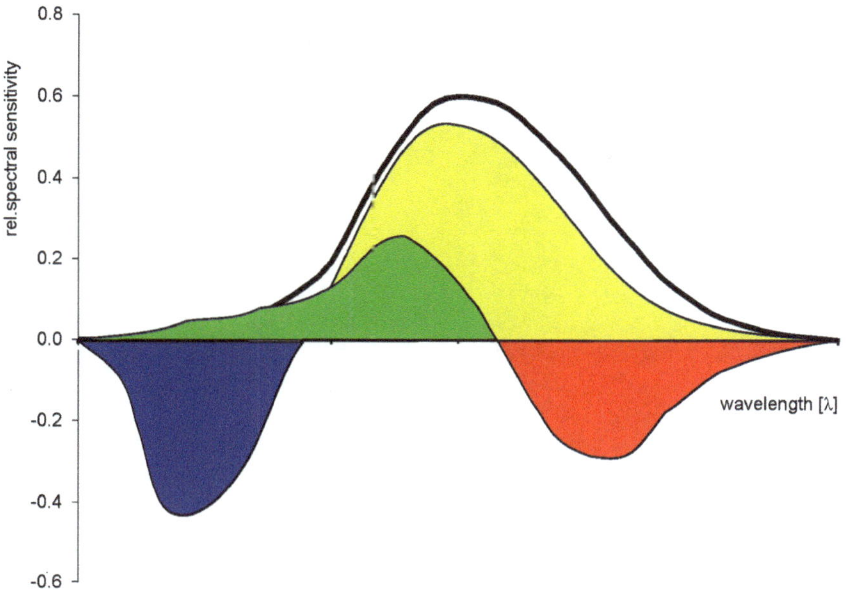

Figure 23.5: Johannes von Kries combined the three-color theory with the four-color theory (counter color theory).

23.2 CIE (Commission Internationale de l'Éclairage)

In 1931, various developments came together which led to the CIE calculating color values. Siegfried Rösch developed a colorimeter and a spectral integrator. He also presented – as early as 1928 – his color body with different planes, which added the brightness of the optimum colors to the CIE diagram. Optimal colors are body colors that cannot be realized. Rösch built his color body with the help of these optimum colors, with all body colors lying within this color body. The optimum colors are the brightest at the same shade of color and the most intensive at the same brightness. The color body is based on the CIE diagram with brightness values of 0, to which layers with brightness values of 20, 40, 60 and 80 were added (Figure 23.6).

In the same years, David Wright and John Guild undertook mixing experiments with some observers, and the results were combined to form the normal observer. This normal observer looks at surfaces with a field of view of 2°. In 1964 the standard observer with a field of view of 10° was introduced. With the 2°-observer one has orientated oneself by the position of the light-sensitive sensors (cones), which are concentrated in the yellow spot and offer the best color vision there. With the 10°-observer the field of view is larger and the number of cones is smaller (Figure 23.7).

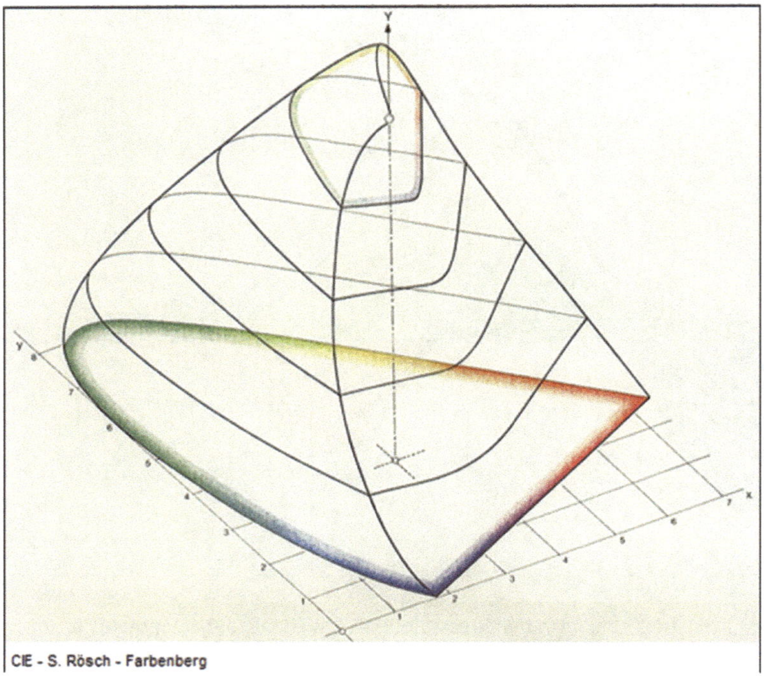

Figure 23.6: Siegfried Rösch created his optimal colors in layers of different brightness.

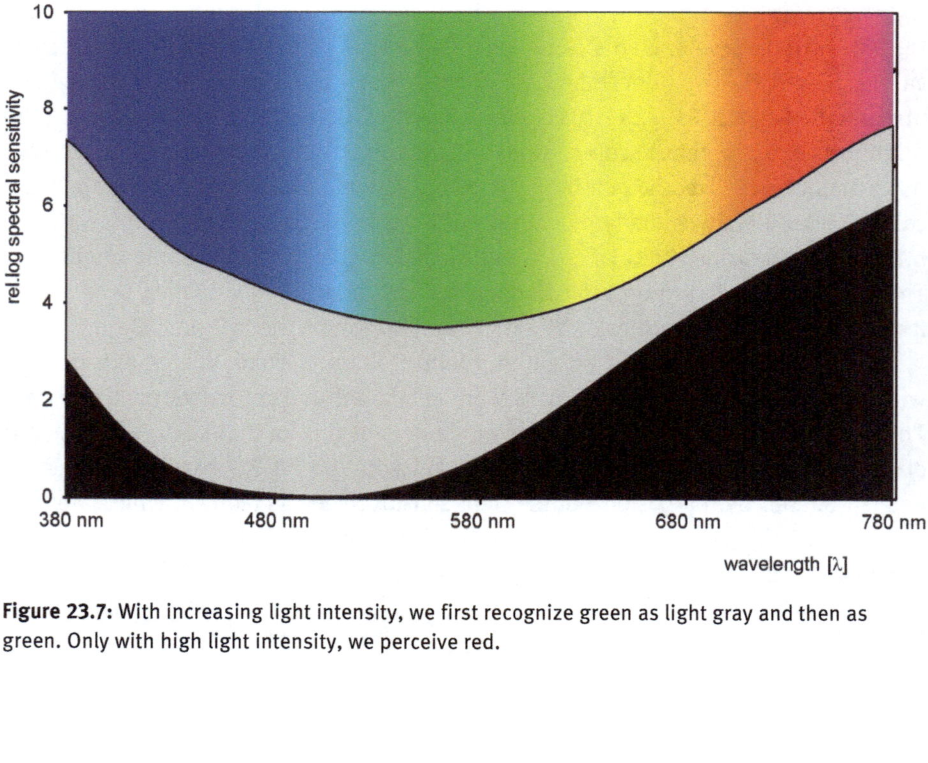

Figure 23.7: With increasing light intensity, we first recognize green as light gray and then as green. Only with high light intensity, we perceive red.

The experiments were carried out with light colors, whereby interference filters were used to generate the spectral lines with 546.1 nm and 435.8 nm the output colors for green and blue. The color red was displayed at 700 nm with incandescent lamps and color filters. In the experiments, a color was projected onto a splitted screen, which was to be simulated on the other part using the light colors mentioned above. The brightness of the light colors was changed by the test persons according to their subjective perception. The scaling of the light colors on the adjustment page resulted in corresponding measurements for the given light color which was to be adjusted. However, some test colors could not be readjusted with these adjustment colors. Instead, the observers could add red light to the given color. Its scale value was listed as a negative red value when readjusting. Correspondingly, the color values were marked with the letters r, g and b and provide the spectral value curves as a function of wavelength (Figures 23.8–23.11).

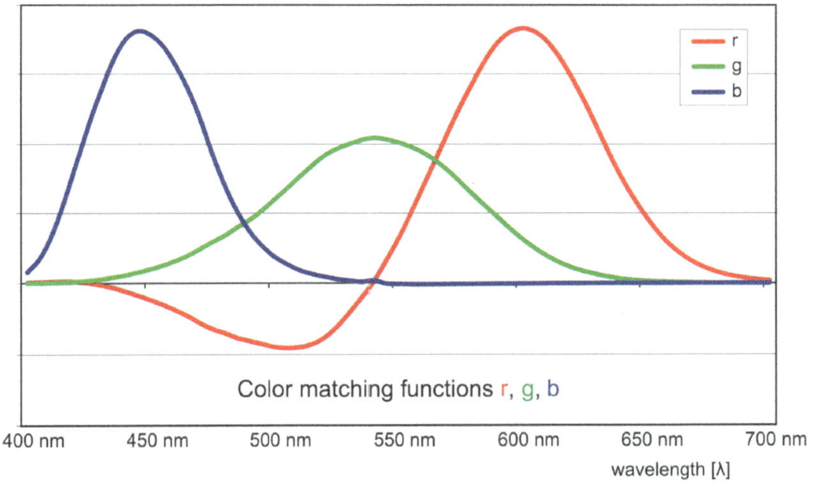

Figure 23.8: Experiments yielding three sensitivity curves for red (r), green (g) and blue (b) with negative red values.

The resulting spectral curves with negative values were converted into positive values at that time in order to calculate the integral better. Since the sensitivity curves of the cones of the human eye do not contain negative values, the spectral curves must also have positive values. The integral under these curves results in the standard color values X, Y and Z, respectively. The standard chromaticity coordinates are calculated accordingly from these values:

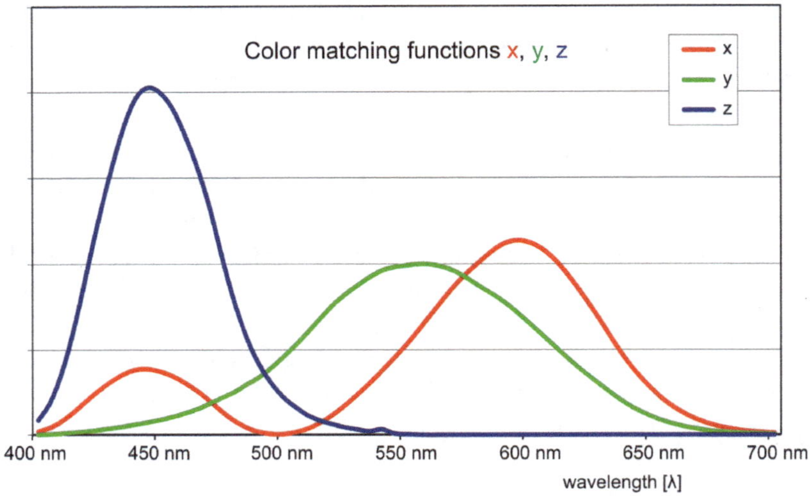

Figure 23.9: Conversion into positive values results in the standard spectral values *x*, *y*, *z*.

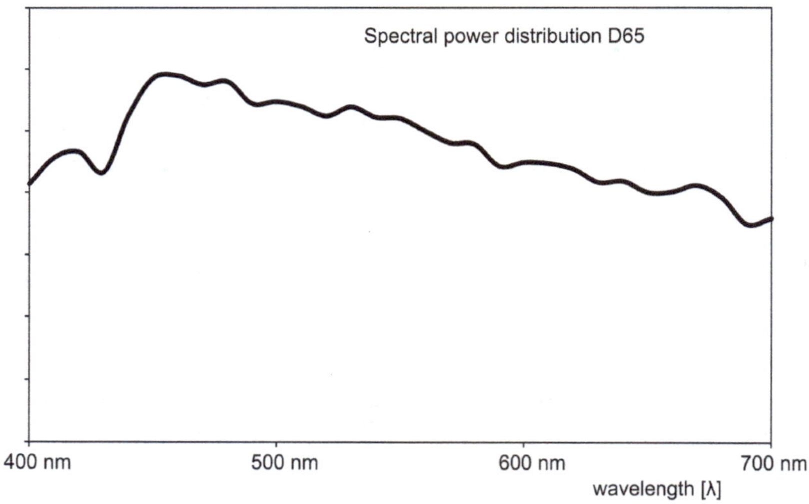

Figure 23.10: Spectral powder distribution D65.

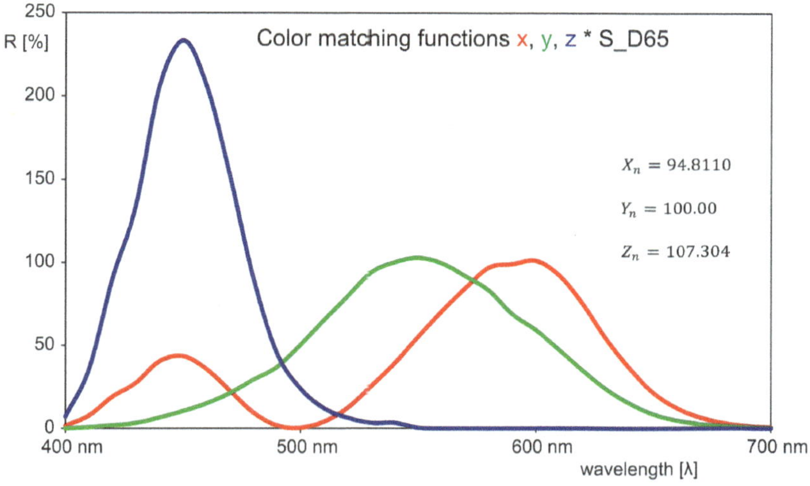

Figure 23.11: By converting the spectra. values related to daylight D65, the standard color values X_n, Y_n and Z_n are obtained.

$$x = \frac{X}{X+Y+Z}$$

$$y = \frac{Y}{X+Y+Z}$$

$$z = \frac{Z}{X+Y+Z}$$

The standard chromaticity coordinates add up to the value 1:

$$x + y + z = 1$$

Since a single value can therefore be calculated from the other two, the standard chromaticity coordinates x and y can be entered in the two-dimensional CIE standard chromaticity diagram with the x-axis and the y-axis perpendicular to it. The shape of this diagram results from the converted chromaticity coordinates for the different wavelengths from 450 nm to 770 nm. The connecting line encompasses this diagram and includes all possible colors. It is also known as the spectral color line with the most saturated colors. The connecting line between the final spectral colors red and blue-violet is called the purple line. On it lie all the purple colors that are not present in the color spectrum and only occur in the human brain (Figure 23.12).

The white point in this standard chromaticity diagram is at the one-third share of the standard chromaticity coordinates:

$$x = y = z = 0.33$$

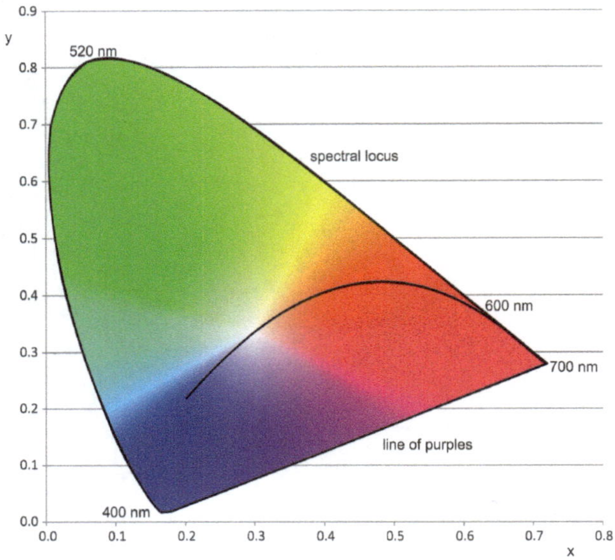

Figure 23.12: Display of the *xy*-values with the outer spectral color line (spectral locus) and the purple line. The black line represents the black-body curve (Planckian locus).

The black-body curve (Planckian locus), on which colors are plotted as a function of temperature (kelvin), also runs through the white point.

If a straight line is laid through the white point, it ends at two points on the boundary line, the spectral color locus. From the white point to the spectral color locus, the colors are identical in hue. Both points are complementary to each other.

The distances on this straight line are also a measure for the saturation: The ratio of the distances between the white point and the color location at hand and between the white point and the location on the spectral color locus plays a role here [2].

The brightness in this representation is defined by the standard color value Y, which is why the color space is also called Yxy color space. For the white point, the following values result for illuminant D65 (today 6504 K instead of 6500 K), which corresponds approximately to noon light in Europe and is referred to as daylight illuminant. For the 10° normal observer, the values (vectors) for the white point are

$$\bar{x}(\lambda) = 031382$$

$$\bar{y}(\lambda) = 0.33100$$

This is used to calculate the standard color values (Y is set equal to 100):

$$X_n = 94.8110$$

$$Y_n = 100.000$$

$$Z_n = 107.304$$

The measurement of a color sample results in the reflection values, which are determined against a white standard (for example $BaSO_4$, Teflon or Opal). These values are multiplied for each wavelength (e.g. at a distance of 10 nm) by the spectral power distribution for D65. These values are then multiplied by the normalized values x, y, and z. The areas under the resulting curves give the color values X_r, Y_r and Z_r for the color panel. The standard chromaticity coordinates can be calculated accordingly (Figures 23.13–23.15).

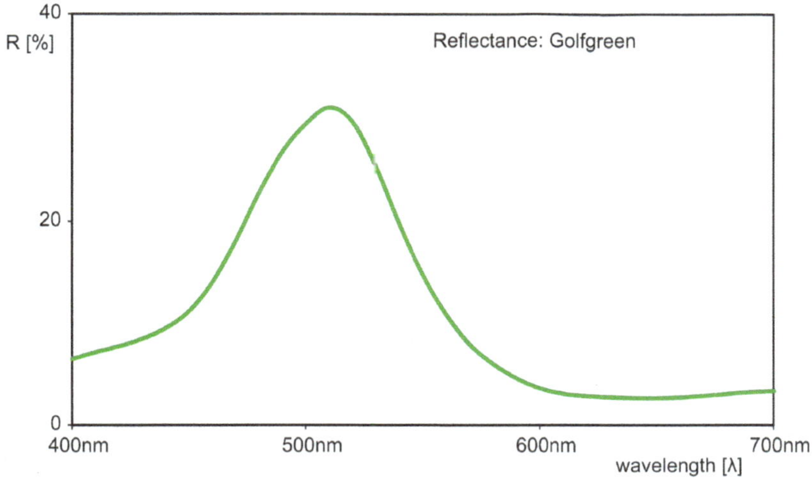

Figure 23.13: Reflection curve of Golfgreen. The maximum is located at the same place as the maximum of the sensitivity of the human eye in daylight. This makes Golfgreen the most striking color.

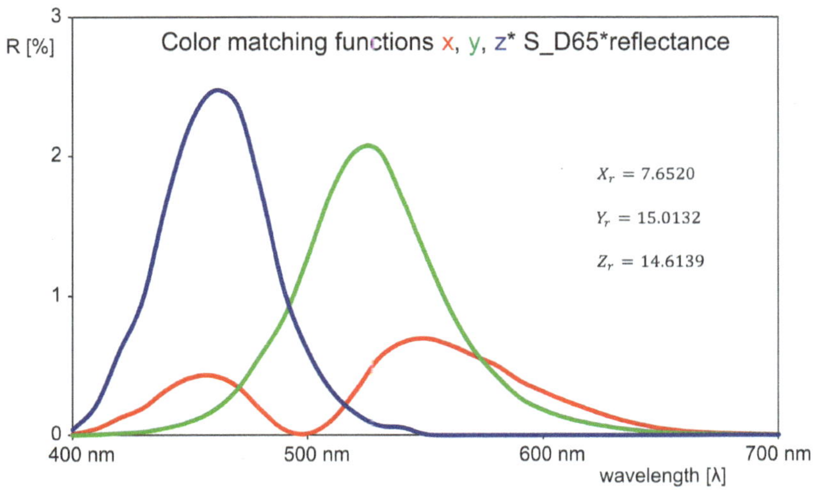

Figure 23.14: Further conversion with the measured reflection values results in the tristimulus values X_r, Y_r and Z_r for the measured color panel.

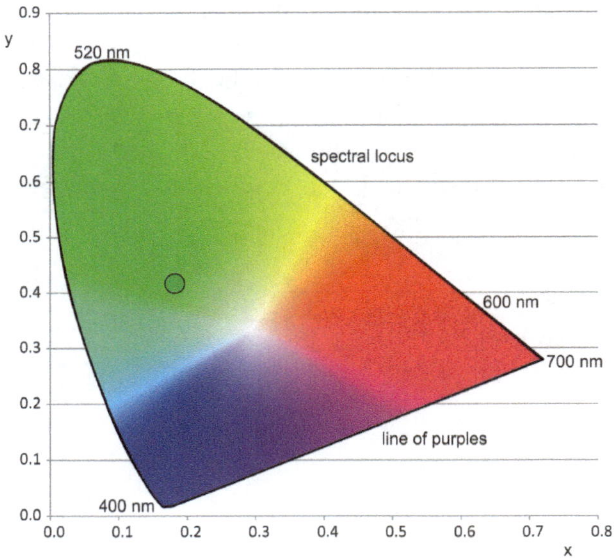

Figure 23.15: Measured value with its calculated *xy*-values.

Standard color values Golfgreen

$$X_r = 7.6520$$

$$Y_r = 15.0132$$

$$Z_r = 14.6139$$

Standard chromaticity coordinates

$$x_r = 0.2053$$

$$y_r = 0.4067$$

$$z_r = 0.3920$$

L*a*b* color space

Richard Sewall Hunter developed a color space in1948, which he obtained by simple conversion of the *XYZ* values. He called the coordinates of this color space L (brightness), a (red–green component) and b (yellow–blue component). Corresponding conversions of the CIE in 1976 were designated as L*, a* and b* in contrast to the Hunter coordinates. The conversion of the *X*-, *Y*- and *Z*-values is carried out according to the following formulas, which are adjusted accordingly for small values:

$$L^* = 116 * \sqrt[3]{\frac{Y}{Y_n}} - 16$$

$$a^* = 500 * \left(\sqrt[3]{\frac{X}{X_n}} - \sqrt[3]{\frac{Y}{Y_n}} \right)$$

$$b^* = 200 * \left(\sqrt[3]{\frac{Y}{Y_n}} - \sqrt[3]{\frac{Z}{Z_n}} \right)$$

$$if \ \frac{X}{X_n} < \frac{216}{24389} \approx 0.008856, \ then \ \frac{1}{116} * \left(\frac{24389}{27} * \frac{X}{X_n} + 16 \right), \ instead \ of \ \sqrt[3]{\frac{X}{X_n}}$$

$$accordingly \ also \ for \ \sqrt[3]{\frac{Y}{Y_n}} \ and \ \sqrt[3]{\frac{Z}{Z_n}}$$

The values published by CIE in 1931 referred to the 2° standard observer, those published in 1964 to the 10° standard observer.

Because of the vectorial representation in the a*b* diagram, it was also suggested to use the values C* for Chroma and h° for Hue instead of the a*b* values. Chroma represents the value of chromaticity. And Hue stands for the hue angle relative to the red axis and has values between 0° and 360°.

$$C^* = \sqrt{a^{*2} + b^{*2}}$$

$$h^\circ = atan2(b^*, a^*)$$

The calculation of the L*a*b* values for Golf green result in the following values. The brightness values L* are shown separately from the color values a*b* (Figure 23.16):

L*a*b*-values Golfgreen

$$L^* = 45.65$$

$$a^* = -59.64$$

$$b^* = 3.47 \ \$$$

$$C^* = 49.78$$

$$h^\circ = 176.00$$

The chromaticity space shows a clear overemphasis in the green-blue area. The CIELUV was introduced with u', v' for a balance in accordance with sensation. The advantage of this system is especially in the representation of mixtures of colors (Figure 23.17).

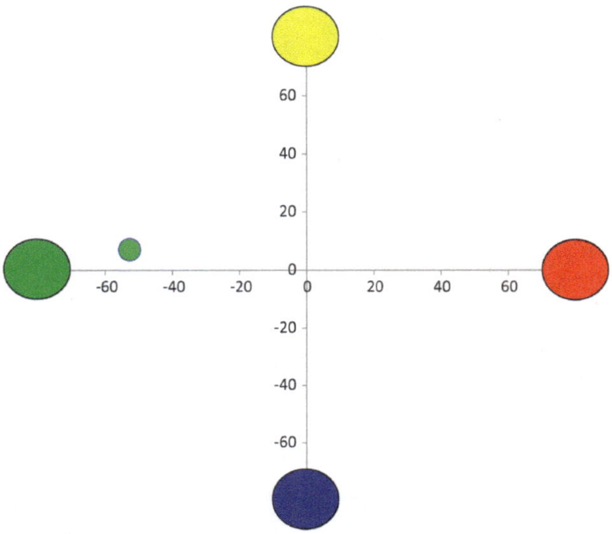

Figure 23.16: In the a*b* system, yellow and blue as well as red and green face each other in a coordinate system. By conversion from the color values X, Y and Z, the L*a*b* values are obtained. Usually the lightness values L* and the color values a*b* are displayed separately.

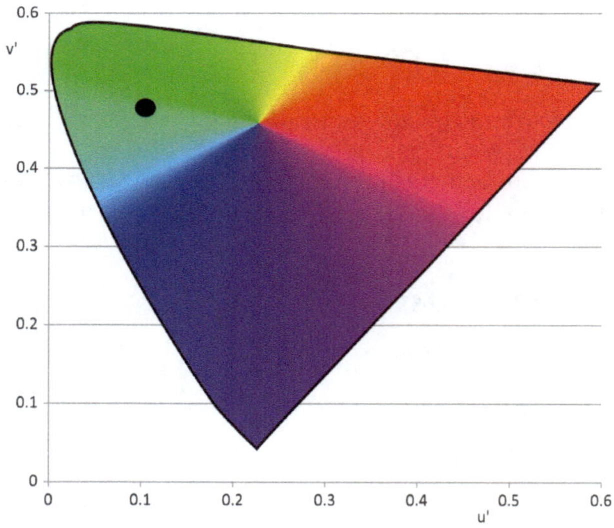

Figure 23.17: u'v'-diagram with measured values of Golfgreen (black dot).

$$u' = \frac{4X}{(X + 15Y + 3Z)} = \frac{4x}{-2x + 12y + 3}$$

$$v' = \frac{9Y}{(X + 15Y + 3Z)} = \frac{9y}{-2x + 12y + 3}$$

u'v'-values Golfgreen

$$u' = 0.11.6$$

$$v' = 0.4883$$

23.3 Color differences

Since the CIELAB space is supposed to be a color distance space in principle, some imbalances regarding color differences appeared. Original formula for color differences:

$$\Delta E^*_{ab} = \sqrt{(L^*_2 - L^*_1)^2 + (a^*_2 - a^*_1)^2 + (b^*_2 - b^*_1)^2} \ \text{(CIE 1976)}$$

$$\Delta E = \sqrt{\left(\frac{\Delta L}{K_L S_L}\right)^2 - \left(\frac{\Delta C}{K_C S_C}\right)^2 + \left(\frac{\Delta H}{K_H S_H}\right)^2} \quad \text{(CIE 1994)}$$

$$\Delta H = \sqrt{(\Delta a^*)^2 + (\Delta b^*)^2 - (\Delta C_{ab}{}^*)^2} \ (not\ \Delta h°!)$$

$$\Delta E = \sqrt{\left(\frac{\Delta L'}{K_L S_L}\right)^2 + \left(\frac{\Delta C'}{K_C S_C}\right)^2 + \left(\frac{\Delta H'}{K_H S_H}\right)^2 + R_T\left(\frac{\Delta C'}{K_C S_C}\right)^2\left(\frac{\Delta H'}{K_H S_H}\right)^2} \quad \text{(CIE 2000)}$$

$$\Delta E = \sqrt{\left(\frac{\Delta L}{l S_L}\right)^2 + \left(\frac{\Delta C}{c S_C}\right)^2 + \left(\frac{\Delta H}{S_H}\right)^2} \quad \text{(CIE:CMC)}$$

In 1994 the CIE published a new distance formula based on the LCh color space with weighting factors. This was followed in the year 2000 by another adjustment to better determine the color differences. The CIE2000 was developed as a tolerancing method which should fix the problem when reversing reference and test specimen.

The color difference E is an important component in the relationship between supplier and customer. It is also used to define the tolerance that describes the deviation from the master sample. (Table 23.1)

The increasing proportion of effect colors has also led to these also being measured. However, geometry is not sufficient here to capture them sufficiently colorimetrically. Aluminum pigments, in particular, and interference pigments reflect more than white near gloss. Although the measuring instruments are calibrated with a white standard, they still accept the L*a*b* values calculated on the basis of the high reflections.

Table 23.1: The tinting colors of a reference were enlarged or reduced. The different methods of deltaE were calculated and compared.

Panel		L*	a*	b*	CIE 1976	CIE 1994 (Gr.Arts)	CIE 1994 (Textiles)	CIE 2000 1:1:1	CMC 1:1	CMC 2:1
M1080	+	41.24	−42.33	13.23	6.65	3.61	3.52	3.44	3.48	3.32
M1079 (Ref.)	Yellow Ochre	42.42	−47.04	8.68						
M1085	–	43.11	−50.01	6.66	3.66	1.79	1.71	1.71	1.81	1.71
M1081	+	43.30	−46.91	19.99	11.34	6.46	6.59	5.84	6.00	5.95
M1079 (Ref.)	Yellow	42.41	−47.04	8.72						
M1086	–	42.27	−47.05	7.09	1.60	0.93	0.95	0.89	0.86	0.85
M1082	+	47.18	−47.65	8.11	4.83	4.78	2.42	4.51	4.81	2.43
M1079 (Ref.)	White	42.41	−47.04	8.72						
M1087	–	37.71	−45.36	9.12	5.02	4.76	2.44	4.20	4.80	2.48
M1083	+	38.56	−49.35	8.29	4.52	3.94	2.10	3.51	4.00	2.17
M1079 (Ref.)	Green	42.41	−47.04	8.72						
M1088	–	45.75	−44.14	9.17	4.44	3.50	1.96	3.29	3.57	2.08
M1084	+	39.96	−43.12	6.66	5.05	2.90	1.94	2.70	3.07	2.20
M1079 (Ref.)	Black	42.41	−47.04	8.72						
M1089	–	43.86	−49.59	9.92	3.18	1.74	1.18	1.63	1.85	1.37

23.4 RGB color space

The CIE system refers to reflective color patterns. In contrast, the RGB color space is based on self-luminous colors that mix additively. The colors can be mixed from the combination of R (red), G (green) and B (blue). The absence of the colors results in black, the full representation of all three colors results in white.

The colors are displayed in a cube with the axes for red, green and blue perpendicular to each other. With $R = G = B = 0$ black is present.

With this representation, not all colors can be built up. Nevertheless, purely arithmetically millions of colors result: With 8 bits per channel, there is a graduation from 0 to 255. Systems with 16 bits per color channel can represent colors from 0 to 65,535 (Figure 23.18)).

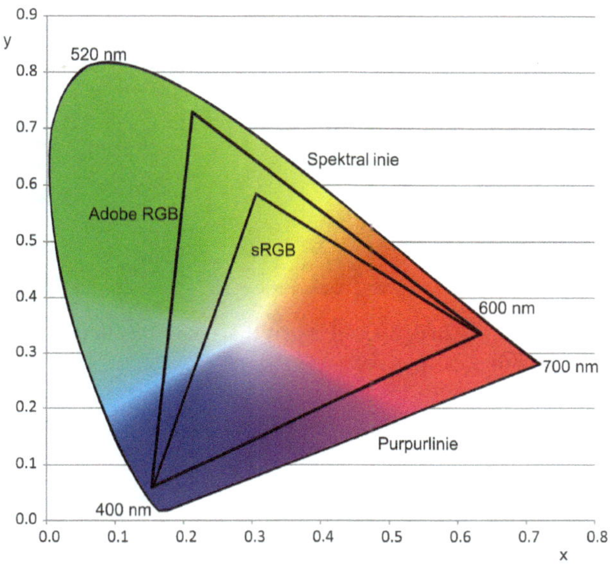

Figure 23.18: Color spaces are different in size as the comparisons with sRGB and AdobeRGB show. sRGB and AdobeRGB form a color triangle.

The RGB system is used for displays and monitors. In order to control the desired color, these are displayed with values from 0 to 255 (0 to 1) or corresponding hexadecimal values. So the most intense red has a representation of rgb(255, 0, 0) or #ff0000. The color can also be described in the HSL system (Hue, Saturation, Lightness) as hsl(0, 100, 50) (also hsl(0°, 100%, 50%)). Correspondingly, for green at full display rgb(0, 255, 0) or #00ff00 or hsl(120, 100, 50) applies. An orange is a mixture of red and green and could have the following example values: rgb(240, 128, 0), #f08000, hsl(32, 100, 47).

The RGB color space is used to translate and display colors on monitors and printers. It is limited by the colors that the device can display. Image processing should also be viewed critically, because, for example, changing the brightness of one color will not affect the other components proportionally. (Figure 23.19)

The RGB color space is a trichromatic color space with red, green and blue components. Its algorithms sometimes lead to non-application-optimal results. This also applies to the conversion to grayscale images: A classification of all three color components to 0.333 (0.333*R + 0.333*G + 0.333*B) only produces a low-contrast greyscale image. A stronger contrast is achieved by considering only

Figure 23.19: Near gloss, interference pigments often have color values that lie outside the RGB color space. They can be adjusted by appropriate conversion.

the green component. And according to the NTSC standard the grayscale intensity is calculated like this:

$$0.2989\ {}^{*}R + 0.5870\ {}^{*}G + 0.1140\ {}^{*}B$$

23.5 Measuring instruments

A parallelism can be shown between the development of modern pigments and measuring instruments. When the CIE drew up the first definitions in 1931, there were already various photometers that could be used to determine the spectral properties of colors and pigments. As late as the 1960s, newly developed measuring instruments (e.g. Elrepho, DMC25) were primarily used for adjustment and production monitoring at paint manufacturers. Due to production at several locations with different paint systems, car manufacturers also began to introduce instrumental as well as visual assessment.

In the eighties, more and more cars were painted with effect paints. This trend was favored on the one hand by the changeover to 2-layer coatings – here the base coat contains the color and effect components and the clear coat sprayed over it protects the base coat – and on the other hand new aluminum and interference pigments were developed. The latter were first used in coatings systems at the end of the seventies. They were initially offered by coatings and car manufacturers as so-called design paints. In 1985, white interference pigments were offered for the first time in series production paints. And the proportion of effect pigments rose sharply thereafter.

It were the companies ZEISS and DuPont who thought extensively about measuring methods for these pigments [3]. It was clear that the previous measuring instruments with only one geometry were no longer sufficient. On the basis of its own experiments, DuPont developed the so-called Dupontmac with three measuring geometries: With 45° illumination, measurements were taken at 15°, 45° and 110° of the gloss angle [4]. ZEISS built the GK111 multi-angle instrument, in which the aspecular angles were adjustable in steps from 20° to 70° [5]. This was followed in 1990 by the GK311/M, whose illumination and observation angles could each be adjusted in 5° steps by motorized software (Figures 23.20 and 23.21).

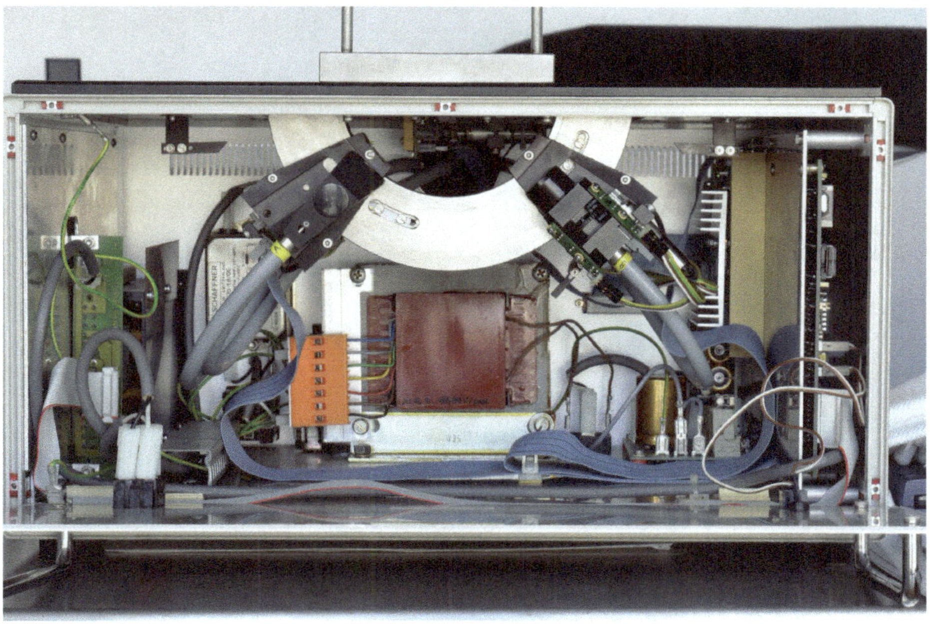

Figure 23.20: With the GK311/M from Zeiss, the illumination and observation heads run on a half arc.

Shortly afterward, the first portable multi-angle measuring instrument appeared: the X-Rite MA58 illuminated the color panel at 45°; measurements were taken at 25°, 45° and 75° off the gloss angle (specular). A few years later, the MA68 was introduced with illumination at 45° and angles of 15°, 25°, 45°, 75° and 110° from the gloss angle (off gloss angle = aspecular) (Figure 23.22).

Konica Minolta built the CM-512m3 with three circular illuminators, which can be used even with slightly curved surfaces. The current CM-M6 has the same geometries as the common portable units. It illuminates at 45° and measures at as-15°, as15°, as25°, as45°, as75° and as110° (aspecular).

Based on many experiments, Datacolor presented its Multi-FX10. This device had three illumination angles at 15°, 45° and 65° from the normal, measured at 15° each

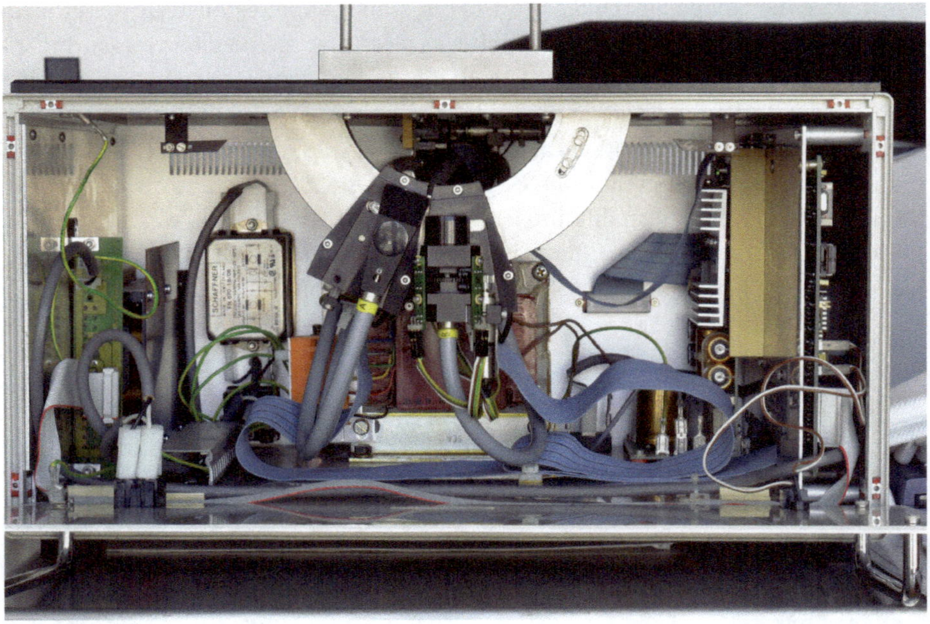

Figure 23.21: Illumination angle and the observation angle can be adjusted independently of each other in 5° steps.

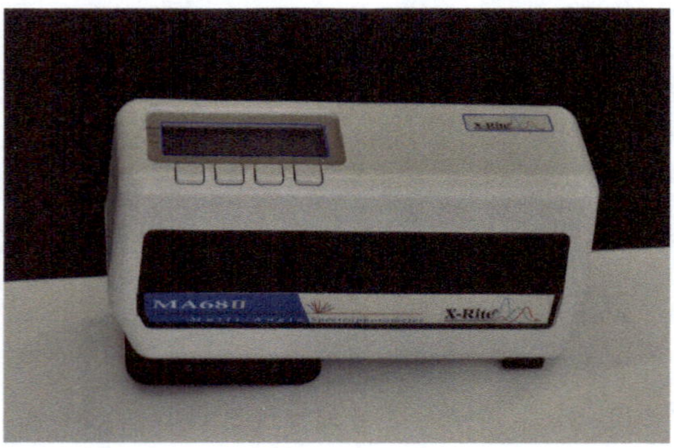

Figure 23.22: X-Rite MA68 was the first portable instrument with multi-angle geometries.

in the cis and trans position from the respective gloss angle. Cis-position means that the measurement takes place between the gloss angle and the illumination angle. The measurement in trans-position takes place on the side of the gloss angle opposite the illumination angle. In addition, the aspecular series was measured at 45° illumination (Figure 23.23).

In addition to color, new pigments required other aspects of appearance for evaluation. AkzoNobel, Merck and BYK Gardner joined forces to use a digital camera in

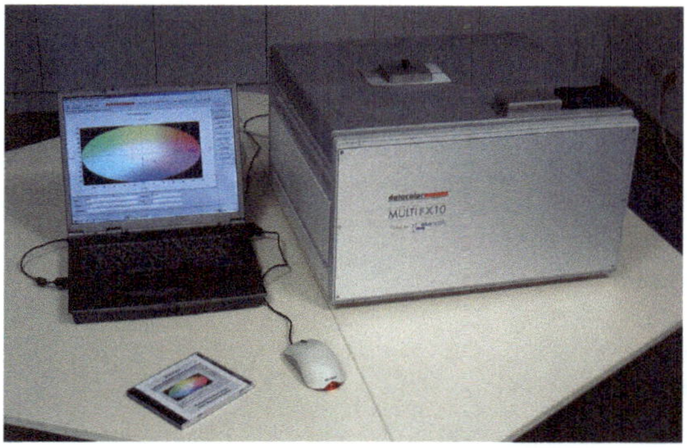

Figure 23.23: Three illumination angles, each with + 15° and −15° aspecular as well as the classic aspecular angles for 45° illumination. This is how the Datacolor MultiFX10 presented itself.

the newly developed BYKmac to determine a Sparkle value. This was intended to record the optical properties of the new Xirallic pigments. In the BYKmac, the −15° observation geometry (aspecular) was taken from the ASTM Test Method and introduced as an additional measurement geometry (Figure 23.24). The newly developed X-Rite MA98 was also based on the ASTM2539 Test Method, which proposes two illumination angles. In addition to the so-called in-plane angles, the instrument had other off-plane geometries. In the current MA-T12 and MA-T6 instruments from X-Rite, the light paths are reversed (Figure 23.25).

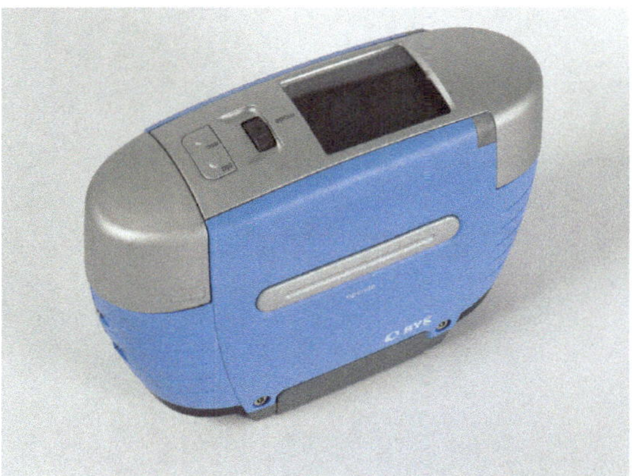

Figure 23.24: With the BYKmac, sparkle values under three geometries are determined in addition to the color values.

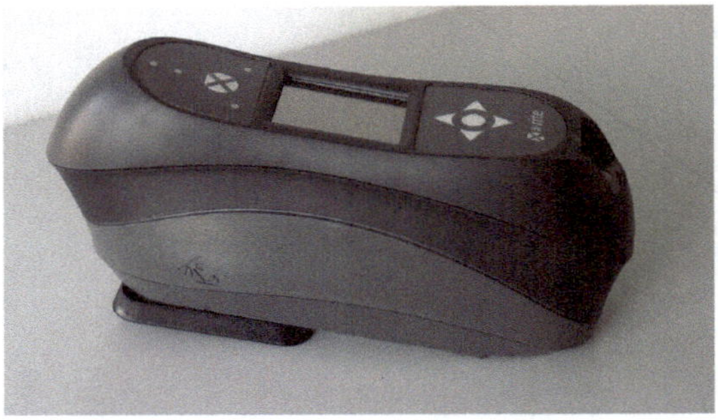

Figure 23.25: X-Rite MA98 offers both in-line and off-line geometries. It has two illumination angles.

23.6 Calibration

The measurement also includes a suitable calibration by a white standard. For a long time $BaSO_4$ (barium sulfate) was used for this purpose. Appropriate tablets were made from the powder material.

Today, white opal is usually used, which is cut to the desired size. In some cases Teflon (e.g. Polytetrafluoroethylene, Spectralon) is also used, which is a synthetic material. However, experiments show that this material is not suitable for flat illumination angles and that there are significant deviations from the calibration with white opal (Figure 23.26).

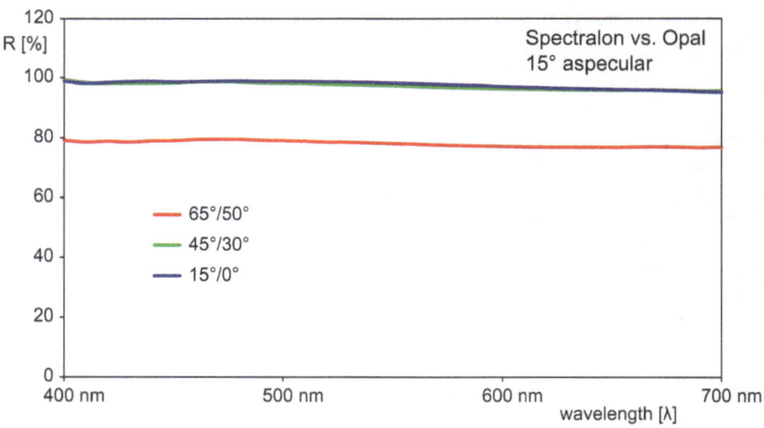

Figure 23.26: Measuring of two calibration materials against each other: Obviously Teflon has problems with flat illumination.

Therefore, the white standard used should also be checked for plausibility. Thus, the reflection curve of an interference pigment shifts toward shorter wavelengths and also rises if the pigment is illuminated flatter at the same difference angle to the gloss angle (e.g. 15° aspecular). If these two criteria are not met, the white standard should be checked (Figures 23.27 and 23.28).

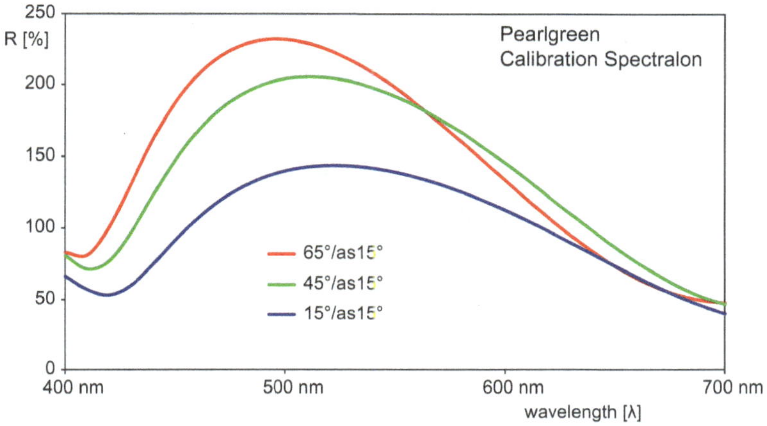

Figure 23.27: Interference pigments show higher reflections when they are illuminated at a lower level. A malfunction due to insufficient calibration material can be seen.

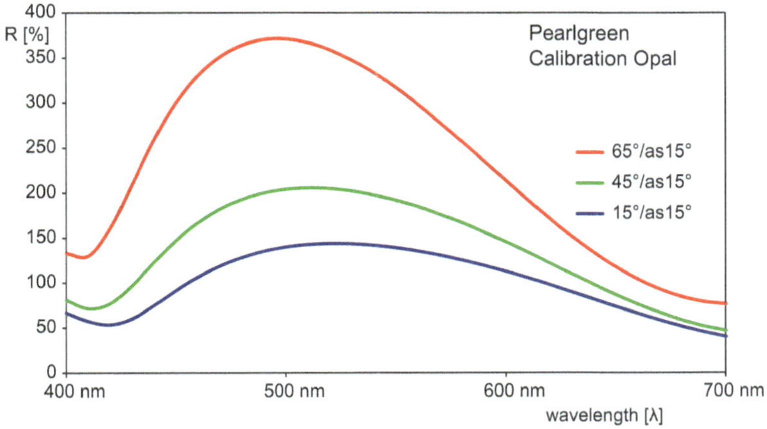

Figure 23.28: Correct calibration produces plausible measured values: The difference between the reflection maxima must increase with flatter illumination.

23.7 Measuring geometries

The exact description and reproducibility of the measurement includes the specification of the measurement geometry. This is particularly important between supplier and customer. Devices with different measuring geometries can deliver different results, so that no reasonable communication between supplier and customer is possible.

For solid colors, i.e. such applications in paint or plastics with colored pigments, for example, measuring instruments with a directed geometry or with a sphere geometry are generally used (Figure 23.29). There are colored pigments with a pronounced reflection maximum (green, blue) and pigments with a reflection plateau (yellow, red). Mixtures with white pigments, with aluminum pigments or interference pigments lead to an increase in chromaticity and brightness up to a turning point for green, blue and blue-violet colored pigments. Thereafter, the chroma decreases and the brightness continues to increase until the achromatic mixing partner [6].

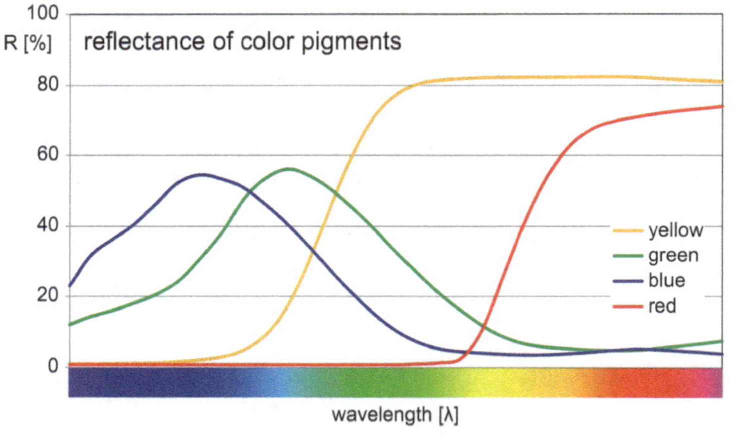

Figure 23.29: Typical reflections of colored pigments (direct illumination). There are pigments with a pronounced reflection maximum (green, blue) and pigments with a reflection plateau (yellow, red).

Since aluminum and interference pigments have angle-dependent colors and brightness, multi-angle measuring devices are used for their measurement. As a rule, measurements are made in the so-called cis position. Here, the angle of observation is between the angle of illumination and the corresponding gloss angle. An observation position on the other side of the gloss angle, i.e. the side of the gloss angle opposite the illumination angle, is called the trans position [7].

23.8 Directed or sphere measurements

A directional measurement geometry means that a color panel is illuminated at a defined angle. The measurement is carried out at a likewise defined angle from the corresponding gloss angle. For solid colors, an illumination of 45° to the panel horizontal and a measurement under 0° (in the normal) has become established. A reverse constellation is also possible with illumination at 0° and measurement at 45° as shown in Figure 23.30.

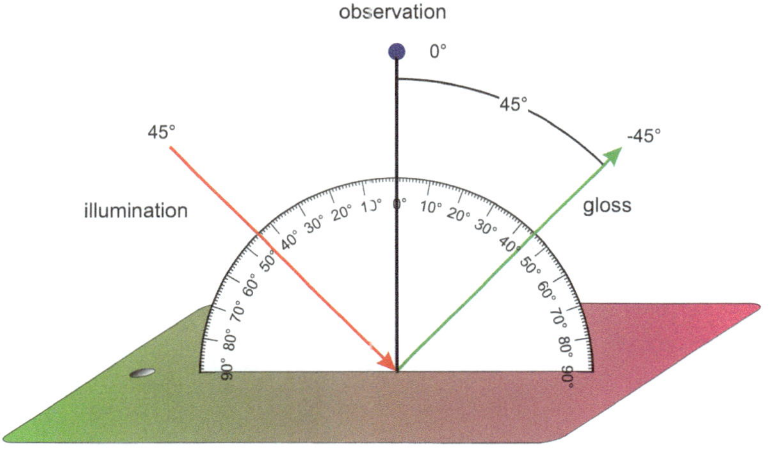

Figure 23.30: In the case of directional geometry, illumination is at an angle of 45°. The gloss angle is then −45°, measured in the normal at 0°. This corresponds to 45° of the gloss. The arrangement can also be reversed.

With the sphere geometry, the color panel is illuminated diffusely. Here, diffuse light is generated in an integrating sphere. The inside of the sphere is coated with a white material such as barium sulphate or ceramic, which reflects strongly. So-called shadows prevent the color pattern from being illuminated directly (Figure 23.31).

This geometry is used in particular for glossy color samples and for structured color samples, because in the latter case the measured values change insignificantly if the surface is of different nature. The light reflected from the color sample is measured at an angle of 8° to the normal (Figure 23.32).

A so-called specular trap in the −8° position is intended to prevent the light reflected at this point from reaching the color panel. The measurement can be carried out either with the specular trap closed (SCI) or with the specular trap open (SCE) (Figure 23.33).

Measurements in SCE mode (Specular Component Excluded) capture the color effect (appearance), while measurements in SCI mode (Specular Component Included)

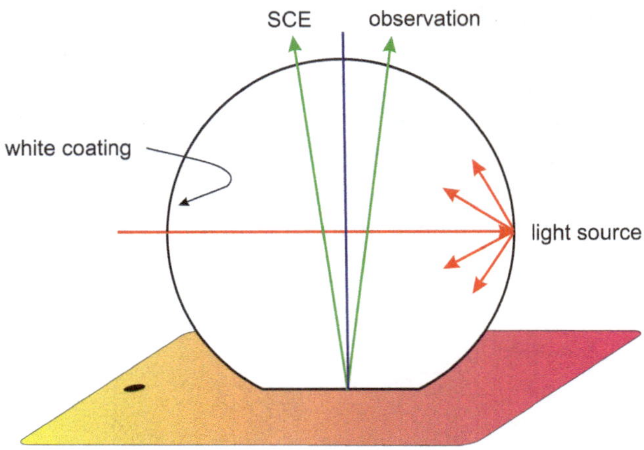

Figure 23.31: With sphere geometry, diffuse illumination is used. With the help of a gloss trap, measurements can be made with (SCI) or without gloss (SCE).

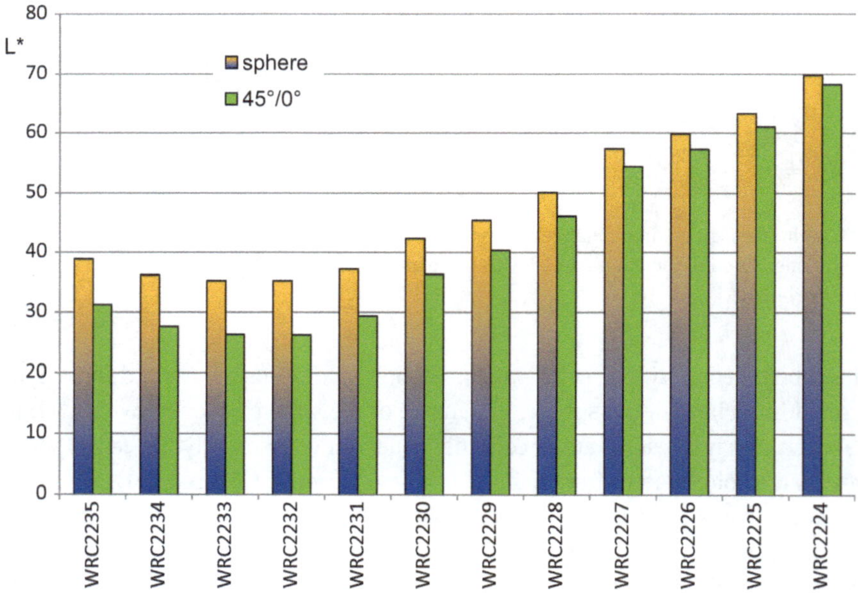

Figure 23.32: Measurement results of the sphere geometry and the directed geometry cannot be compared. For this reason, coordination between different users is always necessary.

capture the pigmentation (color), so that structures, textures and gloss levels are not measured. Here too, however, the agreement with the customer must be observed so that different results are not compared (Figure 23.34).

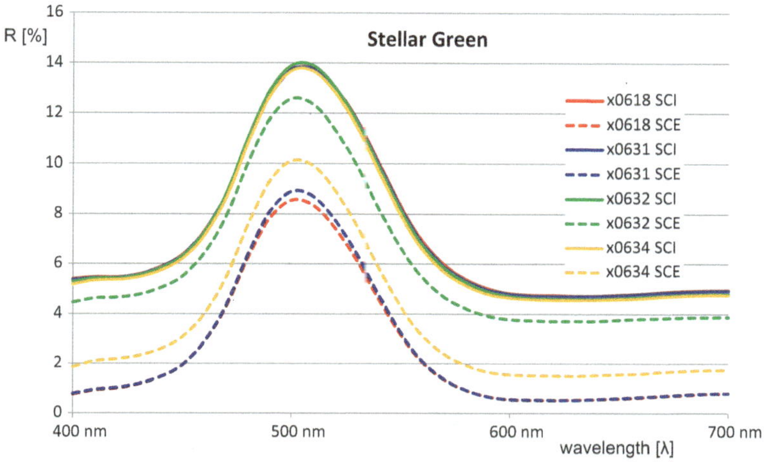

Figure 23.33: Comparative measurements of different color samples show that different gloss levels have little influence on measurements with gloss (SCI).

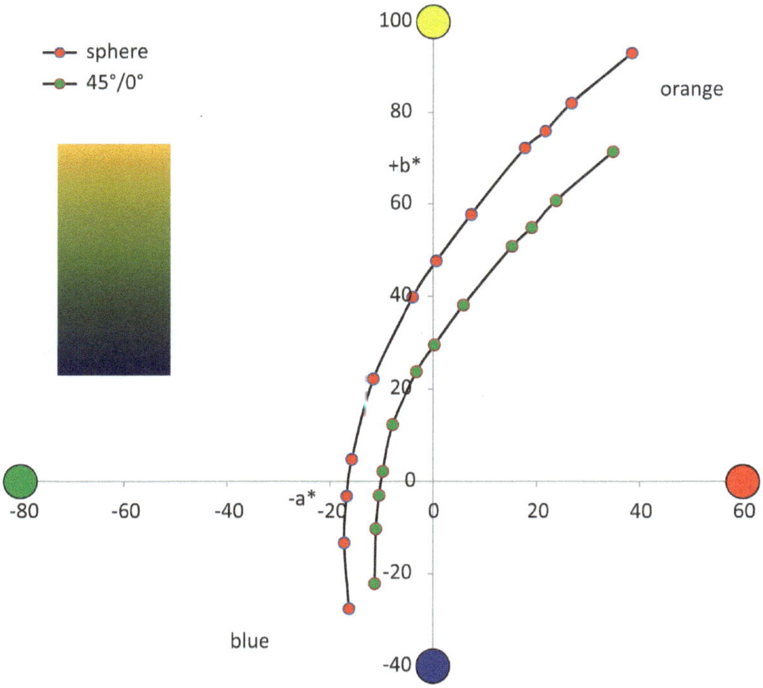

Figure 23.34: Mixing tests between blue and orange colored pigments show different courses of the measurement results with the sphere geometry and the directed geometry.

23.9 Multi-angle measurements

For effect pigments, several geometries are necessary to capture and describe the effect. The current portable devices offer an illumination of 45° and partly additionally one at 15°. The measurements are made at –15°, 15°, 25°, 45°, 75° and 110° off gloss (aspecular) as shown in Figure 23.35. The selection of the geometries is more or less random and historically determined. Proposals for a different selection have not been accepted. It should also be noted that the 110° geometry provides only a very weak electrical signal. In the case of visual matching at this geometry, the color panel is observed below the light source.

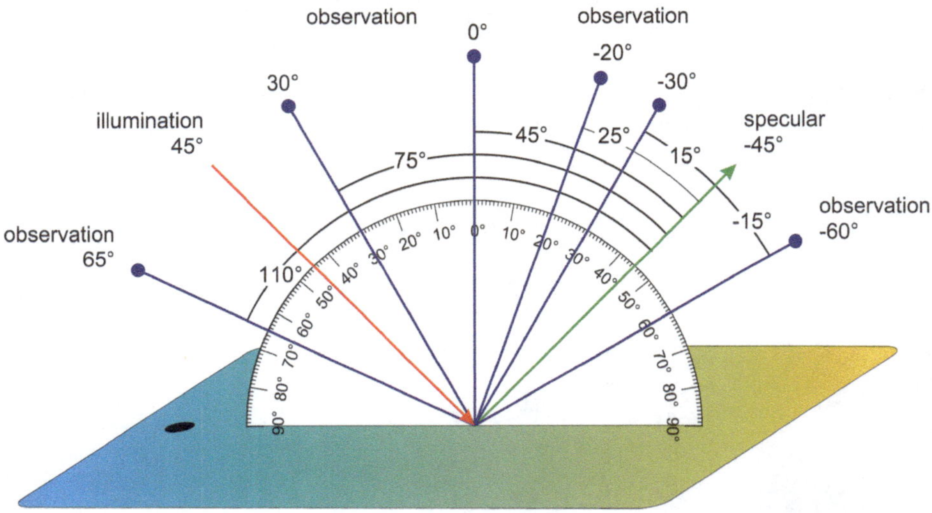

Figure 23.35: Measuring geometries of the current portable measuring instruments: illumination at 45° and measurement at –15°, 15°, 25°, 45°, 75° and 110°.

The purpose of multi-angle measurements is to detect the effect that occurs at a distance from the gloss angle or vice versa. At –15° and 15° these are near-gloss geometries. From 45° on, the geometries are far from gloss.

23.9.1 Aluminum pigments

This type of pigment is characterized by a high reflection near gloss angle, which is also measured at –15° and 15° of gloss. Here the reflection values are sometimes higher than the white values of the calibration standard. Due to the lack of standards for effect pigments, the white standards used are still considered sufficient.

The results of the measurement depend above all on the aluminum pigments themselves and also on their surrounding medium. The orientations can be influenced by appropriate additives. The larger the aluminum pigments are, the stronger their gloss. However, the opacity decreases, which is clearly evident when measuring over black and white backgrounds (Figures 23.36 and 23.37).

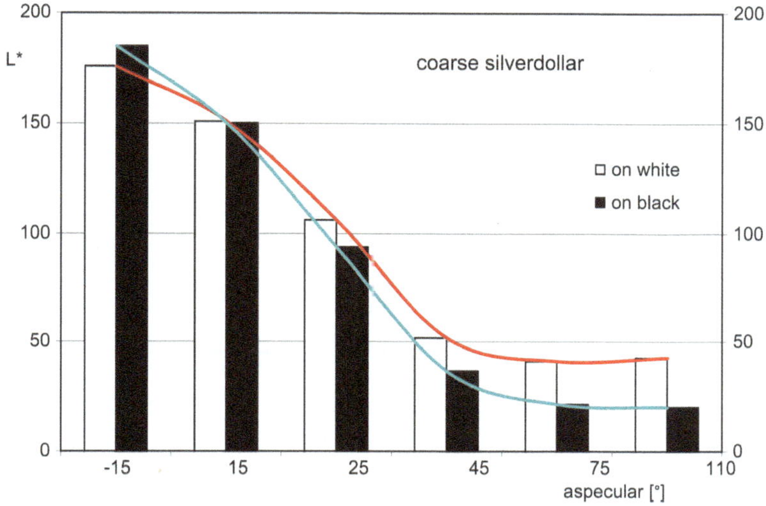

Figure 23.36: If a coarse aluminum pigment is applied over a black and white background, the change in brightness becomes clear when a more lustrous effect is observed.

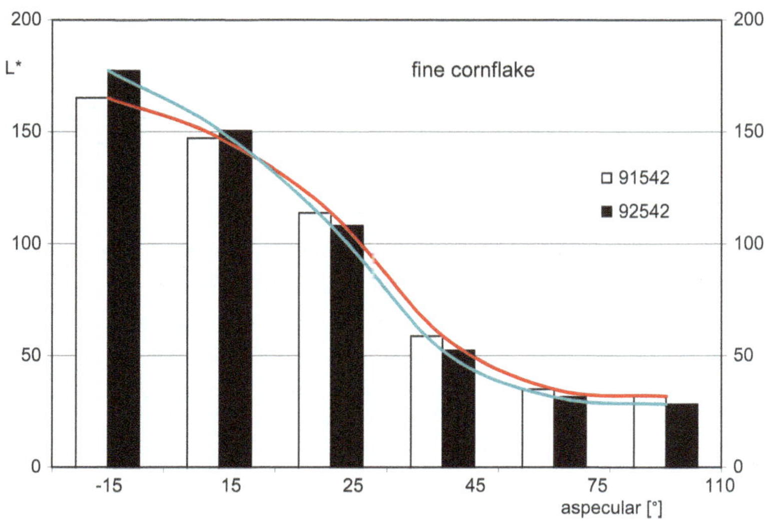

Figure 23.37: With a fine aluminum pigment, the change in brightness is less pronounced when it is far away from gloss observed. In addition, the background is less noticeable.

The so-called flop – i.e. the change in brightness in relation to the angle of observation – depends on the pigment type. In comparison, an aluminum pigment can be brighter than another one near gloss. A change can take place away from the gloss and the previously glossless pigment increases significantly compared to its brightness as shown in Figure 23.38.

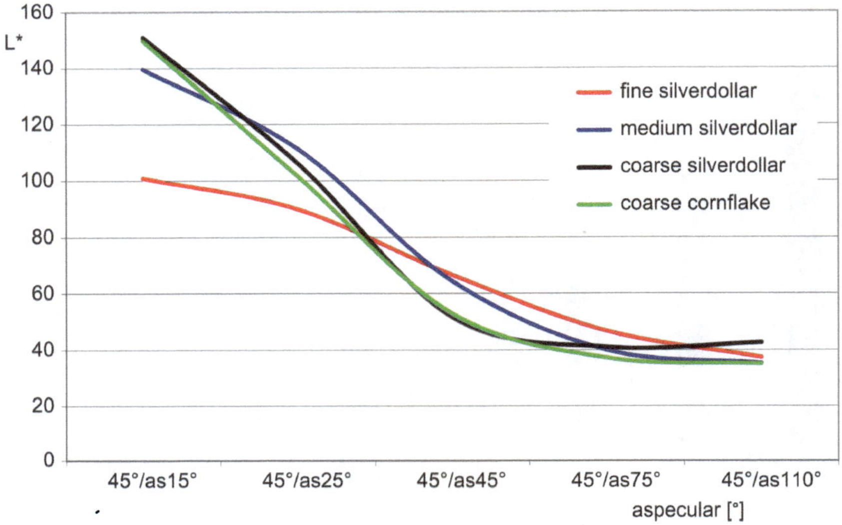

Figure 23.38: Different types of aluminum show different flop behavior.

Since aluminum pigments have no color, they show similar reflection curves to white pigments when mixed with colored pigments. To obtain a more precise statement about their optical properties, a blue colored pigment can be added, for example. The closer the gloss is measured and also observed, the higher the reflection values and thus the reflection curves increase. There is no color shift in aluminum pigments themselves.

In the mixing example with a blue colored pigment, the chroma and brightness initially increase when an aluminum pigment is mixed with blue. From the turning point with the highest chroma, it decreases again, while the brightness increases further when more aluminum pigment is mixed in (Figures 23.39 and 23.40).

Decisive for the flop with effect pigments is their orientation in the medium used. In spray application, this can be influenced by the spray technique (wet or dry) (Figures 23.41 and 23.42).

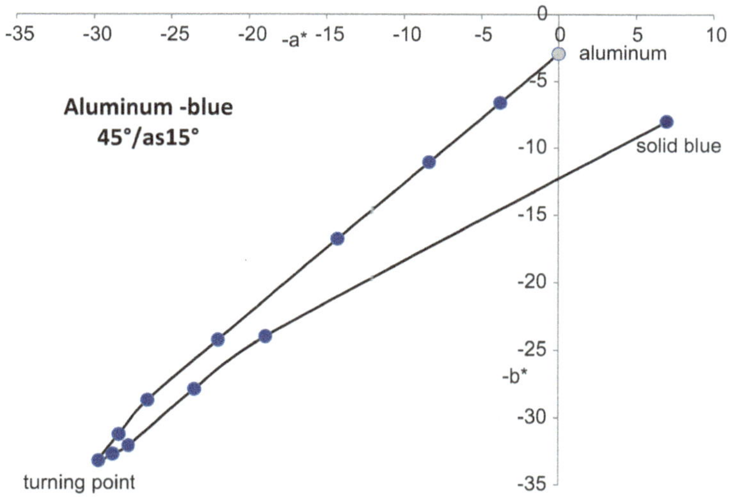

Figure 23.39: Colorfulness of a mixture of an aluminum pigment with a blue colored pigment increases up to a turning point and then decreases again.

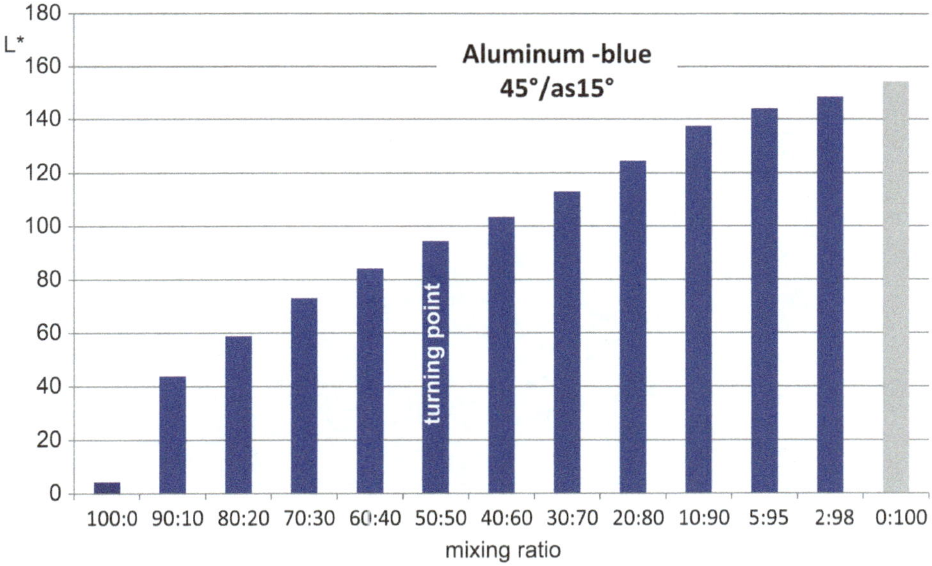

Figure 23.40: Evaluation of the brightness values shows their continuous increase when the proportion of aluminum pigment is increased.

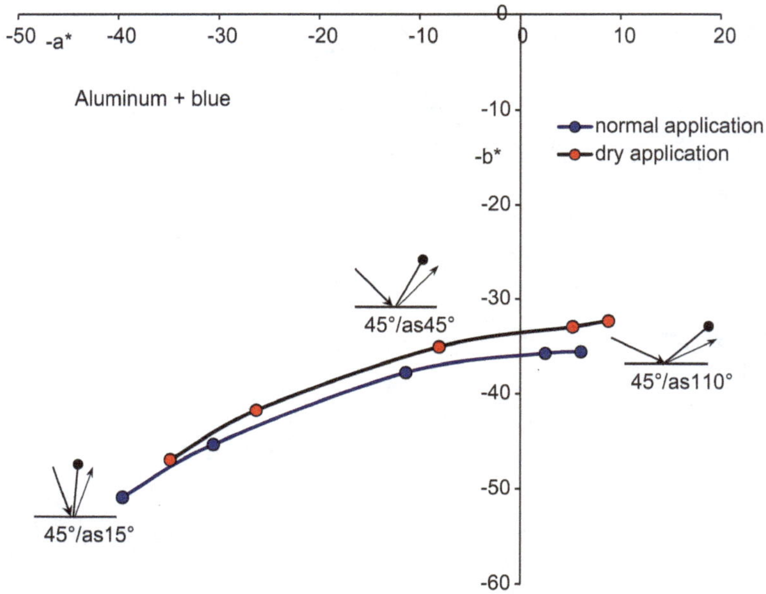

Figure 23.41: Type of spray application influences the result. In this example, the mixture of an aluminum pigment with a blue colored pigment is sprayed wet compared to the normal application conditions.

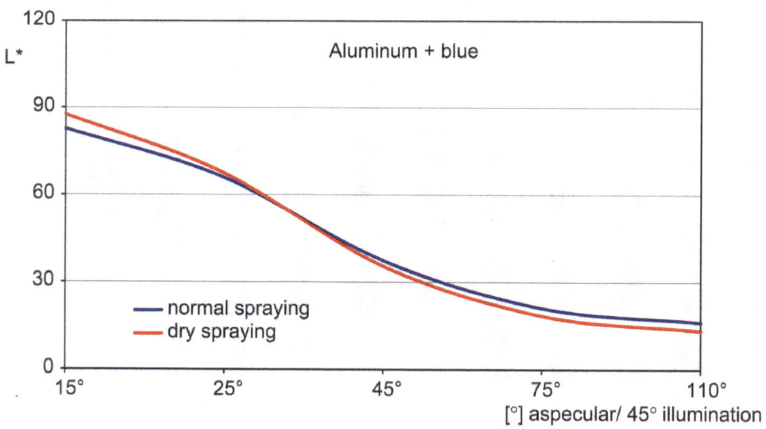

Figure 23.42: Flop behavior is also influenced by the spray application. The brightness can increase close to gloss and decrease further away from gloss when sprayed dry.

23.9.2 Interference pigments

This type of pigment differs from aluminum pigments in a number of essential points: With transparent interference pigments, the color of the background plays a role. Differences therefore also occur when mixed with the two other pigment types: Interference pigments mix with each other additively; this behavior can also be observed colorimetrically (Figures 23.43–23.46).

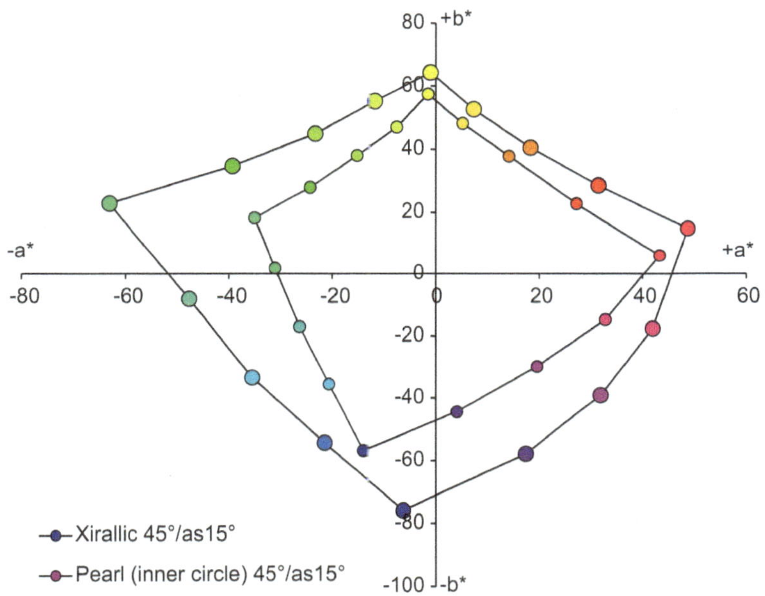

Figure 23.43: If the interference pigments of a series are mixed together, the largest possible color range is obtained.

Mixtures of interference pigments result in reflection curves that run through nodes of the reflection curves of the starting pigments as shown in Figures 23.47 and 23.48.

Interference pigments generate their colors by superimposing light waves. The interference is caused by splitting the incident light and shifting certain light waves in relation to the original light waves. In most cases, interference pigments are structured in such a way that a highly refractive metal oxide such as titanium dioxide coats a transparent platelet (e.g. natural or artificial mica). When light hits this pigment, it is partially reflected. The other part penetrates the metal oxide layer under refraction. At the boundary layer to the substrate, another part is reflected in the direction of the first part and interferes with it. After further refractions and reflections, the remaining part leaves the interference pigment at the back [8].

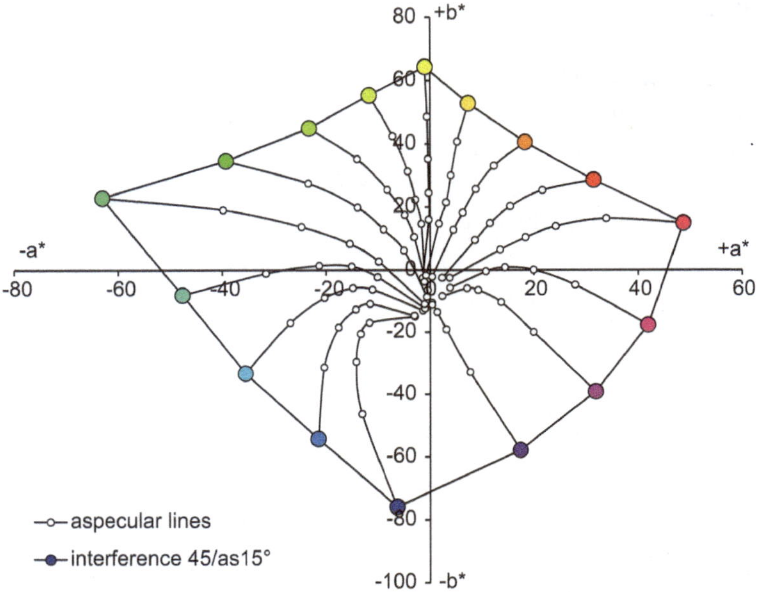

Figure 23.44: Aspecular lines of interference pigments and their mixtures run almost in the same direction.

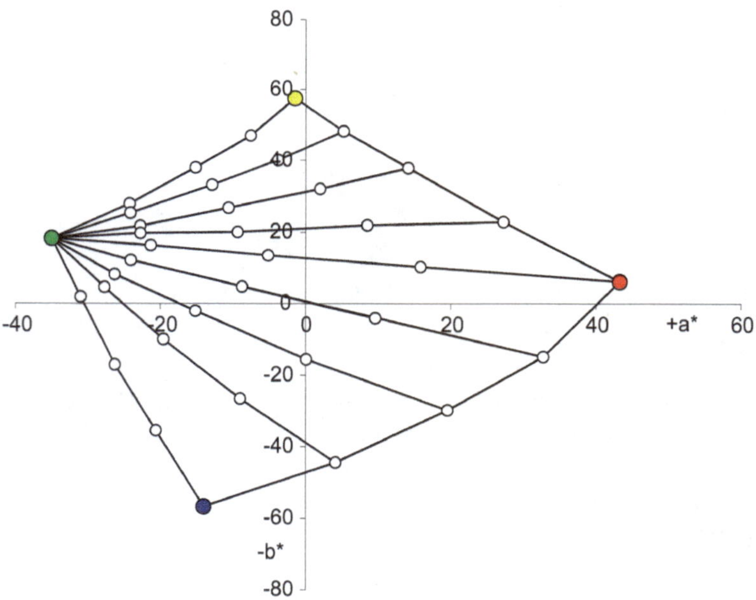

Figure 23.45: Interference pigments mix almost ideally. The lines show triple mixtures of green, red and blue interference pigments.

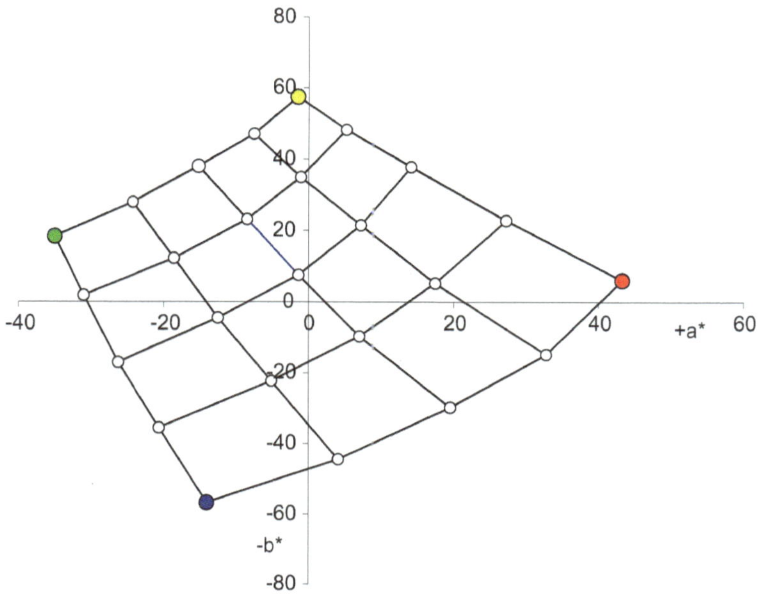

Figure 23.46: Mixtures of four different interference pigments show also an ideal behavior.

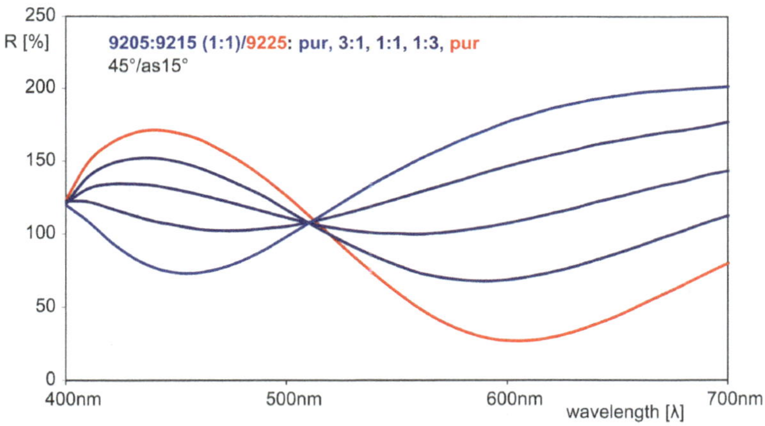

Figure 23.47: Reflection curves of the starting pigments intersect in one or two nodes. The reflection curves of their mixtures run through these nodes.

The simplified interference formula shows qualitative relationships between individual components: The maximum reflection depends on the layer thickness *d* of the coating metal oxide. The higher the layer thickness, the more the maximum shifts toward the longer wavelength. Thinner coatings produce white interference pigments that reflect over the entire spectral range. In the adjacent UV range they have a minimum. With increasing layer thickness, this minimum and the subsequent

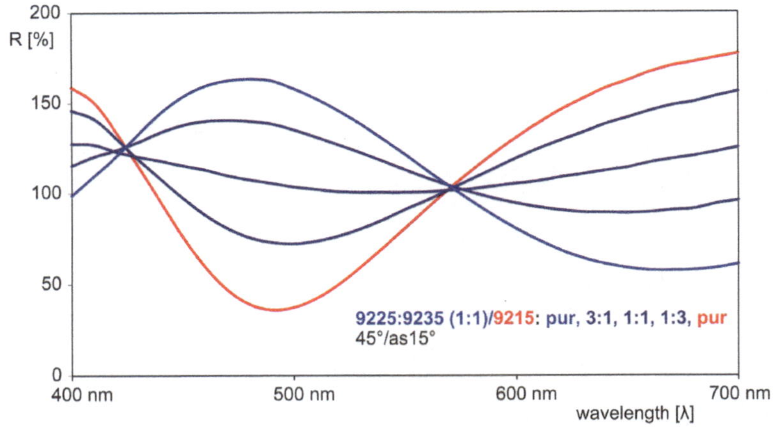

Figure 23.48: Nodes that are formed for multiple mixes.

maximum shifts into the visible range, resulting in yellow interference pigments. With higher layer thickness red interference pigments are formed. If the minimum shifts further into the long-wave spectral range with increasing film thickness, the maximum is in the blue and then in the green spectral range.

The resulting color of the interference pigment depends on the refractive index of the highly refractive material of the coating material. Mostly titanium dioxide TiO_2 is used to produce white and colored interference colors. With iron oxide Fe_2O_3 one obtains red to copper. A combination of both metal oxides produces yellow to golden colors [8].

An important factor for the colors is the angle of the incident light. This angle refers to the normal. The simplified interference formula shows that with increasing angle the reflection maximum shifts toward shorter wavelength, i.e. when the illumination angle becomes flatter. The layer thickness d and the refractive index n cannot be changed by the user. But the user can influence the interference via the angle of incidence λ. This gives the user a possibility to describe and identify an interference pigment.

The formula still needs to be extended by a phase shift of half a wavelength, which happens during the transition from the optically thinner to the optically denser medium. On the other side of the pigment – i.e. from the transition from the optically denser to the optically thinner medium – this phase shift is missing. For this reason, the color of transmission is complementary to the reflection color.

$$\Delta = 2d\sqrt{n^2 - sin^2\alpha} + \frac{\Delta}{2}$$

The following procedure is recommended for measuring interference pigments in order to describe and identify them optically unambiguously: Due to their optical properties, they are measured on the one hand at a constant difference angle from the gloss angle (aspecular) and steep, classical and flat illumination angles as shown in Figure 23.49. On the other hand, the measurements at fixed illumination angle and increasing difference angle are used to determine their optical behavior with respect to the gloss angle [9] [10].

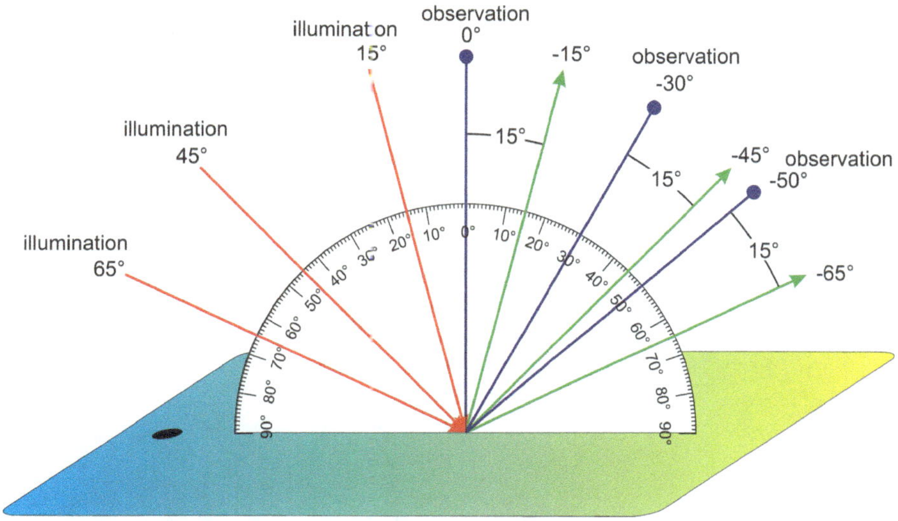

Figure 23.49: Ideally, interference pigments are measured and observed with three illuminations: With steep, classic and flat illumination for close to gloss observation.

After intensive discussions in ASTM International (formerly American Society for Testing and Materials), the geometries for the measurement of interference pigments were defined [11]. At that time, flat illumination was not feasible in portable instruments and was therefore not included in the Standard Test Method ASTM2539. In addition to the classic illumination of 45°, a similar illumination at 15° was defined. For both illuminations an additional measurement geometry of –15° from the gloss angle was introduced [12] [13].

The classical difference angles, originally developed for aluminum pigments, are also defined in this standard test method:

45° illumination/as-15° and 45°/as15°
15° illumination/as-15° and 15°/as15°
45° illumination/as15°, 25°, 45°, 75° and 110°

Two typical lines result from the measured a*b*-values of these geometries: The interference line, based on the measurement geometries 15°/as15°, 45°/as15° and 45°/as-15° as shown in Figure 23.50. This interference line is typical for each individual interference pigment [14]. Its orientation is always counterclockwise when the illumination is flatter. The trans measurement geometry 45°/as-15° is used as a substitute for the cis measurement geometry 60°/as15° (illumination/observation). This application is possible due to the reversibility of the light path as shown in Figure 23.51. The corresponding reflection curves show a clear behavior: The flatter an interference pigment is illuminated, the more the reflection curve shifts toward shorter wavelength [15]. Thus, in case of a yellow or golden interference pigment, it shifts to the green spectral range. Or with a green interference pigment it shifts into the blue spectral range. These color shifts are typical for interference pigments. The shift of the reflection curve toward shorter wavelength is accompanied by an increase in reflections (Figure 23.28).

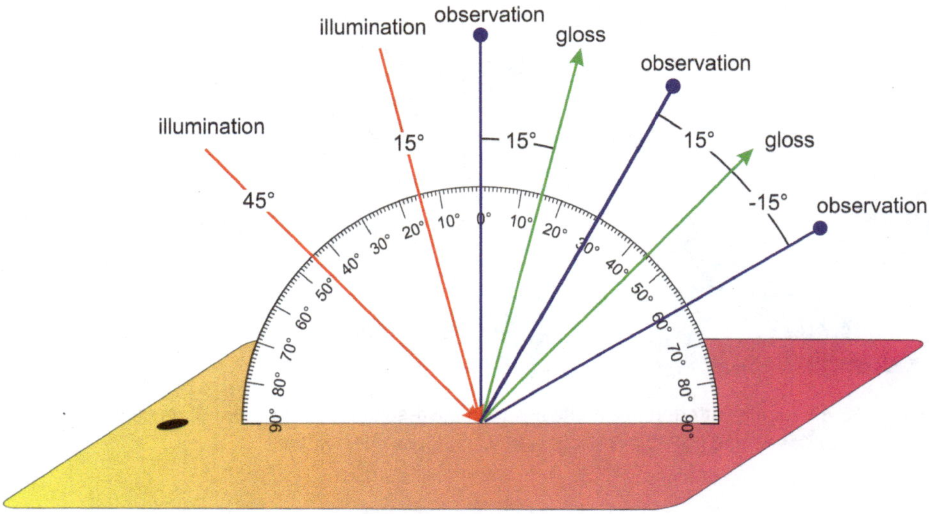

Figure 23.50: ASTM introduced two illumination angles with observations as-15° and as15°.

Correspondingly, the a*b*-values move counterclockwise when illuminated flatter, i.e. pearl green becomes bluer, pearl red becomes more yellow as shown in Figure 23.52.

Many test measurements have shown that the near-gloss geometries as-15° and as15° are most suitable for measuring interference pigments. Measurements closer to gloss – for example at as10° – can prove critical, especially if the color samples are sealed with clear coat. The closer one measures to gloss, the greater the differences between the reflection maxima must become: If the difference between the reflection maxima of 45°/as10° and 45°/as15° is smaller than that of the reflection maxima of 45°/as15° and 45°/as20°, the measurement at 45°/as10° is not plausible and must be classified as an artifact (Figures 23.53 and 23.54).

The second line is composed of the measured values at a fixed illumination angle of 45°. It is called the aspecular line [16]. In these measurements, the color change from reflection color to transmission color can be seen when the interference pigment is applied to a white background as shown in Figure 23.55. The transition range is between 20° and 30° from the gloss angle. The measurement results in this

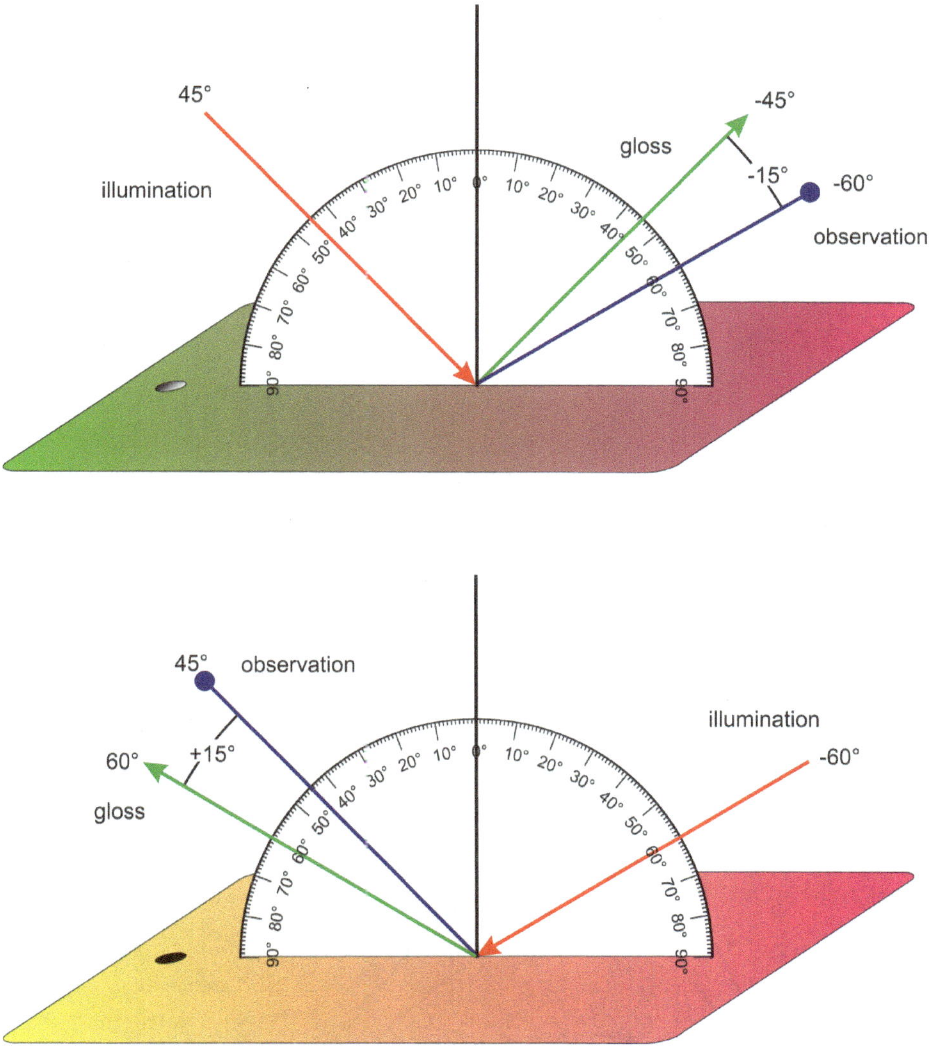

Figure 23.51: Due to the law of light reversal, the geometry 45°/as-15° replaces the cis-geometry 60°/as15°.

range may be critical, as it is not possible to clearly differentiate between the interference pigments (Figure 23.56).

The difference angle to the gloss also plays a role in the aspecular lines. Very close measurement angles to the gloss angle can provide artifacts.

Both lines form an anchor shape which is typical for the respective interference pigment and can be used for its identification and characterization. The anchor shape can be changed by adding colored and aluminum pigments [17]. The aspecular line in particular is influenced, while the interference line remains in its shape and is

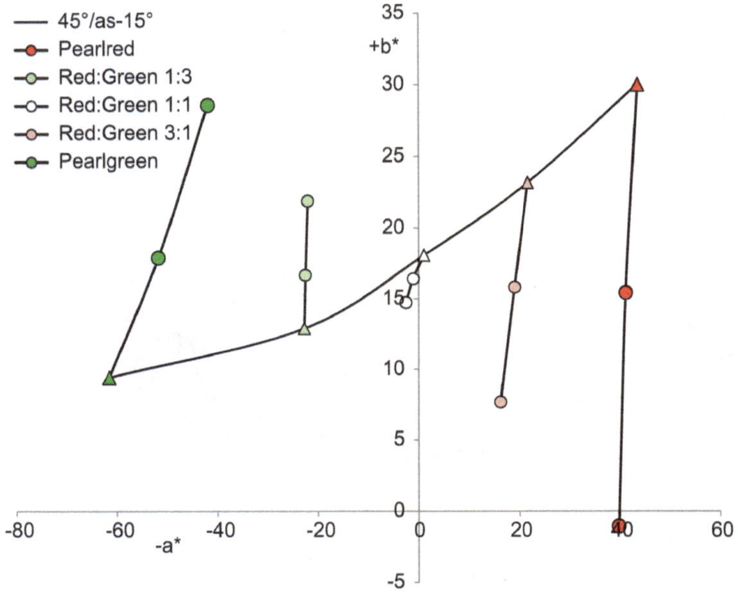

Figure 23.52: When mixing complementary interference pigments, the direction of the interference lines rotates.

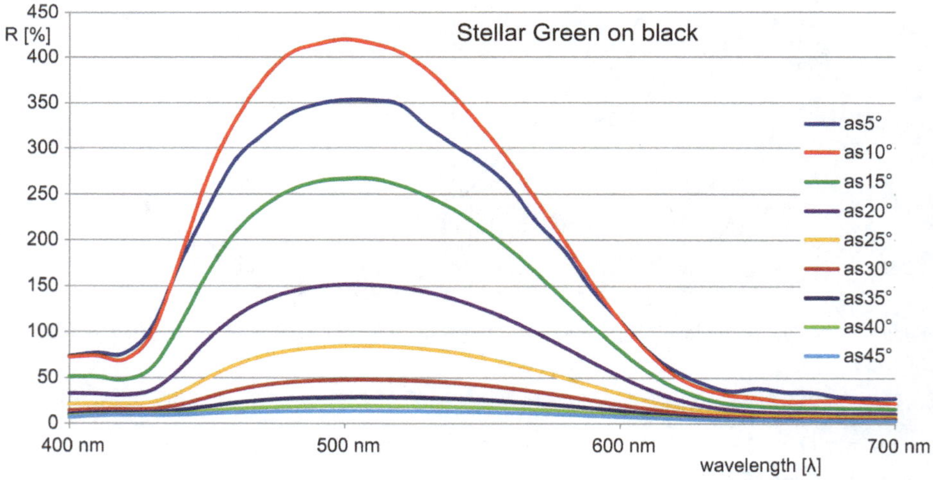

Figure 23.53: Distances between the reflections become larger when illumination is flatter. However, the reflection at 5° from gloss is an artifact.

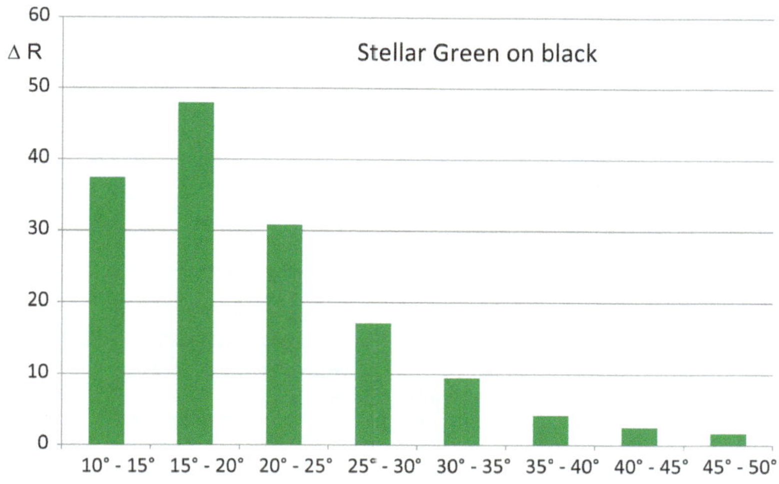

Figure 23.54: Even at 10° from gloss, artifacts are usually obtained. Here the difference between the reflection maxima of 10° – 15° is smaller than the difference of 15° – 20°.

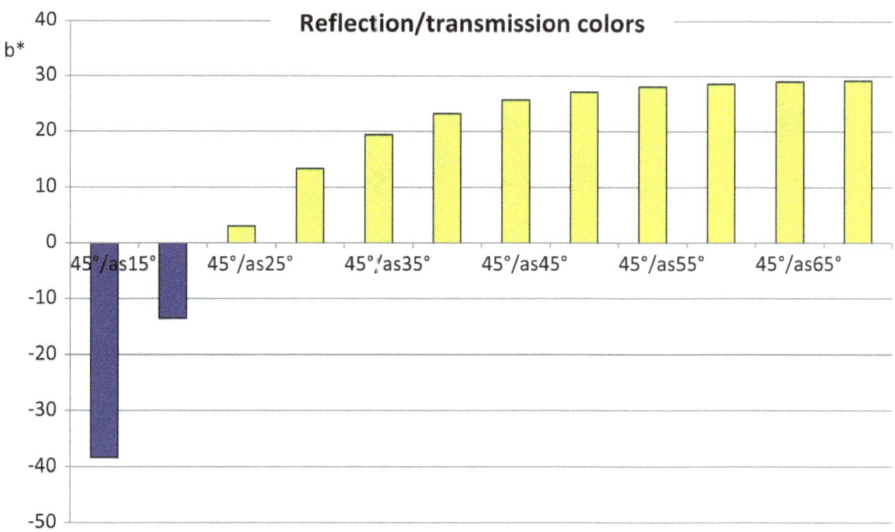

Figure 23.55: Transition from reflection to transmission color: Transition range between 20° and 30° from gloss.

shifted. If it were omitted or completely changed, mixing would not make sense: The optical properties of an interference pigment remain close to the gloss where the interference can be observed. Thus, the blue reflection color of an interference pigment would also be preserved in mixtures with a red colored pigment. If the blue reflection color were no longer visible, mixing would be pointless (Figure 23.57).

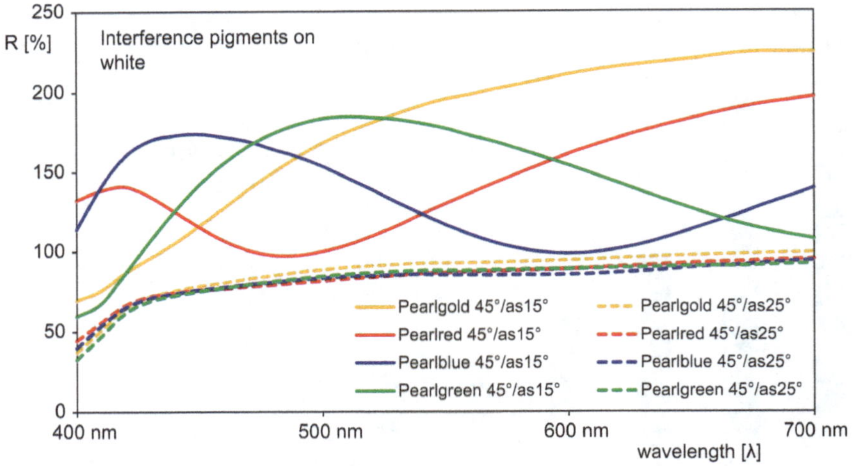

Figure 23.56: Close to the gloss the intensive reflection color can be observed and measured. At 25° from the gloss the reflections of different interference pigments appear the same.

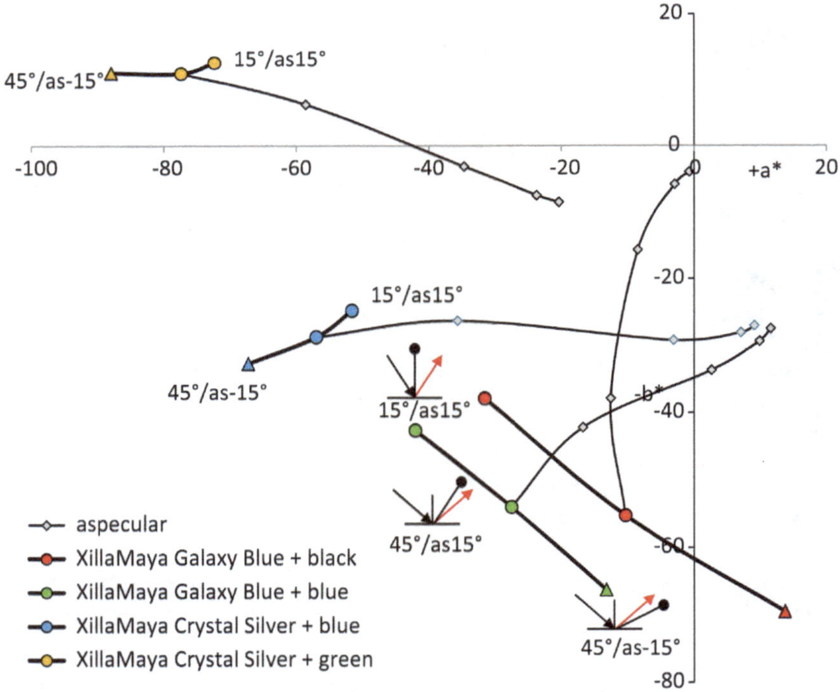

Figure 23.57: Typical interference line of an interference pigment is retained even in mixtures with colored pigments.

The best mixing results can be achieved with rather dark colored pigments such as green and blue. Yellow and red colored pigments have a high brightness and a chroma, which can have a disturbing effect on the interference effect. Blends with aluminum pigments are often critical because aluminum pigments can suppress the interference effect. Nevertheless, mixtures of colored, aluminum and interference pigments are used in automotive coatings today. The aluminum pigments reduce the transparency of the mixture, the interference pigments provide the soft effect and the colored pigments usually determine the color direction.

White interference pigments do not show any shift in the reflection curves under flatter illumination because they do not cause any color shift. However, the reflection values increase with flatter illumination (increasing angle of illumination). This behavior is similar to that of aluminum pigments. For this reason, the interference line runs in an extension of the aspecular line as with aluminum pigments (Figure 23.58).

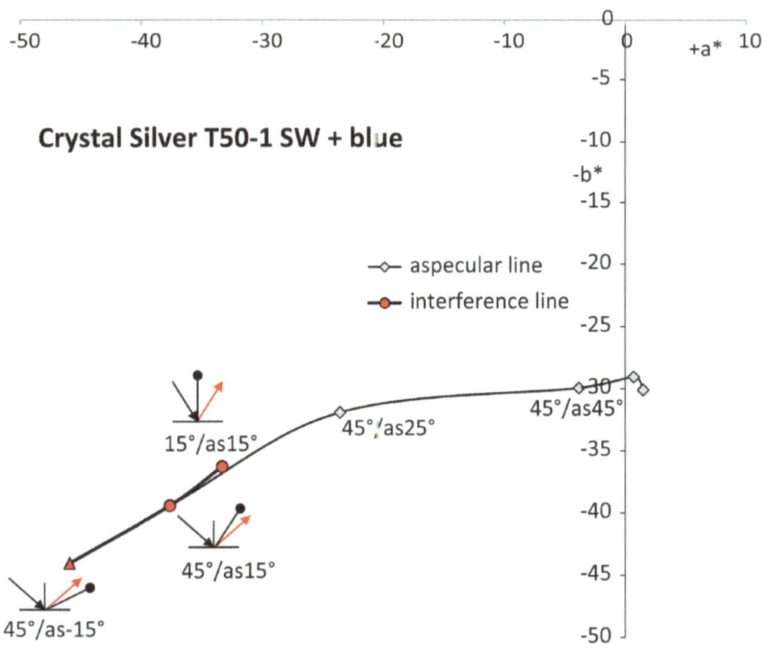

Figure 23.58: Interference line of white interference pigments runs in the extension of the aspecular line.

White interference pigments do not show any coherent measurement results and are therefore difficult to assess. For testing, they are therefore mixed with colored pigments (e.g. blue) (Figures 23.59 and 23.60).

Since the common interference pigments – titanium dioxide coated platelets – are transparent, this is also noticeable in the measurements. If such interference pigments

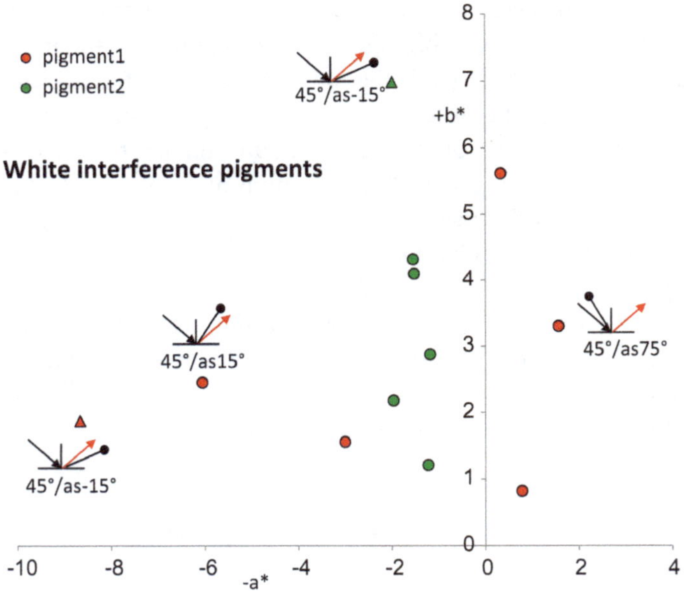

Figure 23.59: Silver–white interference pigments have little chroma. So, it's hard to take plausible measurements.

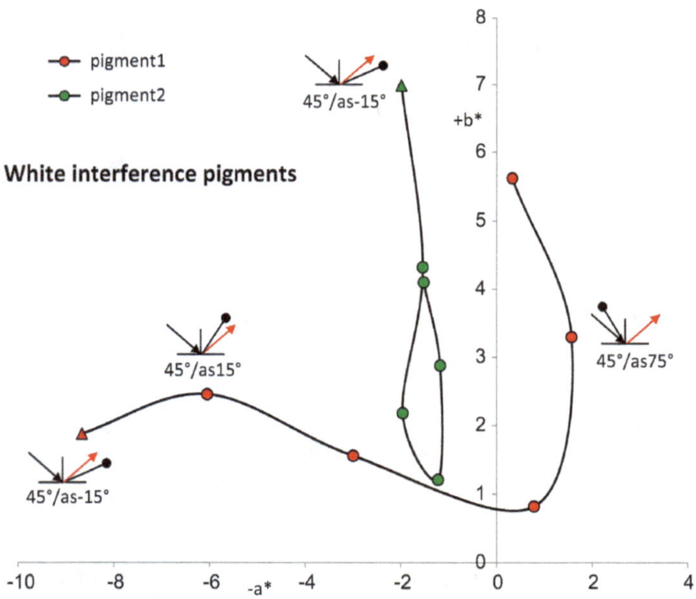

Figure 23.60: Combining of the measured values by a line shows the difficulty to interpret the results, especially when the scaling of the axis is kept in mind.

are applied to a white background, the interference line is comparable as shown in Figure 23.61. However, it shows a lower chroma value. This is due to the white background which shines through the transparent interference pigments and weakens the actual reflection color. The reflection curves of the aspecular geometries "tilt" from the reflection color to the complementary transmission color when measuring further from gloss. Between both positions, there is a transition area between 20° and 30° from the gloss. This transition becomes clear not only with the course of the reflection curves. If one applies the a* or b* values against the aspecular geometries, the change from reflection to transmission color becomes clear. For red–green or green–red changes, the a* values are used, for yellow–blue or blue–yellow changes the b* values.

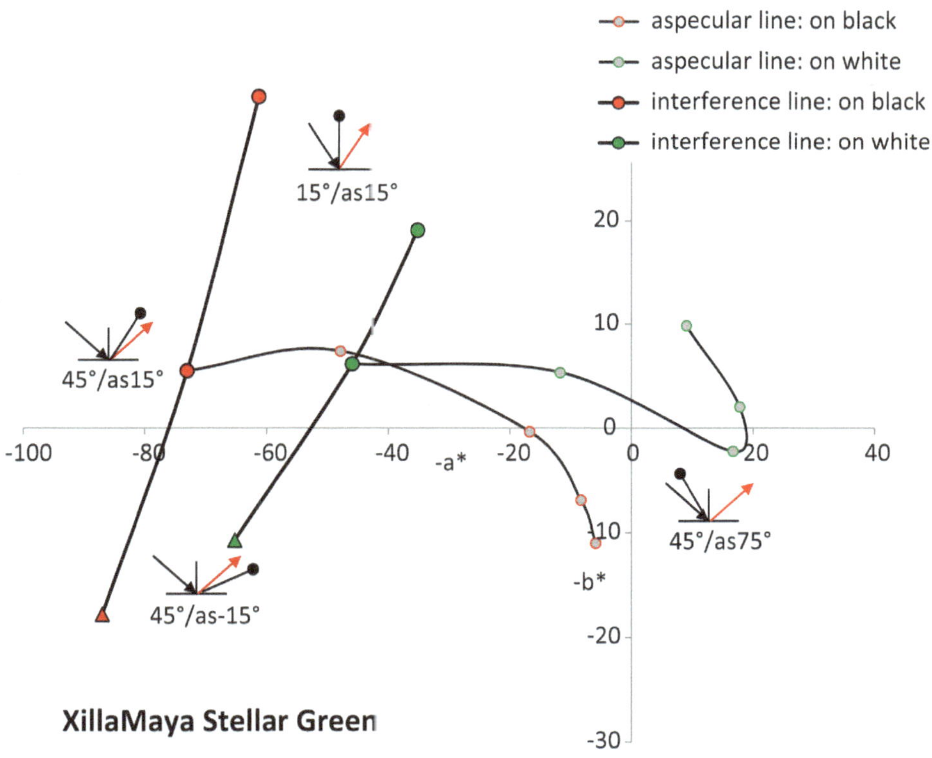

Figure 23.61: Interference pigments on a white background show values at 45°/as75° and 45°/as110° which supposed to be artefacts.

These results show that geometries far away from gloss do not detect the interference. Measurements in the transition region do not necessarily allow differentiation between interference pigments.

For measuring instruments with only one illumination, the part 45°/as15° to 45°/as-15° can be used. This line "bends" counterclockwise relative to the aspecular line. This short interference line behaves in the same way as the described long interference line consisting of three measured values. This also applies to white interference pigments where this section is an extension of the aspecular line.

23.10 Anchor shape

The anchor shape, consisting of interference and aspecular line, clearly characterizes an interference pigment. For description, the zero point of a coordinate system is placed in the measured values of the geometry 45°/as15°. Six parameters can be calculated from the anchor shape, which together describe an interference pigment (Figures 23.62 and 23.63).

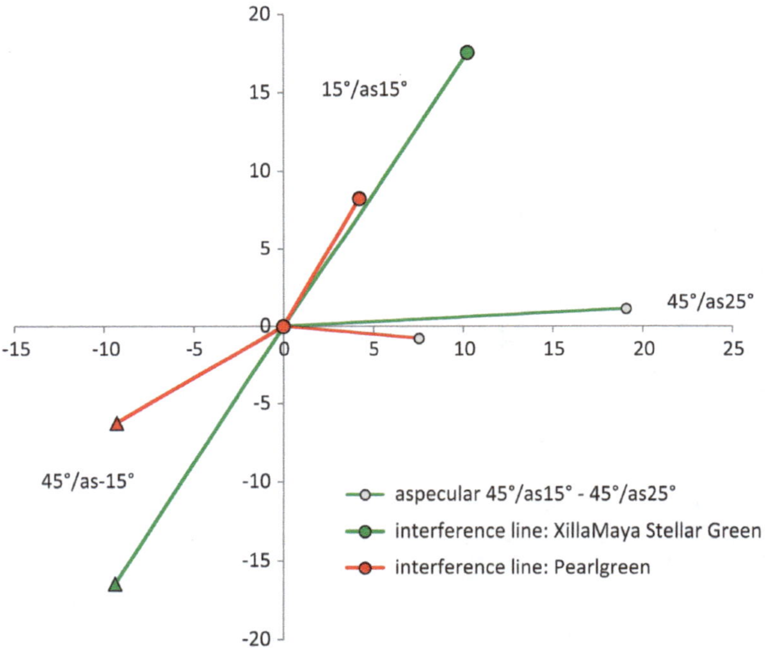

Figure 23.62: A coordinate system that is placed in the measuring point of the geometry 45°/as15°, from which various parameters are calculated.

1. the angle between the partial axis of the interference line from 45°/as15° to 15°/as15° and the *x*-axis
2. the angle between the two parts of the interference line
3. the length of the partial axis of 45°/as15° and 15°/as15
4. the ratio of lengths of both parts of the interference line
5. the angle between the partial axis of the aspecular line 45°/as15° to 45°/as25° and the *x*-axis

6. the angle between the partial axes of the interference line 45°/as15° and 15/as15° and the aspecular line 45°/as15° and 45°/as25°

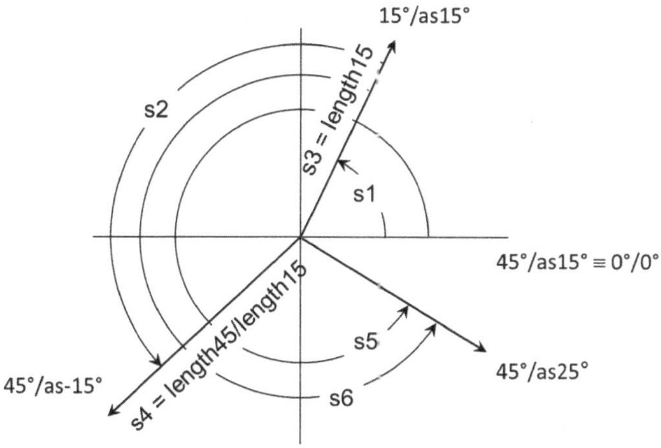

Figure 23.63: Six parameters describing an interference pigment.

These parameters can distinguish between interference pigments that are similarly colored but differently structured, as well as colored and white interference pigments. If the parameters and the corresponding data of the pigments are collected in a database, they can be used to identify interference pigments by query.

23.11 Liquid crystals

Liquid crystals form an intermediate phase between their solid and liquid state with properties of both phases. In this intermediate phase they are colored and change their color depending on the temperature. Today's liquid crystals for screens and displays are not temperature-dependent (Figures 23.64 and 23.65).

23.12 Visual matching

Philosophical aspects can also be cited in the question, do we really see reality. The eye "translates" light rays into optical stimuli, which are converted by the brain into colors and images. Measuring instruments record the light rays, which are physically described. With the help of mathematical formulas, the physical measured values are also "translated" into physiological color values. A camera – digital or analogue – also "translates" the light rays into colors and images. As a result, it can happen that the viewer of the photo thinks he has an exact image of his surroundings in front of

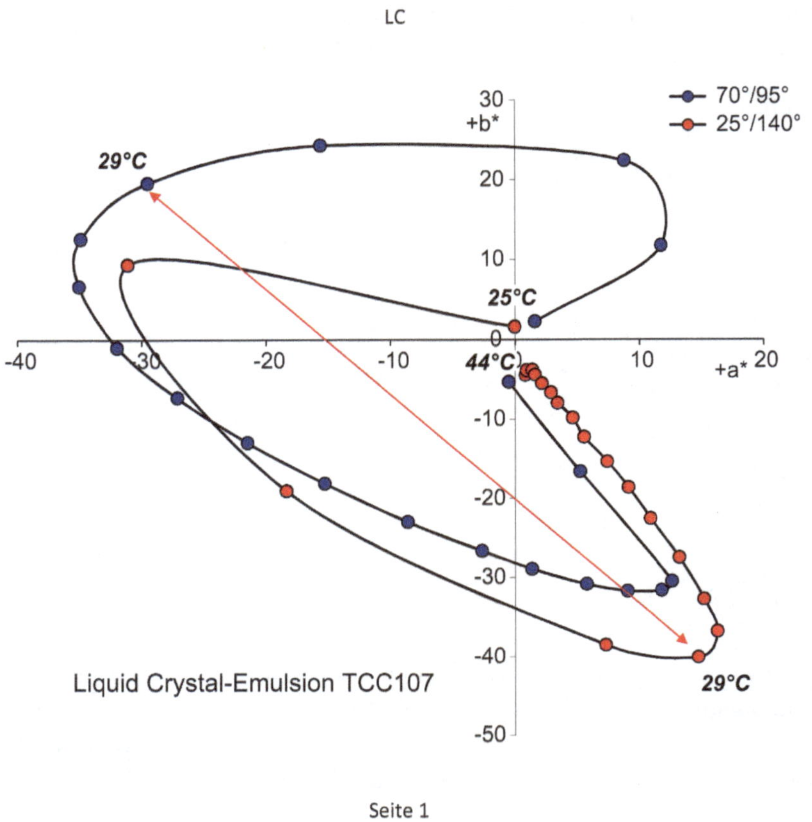

Figure 23.64: Temperature-dependent measurement of a liquid crystal: Shift of the color from red to blue with increasing temperature.

him. One could also speak here of a “retuning”, in which the brain pretends to be identical with the environment.

The different ways of translating the light rays into colors do not necessarily lead to the same result. In the end, however, the visual impression is always decisive. The eye is and remains the yardstick for the representation and evaluation of a color.

There can often be discrepancies between the visual and instrumental assessment of a color. This is especially the case with effect colors, whose color shifts depending on the angle. In most cases, the geometries of both methods do not match and cause color differences.

The current portable measuring instruments illuminate the color sample at a fixed angle. The observation or measurement is carried out at fixed difference angles from the gloss angle at –15°, 15°, 25°, 45°, 75° and 110°.

If a color panel is to be visually inspected, the observer initially holds it so that the light source is reflected in the gloss. Then the observer tilts the color panel downward and then upward or vice versa (Figures 23.66 and 23.67).

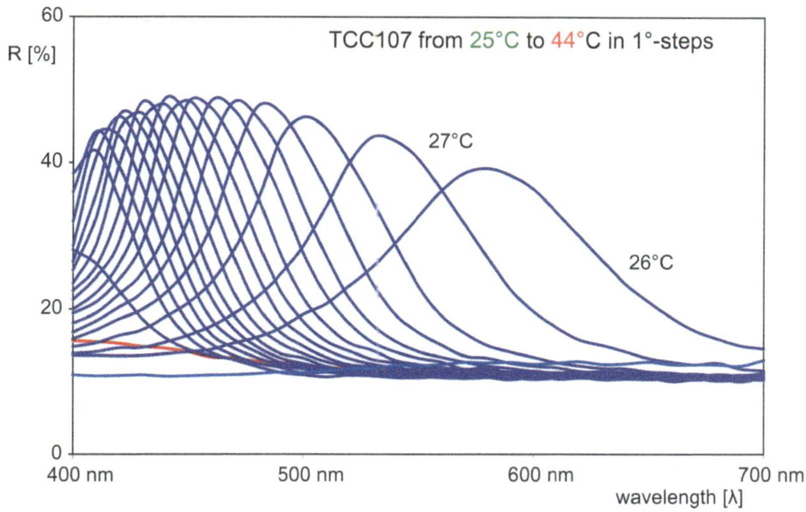

Figure 23.65: Maxima of the reflection curves of a liquid crystal shift to the short wavelength spectral range as the temperature rises.

We start from a starting position where the observer can see the reflection of the light source, for example at the window or in the light booth. If the light source is 20° from the normal, which is thought to be perpendicular to the color panel, the observer also observes at 20°, i.e. the angle between the light source and the observer is 40°. Both positions, that of the light source and that of the observer, do not change and are fixed. If the observer now tilts the panel downward, the light source moves in the direction of the normal, i.e. the angle between the light source and the normal becomes smaller. At the same time, the observer "wanders" to the other side of the gloss angle, i.e. to the side of the gloss angle opposite the illumination. Then the angle between the light source and the normal is 5° for instance. The angle between the normal and the gloss angle is also 5°. Since the position of the observer does not change, his angle to the normal increases to 35°. If the color panel is tilted further, the normal changes to the other side of the illumination angle and the angle of the observer continues to increase (Figures 23.68 and 23.69).

If the observer tilts the color panel toward him, the normal also tilts toward the observer. This increases the angle of illumination, just like the angle of gloss. And the observer moves further away from this angle of gloss.

During the evaluation of the numerical geometries, it turns out that the same geometries are taken up when tilting up and down. So in the end it does not matter if the color panel is tilted up or down. However, it is in the nature of man to perform these movements together (Figures 23.70 and 23.71).

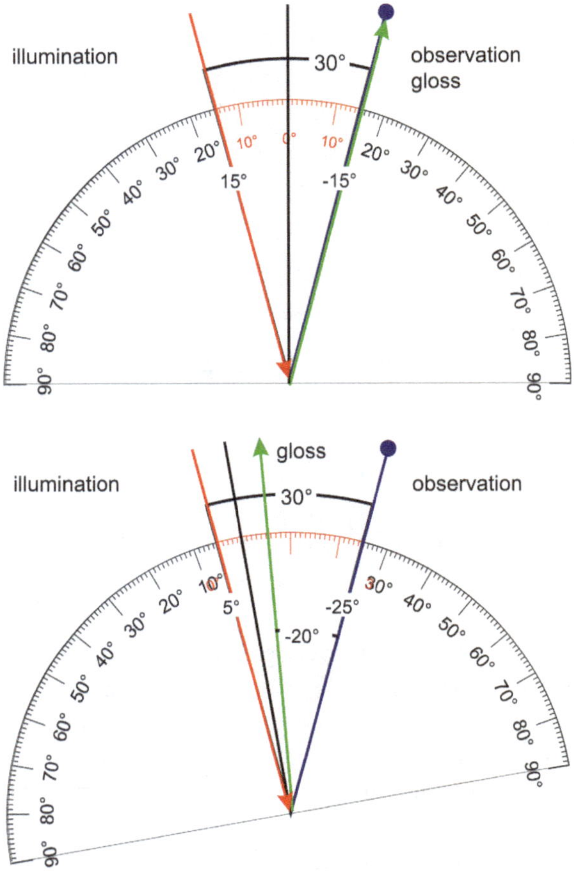

Figure 23.66: If the observer moves the color panel away from himself, the normal also tilts down. The angle between him and the light source always remains the same.

Often the color panel is also held above the head and then moved back and forth. Observation is done with the back to the light source. Here, too, the same geometries result as in the front movements.

The considerations serve to clarify to what extent a comparison between visual and instrumental assessment is possible. Since the measuring geometries play a major role in effect pigments, no agreement is possible for the visual assessment described: In the case of portable measuring instruments, the color panel is illuminated by a fixed light source, whereas in these visual assessments the angle of the illumination changes continuously. The observation angles also change their positions.

In order to occupy the same positions during visual and instrumental assessment, a light source must be placed on the color panel. If this color panel is then tilted, the position of the light source – as with the measuring instruments – always

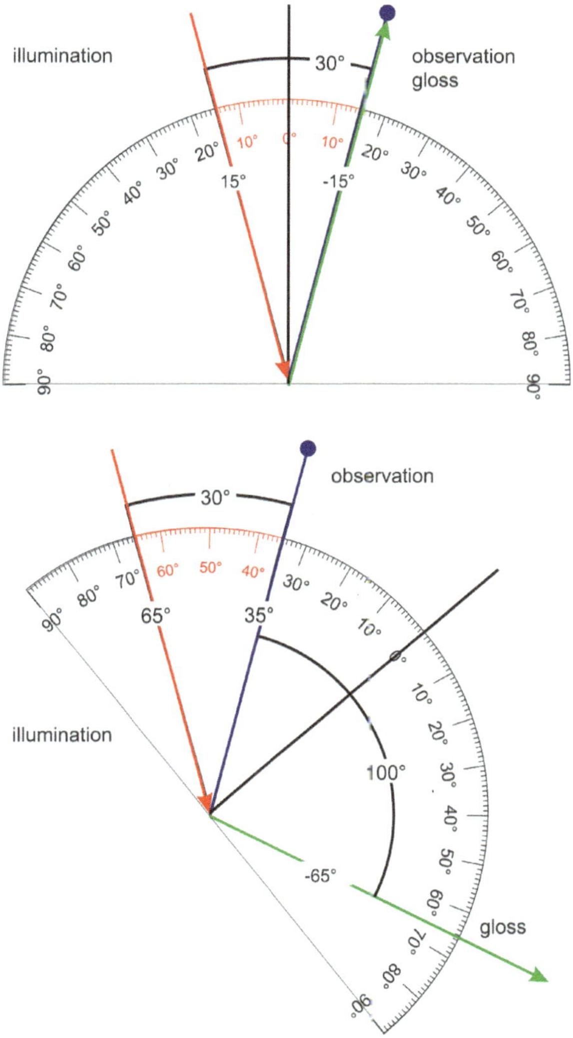

Figure 23.67: When the color panel is tilted toward the observer, the normal also moves toward him.

remains the same. The angle of observation and thus the difference angle (aspecular) changes.

To observe the color change of the interference, the color panel is held at approximately the same height of the light source when the arm of the observer is extended against this light source. Thus the panel is illuminated flat and observed close to the gloss. By lowering the outstretched arm to the bottom and aligning the panel parallel, different illumination angles are passed through. In the lower position, the color panel is also observed close to the gloss under steep illumination. In

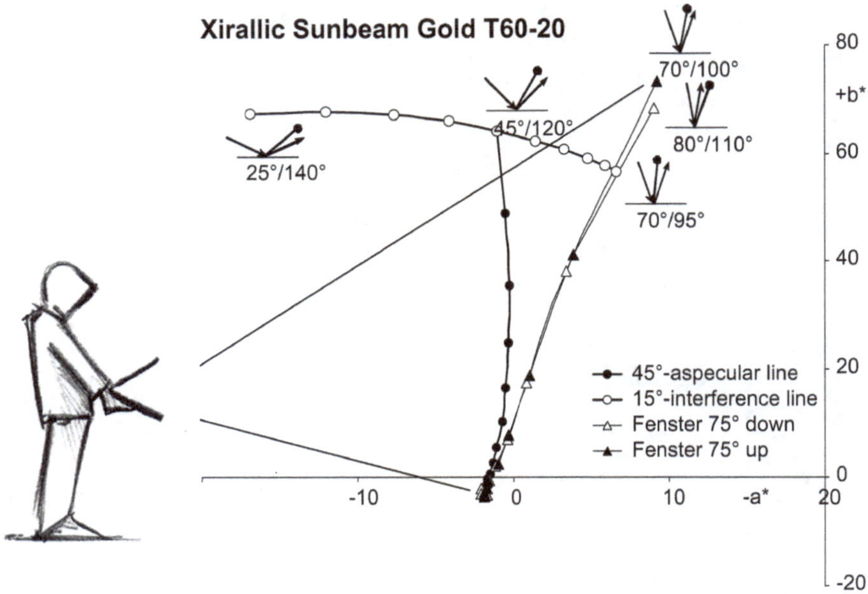

Figure 23.68: Color shift observed during tilting does not correspond to the measured color shift.

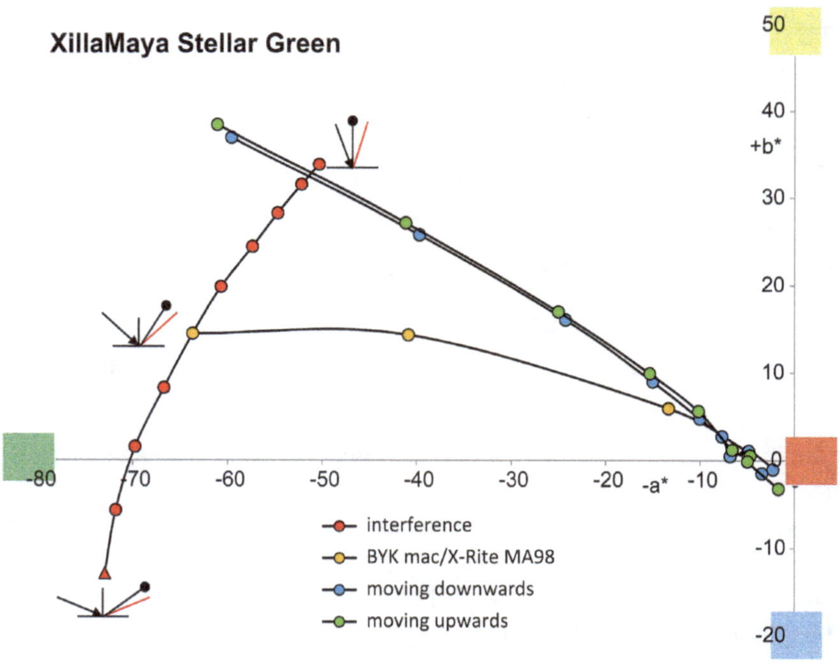

Figure 23.69: Measurements show that the color shift when tilted up corresponds to that when tilted down.

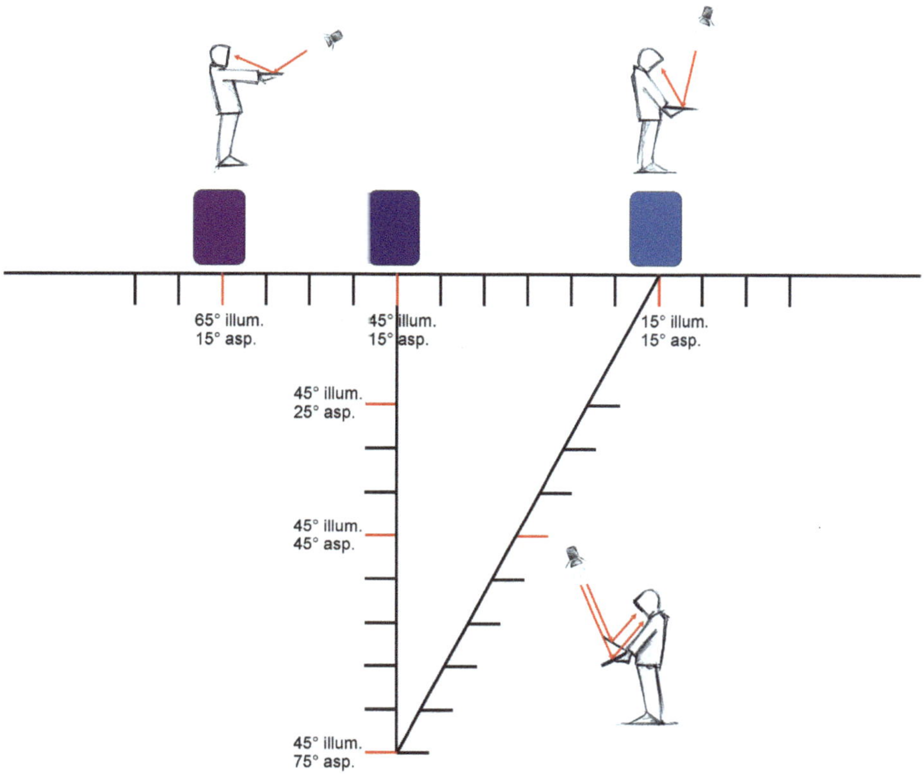

Figure 23.70: Differences between the visual and instrumental assessment are clearly visible.

this way the color change of the interference pigments can be observed visually as shown in Figure 23.72.

The measurement of the colors serves their objective description. This enables communication about colors. Today's demands on the measurement of colors are very high. For this reason, knowledge of the different aspects is necessary to avoid misunderstandings.

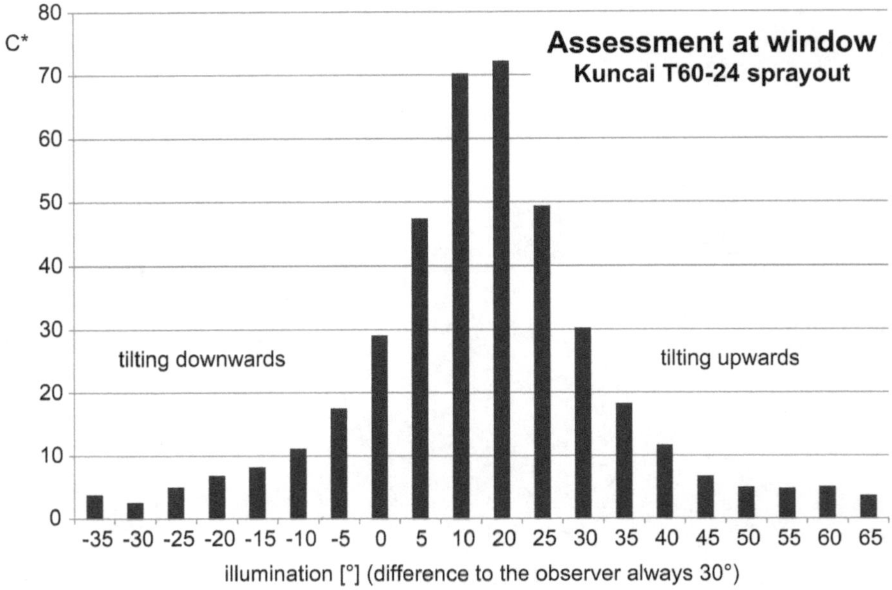

Figure 23.71: If the observer tilts a sample panel up or down, the chroma passes through the same values.

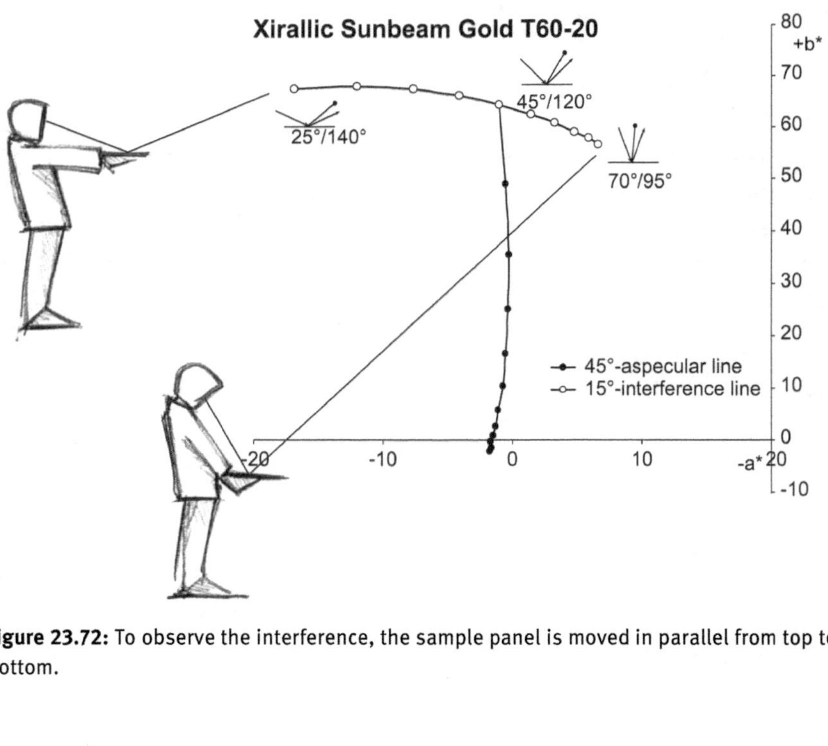

Figure 23.72: To observe the interference, the sample panel is moved in parallel from top to bottom.

References

1. ISO 10527:2007, CIE standard colorimetric observers.
2. Wyszecki G, Stiles WS. Color Science, John Wiley & Sons, 1982.
3. Gerlinger H. Industrielle Farbmesstechnik im Umbruch. Technisches Messen. 1992;59:5.
4. Rodrigues ABJ, Measurement of Metallic and Pearlescent Colors. Die Farbe. 1990;37:65.
5. Gerlinger H. Color Measurement of Pearlescent Paints. Die Farbe. 1990;37:123.
6. McCamy CS. Relating Colorimetry to Visual Observations. Die Farbe. 1990;37:43.
7. Cramer WR. Farbmessung an Glimmerpigmenten. Technisches Messen. 1992;60:229.
8. Gabel PW, Hofmeister F, Pieper H. Interference Pigments a as Focal Point of Color Measurement. Kontakte (Darmstadt). 1992;(2):25.
9. Cramer WR. Magical Mixtures, Paint & Coatings Industry. 1999;9:72.
10. Cramer WR, Gabel PW. Das gewisse Etwas. Farbe + Lack. 2013;109:78.
11. Cramer WR. Examples of Interference and the Color Pigment Mixtures Green with Red and Red with Green. Color Res Appl. 2002;3:276.
12. Kirchner E, Cramer WR. Making Sense of Measurement Geometries for Multi-angle Spectrophotometers. Color Res Appl. 2012;6:186.
13. Cramer WR. Strategisches Farbdesign. Farbe + Lack. 2003;109:30.
14. Cramer WR. Methods for Describing Color and Effect. Paint & Coatings Industry. 2018;6:40.
15. Cramer WR. Undercover Influences. Eur Coat J. 2017;(5):56.
16. Cramer WR. Visual & Instrumental Geometries in Colour Matching. China Coat J. 2012;7:54.
17. Cramer WR. Characterisation of Interference Pigments. China Coat J. 2019;9:130.

Heinz Mustroph

24 Cyanine dyes

Abstract: Cyanine dyes are characterized by an odd number $2n+3$ of π-centers and $2n+4$ π-electrons (where n is the number of vinyl groups –CH = CH–). This special feature has a marked impact on their electronic structure and thus their equilibrium structure in the electronic ground state as well their color and electronic spectrum, respectively. Their first technical application was the use as spectral sensitizers in silver halide photography. Today they have numerous of applications in digital optical data storage, Computer-to-Plate lithographic printing plates, bio-analysis and medical diagnostics.

Keywords: Brunings-Corwin effect, cyanine dyes, cyanine limit, Dewar's Rules, electronic spectroscopy, lithographic printing plates, optical data storage, polymethine dyes, silver halide photography, vinylene shift

24.1 Fundamentals

24.1.1 Basics of the cyanine dye structure

The first synthesized dye of this dye class exhibits a cyan color, which ultimately gave the whole dye class its name. In general *cyanine dyes* consist of a polymethine chain linking two terminal heterocycles, each of which contains a nitrogen atom in conjugation with the polymethine chain. When the two terminal heterocycles are the same the dyes are called symmetrical cyanine dyes, and when these end groups differ the colorants are known as unsymmetrical cyanine dyes. Both subclasses are members of the polymethine family of dyes [1–7].

Cyanine dyes are characterized by an odd number $2n+3$ of π-centers and $2n+4$ π-electrons (where n is the number of vinyl groups –CH = CH–). This special feature has a marked impact on their electronic structure and thus their equilibrium structure in the electronic ground state [1–7]. In all cyanine dyes the π-charge density distribution along the carbon atoms of the polymethine chain alternates. Especially in symmetrical cyanine dyes the differences in the carbon–carbon equilibrium bond lengths are small, which results in narrow absorption bands in the electronic spectrum.

This article has previously been published in the journal Physical Sciences Reviews. Please cite as: H. Mustroph, Cyanine Dyes *Physical Sciences Reviews* [Online] 2020, 5. DOI: 10.1515/psr-2019-0145

https://doi.org/10.1515/9783110587104-024

24.1.2 Experimental properties of the ground state equilibrium structure

From ^{13}C NMR investigations it is well-known that the π-charge density distribution along the carbon atoms of the polymethine chain alternates [3, 4, 6, 8]. This is evidenced by the ^{13}C NMR chemical shifts (δ) along the polymethine chain of cyanine dye **1** (Table 24.1).

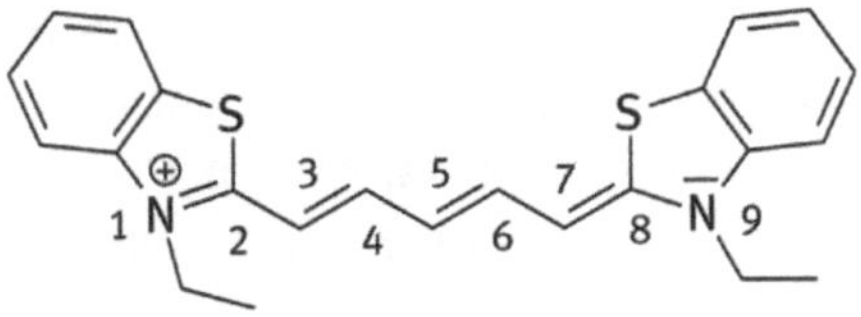

1

Table 24.1: ^{13}C NMR chemical shifts (δ) in ppm along the polymethine chain of cyanine dye **1** measured in [D6]DMSO [8].

C-atom	C2	C3	C4	C5	C6	C7	C8
δ [ppm]	163.2	100.0	150.4	120.8	150.4	100.0	163.2

On the basis of oversimplified theoretical models it is often said that all equilibrium bond lengths in symmetrical cyanine dyes along the conjugated chain are equal [3, 4]. However, in reality the nitrogen–carbon and all carbon–carbon bond lengths do not have identical values. Experimental results show that the differences in the carbon–carbon equilibrium bond lengths in symmetrical cyanine dyes are small, but of significance to this discussion is that they are not identical [9–11] (Table 24.2).

Table 24.2: Equilibrium bond lengths R_e [pm] of cyanine dye **1** along the polymethine chain from crystal structure analysis [9].

Bond	N1-C2	C2-C3	C3-C4	C4-C5	C5-C6	C6-C7	C7-C8	C8-N9
R_e [pm]	138.1	137.6	140.7	137.7	138.3	140.2	139.4	134.7

To demonstrate that the benzothiazole heterocycles are not responsible for the differences in bond lengths, the equilibrium bond lengths for 1,7-bis(dimethylamino) heptamethine streptocyanine **2** are given in Table 24.3.

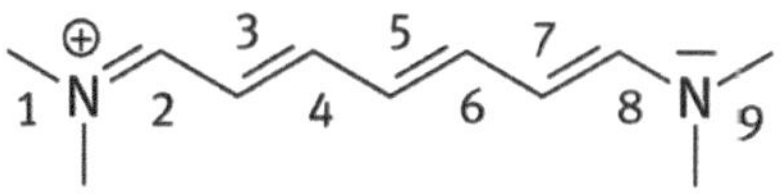

2

Table 24.3: Equilibrium bond lengths R_e [pm] of 1,7-bis(dimethylamino)heptamethine streptocyanine **2** along the conjugated chain from crystal structure analysis [11].

Bond	N1-C2	C2-C3	C3-C4	C4-C5	C5-C6	C6-C7	C7-C8	C8-N9
R_e [pm]	134.5	136.2	137.4	138.2	137.4	140.7	136.4	130.7

Also in the simple streptocyanine **2** there is a small difference between the equilibrium bond lengths. Symmetrical cyanine dyes exhibit an odd alternating conjugated chain, an alternating π-charge density distribution along the carbon atoms of the polymethine chain, and small differences between the carbon–carbon equilibrium bond lengths [3, 4, 6].

In contrast to cyanine dyes polyenes are characterized by an even alternating conjugated chain, equal π-charge density distribution along the carbon atoms of the polymethine chain and substantial differences between the carbon–carbon equilibrium bond lengths [6, 7]. The last point will be illustrated by the carbon–carbon equilibrium bond lengths in the polyene *trans, trans*-1,3,5,7-octatetraene derivative **3**, which are significantly larger than in the streptocyanine **2** (Table 24.4).

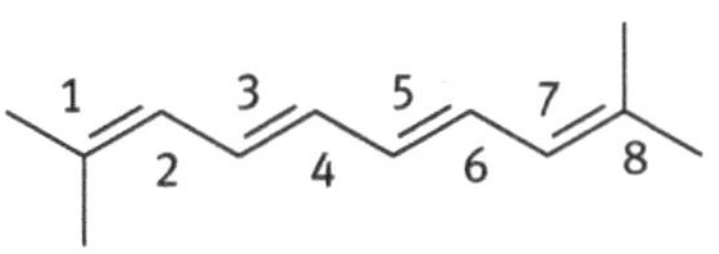

3

Table 24.4: Carbon-carbon equilibrium bond lengths R_e [pm] of polyene **3** along the polymethine chain from crystal structure analysis [12].

Bond	C1-C2	C2-C3	C3-C4	C4-C5	C5-C6	C6-C7	C7-C8
R_e [pm]	133.6	145.1	132.7	145.1	132.7	145.1	133.6

The strongly alternating equilibrium bond lengths in the ground electronic state $R_e(S_0)$ in polyenes lead to large changes of the equilibrium bond lengths in the first excited electronic state $R_e(S_1)$, $[R_e(S_1) \gg R_e(S_0)]$. Therefore, the electronic transition intensity is spread over many members of the vibrational progression and the intensity is not largely concentrated in the 0–0 vibronic transition [7]. A consequence of many vibrational states becoming involved in the electronic transition is that the absorption band broadens, which tends to translate to the colors arising from absorption being duller.

The influence of the molecular equilibrium geometry in the electronic ground state of cyanine dyes on the electronic spectra will be discussed in more detail below. Briefly, if all $R_e(S_O)$ would be equal, the first excited state would have exactly the same equilibrium bond lengths, $[R_e(S_O) = R_e(S_1)]$ and the same potential energy curve. In this case the only vibronic transition is evidently the 0–0 transition. However, the appearance of vibronic bands additional to the 0–0 vibronic band is inconsistent with this model. Owing to the small alternation of equilibrium bond lengths within the polymethine chain of symmetrical cyanine dyes there is a change of $R_e(S_O)$ to $R_e(S_1)$. Indeed it is relatively low $[R_e(S_1) \gtrneqq R_e(S_O)]$ and the absorption intensity is largely concentrated in the 0–0 vibronic transition but this difference explains the appearance of additional vibronic sub-bands [7].

24.1.3 Theoretical models for the description of the electronic structure of cyanine dyes

Regarding the theoretical models describing the electronic structure of symmetrical cyanine dyes the over-simplified model remains very popular. Therefore, a short look into the history of the structural theory is helpful for a better understanding of this development: in the 1860s August Kelulé published some papers suggesting that the benzene molecule has a structure consisting of a six-membered ring of carbon atoms with alternating single and double bonds. In the formalism of the later developed Valence Bond (VB) theory it is said that the electronic structure of benzene can be described by a linear combination of two contributing Kekulé structures. Indeed, with this model it is possible to explain qualitatively the equal carbon–carbon bond lengths and equal π-charge density distribution in the electronic ground state.

This simple model with two contributing structures has been transferred to cyanine dyes (Figure 24.1) and extensively used by Leslie G. S. Brooker et al. [13, 14].

4L **4R**

Figure 24.1: Linear combination of two contributing structures **4L** (with the positive charge on the left hand side) and **4R** (with the positive charge on the right hand side) which describes the resonance hybrid of the Valence Bond theory.

The *resonance hybrid* of both contributing structures **4L** and **4R** leads to the results that all carbon–carbon bond lengths are equal, the positive charge is shared between the two nitrogen atoms only and the π-charge density on all methine groups is equal.

In the 1960s the so-called *mixed valence complexes* started to attract increasing attention due to their interesting optical and electron transfer properties. A mixed valence complex is defined to contain electron donor and acceptor sub-structures separated by a bridge. On the basis of the formal similarity with the simple model for the cyanine dyes alone Melvin B. Robin and Peter Day used it to describe the properties of mixed valence complexes and explained the optical and electron transfer properties with the two contributing structures **4L** and **4R** [15].

In the same manner Seth R. Marder et al. used the VB model with two contributing structures. For the resonance hybrid where the two contributing structures contribute equally they introduced the term *cyanine limit* [16–18]. It follows from this model that all carbon–carbon bonds have equal lengths in the electronic ground and excited state. As is in Brooker´s model the π-charge density distribution on all methine groups is equal and the positive charge is shared between the two nitrogen atoms only.

As mentioned above, from ^{13}C NMR investigations it is well-known that the π-charge density distribution along the carbon atoms of the polymethine chain of cyanines alternates [3, 4, 6, 8]. After due analysis of the model with only two contributing structures, it appears that the positive charge is shared between the two nitrogen atoms only and the π-charge density on all methine groups is equal.

Based on Walter G. König's work concerning polymethine dyes [19] and the knowledge of the results of ^{13}C-NMR measurements Siegfried Dähne introduced the phenomenological model of the *ideal polymethine state* [20, 21]. Without an explanatory theoretical model, he assumed in the ideal polymethine state that π-charge density distribution along the polymethine chain exhibits a maximum of alternation and carbon–carbon bond lengths are identical. This phenomenological model of the ideal polymethine state is illustrated by formulae like **5**. With it the alternating π-charge density and equal carbon–carbon bond lengths shall be outlined.

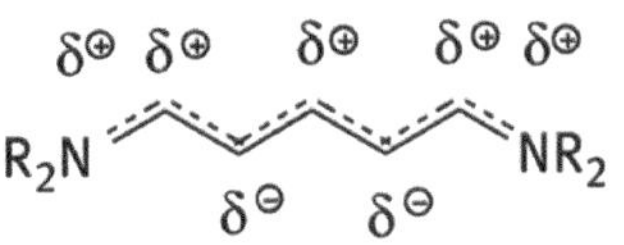

5

It is correct that the π-charge density distribution along the polymethine chain alternates markedly, while the differences in the carbon–carbon equilibrium bond lengths in symmetrical dyes are small, but these lengths are not exactly the same.

It is worth noting that the alternating π-charge density is a substantial difference between the concept of the ideal polymethine state and the model of the cyanine limit.

However, the ideal polymethine state was not derived directly from the theory of the electronic structure: it is a phenomenological description only. By 1939/1940 Linus Pauling [22] and Theodor Förster [23] had already pointed out that the model with two contributing structures is too simple to describe the electronic structure of cyanine dyes. They realized that in addition to these two contributing structures the resonance of positive charge throughout the whole conjugated system must be considered with the other charge separated contributing structures **6S1** to **6S5** (Figure 24.2).

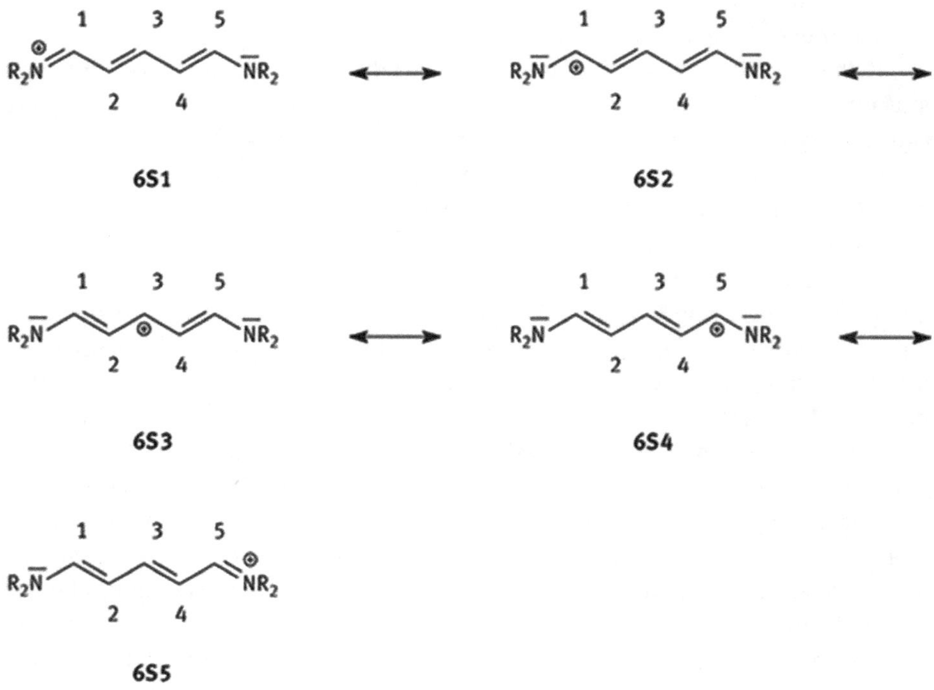

Figure 24.2: Linear combination of five charge separated contributing structures describing the resonance of a positive charge throughout the whole conjugated system within the Valence Bond theory.

Karl F. Herzfeld and Alfred L. Sklar made explicit quantum chemical calculations and demonstrated that the model with two structures is really too simple for the description of cyanine dyes [24–26]. It is therefore all the more surprising that these important evaluations have been ignored, even in recent work, and so this point will be explained in more detail next.

The π-charge density ζ is the difference between π-core charge z and π-electron density q ($\zeta = z - q$). The VB π-core charge on a single N atom is $z = 2$ and the π-electron density is $q = 2$ ($\zeta = 0$), on a N(+) atom is $z = 2$ and $q = 1$ ($\zeta = +1$) and on C it is $q = 1$, $z = 1$

($\zeta = 0$). In the resonance hybrid q is the weighted average of the number of π-electrons occurring in the various contributing structures.

By making the simple assumption that each charge separated contributing structure **6S1–6S5** contributes equally we calculate for N in the resonance hybrid $q_N = 9/5$ and for C1 $q_{C1} = 4/5$ and so on (Table 24.5). Of course, these simple quantitative results do not describe the experimental results exactly (Table 24.1). However, they demonstrate that the VB approach based on just two contributing structures is too simple and leads to wrong conclusions regarding the π-charge density alternation.

Table 24.5: Calculated VB π-electron densities q_i, π-charge densities ζ_i and π-bond orders p_{ij} with the assumption each contributing structure **6S1–6S5** contributes equally.

VB π-electron densities q_i						
N	C1	C2	C3	C4	C5	N
1.8	0.8	1.0	0.8	1.0	0.8	1.8
VB π-charge densities ζ_i						
0.2	0.2	0.0	0.2	0.0	0.2	0.2
VB π-bond orders p_{ij}						
N-C1	C1-C2	C2-C3	C3-C4	C4-C5	C5-N	
0.2	0.6	0.4	0.4	0.6	0.2	

The VB π-bond orders $p_{ij} = 1$ for a double bond and $p_{ij} = 0$ for a single bond in the single contributing structures leads with the model of two structures (Figure 24.1) to $p_{ij} = 1/2$ for all bonds – the ideal polymethine state or the cyanine limit.

In contrast to the simple model, from the Pauling-Förster model follows a second interesting conclusion. In the resonance hybrid Figure 24.2 the VB π-bond order C1–C2 is 3/5 (i. e. > 1/2) and C2–C3 is 2/5 (i. e. < 1/2). By means of this model carbon–carbon bond lengths differences within the polymethine chain are explained, which cannot be described by the model with only two contributing structures.

The theoretical models are discussed here in such detail because, during the last 80 years or so, the VB approach based on just two contributing structures has remained very popular and based on it some further very popular concepts with the same failures were created. The reader should bear in mind that this approach is too simple to describe the electronic structure of cyanine dyes or mixed valence complexes qualitatively.

24.1.4 Reactions of cyanine dyes with electrophiles and nucleophiles

The concept of resonance has been fundamental to our understanding of reactivity in conjugated systems. However, it must be noted that the minimum required number of contributing structures will be taken into account also in qualitative considerations of the electronic structure of molecules. The VB approach based on just two contributing structures does not describe the alternating π-charge density distribution. At least with the simple Pauling-Förster model, it is possible to describe this effect qualitatively, as illustrated in Table 24.5. No question, for an exact correlation with the experimental results like in Table 24.1 more sophisticated theoretical calculations are necessary.

The π-charge density distribution in the polymethine chain has important consequences. For example it explains the behavior of cyanine dyes toward acids and bases. Due to the approach with two contributing structures the π-charge density on all methine groups is $q = 0.5$ and so there would be no preferred orientation for the attack of a proton or a hydroxide anion. Experimentally it is found, in the first step decreasing the pH value a proton attacks the C2-atom in the polymethine chain and increasing the pH value a hydroxide anion attacks the C1-atom leading to a discoloration of the dye-solution [27, 28]. This is in perfect accord with the π-charge density distribution experimentally derived by the ^{13}C NMR chemical shifts (Table 24.1).

The electronic structure of polymethines determines not only their interactions with protons and hydroxide anions. In a similar manner it influences the reactivity with electrophiles and nucleophiles.

A wide range of cyanine dyes substituted at the *meso* position by an alkyl group or a halogen atom are known. Especially in silver halide sensitizers the alkyl groups were introduced in trimethine cyanine dyes to control the aggregation properties and to adjust the absorption maximum.

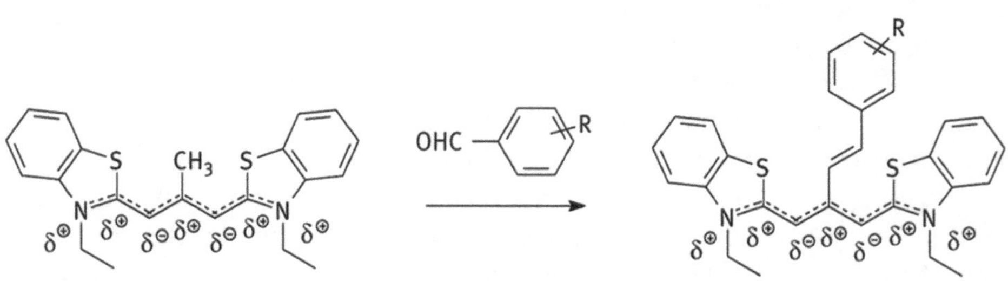

7 8

In *meso*-substitued trimethine cyanine dyes the alkyl group is linked with a partial positive charged methine group, which increases the CH-acidity of the adjacent CH_2R group and so the thermodynamic stability of the resulting carbanion. This effect enables the reaction of **7** with a benzaldehyde derivative under basic conditions to furnish the new cyanine dye **8** [29].

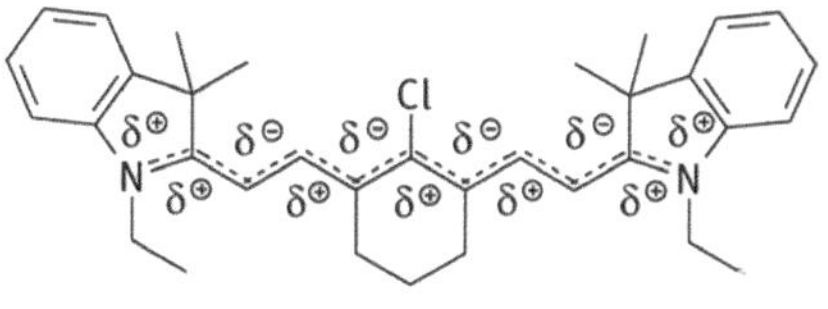

9

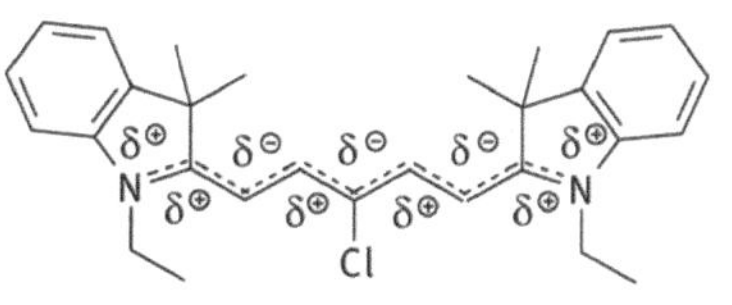

10

The influence of a partial positive charged methine group on the reactivity is demonstrated by **9** and **10**. The positive partial charge of the central methine group in **9** causes its chlorine substituent to be a good leaving group which can be easily replaced with a wide range of nucleophiles [30, 31]. On the other hand, owing to the negative partial charge of the central methine group in **10** the chlorine atom of this dye cannot be substituted by nucleophiles in the same simple manner. Note that this difference in synthetic behavior cannot be explained with the VB approach based on just two contributing structures.

24.1.5 Fine structure of the electronic spectra

The most important property of the dyes is their color and thus their absorption spectra in the visible electromagnetic region. Numerous models have been developed that seek to predict and/or understand spectra of cyanine dyes, but they have shortcomings – an example is the well-known cyanine limit model [16–18]. With reference to the above discussed simple model with two contributing structures in the cyanine limit model it is assumed that the two equivalent limiting forms contribute equally to the electronic ground and excited state of cyanine dyes, which results in equal π-bond orders in S_0 and S_1, respectively. That means that the experimental equilibrium bond lengths in S_0 and S_1 are exactly the same $[R_e(S_0) = R_e(S_1)]$ and the potential energy curves of the two electronic states have the same shape, i. e. both force constants k are the same $[k(S_0) = k(S_1)]$.

With the assumptions the potential energy curves are described by the harmonic oscillator model and $k(S_0) = k(S_1)$ the intensities $I_{0\text{-}v}$ of vibrational-electronic transitions between the vibrational ground state ($v = 0$) in S_0 and vibrational states v in S_1 can be calculated with eq. (24.1) in a simple manner. For background behind the relationship described in this Equation, please see [7] and [32].

$$I_{0-v} = e^{-S} S^v / v! = I_{0-0} S^v / v! \tag{24.1}$$

The parameter S is called the *coupling strength*,

$$S = \tfrac{1}{2}\, k/h\nu_{vi}[R_e(S_1) - R_e(S_0)]^2 \tag{24.2}$$

where ν_{vi} is the vibrational frequency of the symmetric harmonic valence vibration of a diatomic molecule in S_0 and in S_1.

From $R_e(S_1) - R_e(S_0) = 0$ follows $S = 0$ (per definition holds $0^0 = 1$ and $0! = 1$). In all other cases $v > 0$ it follows $S^v = 0^v = 0$. So, for $R_e(S_0) = R_e(S_1)$ the electronic absorption band will consist only of the 0–0 vibronic transition and should show no vibrational fine structure [7, 32].

However, the electronic spectra of most symmetrical cyanine dyes exhibit a clear-cut fine structure (Figure 24.3) with an energy spacing between the first two sub-bands of $1{,}200\,cm^{-1} \pm 200\,cm^{-1}$ [32–37]. Thus, the appearance of vibronic bands in the spectra of many symmetrical cyanine dyes is inconsistent with the cyanine limit model [32–37]. The S_0–S_1 electronic transition in cyanine dyes can be ascribed mainly to an electronic transition from the highest occupied molecular orbital (HOMO) to the lowest unoccupied molecular orbital (LUMO). The HOMO has bonding character whereas the LUMO has antibonding character. An electron in an antibonding molecular orbital results in a lengthening of the bond, that is $R_e(S_1) > R_e(S_0)$. It

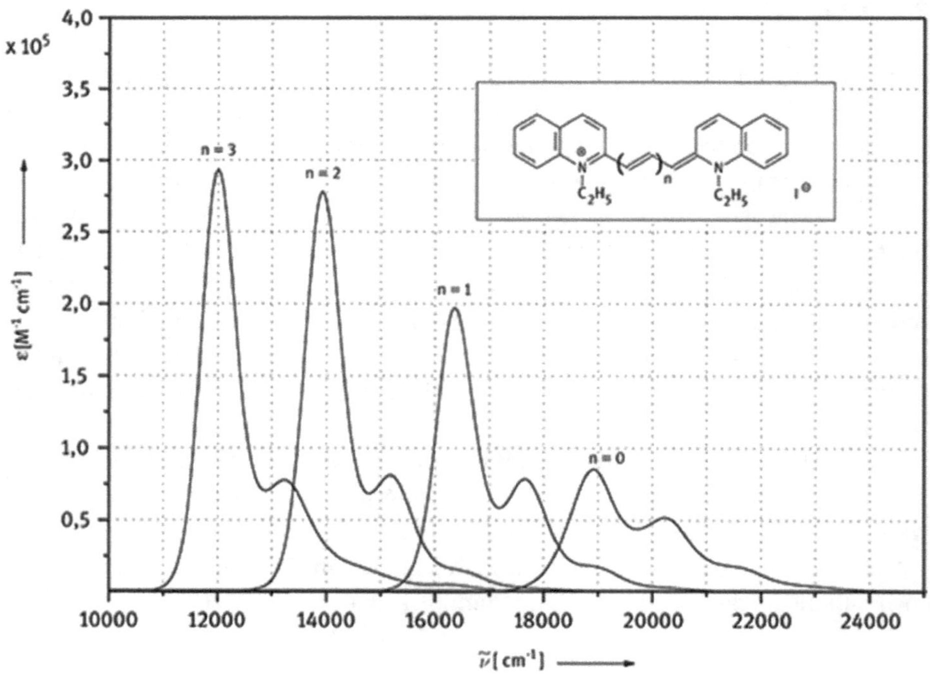

Figure 24.3: Electronic absorption spectra of a series of carbo-cyanine dyes in dichloromethane solution which show the influence of polymethine chain length n.

follows that the assumption R_e (S_0) = R_e (S_1) is wrong. The observed sub-bands correspond mainly to vibronic transitions from $v = 0$ in S_0 to $v = 0, 1, 2, \ldots$ in S_1, where v is the vibrational quantum number of the respective vibrational state of the symmetric carbon–carbon valence vibration of the polymethine chain [32–37].

However, one should pay attention to unsymmetrical cyanine dyes. With increasing difference of the electronic structure between ground and excited state the difference $R_e(S_1) \gg R_e(S_0)$ increases substantially. This causes that the relative intensity of vibronic transition 0–v to increase and that of the 0–0 transition to decrease. In addition, the clear-cut fine structure of the absorption band could be blurred (Figure 24.4) [38].

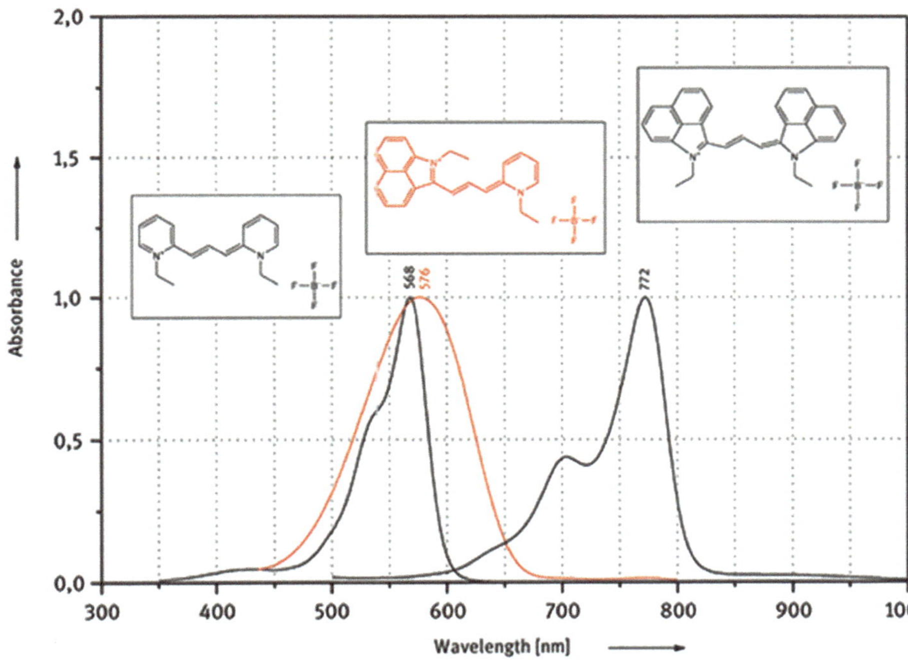

Figure 24.4: Normalized electronic absorption spectra of two symmetrical cyanine dyes and the corresponding unsymmetrical dye in DMSO.

24.1.6 Vinylene shift and intensity

In the formula in Figure 24.3, n is the number of vinylene groups in the polymethine chain (n = 0, 1, 2, etc.). As the value of n increases by one, the maximum of the 0–0 vibronic transition gives a bathochromic shift of about 100 nm (the so-called *vinylene shift*) [1–6, 19, 34, 37] as illustrated in Figure 24.3 and Table 24.6.

Table 24.6: Absorption maxima λ (nm) and molar absorption coefficients ε (M^{-1} cm^{-1}) of the 0–0 and 0–1 sub-bands in the electronic absorption spectra of cyanine dyes in dichloromethane and the wavenumber difference $\tilde{\nu}(0–1) - \tilde{\nu}(0–0)$ in cm^{-1} of both sub-bands showing dependence on the number of vinylene groups n.

n	λ (0–0)	ε	λ (0–1)	ε	$\tilde{\nu}(0–1) - \tilde{\nu}(0–0)$
0	529	85,000	495	51,600	1,298
1	611	197,000	567	78,300	1,270
2	718	278,000	659	80,700	1,251
3	833	293,000	757	79,000	1,205

Second, these experimental values illustrate the overall effect, as the absorption maximum of the 0–0 sub-band increases, the molar absorption coefficient ε rises [37]. However, beginning with $n = 4$ and more noticeably at chain lengths $n > 4$, the absorption band becomes increasingly broad and ε falls. The reason is the presence of a complex mixture of geometric isomers and loss of molecular planarity, which cannot be described by simple models.

24.1.7 Dewar´s Rules

Based on Perturbational Molecular Orbital (PMO) theory Dewar drew up a set of rules for predicting the effect of structural changes on the 0–0 vibronic transition energy of odd alternant conjugated systems [39]. He labelled the terminal atoms of the polymethine chain with a star and did also for every alternate atom between them. So, in odd alternant conjugated systems it is characteristic that they have one more "starred" than "un-starred" atoms (Figure 24.5).

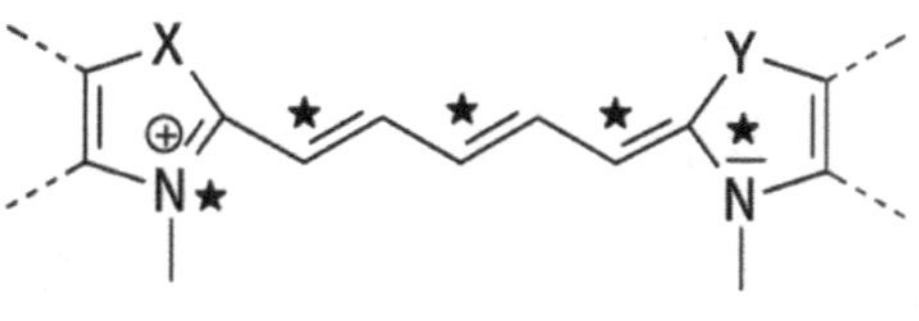

Figure 24.5: Illustration of Dewar´s Rules for cyanine dyes.

Dewar´s Rules predict an alternating influence of substituents on the chain or heteroatoms in the chain on the 0–0 vibronic transition depending on their electronic properties. Very briefly, they apply as follows:

1. Replacement of a carbon atom at a "starred" position by a more electronegative heteroatom or introduction at a "starred" position of an electron-withdrawing substituent causes a hypsochromic shift and vice versa a bathochromic shift at an "un-starred" position.

2. Attaching an electron donating substituent at a "starred" position causes a bathochromic effect and vice versa a hypsochromic effect at an "un-starred" position.

The original rules [39] have been subsequently modified and extended [40]. There are a lot of experimental results, which support these rules [41–43].

24.1.8 Brunings-Corwin effect

In addition, in even alternant conjugated systems (e. g. polyenes, azobenzenes) steric effects which result in non-planarity cause a reduction of intensity and a hypsochromic shift. Also in odd alternant conjugated systems steric effects cause a decrease in intensity. However, in complete contrast to even systems, deviations from planarity in odd systems lead to a bathochromic effect, which was first observed by Brunings and Corwin [44]. They observed that the N,N´-di-methyl dye **12** absorbed at longer wavelengths than the di-NH dye **11** [44]. The planarity along the polymethine chain of dye **12** is diminished compared to dye **11** in order to relieve the steric crowding caused by the introduction of methyl groups, and this increased deviation from co-planarity in an odd-alternant system results in a red-shifted absorption maximum of lower intensity.

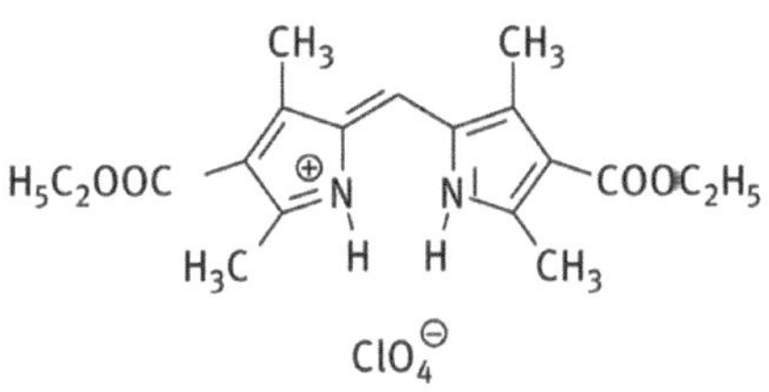

11

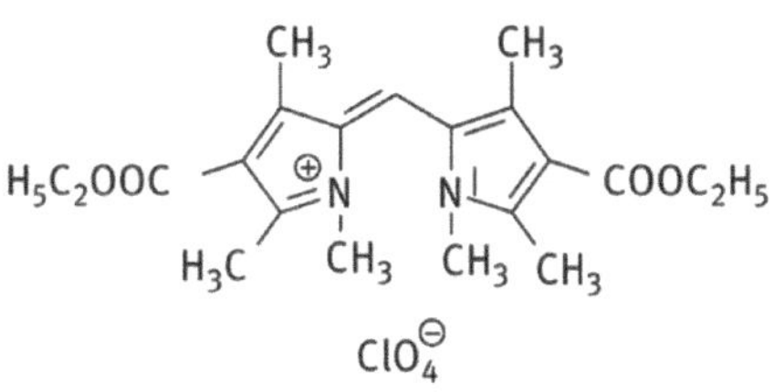

12

λ_{max} = 470 nm (ε_{max} = 150,000 M^{-1} cm^{-1}) λ_{max} = 510 nm (ε_{max} = 85,000 M^{-1} cm^{-1})

The general validity of this effect was confirmed by extensive further experiments [45].

24.1.9 Cyanine dye aggregates

Cyanine dyes are well-known to undergo pronounced spectral changes in water which are dependent upon the chemical structure of the dye, concentration, temperature and the presence of electrolytes or surfactants. These spectral changes have long been attributed to dye molecule aggregation. The phenomenon of cyanine dye aggregation has implications for application of these dyes in fields such as color photography and fluorescent dyes for bio-analysis and medical diagnostics – this aspect will be discussed in the corresponding chapters. Progressive increases in dye concentration results in an initial loss of the monomer M-band intensity at the expense of the coexisting blue-shifted dimer D-band, followed by the H-band (hypsochromically shifted compared to the monomer) and in some cases followed by the sudden appearance of an additional, red-shifted, characteristically sharp, J-band (J for Edwin E. Jelley) above a critical dye concentration. Aggregation (self-association or self-organisation) of cyanine dyes has been studied extensively since these are the best known self-aggregating dyes and was extensively reviewed [3–5, 46–48]. This type of behavior in water is due to the strong intermolecular van der Waals attractive forces (dispersion forces), associated with the high π-electronic polarizability of the polymethine chain, in combination with the hydrophobic effect [46–51].

The relationship between the number and the relative orientation of the dye molecules and the spectral shift of a dye aggregate has been explained in terms of molecular exciton theory [52]. In reality the shape and splitting of electronic absorption bands are influenced by molecular vibrations and lattice vibrations [32, 52]. To make it easier to explain the basics of the molecular exciton theory both types of vibrations are not considered in this chapter. In addition, for the present purpose it will be sufficient to consider the theory in the simple point dipole approximation proposed by McRae and Kasha and to start with a dimer [53]. Within this model the energy of the electronic ground state of the dimer is derived from the solution of the Schrödinger equation (SE). It follows that the intermolecular van der Waals interaction energy lowers the energy of the ground state of the molecules in the dimer in comparison with the monomers. For simplicity in the following the van der Waals interaction energy is neglected and so all electronic transitions start from the electronic ground state of the monomers [S_0(M)] in Figure 24.6].

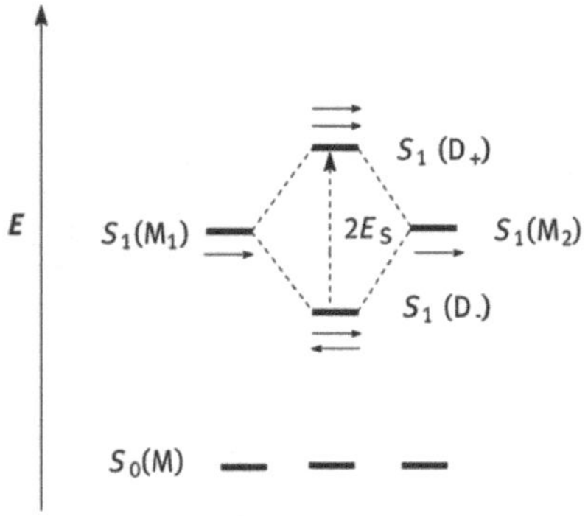

Figure 24.6: Interaction diagram of the two excited monomer states $S_1(M_1)$ and $S_1(M_2)$, respectively, to both dimeric states $S_1(D_+)$ and $S_1(D_-)$ with the splitting energy $2E_S$ and the electric transition dipole moments symbolized by the arrows.

The solution of the SE of the dimer excited electronic state results in a splitting off into the two excited electronic states $S_1(D_+)$ and $S_1(D_-)$ (Figure 24.6). Because of electronic degeneracy of both single molecules, that is, they have equal electronic energies, the energy of $S_1(D_+)$ and $S_1(D_-)$ differ energetically by the *splitting energy* $2E_S$. The electronic transition energy of the monomer $\Delta E[S_1(M)]$, respectively, the wavenumber $\tilde{\nu}_M$ is

$$\Delta E[S_1(M)] = E[S_1(M)] - E[S_0(M)] = \tilde{\nu}M \tag{24.3}$$

and the respective transition energies of the dimers arise in terms of $\Delta E[S_1(M)]$ to

$$\Delta E[S_1(D_+)] = \Delta E[S_1(M)] + E_S \tag{24.4}$$

$$\Delta E[S_1(D_-)] = \Delta E[S_1(M)] - E_S \tag{24.5}$$

Therefore, $\Delta E[S_1(D_-)]$ should be shifted bathochromically to the monomer absorption and $\Delta E[S_1(D_+)]$ hypsochromically. In cyanine dyes the electric transition dipole moment lies along the axis of the molecule. Because the two transition moments are parallel to each other and are equal in magnitude it follows that the vectorial addition of the two moments to $S_1(D_-)$ gives zero, so that only the transition to $S_1(D_+)$ can be measured experimentally.

The solution of the SE for E_S leads to an extensive term [53]. For parallel orientated molecules in a dimer, E_S becomes

$$E_S = \boldsymbol{M}^2 r^{-3}(1 - 3\cos^2\alpha) \tag{24.6}$$

where $\boldsymbol{M}$ are the electric transition dipole moment in the monomers, r the distance between centers of adjacent molecules and α the tilt angle between the long molecule axes and the line connecting the centers of the parallel orientated molecules [53] (Figure 24.7).

Figure 24.7: Schematic representation of the tilt angle of the long molecule axes and the line connecting the centers of the parallel orientated molecules in the aggregate, where the arrows symbolize the electric transition dipole moments of the monomers.

Therefore, E_S in a dimer depends on $\boldsymbol{M}$, r and α. In addition, in N-mer aggregates the splitting energy depends on the number of molecules in the aggregate N [eq. (24.7)].

$$E_{SN} = (N-1)/N \cdot \boldsymbol{M}^2 r^{-3}(1-3\cos^2\alpha) \tag{24.7}$$

Experimentally it was established that the shift of the transition energy $\Delta\tilde{\nu}_{\mathrm{N}}$ of the *N*-mer aggregate $\tilde{\nu}_{\mathrm{N}}$ from the monomer transition energy $\tilde{\nu}_{\mathrm{M}}$ [eq. (24.3)], can governed by eq. (24.8) [54].

$$\Delta\tilde{\nu}_{\mathrm{N}} = \tilde{\nu}_{\mathrm{N}} - \tilde{\nu}_{\mathrm{M}} = (N-1)/N \cdot \Delta\tilde{\nu}_{\infty}(1-3\cos^2\alpha) \tag{24.8}$$

Here $\Delta\tilde{\nu}_{\infty}$ is the shift of the transition energy of an infinite aggregate. Equation (24.8) provides a powerful tool for experimental investigations into the kind and size (J-aggregate, H-aggregate, dimer) of the aggregates, which will be illustrated in the following.

Whether a dye aggregate exhibits a bathochromic or hypsochromic shift depends on the tilt angle α.

In H-aggregates it is assumed that they have a tilt angle $\alpha = 90°$ ($\cos 90° = 0$). Because the term $(N\text{-}1)/N$ converges rapidly, it follows that the separation of the transition energy of the H-aggregate $\Delta\tilde{\nu}_{\mathrm{H}}$ is roughly equal to the infinitely large aggregate $\Delta\tilde{\nu}_{\infty}$.

$$\Delta\tilde{\nu}_H = \tilde{\nu}_H - \tilde{\nu}_M \approx \Delta\tilde{\nu}_{\infty} \tag{24.9}$$

The next interesting conclusion from this model is that the separation between monomer and parallel dimer transition $\Delta\tilde{\nu}_{\mathrm{D}}$ is half the separation between monomer and infinitely large aggregate.

$$\Delta\tilde{\nu}_D = \tilde{\nu}_D - \tilde{\nu}_M \approx {}^1\!/_2\Delta\tilde{\nu}_{\infty} \tag{24.10}$$

For the dimeric transition energy it was found experimentally $\Delta\tilde{\nu}_{\mathrm{D}} \approx 1{,}400\ \mathrm{cm}^{-1}$ which is similar in magnitude to the energy of the 0–1 vibronic transition. Therefore, when discussing the electronic spectra of monomeric cyanine dyes, it must be carefully considered whether the Lambert–Beer law applies. The formation of dimers in solution leads to deviations from the Lambert–Beer law, enabling their presence at various concentrations to be tested for spectroscopically through determination of molar absorptivity within that range of concentration.

Due to the approximated value for $\Delta\tilde{\nu}_{\mathrm{D}}$ we can deduce $\Delta\tilde{\nu}_{\mathrm{H}} \approx 2{,}800\ \mathrm{cm}^{-1}$ as the largest hypsochromic shift of infinite H-aggregates.

In J-aggregates the dye molecules are disposed head-to-tail, i. e. α decreases Figure 24.7 b). With decreasing α the energy of the upper excited state decreases and hence the absorption maximum of a J-aggregate is bathochromically shifted compared to the monomer [3–5, 46–48, 53]. For the angle $\alpha = 54.74°$ ($\cos^2 54.74° = 1/3$) the splitting energy is zero and so $\tilde{\nu}_{\mathrm{N}} = {}_{\mathrm{M}}$.ny tilde subscript M as one symbol With further α decreasing $\tilde{\nu}_{\mathrm{N}}$ is bathochromically shifted in comparison with $\tilde{\nu}_{\mathrm{M}}$. This shift could be up to several hundred nm.

This model explains in a simple manner the basic behavior of cyanine dye aggregates. More sophisticated models were recently reviewed by Nicholas J. Hestand and Frank C. Spano [55].

24.2 History of the cyanine dyes

It has been well documented that in the Easter holidays of 1856 William H. Perkin was following up on August Wilhelm Hofmann's suggestion that quinine might be prepared synthetically from allyl toluidine. As a result of the reaction he got a black precipitate. Extracting it with ethanol he obtained purple crystals. He recognized that this new material might serve as a dye and patented it that summer [56]. It proved particularly popular in France and acquired the name *Mauveine* there. With this trade name the dye became very popular and in general it is said, *Mauveine* was the first synthetic dye.

However, at a meeting of the Royal Society of Edinburgh on April 7th 1856 Charles Greville Williams reported that quinoline obtained by the distillation of cinchonine with potash, after quarternization with methyl (**13a**), ethyl (**13b**) and amyl iodide (**13c**) gave rise to beautiful blue dyes on treatment with ammonia [57]. The paper describes many experiments in considerable detail together with analytical results that one can expect that William´s dye was the first synthetic dye of commercial value. Nevertheless, this dye is hardly mentioned in papers or book chapters describing the history of synthetic dyes.

Williams did not suggest a name for the blue dye. In the same year also von Babo examined decomposition products of cinchonine and synthesized exactly the same blue dye type which he named *Irisin* [58].

It is interesting neither Perkin nor von Babo or Williams had the goal to synthesize dyes. Both synthetic dyes arose out of doomed efforts to synthesize quinine and cinchonine, respectively. Quinine was used chiefly in the treatment of malaria. Administration of quinine dramatically improves the condition of a person with malaria. In addition quinine also played a significant role in the colonization of Africa by Europeans. Therefore, it was very important to find alternative sources based on chemical synthesis.

Also the first synthesis in a large scale had to do with quinine. Chinoidin, a mixture of quinine, cinchonine and other alkaloids, was a by-product of quinine extraction and thus commercially available. In 1861, Guido Schnitzer described the synthesis of the blue dye in a larger scale starting from chinoidin. For the quarternization he used amyl iodide which has a lower volatility than methyl or ethyl iodide and was easier to handle at that time. He called this dye *Chinolinblau* (Quinoline Blue) [59].

However, neither *Irisin* nor *Chinolinblau* prevailed as a name for this dye type. The company of Ménier in Paris manufactured larger quantities of **13c** using amyl iodide and exhibited the dye on the *London International Exhibition on Industry and Art* from 1 May to 15 November 1862. Here, it was Ménier, who gave the name *cyanine* to this blue dye which later became widely synonymous with the whole class of colorants rather than just this single dye.

Later it was found, that Williams´ "quinoline" was a mixture of quinoline and lepidine (4-methylquinoline) that gave the dyes **13**.

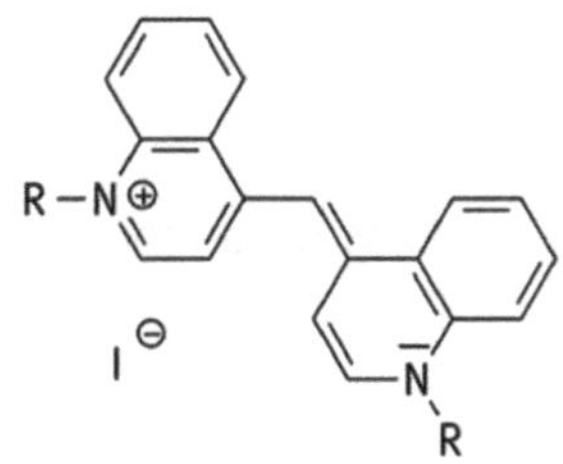

13a: R = CH_3
13b: R = C_2H_5
13c: R = C_5H_{11}

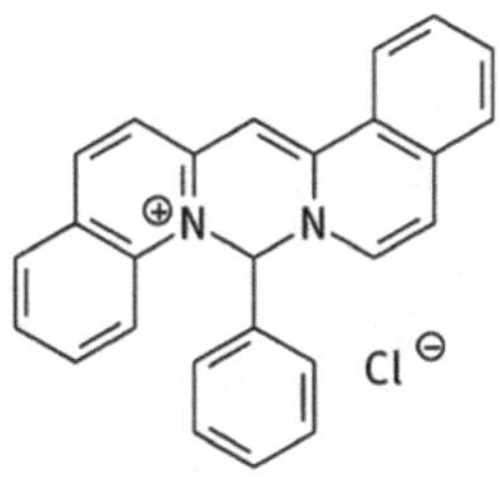

14

The next "cyanine dye" was synthesized by Emil Jacobsen in1882 and called "Chinolinroth" (Quinoline Red) **14** [60]. Just one year later *isocyanine* (2:4´-cyanine) **15** was synthesized [61]. Only in 1920 was the third type of the monomethine cyanines, the *pseudo-isocyanine* (2:2´-cyanine) **16** described [62].

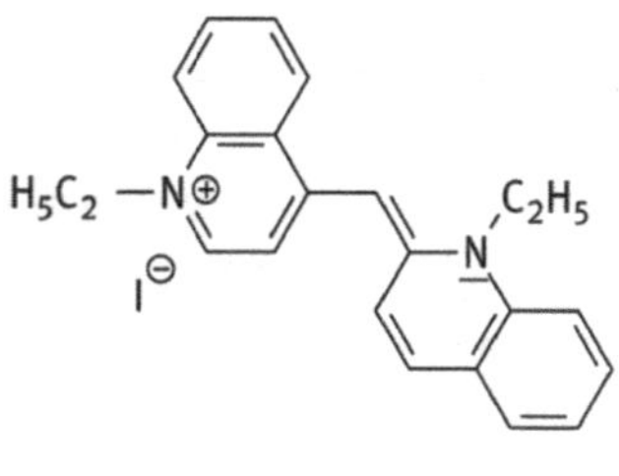

15

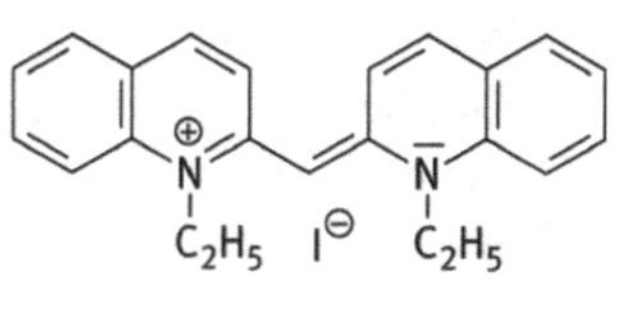

16

Already in 1905 Homolka carried out the condensation of two 2-methylquinoline molecules with one molecule formaldehyde [63]. The reaction product **17** was called *pinacyanol* (2:2´-trimethine-cyanine). Also the trimethine cyanine analogues of **15** (2:4´-trimethine-cyanine) and **13b** were synthesized. The latter (4:4´-trimethine-cyanine) **18** was called *Kryptocyanine*. Walther König discovered that this dyes could be obtained more smoothly using ethylorthoformate [64]. This reaction enables the synthesis of cyanine dyes from other heterocycles (imidazoles, indoles, oxazoles, thiazoles) which went on to be of increasing importance as sensitizers in silver halide photography.

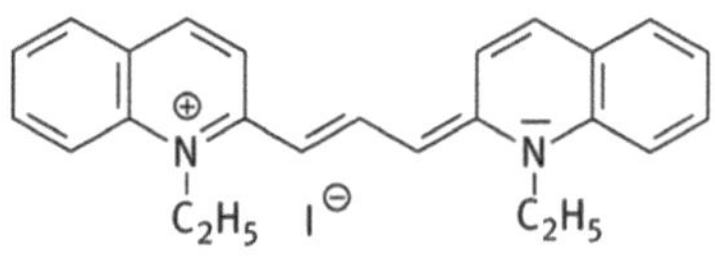

17

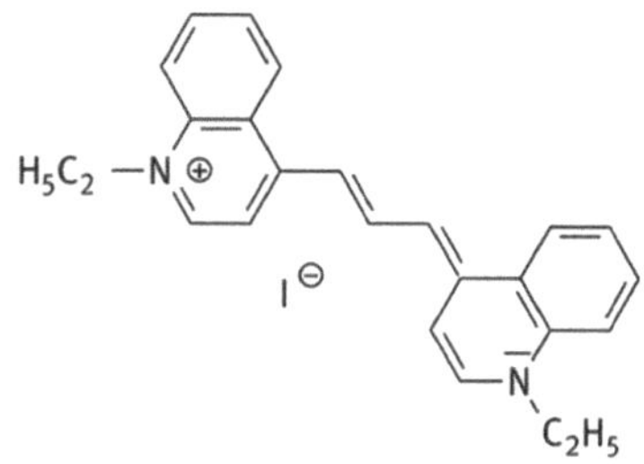

18

After these beginnings, the chemistry of cyanine dyes increased dramatically [65]. The first and most important technical application was the spectral sensitization of silver halide photographic materials. The further development of the chemistry of cyanine dyes is described in connection with the respective technical applications.

24.3 General synthetic routes to cyanine dyes

As was defined at the start of this chapter, cyanine dyes consist of a polymethine chain linking two nitrogen atoms which are part of terminal heterocycles. So, in general the synthetic principle is to combine two heterocyclic quaternary salts with the polymethine chain. The main principles to synthesize cyanine dyes were developed up to the 1960s. The classical standard for the synthesis is the comprehensive book of Frances M. Hamer [1]. Over the time the synthesis methods have been summarized in further review papers and books [2, 66–72].

Williams' "quinoline", besides quinoline, contained fortuitously lepidine (4-methylquinoline), which was an essential second precursor for the dyes. Quaternization of this mixture with methyl, ethyl or amyl iodide, and subsequent treating of the brown oil with ammonia gave the beautiful blue dyes [57]. Both *Irisin* [58] and *Chinolinblau* [59] were synthesized in a similar manner starting from chinoidin. Von Babo used methyl sulfate for the quarternization and treated the obtained oil with caustic potash while Schnitzer used amyl iodide and treated the oil with dilute sodium hydroxide solution. However, this synthetic method is not very efficient.

For the synthesis of monomethine cyanine dyes **21** the standard method is to react a heterocyclic quaternary salt with a leaving group R in **19** with a heterocyclic quaternary salt with a methyl group in basic solution **20** (Figure 24.8).

Longer-chain cyanine dyes are synthesized with the aid of an additional chain-builder. It is an electrophilic reagent which reacts with the nucleophilic base of a heterocyclic quaternary salt.

The first trimethine cyanine dye was synthesized by Homolka using formaldehyde as chain-builder [63]. The application of formaldehyde as chain-builder for the synthesis of cyanine dyes is very limited. In some cases chloroform or dimethylformamide

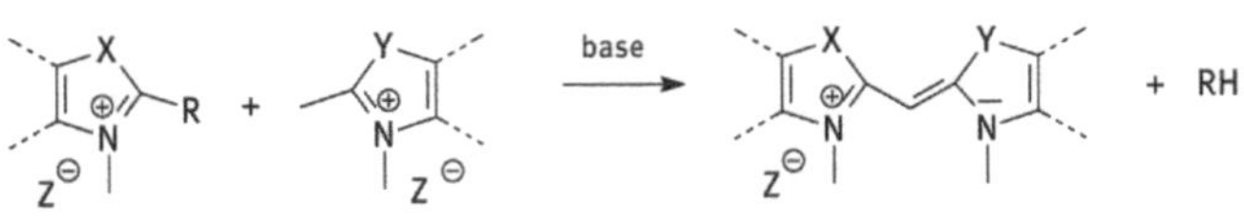

19 20 21

Figure 24.8: Reaction Scheme.

are used, too. Only when König discovered that trimethine cyanine dyes could be obtained more smoothly using ethylorthoformate wider synthetic possibilities were opened [64]. Furthermore, he found that by using triethoxypropene it will be possible to synthesize pentamethine cyanine dyes.

Today in most cases the corresponding streptocyanine diphenylformamidine hydrochloride **23** (a reaction product of ethylorthoformate) in combination with a variety of heterocyclic quaternary salts in the presence of acetic anhydride and a suitable base in a suitable solvent is used (Figure 24.9). With **23** it is possible to stop the reaction at the level of the "half" dye **24**, which enables the synthesis of unsymmetrical cyanine dyes **26.** In the case of the synthesis of symmetrical dyes simple two molecules of the same heterocyclic quaternary salt react with **23** in one step.

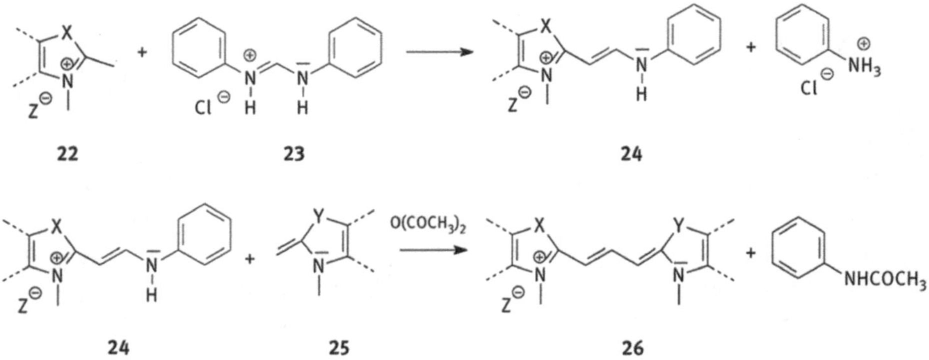

Figure 24.9: Reaction Scheme.

The longer-chain streptocyanine **27a** (malondialdehyde dianil hydrochloride) and **27b** (glutacondialdehyde dianil hydrochloride) are useful synthetic reagents in the preparation of pentamethine and heptamethine cyanine dyes, respectively. Longer chain streptocyanines (**27**, n = 3, n = 4) can be synthesized. However, neither they, nor the resultant nonamethine and undecamethine dyes, are very stable, which is why they are of academic interest mainly.

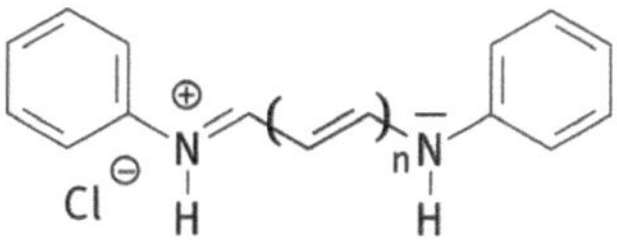

27a (n = 1)
27b (n = 2)

For fine tuning of the absorption and aggregation properties especially for trimethine and pentamethine cyanine dyes the chain-builders are substituted in their meso position. Often heptamethine cyanine dyes are synthesized with five- or six-membered carbocyclic dialdehyde dianils like e. g. **28** and **29**.

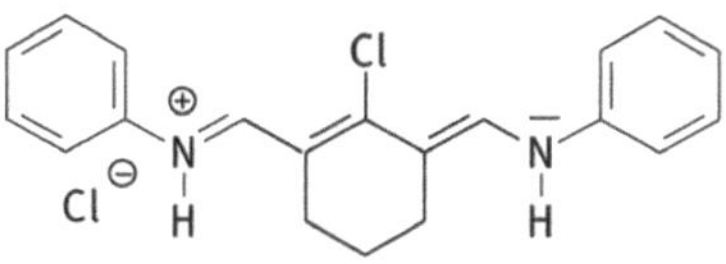

28

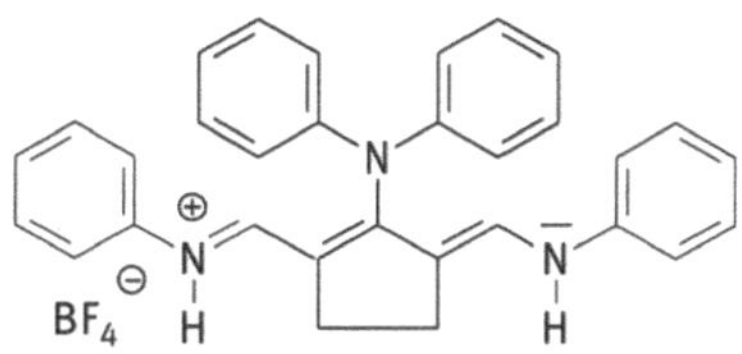

29

Both chain-builders **28** and **29** are widely used at an industrial scale. Finally, it should be mentioned, in addition to aldehydes, cyclic ketones such as isophorone and dimedone are used. Owing to the low synthetic yields associated with their use, they are employed for very special syntheses only.

24.4 Commercial uses of cyanine dyes

24.4.1 Textile dyeing

Initially, synthetic dyes were developed for dyeing textiles but also leather, fur and paper. However, in most cases cyanine dyes were too unstable during application for textile dyeing. In 1924, the synthesis of the indole trimethine dye Astraphloxin FF **30** was described [73]. Due to its attractive hue this cyanine dye found application in textile dyeing.

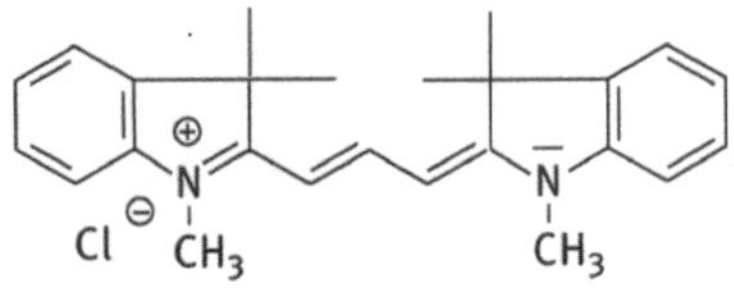

30

24.4.2 Silver halide photographic materials

The driving force was the application of cyanine dyes as *spectral sensitizers* in silver halide photography [74–77]. AgCl is white, AgI is yellow and therefore the wavelength limit of spectral sensitivity lies at about 420 nm for AgCl, 470 nm for AgBr and 520 nm for AgI. Due to its own color a photographic silver halide material is highly sensitive in the ultraviolet region and falls steadily throughout the blue and drops to zero beyond 520 nm. So, these systems are "color blind" in the sense that they are not sensitive to green, yellow, orange and red light. In 1873 Hermann W. Vogel discovered during a study of the spectral sensitivity of dry collodium plates the *spectral sensitization*. It opened the possibility of extending the sensitivity to longer wavelengths by the addition of dyes (spectral sensitizers) that adsorb on the surface of the silver halide crystals. Certain plates with dyes added for halation protection gave spectral sensitivity maxima in the green. As part of his experiments he added Corallin (Korallin) to a photographic emulsion and extended the sensitivity to yellow light. Some years later he found that with the "cyanine" (Chinolinblau) **13c** he could extend the sensitivity to orange. In 1886, Vogel patented Chinolinblau as an orange sensitizer in combination with quinoline red (Chinolinroth) **14** as green sensitizer.

The sensitizers of the first generation were derivatives of quinoline like isocyanine **15**, pseudo-isocyanine **16**, pinacyanol **17** and Kryptocyanine **18**. König´s discovery that the bases of 2-methylsubstituted heterocyclic quaternary salts react with ethylorthoformate to cyanine dyes allowed the synthesis of cyanine dyes from other heterocycles which so became of importance as sensitizers. In addition, using triethoxypropene it was possible to synthesize dyes with longer polymethine chain like e. g. **17** and **18**. With these kinds of sensitizer dyes it was possible to manufacture black and white photographic materials that were sensitive to all visible wavelengths.

Shortly after the invention of photography, the interest for color images arose. In the following years many ingenious ideas have been suggested and tried for utilizing color photography. In the nineteenth century the additive color mixture of the three primary colors blue, green and red was technically realized by different systems. However, it was also known that the subtractive system with the complementary colors yellow, magenta and cyan is more apparent, but it was not so easy to realize. The most successful process was the so called *chromogenic development*, which was discovered in 1911 by Rudolf Fischer and Hans Siegrist. In this process the exposed silver halide acts as the oxidizing agent to for *p*-phenylendiamine, the "developer". The oxidized developer then reacts with color "couplers" creating indoaniline and azomethine dyes resulting in the formation of colored images.

Very roughly, if one considers the visible range of the spectrum approximately from 400 to 700 nm and divide into three equal areas with the respective three mean values (450, 550, 650 nm) three sensitizer dyes are needed. Of course, in the practice of color photography, it is not that easy. Dye mixtures are used for good color reproduction, and spectral sensitization also depends on whether it is a recording or copying material. For the details please refer to the literature [2, 4, 74–77].

For color photography sensitizer dyes with a narrow absorption band and, therefore, narrow sensitization area are necessary to avoid color distortions. These requirements are fulfilled by J-aggregates. So, excellent color sensitizers are dyes with a strong tendency to form J-aggregates. Furthermore, in order to prevent residual coloration after the development process dyes with sulfoalkyl or carboxyalkyl substituents on the heterocyclic nitrogen are preferred for their solubility in water.

As mentioned above, J-aggregation of cyanine dyes is determined by strong intermolecular van der Waals attractive forces, associated with the high π-electronic polarizability of the dyes and steric effects. Substitution with, for example, 5-CN, 5-Cl, 5-CH_3, 5-phenyl functions or 4,5-benzannelation in the benzothiazole or benzoxazole heterocycle increase the π-electronic polarizability and with it the tendency to form J-aggregates. From the steric side, especially in trimethine cyanine dyes an ethyl group in meso-position increases the tendency to form J-aggregates.

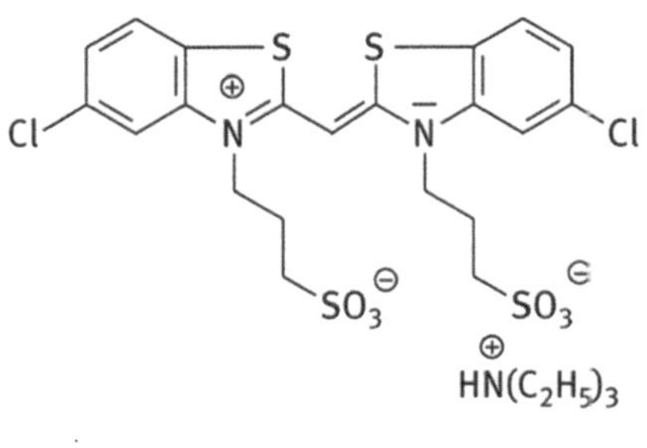

31

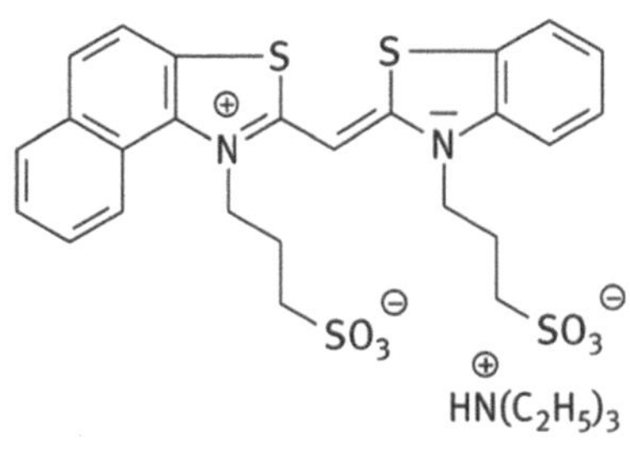

32

The most important blue sensitizes are monomethine cyanine dyes. Both example dyes **31** and **32** are excellent blue sensitizers and were used on a large technical scale in materials for color photography.

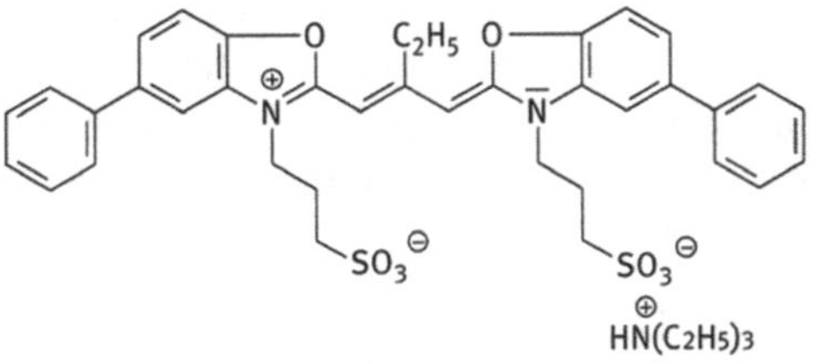

33

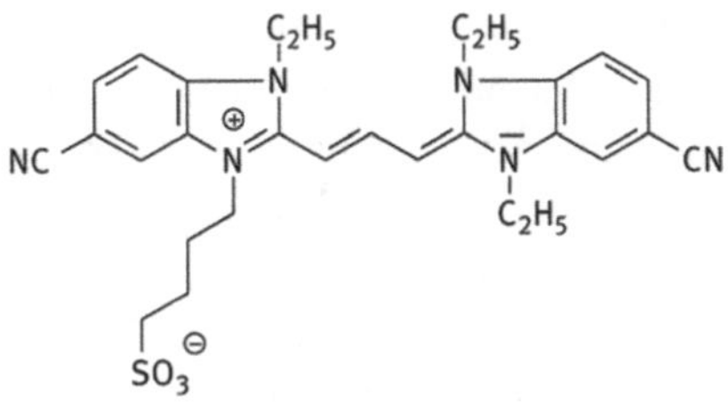

34

Besides other dye types, the most important green sensitizers are trimethine cyanine dyes. Especially in the combination of an ethyl group in meso-position with 5-phenylbenzoxazole like in **33** they produce narrow J-bands and are used in all color materials (paper, films). In color recording films **33** is combined with types like **34** to sensitize over a wider spectral region.

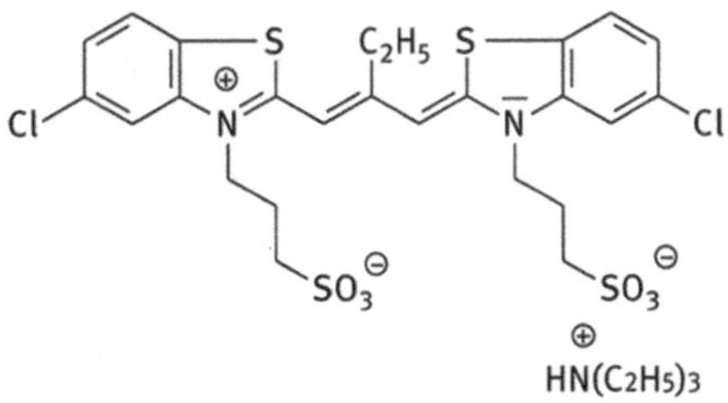

35

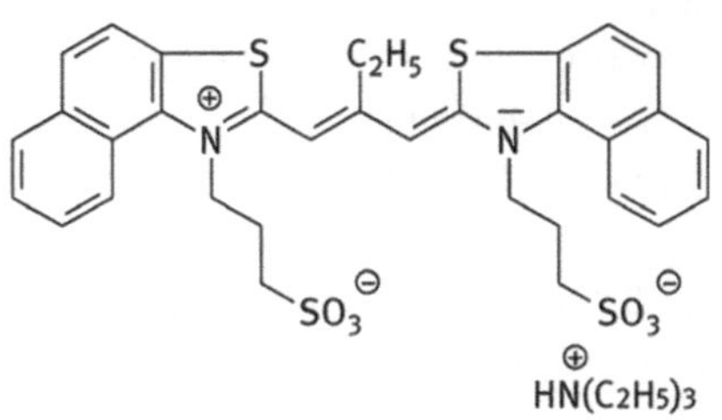

36

The same structural basic principles are used for red sensitizers. The dye **35** is characterized by intense, narrow J-aggregate sensitization at about 650 nm. A dye such as **36** is used for sensitization in the longer-wavelength red region.

The action of sensitizer dyes and their chemistry is a very wide field and for further reading the excellent refs [1, 2, 75–78]. are suggested.

24.4.3 Digital optical data storage with dyes

Classical silver halide photography is one of the oldest optical data storage systems. Since it works in an analogue manner, it suffers from all the problems of such systems like deterioration on repeated playing, which lead to hissing and crackling during play-back. This wear does not occur with digital storage, and so the improvement in play-back quality was a significant driving force for the development of digital data storage. The second important reason for the introduction of digital media was the application software and data for computers which are stored digitally. Optical storage media have played a major part in this development.

In 1984, the first high-power *laser diodes* (LDs) with emission wavelength maxima of around 810 nm were introduced commercially. The subsequent appearance of LDs with emission maxima in the region of 780 and 830 nm opened up many new technical possibilities which in turn required new materials that efficiently absorb in the 780, 810 and 830 nm ranges.

The start of this technology began with a small silvery disc 16 g in weight, with a diameter of 12 cm and a thickness of 1.2 mm – the *compact disc* (CD). Philips and Sony began the development of a digital recording system for sound at the end of the 1970s in the form of the *compact disc digital audio* system (CD-Audio). Within a few years, CD-Audio had won through dramatically. Important advantages compared to analogue phonograph records were the extraordinary purity of sound and simple handling. The data are stored bit by bit (binary digit) as depressions (pits) and regions without depressions (lands). The used LD exhibited laser emission about $\lambda = 780$ nm.

Due to the success of *compact disc digital audio* system, digital data storage has become a mainstream technology. In addition to the digital storage of music and speech, this form has also allowed the digital storage of application software and data for computers. These data stores are called CD-ROM (*Read Only Memory*) and came onto the market 3 years later.

Both, CD-Audio and CD-ROM, are pre-recorded media and cannot be recorded by the user. Therefore, a lot of work was being undertaken to develop an optical recording technique for recordable digital media. An early magneto-optical version of the CD was not a success, because of incompatibility with the CD. Sony used the technology of magneto-optical CD to introduce the minidisc (MD) for sound and data recording in 1992. The next trial, a disc based on organic dye layers, was the so called WORM (write once read many) [79–81]. In this system laser diodes with $\lambda = 830$ nm were used due to the higher power of these LDs in comparison with LDs $\lambda = 780$ nm. The laser radiation is absorbed by an amorphous dye layer. The absorbed laser radiation is converted into heat, and the dye is destroyed by the generated heat. A typical used cyanine dye in WORM technology was **37**. The resulting differences in reflection can then be read by a laser of lower intensity [79–81].

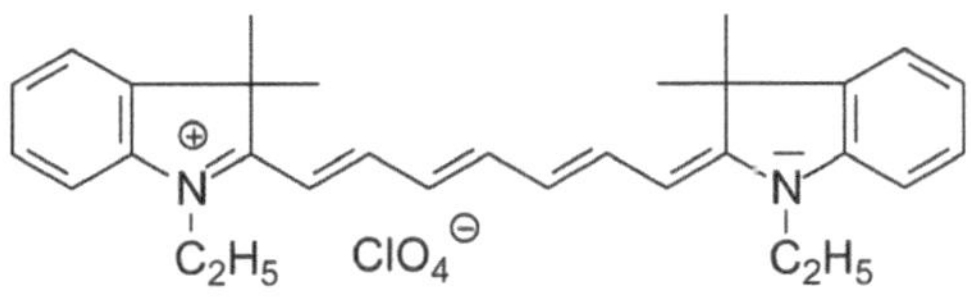

37

Despite all the effort, WORM was not a commercial success, because there was no generally binding standard: WORM media of different manufacturers were not compatible and special hardware was required in each case for burning and play-back and they were not compatible with the CD standard. For these reasons the system and the apparatus remained very expensive, and despite an actually existing demand, it did not win through.

Surprisingly, in 1989 the Japanese company Taiyo Yuden introduced a recordable CD, the CD-R (recordable) that was completely compatible with the CD standard and could be read in normal CD-ROM and CD-Audio drives. The difference compared to WORM was that the reflection at a metal surface was exploited, in the same way as with CD with the same laser wavelength $\lambda = 780\,\text{nm}$. Also in CD-R a very thin amorphous dye layer is used. During the recording process the dye layer absorbs the laser radiation and converts it into heat with destruction of the dye molecules. In addition, the polycarbonate melts through the heat produced and the silver mirror is deformed. Thus, differences in optical wavelengths are produced in the recorded regions (written marks). The deformed region corresponds to the pit in the molded disc and has a lower reflection because of the destructive interference of the laser beam [82, 83].

One of the first used dyes was the cyanine dye **38**. In the early CD-R sputtered gold was used as reflection layer. Later it was replaced by a silver alloy and, therefore, **38** with iodide anion were replaced by **39** and **40** since reactions of the iodide anion with the surface of the silver layer occurred, resulting in poorer reflection. With increasing recording speed the power of the LDs was increased, which needed cyanine dyes without perchlorate anion and higher thermal stability for example, **41**.

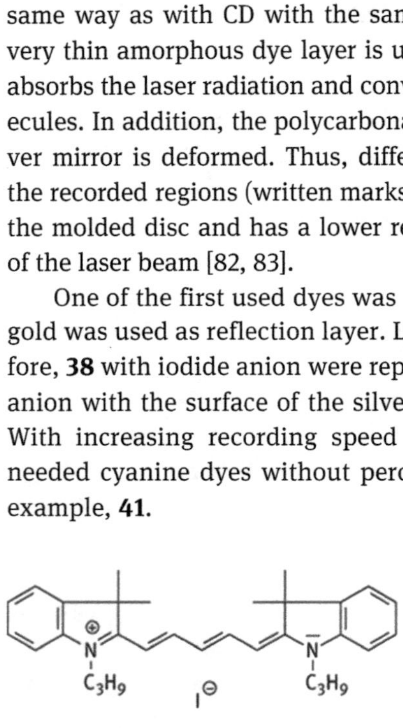

38

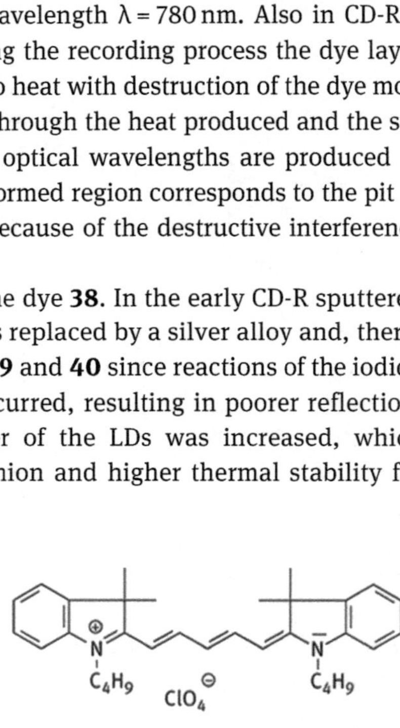

39

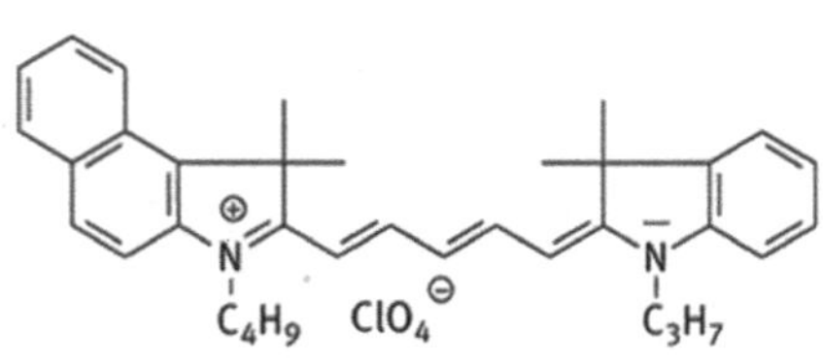

40

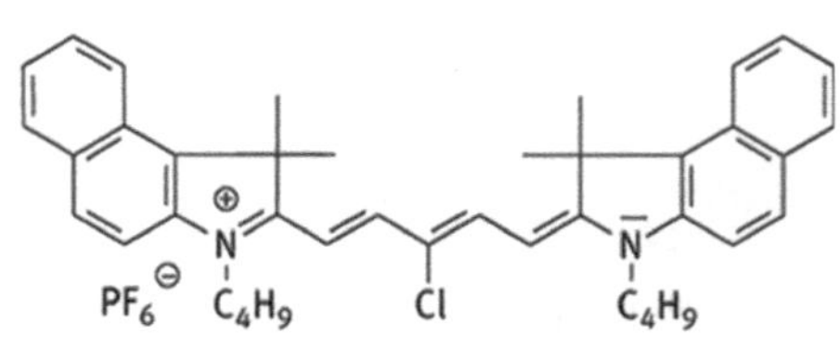

41

An increased information density for video recording was dependent upon the employment of LDs emitting shorter wavelength radiation. After development of such LDs and intensive debates a standard for video recording was laid down for the DVD (digital versatile disc) in 1996. A DVD can store 4.7 GB data on a single information layer. As with CDs, developments followed the pre-recorded DVD format that also led to being able to record a DVD [82, 83]. After a longer development the last version of the DVD/R uses LDs with emission wavelengths 650 nm and corresponding new dyes had to be developed. Also here, cyanine dyes were used in a wide range. One of the most used dyes was **42**. For solubility reasons the N-alkyl groups are unsymmetrical. With increasing recording speed the absorption of the amorphous dye layer at 650 nm has to be increased. For this purpose, for example, **43** was introduced.

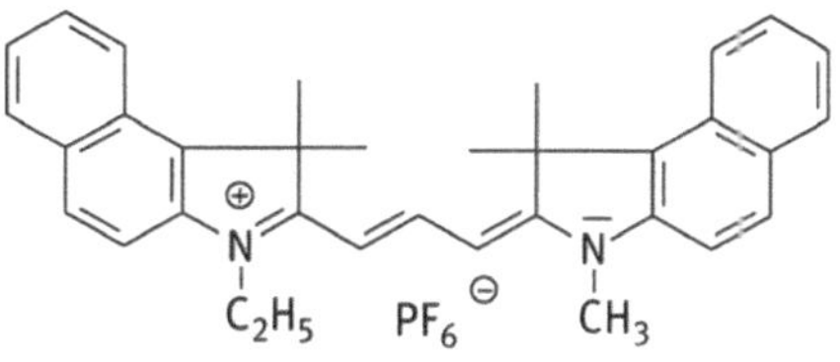

42

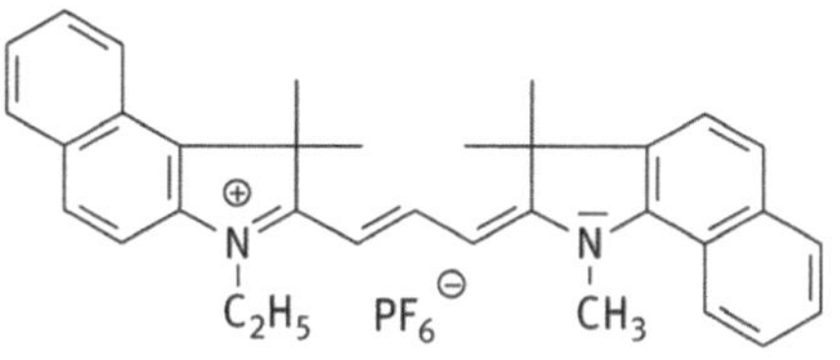

43

For further reading, the technology of the WORM, CD-R and DVD/R, respectively, is described in detail in refs [79–83].

24.4.4 Computer-to-Plate lithographic printing plates

The first economic printing technique was developed around 1450 by Johannes Gutenberg and was based on a relief printing process with letter types, the letterpress printing. Today besides letterpress, gravure, screen and lithographic printing is used. In lithographic printing the printing and nonprinting areas are effectively in the same plane, but they differ in their physicochemical properties. The printing areas are oleophilic, the nonprinting hydrophilic. So, in principle platemaking is simple, and lithographic printing has had the greatest potential for a quick and economic manufacture and storage process of the printing plates [84–86]. The quick and economic manufacture of lithographic printing plates is realized by a light sensitive coating on a flexible carrier. Currently aluminum is used as the predominant substrate. Plates based on a polyester substrate are now only rarely found.

In the 1940s, the Kalle company started selling the first industrially manufactured pre-sensitized plates, that is, with a light-sensitive coating based on diazo- or bisazide chemistry. The printing plate is produced by exposure with UV light through a film, held in contact with the plate UV light-sensitive coating under vacuum. By this selectively exposure the latent image is produced. At the end of this process the printing plate is developed and ready to print. This is the conventional, or analogue, platemaking processes. In the course of the 1950s, offset printing emerged as the world's dominant printing technique.

The first half of the 1990s saw the beginning of the development of a complete new technology for lithographic printing plates. Most important, in this technology the image of the page from a digital file created in a desktop publishing (DTP) application is recorded by laser beam directly from a computer to a printing plate. The direct transfer from computer to offset plate is called Computer-to-Plate (CtP). The companies Kodak and Creo developed image plate setter with high power laser diodes of emission wavelength 830 nm. These technologies depended upon functional colorants absorbing NIR radiation in the region of 830 nm.

Today in the coating of the plate different working principles are used. Roughly, one can distinguish between the alkaline processing plates and on-press processing plates or between heat-sensitive plates based on physical modifications, heat-sensitive plates based on chemical modifications and photo-sensitive plates based on radical photo-polymerization [86].

The most used NIR-sensitizer for heat-sensitive plates is the cyanine dye **44**, which is manufactured in ton-scale per year. The function of the NIR-sensitizer is to convert laser rays into heat and the produced heat leads to physical or chemical changes in the printing plate, which produces the latent image.

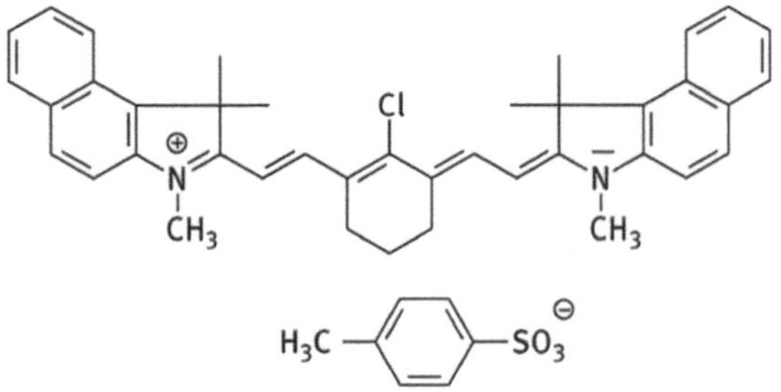

44

In the photo-sensitive plates the function of the NIR-sensitizer is to absorb laser rays and produce reactive radicals which start radical photo-polymerization. Here, the two basic systems

- radical photo-polymerization of monomers (oligomers) in a matrix of polymers
- radical photo-polymerization of monomers (oligomers) in a matrix of particles

are used. Photo-sensitive plates are the most sensitive plate types and this approach has evolved into the world's dominating CtP printing technique. Proven NIR-sensitizers for radical photo-polymerization are e. g. the two cyanines **45** and **46**.

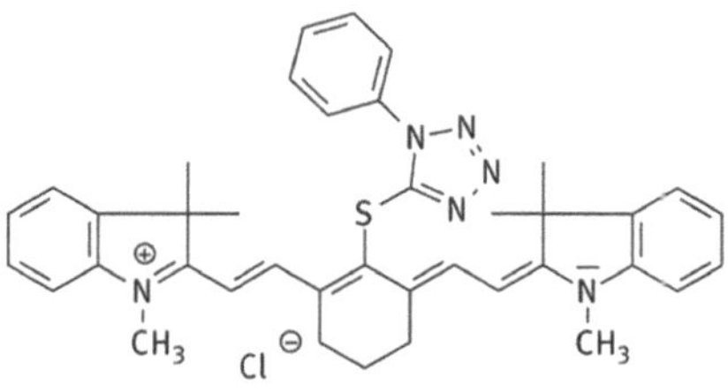

45

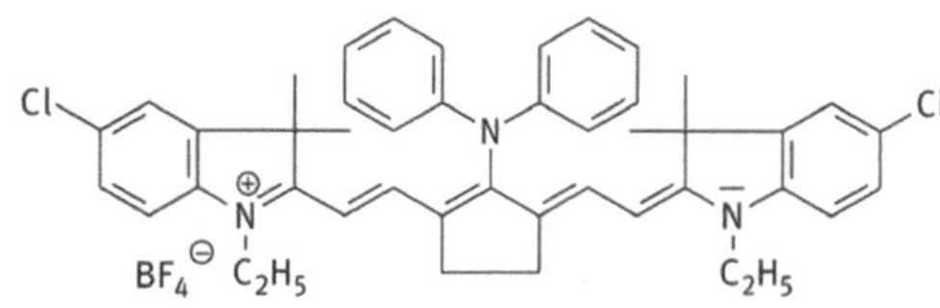

46

The printing process and the manufacture of printing plates are highly sophisticated technologies and for further reading refs [84–86] are recommended.

24.4.5 Fluorescent dyes for bio-analysis and medical diagnostics

The so-called fluorescent marker technology is a process that fluorescent dyes combining with a substrate (analyte) by physical or chemical action to form a complex. The fluorescence signal or the fluorescence intensity of the marked object is detected and this signal is transferred qualitatively or quantitatively. With the development of modern medicine, molecular biology and various advanced fluorescence detection technologies and instruments fluorescent marker technology has been appreciated extensively.

Since the structures of cyanine dyes can be readily modified to tune desirable properties such as fluorescence wavelength and quantum yield, solubility, permeability, and physical (non-covalent) or chemical binding (covalent), an extensive number of cyanine dyes have been synthesized and developed as fluorescent marker for bio-analytical marking and imaging [87–92].

24.4.5.1 Non-covalent protein markers

The great specific and structural diversity of proteins and peptides, respectively, has necessitated the development of a wide range of dye markers, possessing the necessary properties to interact optimally with a substrate. Because of the enormous possibilities for their structural modification, cyanine dyes have long served as such markers.

Often the non-covalent binding of cyanine dyes to proteins is used to detect and analyze biomolecules in living organisms. In a simple manner an aqueous solution of a marker dye is introduced into the living organism. Here the most commonly used dye is Indocyaninegreen ICG (Cardiogreen) **47**. ICG has low toxicity and the sterile lyophilisate of an aqueous ICG solution is approved in many European countries and USA. Therefore, this fluorescent dye is widely used in medical diagnostics, such as monitoring cardiac output, hepatic function, and liver blood flow and for ophthalmic angiography.

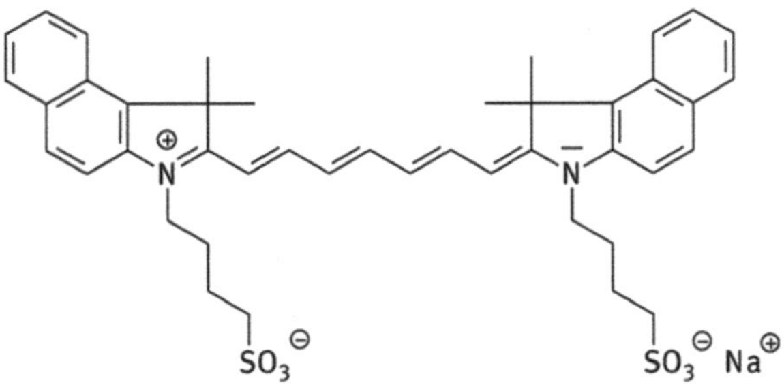

47

ICG is also used in gel electrophoresis to separate various proteins stained with it [87, 88]. A lot of other non-covalent protein marker dyes and applications are described in the literature [87, 88].

24.4.5.2 Covalent protein markers

In contrast to non-covalent staining, dyes for covalent binding (labeling) employs fluorescent dyes that feature reactive groups. A spacer group is usually placed between the fluorescent dye and the reactive group with the intention of preventing, or at least attenuating, undesirable steric interactions between the signal-generating dye and the labeled component as illustrated on **48**. In addition, covalent protein marker molecules typically bear a number of sulfonic acid groups to increases their water solubility and reduce the tendency of the dye to aggregate in aqueous solution.

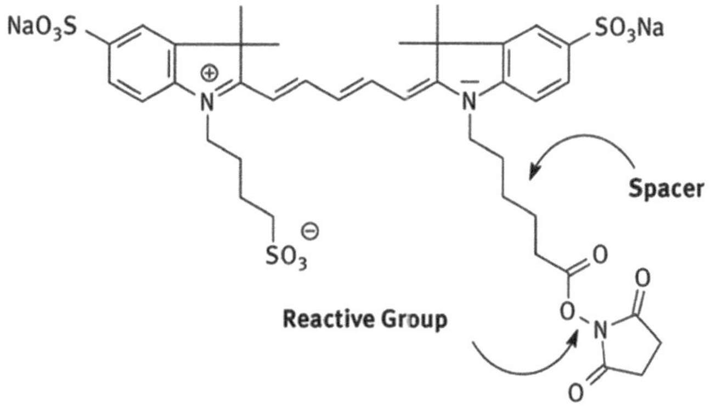

48

The most commonly used reactive group to label proteins and peptides, respectively, is the N-hydroxysuccinimide ester, as in **48**.

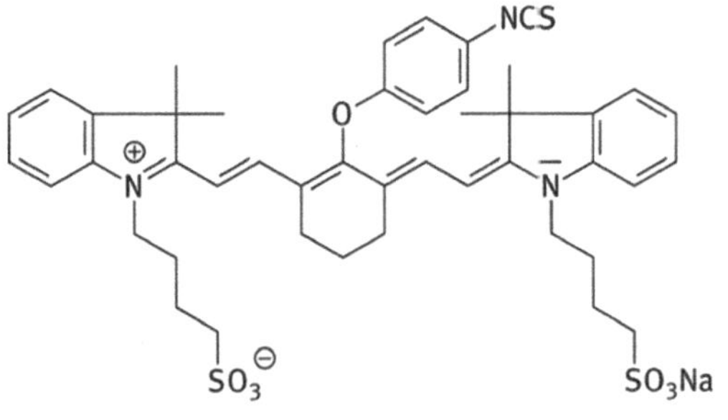

49

Besides this reactive function the isothiocyanate group (–NCS) as in **49** is sometimes used in labelling; this motif reacts with a primary amino group in an amino acid to form a thiourea linkage.

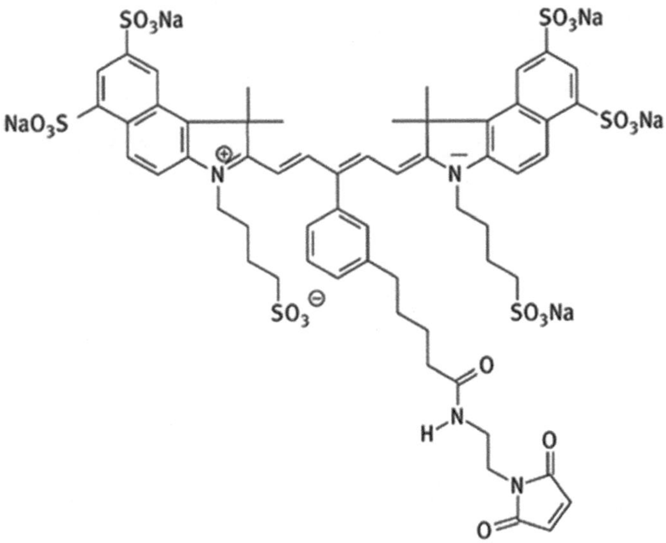

50

Cysteine and methionine are sulfur containing amino acids. For the reaction with a thiol (mercapto) group of an amino acid the maleimido group as in **50** is the standard reactive function.

In the 400–600 nm region of the electronic spectrum derivatives of coumarins, fluorescein and rhodamines are the most used fluorescent labels, whereas in the range 600–800 nm cyanine dyes dominate e. g. **48** for laser excitation around 640 nm, **50** for 680 nm and **49** for 780 nm.

This dye chemistry is a broad field and further covalent protein marker cyanine dyes and applications are described in the literature [87, 89–91].

24.4.5.3 Intercalating cyanine dyes

In biochemistry the insertion of a chemical compound between adjacent base pairs of the deoxyribonucleic acid (DNA) helix is called intercalation. This process is used as a method for analyzing DNA and it is also the basis of certain therapeutic treatments to inhibit DNA replication in rapidly growing cancer cells.

The fluorescence quantum yield of intercalating dyes in molecular disperse solution is rather low due to torsional vibrations of the heterocycles. Therewith, it depends on the environment. The DNA helix interacts through stacking forces with the intercalating dyes reducing the torsional vibrations which increase the intensity of fluorescence. This allows direct quantitative DNA measurements.

There is a large family of intercalating dyes. The most important are cyanine dyes based around Thiazole orange (TO) and Oxazole yellow (YO). The modified derivatives TO-PRO-1 **51a** and YO-PRO-1 **51b** are cyanine dyes with a second positive charge in the chain and so consequently stick very strongly to DNA.

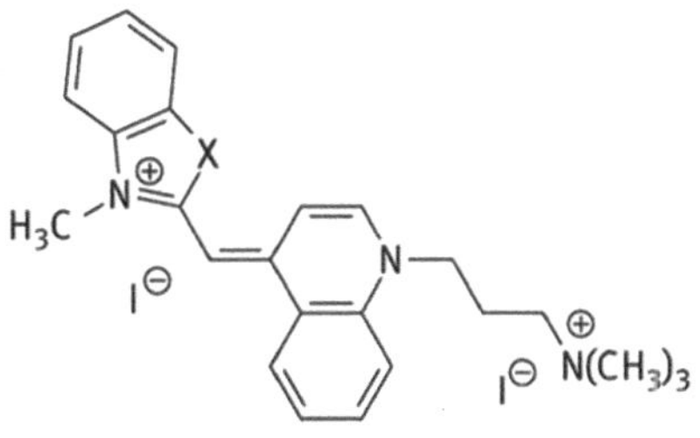

51a: X = S
51b: X = O

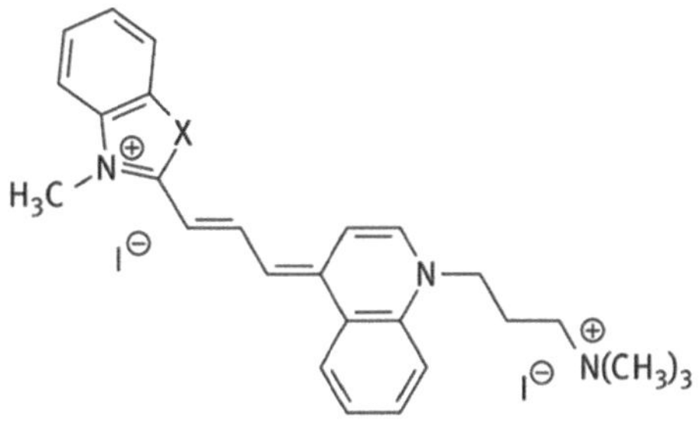

52a: X = S
52b: X = O

To extend the fluorescence into the red region the conjugated system of the monomethine dyes **51** was lengthened to furnish to trimethine dyes TO-PRO-3 **52a** and YO-PRO-3 **52b**. Increasing chain length even further, for example to create dyes such as TO-PRO-5, enables emission to be pushed beyond the visible region and into the near infrared.

If a single intercalating dye bearing two positive charges sticks very strongly to DNA, its dimer should be even better. So dimers of these compounds are also synthesized and known as TOTO-1 and TOTO-3, as well YOYO-1 and YOYO-3 [87, 92].

24.4.5.4 Covalent DNA sequencing labels

Genetic information is stored in DNA in the nuclei and mitochondria of cells. DNA sequencing is the technology to determine the nucleotide sequence of DNA. The nucleotide sequence is the most fundamental level of knowledge of a gene or genome. DNA sequencing techniques include any method or technology that is used to determine the order of the four bases.

The original methods of DNA sequencing were implemented through the use of radioactive labels. Radiolabelling remains popular owing to its high sensitivity and ease of use. However, the dangers of radioactivity, paved the way for fluorescence labels, which are simple, sensitive and easy to automate.

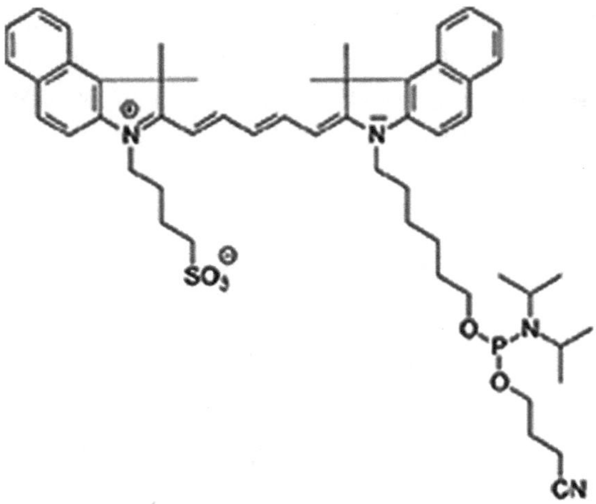

53

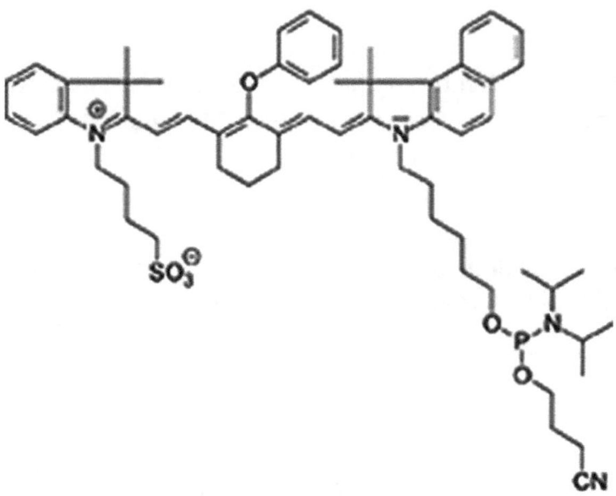

54

Sometimes the isothiocyanate or N-hydroxysuccinimide ester reactive group is used to couple the dye to a primary amine connected with a phosphor amidite functionality. The phosphoramidite group couples the dye to the 5´-OH nucleotide of an oligonucleotide. Introducing the amidite functionality in a direct way like e. g. **53** for 680 and in **54** for 780 nm allows automated DNA sequencing.

In summary, cyanine dyes are among the most widely used fluorescent dyes for protein and nucleic acid detection. They have numerous of applications in bio-analysis and medical diagnostics. During the last few decades many new cyanine dyes have been synthesized and commercialized to improve application properties. Their number is so great that they cannot discussed here in detail and so for further reading refs [87–92] are recommended.

References

1. Hamer FM. The cyanine dyes and related compounds. In: Weissberger A, editor. The chemistry of heterocyclic compounds, vol. 18. New York: Interscience, 1964.
2. Sturmer DM, Diehl DR. Polymethine dyes. In: Kirk-othmer encyclopedia of chemical technology, vol. 18, 3rd ed. New York: Wiley, 1982:848–74.
3. Tyutyulkov N, Fabian J, Mehlhorn A, Dietz F, Tadjer A. Polymethine dyes – Structure and properties. Sofia: St. Kliment Ohridski University Press; 1991.
4. Bach G, Daehne S. Cyanine dyes and related compounds. In: Sainsbury M, editor(s). Second supplements to the 2nd edition of Rodd's chemistry of carbon compounds Vol. IV B. Amsterdam: Elsevier, 1997:383–431.
5. Mishra A, Behera RK, Behera PK, Mishra BK, Behera GB. Cyanine dyes during the 1990s: a review. Chem Rev. 2000;100:1973.
6. Mustroph H. Polymethine dyes. Phys Sci Rev. 2020;5. doi:10.1515/psr-2019-0084.
7. Mustroph H. Dyes: quantum chemical calculation of electronic spectra. Phys Sci Rev. 2019;4. doi:10.1515/psr-2019-0040.
8. Grahn W. ^{13}C-NMR-Spektren von di- und trinuclearen Polymethincyanin- Farbstoffen. Tetrahedron. 1976;32:1931.
9. Shklyarevskiy IO, Christianen PC, Aret E, Meekes H, Vlieg E, Deroover G, et al. Determination of the molecular arrangement inside cyanine dye aggregates by magnetic orientation. J Phys Chem B. 2004;108:16386.
10. Cambridge Crystallographic Data Center (CCDC). https://www.ccdc.cam.ac.uk/
11. Kulpe S, Kuban RJ, Schulz B, Dähne S. Crystal structure determination of bis-dimethyl heptamethine cyanine chloride, $C_{11}N_2H_{19}Cl \cdot 4\ H_2O$. Cryst Res Technol. 1987;22:375.
12. Baughman RH, Kohler BE, Levy IJ, Spangler C. The crystal structure of *trans,trans*-1,3,5,7-octatetraene as a model for fully-ordered trans-polyacetylene. Synth Met. 1985;11:37.
13. Brooker LG. Absorption and resonance in dyes. Rev Mod Phys. 1942;14:275.
14. Brooker LG. Some recent developments in the chemistry of photographic sensitizing dyes. Experientia Suppl. 1955;2:229.
15. Robin MB, Day P. Mixed valence chemistry - a survey and classification. Adv Inorg Chem Radiochem. 1968;10:247.
16. Marder SR, Perry JW, Tiemann BG, Gorman CB, Gilmour S, Biddle SL, et al. Direct observation of reduced bond-length alternation in donor/acceptor polyenes . J Am Chem Soc. 1993;115:2524.
17. Gorman CB, Marder SR. An investigation of the interrelationships between linear and nonlinear polarizabilities and bond-length alternation in conjugated organic molecules. Proc Natl Acad Sci USA. 1993;90:11297.
18. Marder SR, Perry JW, Bourhill G, Gorman CB, Tiemann BG, Mansour K. Relation between bond-length alternation and second electronic hyperpolarizability of conjugated organic molecules. Science. 1993;261:186.

19. König W. Über den Begriff der „Polymethinfarbstoffe" und eine davon ableitbare allgemeine Farbstoff-Formel als Grundlage einer neuen Systematik der Farbenchemie. J Prakt Chem. 1926;112:1.
20. Dähne S. Systematik und Begriffserweiterung der Polymethinfarbstoffe. Z Chem. 1965;5:441.
21. Dähne S. Color and constitution: one hundred years of research. Science. 1978;199:1163.
22. Pauling L. A theory of the colors of dyes. Proc Natl Acad Sci. 1939;25:577.
23. Förster T. Quantenmechanische Rechnungen zur Theorie der organischen Farbstoffe. Z Phys Chem B. 1940;47:245.
24. Herzfeld KF. On the absorption spectrum of some polymethine dyes. J Chem Phys. 1942;10:508.
25. Sklar AL. Energy levels and color of polymethine dyes. J Chem Phys. 1942;10:521.
26. Herzfeld KF, Sklar AL. Colour and constitution of polymethine dyes. Rev Mod Phys. 1942;14:294.
27. Beretta P, Jaboli A. Correlations between the acid-base equilibria and the electrochemical properties of some carbocyanine dyes. Photogr Sci Eng. 1974;18:197.
28. Strekowski L, Mason JC, Britton JE, Lee H, Van Aken K, Patonay G. The addition reaction of hydroxide or ethoxide ion with benzindolium heptamethine cyanine dyes. Dyes Pigm. 2000;46:163.
29. Kimura M, Mitekura H, No T, Suzuki K. Synthesis of *meso*-substituted trimethine cyanine dyes and evaluation of their sensitivities in sensitized photo-polymerization. Bull Chem Soc Jpn. 2002;75:2655.
30. Strekowski L, Lipowska M, Patonay G. Facile derivatizations of heptamethine cyanine dyes. Synth Commun. 1992;22:2593.
31. Strekowski L, Lipowska M, Patonay G. Substitition reactions of a nucleofugal group in heptamethine cyanine dyes. J Org Chem. 1992;57:4578.
32. Mustroph H, Towns A. Fine structure in electronic spectra of cyanine dyes : Are sub-bands largely determined by a dominant vibration or a collection of singly-excited vibrations? ChemPhysChem. 2018;19:1016.
33. Pouradier J. Remarque sur le spectre d´absorption des cyanines en solution. J Chim Phys. 1964;61:1107.
34. Scheibe G. Ergebnisse der Absorptionsspektroskopie hinsichtlich Lage, Intensität und Struktur der Banden. In: Jung W, editor. Optische Anregung organischer Systeme. Weinheim: Verlag Chemie; 1966:109–59.
35. Mustroph H, Reiner K, Mistol J, Ernst S, Keil D, Hennig L. Relationship between the molecular structure of cyanine dyes and the vibrational fine structure of their electronic absorption spectra. ChemPhysChem. 2009;10:835.
36. Meguellati K, Ladame S, Spichty M. A conceptually improved TD-DFT approach for predicting the maximum absorption wavelength of cyanine dyes. Dyes Pigm. 2011;90:114.
37. Mustroph H, Ernst S, Senns B, Towns AD. Molecular electronic spectroscopy: from often neglected fundamental principles to limitations of state-of-the-art computational methods. Color Technol. 2015;131:9.
38. Mustroph H, Reiner K, Senns B. Bond length alternation in unsymmetrical cyanine dyes and its influence on the vibrational structure of their electronic absorption spectra. Color Technol. 2017;133:469.
39. Dewar MJ. Colour and constitution. Part 1. Basic dyes. J Chem Soc. 1950:2329.
40. Dewar MJ. Modern theories of color. Chem Soc (London) Spec Publ. 1956;4:64.
41. Knott EB. The colour of organic compounds. Part I. A general colour rule. J Chem Soc. 1951:1024.
42. Griffiths J. Colour and constitution of organic molecules. London: Academic Press; 1976.

43. Fabian J, Hartmann H. Light absorption of organic colorants. Berlin: Springer, 1980.
44. Brunings KJ, Corwin AH. Steric influences on the aromaticity of dipyrrylmethenes. The synthesis and study of the properties of a di-N-methyldipyrrylmethene. J Am Chem Soc. 1942;64:593.
45. Brooker LG, White FL, Sprague RH, Dent SG, Van Zandt G. Steric hindrance to planarity in dye molecules. Chem Rev. 1947;41:325.
46. Nys J. Environmental influences on the spectral absorption of sensitizing dyes. In: Berg WF, Mazzucato U, Meier H, Semerano G, editors. Dye sensitization, symposium: Bressanone. London, New York: Focal Press, 1970:26–43.
47. Dietz F. Die Aggregation der Cyaninfarbstoffe und ihre Bedeutung für die spektrale Sensibilisierung. J Signalaufzeichnungsmater. 1973;1:157, 237, 381.
48. Herz A. Aggregation of sensitizing dyes in solution and their adsorption onto silver halides. Adv Colloid Interface Sci. 1977;8:237.
49. Blokzijl W, Engberts JB. Hydrophobic effects. Opinions and facts. Angew Chem Int Ed. 1993;32:1545.
50. Breslow R. Hydrophobic effects on simple organic reactions in water. Acc Chem Res. 1991;24:159.
51. Pirrung MC. Acceleration of organic reactions through aqueous solvent effects. Chem Eur J. 2006;12:1312.
52. Davydov AS. Theory of molecular excitons. New York: Plenum Press, 1971:23–47.
53. McRae EG, Kasha M. The molecular exciton model. In: Augenstein L, Mason R, Rosenberg B, editors. Physical processes in radiation biology. New York: Academic Press, 1964:23–42.
54. Rosenoff AE, Walworth VK, Bird GR. The stability of sensitizing dye aggregates on silver bromide. Photogr Sci Eng. 1970;14:328.
55. Hestand NJ, Spano FC. Expanded theory of H- and J-molecular aggregates: the effects of vibronic coupling and intermolecular charge transfer. Chem Rev. 2018;118:7069.
56. Perkin WH. GB 1984 (26th August 1856).
57. Williams CG. Researches on chinoline and its homologues. Trans Roy Soc Edinburgh. 1857;21:377.
58. von Babo. Ueber einige Zersetzungsprodukte des Cinchonins. J Prakt Chem. 1857;72:73.
59. Schnitzer G. Ueber Anilinroth und Chinolinblau. Chem Centr NF. 1861;6:636.
60. Jacobsen E. DRP 19 306 (14th February 1882).
61. Spalteholz W. Ueber Farbstoffe aus dem Steinkohlentheerchinolin. Ber Dtsch Chem Ges. 1883;16:1847.
62. Fischer O, Scheibe G. Beitrag zur Kenntnis der Chinocyanine. J Prakt Chem. 1920;100:86.
63. Dieterle W. Der Anteil der deutschen Teerfarbenindustrie am Ausbau und an der Weiterentwicklung der Entdeckung H. W. Vogels. Veröff Agfa Wolfen. 1939;6:1.
64. König W, US 1,524,791 (5th May 1923).
65. Doja MQ. The cyanine dyes. Chem Rev. 1932;11:273.
66. Ficken GE. Cyanine dyes. In: Venkataraman K, editor. The chemistry of synthetic dyes, vol. 4. New York: Academic Press, 1971:211–340.
67. Berlin L, Riester O. Methoden zur Herstellung von Cyaninen (Polymethinen). In: Müller E, editor. Methoden der organischen Chemie (Houben-Weyl), vol. 5/1d. Stuttgart: Thieme, 1972:231–98.
68. Sturmer DM. Syntheses and properties of cyanine and related dyes. In: Weissberger A, Taylor EC, editors. The chemistry of heterocyclic compounds, vol. 30. New York: Wiley, 1977:441–587.
69. Sturmer DM. Cyanine dyes. In: Kirk-Othmer encyclopedia of chemical technology, vol. 7, 3rd ed. New York: Wiley, 1979:335–58.

70. Raue R, Riester O. Methinfarbstoffe. In: Ullmanns Enzyklopädie der Technischen Chemie, vol. 16, 5th ed. Weinheim: Verlag Chemie, 1979:635–69.
71. Berneth H. Methine dyes and pigments. In: Ullmann´s encyclopedia of industrial chemistry. Weinheim: Wiley-VCH, 2008.
72. Panigrahi M, Dash S, Patel S, Mishra BK. Syntheses of cyanines: a review. Tetrahedron. 2012;68:781.
73. König W. Über Indolenino-cyanine (Indocyanine). Ber Dt Chem Ges. 1924;57:685.
74. Kainrath P. Fortschritte auf dem Gebiete der Sensibilisierungsfarbstoffe für photographische Halogensilber-Emulsionen. Angew Chem. 1948;60:36.
75. Brooker LG, Vittum PW. A century of progress in the synthesis of dyes for photography. J Photogr Sci. 1957;5:71.
76. Brooker LG. Sensitizing and desensitizing dyes. In: Mees CE, James TH, editors. The theory of the photographic process. 3rd ed. New York: Macmillan, 1966:198–232.
77. Sturmer DM, Heseltine DW. Sensitizing and desensitizing dyes. In: James TH, editor. The theory of the photographic process. New York: Macmillan, 1977:194–234.
78. Keller K, Kampfer H, Matejec R, Lapp O, Krafft W, Frenken H, et al. Photography. In: Ullmann's encyclopedia of industrial chemistry. Weinheim: Wiley-VCH, 2000.
79. Nakazumi H. Organic colorants for laser disc optical data storage. J Soc Dyers Colour. 1988;104:121.
80. Emmelius M, Pawlowski G, Vollmann HW. Materials for optical data storage. Angew Chem Int Ed. 1989;28:1445.
81. Matsui F. Optical recording systems. In: Matsuoka M, editor. Infrared absorbing dyes. New York: Plenum Press, 1990:117–40.
82. Mustroph H, Stollenwerk M, Bressau V. Current developments in optical data storage with organic dyes. Angew Chem Int Ed. 2016;45:2006.
83. Wochele RE, van Houten H, Duchateau JP, Kloosterboer HJ, Verhoeven JA, van Vlimmeren R, et al. Information storage materials, 2. Optical recording. In: Ullmann's encyclopedia of industrial chemistry. Weinheim: Wiley-VCH, 2011.
84. Vollmann HW. New technologies for the filmless manufacture of printing forms. Angew Chem Int Ed. 1980;19:99.
85. Taggi AJ, Walker P. Printing processes. In: Kirk-Othmer encyclopedia of chemical technology. Weinheim: Wiley-VCH, 2000.
86. Baumann H, Hoffmann-Walbeck T, Wenning W, Lehmann HJ, Simpson CD, Mustroph H, et al. Imaging technology, 3. Imaging in graphic arts. In: Ullmann's encyclopedia of industrial chemistry. Weinheim: Wiley-VCH, 2015.
87. Daehne S, Resch U, Wolfbeis OS, editors. Near-infrared dyes for high technology applications. Dordrecht: Kluwer Academic, 1998.
88. Tatikolov AS. Polymethine dyes as spectral-fluorescent probes for biomacromolecules. J Photochem Photobiol C: Photochem Rev. 2012;13:55.
89. Gonçalves MS. Fluorescent labeling of biomolecules with organic probes. Chem Rev. 2009;109:190.
90. Pansare VJ, Hejazi S, Faenza WJ, Prud'homme RK. Review of long- wavelength optical and NIR imaging materials: contrast agents, fluorophores, and multifunctional nano carriers. Chem Mater. 2012;24:812–27.
91. Della Ciana L. New trends in fluorescent reporters in biology. In: Pedras B, editor. Fluorescence in industry. Springer series on fluorescence (Methods and applications), vol. 18. Cham: Springer, 2019:321–39.
92. Deligeorgiev TG, Kaloyanova S, Vaquero JJ. Intercalating cyanine dyes for nucleic acid detection. Recent Pat Mater Sci. 2009;2:1.

Andrew Towns

25 Diarylethene dyes

Abstract: This article introduces the general characteristics of the diarylethene class of photochromic dye and the structural features that make photochromism possible. It touches on the methodologies employed to synthesize these compounds as well as the influences that typical substitution patterns exert on photocoloration. A demonstration is then given of the great diversity pertaining to the potential applications in which researchers are seeking to exploit them as functional colorants.

Keywords: colorant, dye, functional, photochromic, photochromism

The diarylethene (DAE) class is one of the most intensely scrutinized types of photochromic dye. The serendipitous discovery of its relatively unusual photochromism [1] attracted great attention from researchers seeking to exploit light responsive switches in a wide variety of fields. As functional colorants, DAEs continue to be the subject of many avenues of research. Despite the interest shown in them, the use of these dyes does not currently extend beyond low-volume niche applications. This Chapter will describe the characteristics that make them such appealing tools for a diverse range of technologies. As well as taking a brief look at the influences on these properties and the chemistry of DAE dyes, it will also give a flavor of the kinds of potential use to which the photochromism of this class may be put.

25.1 Diarylethene dye characteristics and molecular structure

With the right molecular design, the class offers unusual photochromic properties that are of special interest to researchers:

- P-type photochromism, whereby colorants can be switched between states possessing different light absorption properties through irradiation with light of specific wavelength ranges [2]. Each state is thermally stable. Most photochromic classes, for example naphthopyrans [3], exhibit T-type behavior, whereby photoactivated forms revert in the absence of light to their original state. Consequently, P-type systems are more attractive as functional dyes because they may be switched back and forth on demand between states that would otherwise persist indefinitely. Many DAE dyes behave in this manner for practical purposes. Researchers estimate lifetimes of states of certain DAEs at 30°C in terms of millennia [4]!

This article has previously been published in the journal Physical Sciences Reviews. Please cite as: A. Towns, Diarylethene Dyes *Physical Sciences Reviews* [Online] 2020, 5. DOI: 10.1515/psr-2019-0146

https://doi.org/10.1515/9783110587104-025

- exhibit photochromism in solid state form. Members of most photochromic dye classes do not undergo changes that are visible to the naked eye unless they are in solution, either in a polymeric matrix or a solvent. In contrast, many DAEs switch color in crystalline form [5].
- fatigue resistance enabling cycling between states thousands of times. DAEs are not as hardy as members of other colorant classes, such as those of the naphthopyran and spirooxazine families. However, with the right design they are sufficiently robust to remain of interest as functional colorant switches. Significantly, diarylethene dyes are in general more resilient than members of the other major P-type class (fulgide) [6].

The discovery of these behaviors in DAE compounds during the mid-1980s [1] ignited great interest from technologists working in fields such as optical data storage, nanotechnology, and photonics. During 2000–2016, researchers authored over two thousand academic publications dedicated to DAEs [7]. This amount is more than the total that appeared during the same period which featured the T-type spiropyran, spirooxazine, and naphthopyran classes. The key characteristics that make DAEs attractive as functional colorants originate from specific molecular motifs. The general structure of members of this class is shown in Figure 25.1.

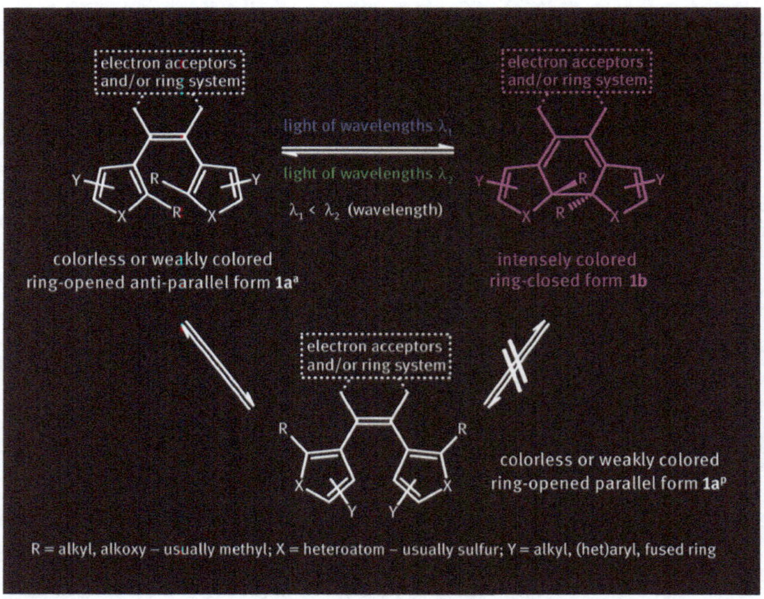

Figure 25.1: Generalized structure and photochromism of DAE dye class.

An enormous number of variants have been reported [8, 9], but the most commonly used DAEs may be thought of as comprising two crucial design elements:

i. an ethene bridge. It is usually substituted in a way that prevents *cis–trans* photoisomerization from occurring in order to stop this process competing with photocoloration. By far the most well-used strategy is to employ a cycloalkene motif. However, one and even two aromatic rings have been utilized as the unsaturated linking function to produce "terarylene" [10] and 'tetraarylene"' [11] P-type photochromic colorants.
ii. two hetaryl rings. They are linked by the ethene bridge. Not all heterocycles furnish robust P-type photochromism, so these ring systems and their substituents must be selected with care. Nevertheless, a huge number of permutations have been synthesized, which include the insertion of second heteroatoms into the heterocycles [11] as well as attaching or fusing aromatic rings onto them [8, 9].

Like the most industrially important photochromic colorant classes, light-driven color changes in DAE dyes involve photoisomerization by means of pericyclic rearrangement [2]. Decorating a central double bond with two heterocyclic rings, usually thiophene-based (see Figure 25.1, X = S), and introducing compact groups to block photochemical side reactions (see Figure 25.1, R ≠ H), affords molecules capable of absorbing light energy to transition reversibly between thermally stable states. The ring-opened forms **1a** of DAEs ("o-DAE") are generally colorless, or only weakly colored, because their non-planarity reduces conjugation between the π-systems of the heterocycles. The parallel conformer $\mathbf{1a^p}$ is not photoactive owing to its geometry, whereas the anti-parallel conformer $\mathbf{1a^a}$ is capable of cyclisation upon photoexcitation [12]. The latter absorbs ultraviolet and/or short wavelength visible light (see Figure 25.1, λ_1) with the net result that a thermally irreversible conversion to the ring-closed form **1b** ("c-DAE") occurs: it is intensely colored owing to the presence of an extended planar conjugated π-system, its longest wavelength absorption maximum typically being 200–300 nm longer than that of the corresponding o-DAE. (With reference to non-photochromic conventional colorants, the extinction coefficients of c-DAEs are comparable to anthraquinonoid and simple azo dyes.) Reversion of **1b** to **1a** only occurs when the chromophore of the former absorbs light energy corresponding to wavelengths further into the visible region (see Figure 25.1, λ_2), which the o-DAE **1a** cannot absorb.

The key to the thermal stability of c-DAE **1b** is the relatively low aromatic stabilization energy of the two heterocycles. It minimizes the tendency of the c-DAE form to ring open thermally to **1a**, thereby restoring the aromatic nature of the heterocyclic rings. (Replacement of the heterocycles with phenyl rings destroys P-type character. The high aromatic stabilization energy of the carbocycles widens the gap in energy between ring-opened and ring-closed forms. Consequently, the corresponding c-DAE form readily isomerizes thermally to the much lower energy o-DAE photoisomer, thereby leading to T-type photochromism.) It is important that substitution patterns on the heterocycles do not weaken the central C-C bond linking them in the ring-closed form to cause thermal instability [13, 14], e. g. through steric hindrance in case of R = isopropyl [15]. An appropriately substituted c-DAE can be stable for years at 20–30°C when kept away from visible light that would cause it to cyclorevert to the corresponding ring open form [6]. The thermal irreversibility makes it possible to resolve and isolate the photoisomers by high performance liquid chromatography. Much more detail concerning design principles and the theory behind the mechanics of photoisomerization can be found in [8] and [9].

The change in molecular geometry that accompanies the photoisomerization of DAEs is modest. For instance, the long dimension of the simple o-DAE **2** (X = Y = 5-Me) shrinks by just 10% on ring-closure while its short dimension grows only 10% and overall planarity increases [8]. Such small alterations enable many DAEs to exhibit pronounced photochromism in environments that completely suppress the photochromic properties of conventional industrial T-type dyes, which undergo greater changes in geometry upon photoisomerization. Numerous DAE colorants transition in polymer substrates with high Young's moduli and relatively inflexible chains that offer too little free volume for commercial T-type naphthopyrans and spirooxazines to operate. Plenty of examples also photoisomerize in their crystalline solid state – an unusual property, which opens up numerous potential applications that are not possible with conventional industrial photochromic colorants. The rate of change is very rapid for DAEs irrespective of whether they are in the form of solutions or crystals: typically switching takes just a matter of picoseconds (10^{-12}s) in either case [6]. The photochemistry of cyclization, as well as cycloreversion, has been much studied: for a good summary, see [8].

By far the most extensively investigated variety of DAE is the dithienylethene (DTE) subclass, specifically dithienylhexafluorocyclopentene derivatives **2** as shown in Figure 25.2. Both the perfluoro substitution and sizing of a five-membered ring increase fatigue resistance compared to that of analogues whose ethene bridges comprise non-fluorinated rings and/or cyclic structures of other sizes. The remainder of this Chapter will therefore tend to focus on this kind of colorant. DTEs are the sole type of DAE available off-the-shelf on the open market [16]. Only research quantities are offered which is reflected by pricing that is pitched at the level of hundreds of dollars per gram, an order of magnitude or two greater than that for commercial volumes of industrial T-type colorants.

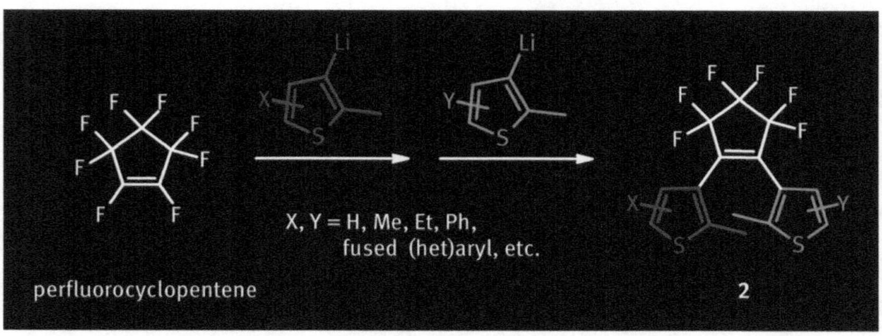

Figure 25.2: Typical synthetic route to DTE derivatives **2** from perfluorocyclopentene.

25.2 Synthesis of diarylethenes

First reported in the early 1990s [1], DTEs **2** are typically prepared by reaction of perfluorocyclopentene, a low boiling (27°C) liquid, with thienyllithium compounds. Most DTE derivatives reported tend to be symmetrical for synthetic expediency; only a single step, involving reaction with two equivalents of a single lithiothiophene (see Figure 25.2, X = Y), is needed. Many asymmetric diarylethene dyes are accessible by modifying this strategy to include sequential reaction of one equivalent each of two different lithioheterocycles (see Figure 25.2, X ≠ Y). Other coupling chemistries can be employed, e. g. use of hetarylboronic acids instead of organolithium derivatives. Another means of DTE synthesis is construction of the cyclopentene ring through cyclisation by intramolecular McMurry coupling of a 1,5-dithienyl-1,5-diketone [17]. For a comprehensive review of pathways to symmetric and asymmetric DTEs, see [18].

A plethora of unsaturated bridging fragments other than hexafluorocyclopentenyl have been explored [19]. Again, synthetic strategies to introduce the ethene bridge include coupling heterocycles to a ring system, or employing cyclisation reactions, creating many DAE examples that comprise heteroatoms and/or a ring of different size. Early instances are the P-type maleic anhydride derivative **4a** and its maleimide analogue **4b** (see Figure 25.3). The former is made through alkaline hydrolysis of the dicyano derivative **3** [8, 20], which itself exhibits P-type photochromism.

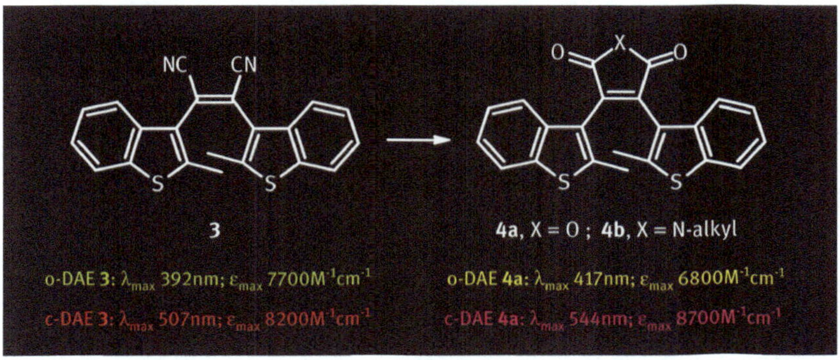

Figure 25.3: Properties of dicyano- and maleic acid-derived DTEs in benzene solution [12].

25.3 Color and constitution of diarylethene colorants

Despite the restrictions imposed on molecular structure to ensure photochromism is P-type (i. e. that ring-closure is thermally irreversible), considerable scope remains for variation in design. The pronounced influence of heterocycle type and substitution on the absorption properties of c-DAEs, coupled with the ingenuity of chemists, means that examples range the whole way across the visible spectrum and beyond. Within the DTE class alone, manipulation of structure generates P-type photocoloration from yellow through to red through to blue and even green (see Table 25.1).

Table 25.1: Influence on absorption properties by substitution pattern of ring-closed forms of dihetarylhexafluorocyclopentene-derived dyes **5** in hexane.

	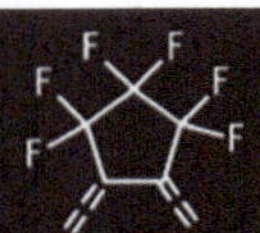	λ_{max} (nm)	ε_{max} ($M^{-1}cm^{-1}$)
5a [21]		425	5800
5b [21]	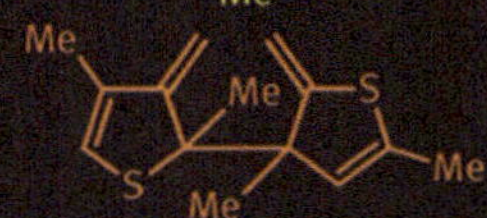	469	4500
5c [22]	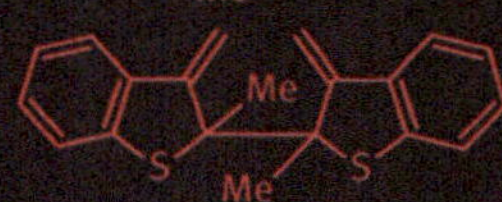	517	9100
5d [21]	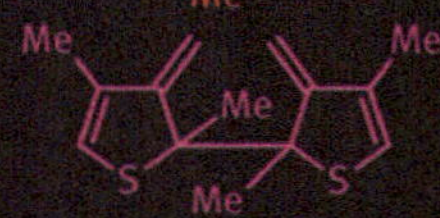	534	5000
5e [23]	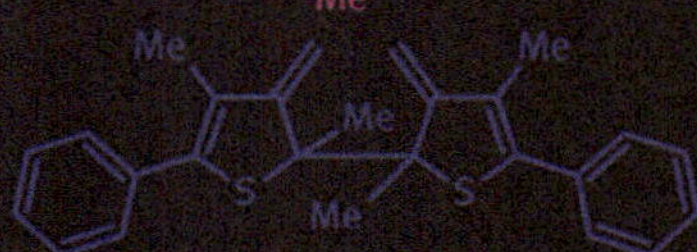	562	11000
5f [23]	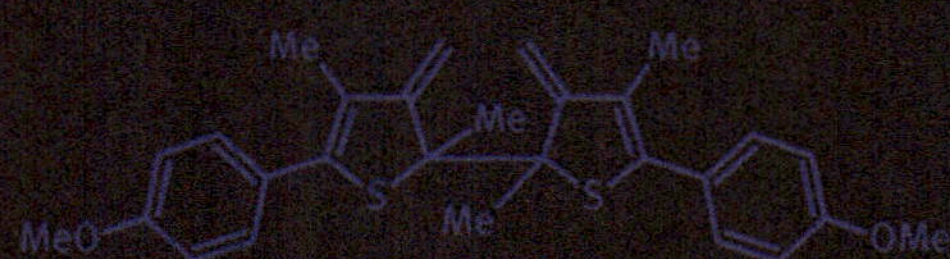	570	14000
5g [23]	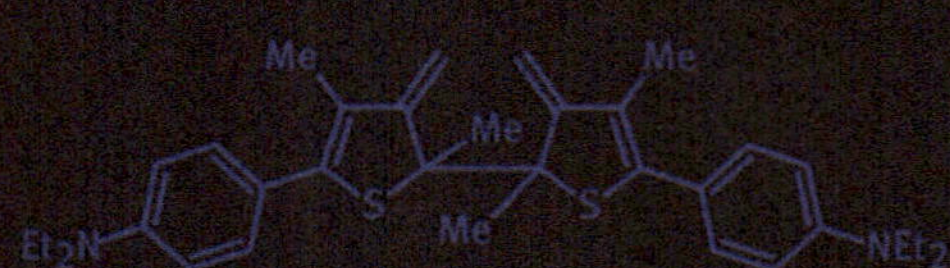	597	18000
5h [8]	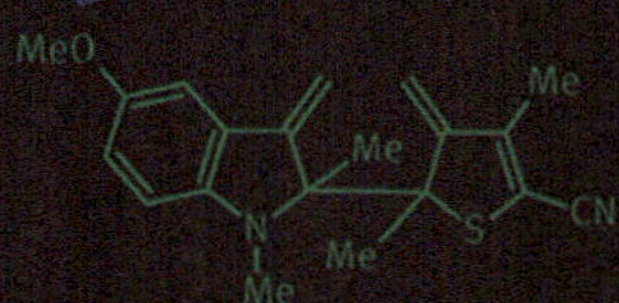	665	not reported
5i [24]	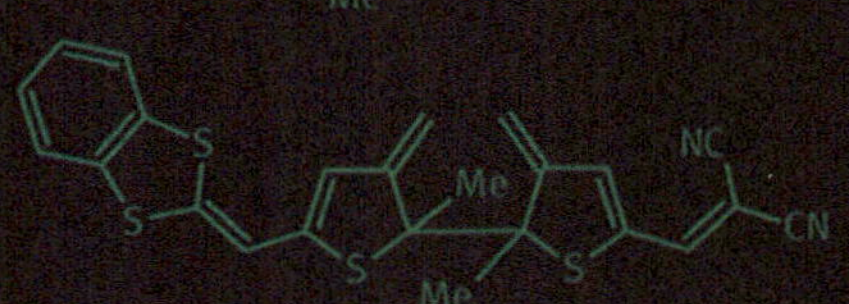	828 (benzene-d6)	23000 (benzene-d6)

Expanding the conjugated system of c-DTEs, e. g. fusing (**5c**) or appending (**5e**) phenyl rings to the parent structure produces bathochromic shifts. Introduction of electron donating functions onto pendant phenyl rings (e. g. methoxy groups of **5f**) gives further red shifts in absorption as does increasing their strength (e. g. diethylamino groups of **5 g**). These changes also increase intensity of coloration: for instance, the extinction coefficients of c-DTEs **5d**, **5e**, **5f** and **5 g** grow progressively (see Table 25.1). Further developing the push–pull nature of the chromophore by incorporation of electron-accepting motifs leads to pronounced bathochromism (**5 h-i**): however, this comes at a price – the central C-C bond of c-DAE **1b** in the case of **5i** is weakened sufficiently that this isomer is no longer thermally stable. While the benzodithiole-based donor system opposite the dicyanoethylene-derived acceptor fragment pushes absorption of **5i** into the near infrared, the DTE exhibits T-type rather than P-type photochromism. The half-life of its ring-closed form is just ~ 3 h at 60°C whereas that for **5c** is > 12 h at 80°C [8]. (Loading electron acceptor groups directly onto the thienyl rings of DTEs tends to lessen their P-type character.) The absorption maxima and extinction coefficients of o-DAEs respond in a qualitatively similar manner to the aforementioned structural changes in c-DAEs, shifting toward the visible region and becoming larger, respectively. As shall be discussed later, DAEs that can be ring-closed with visible light (i. e. avoiding use of UV radiation) are of much interest as functional colorants for certain applications, provided that P-type behavior and robustness are not compromised. DAEs of this kind remain a focus of research [25–27].

Attaching both thienyl rings to the bridge by their 2-positions, as in the case of **5a**, leads to yellow photocoloration. In contrast, the isomeric colorant **5d** in which both thiophene systems are linked to the bridge by their 3-positions becomes red upon exposure to UV. The hypsochromism of **5a** is ascribed to the orientations of the hetaryl systems in its ring-closed form largely restricting π-conjugation to within the cyclohexadiene fragment [21]. Greater delocalization over the thienyl rings relative to **5a** occurs in the case of **5d** upon ring closure, resulting in red-shifted absorption. The absorption maximum of the ring-closed form of analogue **5b**, in which one pendant hetaryl ring is 3-thienyl and the other 2-thienyl, lies between these two extremes. P-type photochromism is not sacrificed: the ring-closed forms of each of these three colorants are thermally stable with absorbance in oxygen-free heptane remaining essentially constant for > 500 h at 80°C [21].

Replacement of the thienyl rings with other heterocycles produces major changes in photocoloration in terms of hue, thermal stability, quantum yield, and other properties. Introduction of a nitrogen heteroatom into DTE **6a** to create the dithiazolyl analogue **6b** leads to pronounced hypsochromism (see Table 25.2). It also reinforces the thermal stability of its c-DAE photoisomer given the reduced aromatic stabilization energy of the 4-thiazolyl fragments in the o-DAE form. (Note that figures of 525 nm and 10,000 $M^{-1}\,cm^{-1}$ are reported for **6b** in [32].) Substitution of the sulfur heteroatom

Table 25.2: Influence on absorption properties of the substitution pattern of ring-closed forms of dihetarylhexafluorocyclopentene-derived dyes **6** in hexane solution.

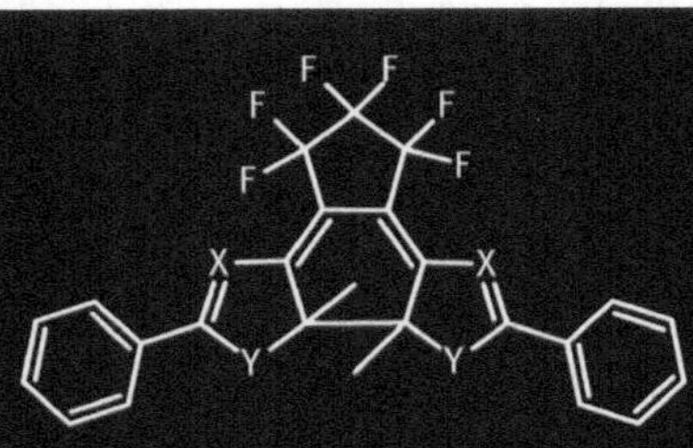

DAE	X	Y	λ_{max} (nm)	ε_{max} ($M^{-1}cm^{-1}$)
6a [28]	CH	S	575	15600
6b [29]	N	S	534	14500
6c [30]	CH	O	525	16500
6d [31]	N	O	462	not reported

of **6a** and **6b** with oxygen yields difuryl and dioxazolyl colorants **6c** and **6d**, respectively, with even less stabilization energy and greater hypsochromism. Depending upon the nature of the ethene bridge, insertion of nitrogen heteroatoms into the thiophene rings of DTEs brings about other benefits such as improved quantum yield for photocyclization by increasing the proportion of the photoactive anti-parallel isomer **1aa** in the o-DAE form [11].

As well as the nature of the pendant heterocycles, the other substituents on the ethene bridge profoundly influence the color of the ring-closed forms. Table 25.3 illustrates how modifying the hexafluorocyclopentyl fragment of DTE **5c** alters shade: breaking the ring to create **7** results in a substantial hypsochromic shift, while adjusting ring size (**8**; n = 0, 2) exerts a more modest effect. Ring size affects planarity of the ring-closed form, and thereby π-conjugation within it, which in turn impacts upon absorption band maximum and shape, and ultimately color [22, 33]. Use of stronger electron acceptors as demonstrated by the dicyano derivative **3** and maleic acid derivatives **4** results in bathochromism (see Figure 25.3) in both ring-opened and -closed forms: the latter are yellowish rather than colorless and thus short wavelength visible light of wavelength ~400 nm as well as near UV triggers color generation. In contrast, the open form of simple hexafluorocyclopentyl DTE **5e** responds only to shorter wavelengths, e. g. mid-UV radiation of ~313 nm. Visible light of >500 nm wavelength drives the return of its blue photoisomer **5e** back to the colorless form.

Table 25.3: Influence of cycloaliphatic ring system on absorption maximum in benzene solution of ring-closed forms of DTE dyes [22].

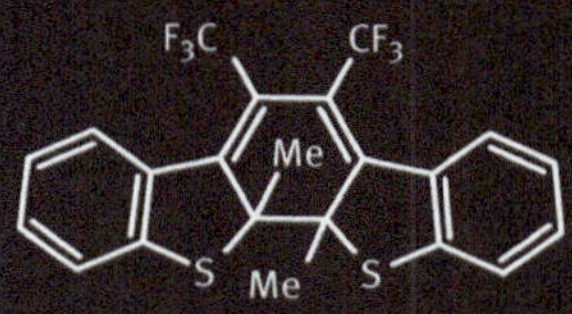

7

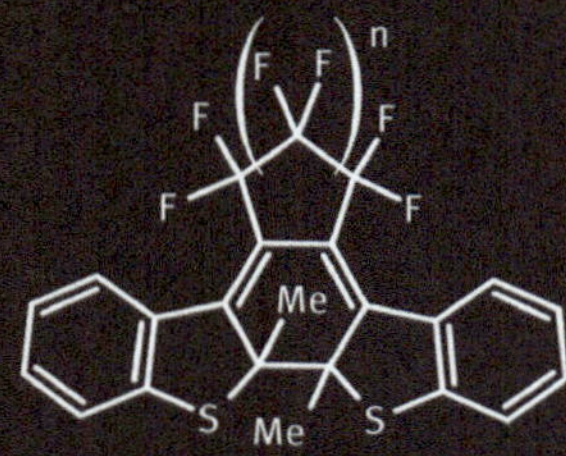

8

DTE	λ_{max} (nm)
7	449
8 (n = 0)	532
5c [8 (n = 1)]	526
8 (n = 2)	510

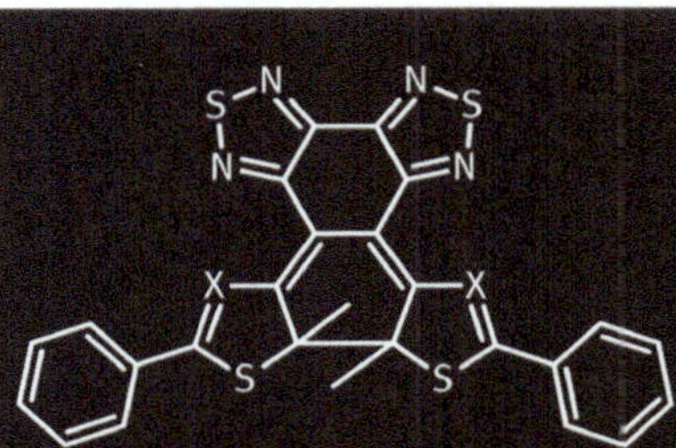

9a X = CH λ_{max} ~570, ~455 [34]

9b X = N λ_{max} 572, 426 [35]

Figure 25.4: Two examples of neutral-colored DAEs and the absorption maxima (in nm) of their ring-closed photoisomers **9** in acetonitrile solution.

Examples of single-molecule P-type DTEs affording dull tertiary shades are known. Dye **9a** is brownish when ring-closed [34] owing to two overlapping absorption bands in the visible region. Remarkably its dithiazolyl analogue **9b** transitions from colorless to black when irradiated with UV [35] (see Figure 25.4). This neutral coloration of the closed ring form, arising from reasonably uniform absorption across most of the visible spectrum, also means that cycloreversion can be effected by irradiation with light of a wide range of wavelengths in the visible region.

Relationships between DAE structure and quantum yields for photoisomerization, thermal stability of photoisomers, and fatigue resistance are well understood [8]. Removing vulnerable unsubstituted positions on thiophene rings of DTEs markedly improves the robustness of dye solutions towards repeated cycling between ring-opened and -closed forms. For exampling, annellation with phenyl rings in the case of **5c** greatly curbs loss of photochromism by suppressing reaction with oxygen to create endoperoxide side photoproducts [22]. Over 90% of photocoloration intensity is retained by this dye even after 14,000 cycles of coloration/decoloration in methylcyclohexane solution. The 4-methyl groups of **5e** inhibit photochemical cyclisation to non-photochromic colored side products, so fatigue resistance is greatly enhanced compared to the analogous DTE **10** lacking substitution at these positions [36] (see Figure 25.5). Hexafluorocyclopentene derivatives are relatively hardy DAEs: for example, **5c** is over an order of magnitude more fatigue resistant in solution than its maleic anhydride analogue **4a** [8]. The state of a DAE often makes a big difference to its resilience: the photochromism of such a dye in its solid state is markedly more robust compared to when it is solution. The crystal lattice sterically hinders side reactions and impedes diffusion of oxygen (for example, see Figure 25.5).

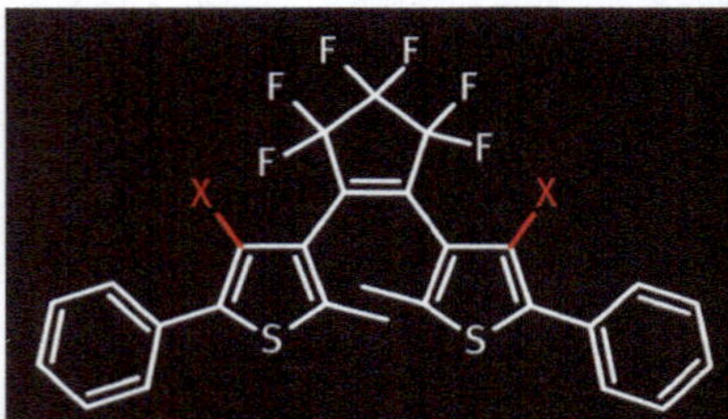

		Hexane	Crystal
X = H	**10**	200	>10000
X = Me	**5e**	2000	>30000

Figure 25.5: Influence of 4-methyl substitution on number of cycles after which 20% loss of DTE molecules occurs in hexane solution under vacuum or in crystalline state [36, 37].

However, since the draw of DAEs lies in them being functional colorants, P-type color change is not necessarily of primary interest: variations in properties unrelated to color that accompany the photoisomerization often more greatly concern researchers. These attributes, and the potential applications which exploit them, are covered in the next section.

25.4 Applications of diarylethene dyes

A report of P-type photochromism in DTEs appeared during 1967 [38]. Over twenty years passed before researchers seized upon the DAE unit as a tool in applications which required light-activated bistable switches. This interest was prompted by publication of research which demonstrated the high thermal stability and resilience of

DTEs in the late 1980s and early 1990s [8, 39]. Very soon afterwards the first fruits of work seeking to exploit DAEs as functional colorants were disseminated in the academic literature [40]. The publication of work relating to DAEs, and especially DTEs, blossomed with the annual number of research papers growing year on year over the next two decades [7, 41]. Numerous review papers and books document these extensive efforts: excellent recent accounts giving general overviews include [9] and [42], as well as individual chapters within [43].

Table 25.4 gives a flavor of the fields in which the exploitation of DAEs has been attempted. It lists a few example applications as well as the nature of the property changes that accompany photoisomerization upon which they depend. The remainder of this section explains some of these entries in more detail. Those outlets involving the solid-state photochromism exhibited by certain DAEs will only be touched on briefly because in such applications these colorants are being employed as pigments rather than dyes [60].

Table 25.4: Some examples of switchable properties of DAEs and instances of applications in which their exploitation has been attempted.

Photoreversible Property	Example application
Absorbance in mid-infrared	Memory element for data storage [44]
Absorption of visible light	Memory elements [45]; focal plane masks [46]; optical masks [46]
Chirality [47, 48]	Liquid crystal displays [49], control of reaction stereospecificity [50]
Conductance	Molecular electronics [45]
Crystal dimensions	Nanotechnology [51]
Cytotoxicity	Photodynamic cancer therapy [52]
Electronic structure	Reaction control [50]; polymerization catalysis [53]
Geometry	Switchable guest-host receptors [51]; adhesion promoter [54], reaction control [50]
Ion affinity	Biological probe [55]
Luminescence	Super-resolution microscopy [56]; memory element [45, 57]
pKa	Multi-stimulus logic gate construction [55]
Refractive index	Waveguide optoelectronic component [46]; holography [46]; memory
Wettability/surface tension [58]	Microfluidics [59]

25.4.1 Information technology and optoelectronics

Photoreversibility of various optical properties – not just color – serves as the basis for potentially very high-density storage systems consisting of DAEs as binary memory elements. Irradiation with UV or visible light alters the state of an element, i. e. writes data. Reading of the data non-destructively (i. e. without altering it) requires that the state of each element be determined by means of another property which varies between open ring and closed ring forms. For example, absorption in the mid-infrared [44], refractive index [46] or chirality [47] of DAE-doped materials can be probed using electromagnetic radiation of wavelengths that do not induce photoisomerization. Fluorescence is another means of non-destructive readout. An example of this approach relies on a DTE whose fluorescent open ring form is switched to a non-fluorescent closed ring photoisomer by irradiation with light of <400nm and vice versa with visible radiation of >600nm [57]. Probing of the state of an individual element is performed by irradiation with light of ~460nm. An element in open ring form will fluoresce at ~590nm. Neither of these two wavelengths lead to photoisomerization, thereby preserving data. A variation on this strategy employs metal complexes linked to DTEs, enabling non-destructive readout of phosphorescence [61]. DAE-based switches have been designed that are not just bistable, but which can adopt more than the two binary states of a bit. For example, the panchromatic absorption of **9b** enables ternary states to be constructed through irradiation with different wavelengths of visible light [35]. In another elegant proof-of-concept study, three DTE units, each of a different structure, were incorporated into a single photochromic colorant that was capable of being switched between yellow, red, and blue hues by particular combinations of UV and visible light for potential use as a multiple state memory element [62].

Photocontrolled switching of electrical current is also of interest for the construction of memory elements that can be read non-destructively by means of current detection, as well as for the manufacture of light-programmable electronic components [63]. Approaches using DAEs may be at bulk or single-molecule scales. In the former case, polymer films doped with appropriately substituted DAEs are sandwiched between electrodes and their conductivity switched by irradiation with light – devices have been constructed that yield higher conductivity when the concentration of colored c-DAE form is raised. Such a device has potential in optical memory and electrical circuitry. Studies of photoelectrochemical switching at the molecular scale have a similar application focus [40]. In conductive organic compounds, electrons flow through their conjugated π-systems. By controlling the breaking and re-making of the chain, conductivity can be switched off and back on. The use of DTEs in this manner was investigated only a year or so after their P-type behavior was first reported in the early 1990s [64]. In o-DAE form **1a**, π-electrons are localized in the hetaryl rings so the photoisomer is "non-conductive" or "insulating". In c-DAE form **1b**, they are delocalized across the molecule, which is thus "conductive". Incorporation of DTE units as monomers into conductive polymer chains was undertaken as long as two decades ago [65].

Despite the ingenuity of the work described above, DAE-based memory devices and optoelectronics have not supplanted non-photochromic technologies, nor are they likely to do so in the near future given the pace of advances in current conventional storage and electronics.

DAEs have also been at the heart of attempts to explore photochromic materials as components in neuromorphically engineered constructs to process information, i. e. computing based on biological nervous systems. The authors of a pioneering study [66] use DTE **5c** as a memory element in an artificial neural network, likening it to an optical "memristor" analogue whose absorbance, rather than resistance, is a function of its history. Such systems are many years away from being a practical proposition. Conventional technologies that do not rely on photochromism are unlikely to be displaced any time soon from their position of dominance.

25.4.2 Imaging

Super-resolution microscopy (SRM) is a clever optical technique that enables features of biological and material science-related objects to be imaged despite their size being smaller than the so-called "diffraction barrier" of ~250 nm [67]. The key figures in the development of this "nanoscopy" technique earned the Nobel Prize for Chemistry in 2014 [68]. SRM's use of fluorescent molecules visualizes living systems and enables the highlighting of component parts such as specific proteins [69] – something that is out of reach of electron microscopy. Resolution can be even further refined using photoswitchable fluorophores [70]. DAEs which photoisomerize between non-emissive and luminescent states have proved useful for SRM [56], although improvements in performance are needed. Researchers are now designing DAE colorants specifically targeted for use in nanoscopy. A priority is enhancing fatigue resistance within the aqueous environments encountered in biological substrates [71, 72]. While DAEs could become an entrenched fixture in SRM, their usage will remain high-value and low-volume.

An ingenious, yet non-commercialized, macroscopic image creation system relies on DAEs acting as dopants in cholesteric liquid crystals (CLC) [49]. The chirality change that occurs upon irradiation of a dopant DAE affects CLC structure which in turn determines the hue of light reflected by the crystalline phase. In principle, this enables the construction of thermally stable flat displays whose color is photocontrolled – no auxiliary electronic drive circuitry is required unlike conventional LC displays.

25.4.3 Process technology

Microfluidic technologies, as used for example in bioassays, depend upon accurate manipulation of liquid flowing in microscopic channels of devices [59]. Localized control of this flow with light is of interest since it enables remote operation and is non-invasive. DAE dyes have been explored as the photoresponsive units in such systems. When incorporated into hydrogels, the hydrophobicity/hydrophilicity of these materials can be manipulated by irradiation with light – the accompanying variation in size acts to channel flow. An alternative is to manipulate droplet movement by locally modifying the wettability of the surface on which they reside.

DAEs are also being explored as tools to enhance control of chemical synthesis. Using the change in electronic structure and, in some cases, chirality that occurs upon photoisomerization, forays into controlling chemical reactions have been made with DTEs and dithiazolylethenes. They play the role of reagents or catalysts, in conjunction with light [73]. The radiation does not serve as an energy source, as it would in traditional photochemistry. Instead it dictates when and where chemical transformations occur. In principle, light allows precise remote control in terms of both space and time by influencing reactivity of starting materials, position of reaction equilibria, catalytic activity and product stereochemistry [50]. Dynamic control of polymerization by means of photoswitched catalysts based on DTEs is being explored [53]. The ability to manipulate the conversion of monomers with precision in space and time is attractive, enabling unprecedented control over polymer composition, but it remains far from commercial realization for technical and economic reasons.

25.4.4 Biotechnology

Just as DAEs have formed part of efforts to harness photochromism in the controlled synthesis of small molecules and polymers, this class often finds itself at the center of research seeking to manipulate the course of biochemistry through light [74–76]. DAEs have been explored in connection with photopharmacology (i. e. changing the activity of biological activity of molecules using light [52, 77]), specifically to modulate the cytotoxicity of drugs by optical means only where required [78, 79]. For example, a photoswitchable analog of the cancer drug cisplatin, modified to incorporate an open-ring DTE unit, is more toxic to cancer cells when it comprises the ring-closed photoisomer [80]. Conversion to the latter form increases the likelihood of interference with DNA repair, accelerating cell death. Localized application of UV would accomplish the photoisomerization only in the region of a tumor, thus minimizing damage to surrounding healthy tissue. However, UV is itself toxic, rendering its use clinically unacceptable [79][81]. Developing therapies that respond to light in the 600–1200 nm window is thus of much interest since such wavelengths penetrate biological tissue more deeply than shorter wavelengths and without causing damage. Compatibility towards aqueous environments in terms of solubility and stability is also important in small molecule-based photoswitches. While these goals present major challenges for DAE dye design, avoiding use of UV does appear feasible. An approach that has the big advantage of utilizing visible light depends upon peptides that resemble the highly cytotoxic antibiotic gramicidin S: incorporation of a ring-closed DTE unit into their backbone reduces cytotoxicity by around an order of magnitude [82, 83]. The premise is that this form has sufficiently low toxicity to enable its systemic application. Irradiation with visible light only in the vicinity of a tumor leads to ring opening of the DTE unit and a localized increase in membrane disruption. Cell death is thus confined to tumor tissue.

As well as switching pharmacodynamic characteristics (i. e. the interaction of a drug with its target), DAE-based compounds have also been investigated in relation to pharmacokinetic properties. The latter term refers to how a drug becomes distributed within living tissue, which has a bearing on its efficacy. DAE photochemistry has been employed to regulate uptake of biomolecules into cells. Peptides containing DTE units embedded into their backbones more readily penetrate cell membranes when the photochromic species are ring-closed owing to the less rigid o-DTE form promoting greater overall chain flexibility [84]. Ingress of such peptides can thus be controlled by irradiation with tissue-friendly red light. DAEs are regarded as one the most promising types of functional colorant for modulating biological activity [81][85]. Nonetheless, much remains to be done in order to realize their potential in creating efficient photocontrollable drugs [86].

DAEs are of interest in “nanocarrier” technology to photocontrol drug delivery [26]. For instance, a DTE bearing tri(ethylene glycol) and nonyloxy side-chains in an aqueous environment spontaneously forms colorless submicron-diameter spheres capable of encapsulating bioactive materials [87]. Upon UV irradiation, the relatively flexible o-DTE photoisomerizes into a more rigid c-DTE, leading to conversion of spheres into blue fiber-like aggregates. However, controlled release from DAE-based nanocarriers by photoirradiation has yet to be realized.

25.4.5 Photomechanical technologies

One of the most striking features of DAE photochromism is the occurrence not only of photochromism in crystals, but also the macroscopic changes in shape and orientation which accompany it. The first reports were of square crystals morphing into rhomboid form when irradiated with UV, as well as crystalline rods bending in response to the stimulus [88]. The latter was used to demonstrate the conversion of light energy into mechanical work: the photoinduced motion led to microscopic beads in proximity being batted away. These deformations occur because of planarization of the DAE structure upon ring-closure, driving contraction along one crystallographic axis and expansion along another. These changes at a molecular level become manifest macroscopically because of the precisely arranged DAE units of a crystal lattice exhibiting a concerted photoresponse. (Certain DAE crystals exhibit the rare phenomenon of photosalience [89]: sudden shattering of a solid upon exposure to intense light.) Pronounced photomechanical responses are more difficult to achieve with polymeric matrices doped with photochromic colorant owing to a lack of long-range ordering of dye molecules in sufficiently high concentration. One means of tackling this difficulty is to employ DAEs to modulate the properties of liquid crystalline polymers. Photomobile materials can be constructed in this way, for example, films that bend toward UV sources and return to planarity when exposed to visible light [90]. By careful arrangement of DAE molecules (and in some cases angle of irradiation), photoinduced twisting is also possible [91]. Bi-directional, thermally-stable light-driven geometry changes typical of DAE units are also of great interest to researchers exploring their potential as components of supramolecular systems [76] and nanomachines [51].

Another potential application reliant upon nanoscale interactions concerns use of DAEs as adhesion promoters. DTE **5e** was found to improve bonding between glass and the polystyrene film into which it had been doped and then activated with UV radiation [54]. The effect remained largely intact following cycloreversion. It was hypothesized that "photochromic annealing" of the polymer had occurred as a result of the geometry change accompanying ring-closure, leading to localized polymer motion reducing voids in the interface. The overall more intimate contact between glass and polymer promoted adhesion.

25.5 Summary

Like the discovery of several dye and pigment classes of key industrial importance, serendipity played a role in the genesis of DAE colorants. The news of their P-type photochromism, which operated in the solid state for thousands of cycles, was seized upon by technologists and material scientists. They have synthesized many thousands of examples [55], as well as incorporating them into polymeric and supramolecular systems, in an effort to develop applications based upon light-responsive switches. The thermal irreversibility of the photochromic transitions exhibited by suitably designed DAE-based compounds has led to their investigation in many spheres of scientific endeavor, such as cancer therapy and information technology. Some of these applications could revolutionize their fields, making an impact on daily life and well-being of the general population. Whether they will successfully compete with non-DAE approaches on a technical and economic footing remains to be seen. Despite the ingenuity and hype surrounding the research dedicated to their exploitation, use of DAE dyes will continue to be restricted to a few low-volume niche end-uses in the short-term at least.

References

1. Irie M. Discovery and development of photochromic diarylethenes. Pure Appl Chem. 2015;87:617–26.
2. Towns A. Photochromic Dyes. to be submitted.
3. Towns A. Naphthopyran dyes. Phys Sci Rev. 2020;5:20190085.
4. Takame S, Kobatake S, Kawai T, Irie M. Extraordinarily high thermal stability of the closed-ring isomer of 1,2-Bis(5-methyl-2-phenylthiazol-4-yl)perfluorocyclopentene. Chem Lett. 2003;32:892–3.
5. Animoto K, Kawato T. Photochromism of organic compounds in the crystal state. J Photochem Photobiol C Photochem Rev. 2005;6:207–26.
6. Irie M. Photoswitchable Molecular Systems based on Diarylethenes. In: Feringa BL, editor(s). Molecular switches, 1st ed. Weinheim: Wiley-VCH. Chapter 2. 2001:37–62.
7. Nakatani K, Piard J, Yu P, Métivier R. Chapter 1 Introduction: organic photochromic molecules. In: Tian H, Zhang J editors, Photochromic materials: preparation, properties and applications, 1–45. Weinheim: Wiley-VCH, 2016.
8. Irie M. Diarylethenes for memories and switches. Chem Rev. 2000;100:1685–716.
9. Irie M, Fukaminato T, Matsuda K, Kobatake S. Photochromism of diarylethene molecules and crystals: memories, switches and actuators. Chem Rev. 2014;114:12174–277.
10. Nakashima T, Kawai T. Chapter 10 Photochromic terarylenes. In: Irie M, Yokoyama Y, Seki T editors, New frontiers in photochromism, 183–204. Tokyo: Springer, 2013.
11. Lvov AG, Khusniyarov MM, Shirinian VZ. Azole-based diarylethenes as the next step towards advanced photochromic materials. J Photochem Photobiol C Photochem Rev. 2018;36:1–23.
12. Uchida K, Nakayama Y, Irie M. Thermally irreversible photochromic systems. reversible photocyclisation of 1,2-Bis(benzo[b]thiophen-3-yl)ethene derivatives. Bull Chem Soc Jpn. 1990;63:1311–15.
13. Kitagawa D, Sasaki K, Kobatake S. Correlation between steric substituent constants and thermal cycloreversion reactivity of diarylethene closed-ring isomers. Bull Chem Soc Jpn. 2011;84:141–7.
14. Kitagawa D, Kobatake S. Strategy for molecular design of photochromic diarylethenes having thermal functionality. Chem Rec. 2016;16:2005–15.
15. Uchida K, Tsuchida E, Aoi Y, Nakamura S, Irie M. Substitution effect on the coloration quantum yield of a photochromic bisbenzothienylethene. Chem Lett. 1999;28:63–4.
16. For example, Tokyo Chemical Industry. Photochromic dyes. www.tcichemicals.com/eshop/en/gb/category_index/12989/ Accessed: 25 Feb 2020.
17. Lucas LN, de Jong JJ, van Esch JH, Kellogg RM, Feringa BL. Syntheses of dithienylcyclopentene optical molecular switches. Eur J Org Chem. 2003;155–66.
18. Szalóki G, Pozzo J-L. Synthesis of symmetrical and non-symmetrical bisthienylcyclopentenes. Chem Eur J. 2013;19:11124–32.
19. Chen S, Li W, Zhu W-H. Chapter 2 Novel ethene-bridged diarylethene photochromic systems: self-assembly, photoswitcher, and molecular logic gates. In: Yokoyama Y, Nakatani K editors, Photon-working switches, 37–62. Tokyo: Springer, 2017.
20. Nakayama Y, Hayashi K, Irie M. Themally irreversible photochromic systems. reversible photocyclisation of non-symmetric diarylethene derivatives. Bull Chem Soc Jpn. 1991;64:789–95.
21. Uchida K, Irie M. A photochromic dithienylethene that turns yellow by UV irradiation. Chem Lett. 1995;24:969–70.

22. Hanazawa M, Sumiya R, Horikawa Y, Irie M. Thermally irreversible photochromic systems. reversible photocyclization of 1,2-bis (2-methylbenzo[b]thiophen-3-yl)perfluorocycloalkene derivatives. J Chem Soc Chem Commun. 1992;206–7.
23. Irie M, Sakemura K, Okinaka M, Uchida K. Photochromism of dithienylethenes with electron-donating substituents. J Org Chem. 1995;60:8305–9.
24. Gilat SL, Kawai SH, Lehn J-M. Light-triggered electrical and optical switching devices. J Chem Soc Chem Commun. 1993;1439–42.
25. Zhang Z, Wang W, Jin P, Xue J, Sun L, Huang J, et al. A building-block design for enhanced visible-light switching of diarylethenes. Nat Commun. 2019;10:4232.
26. Jia S, Fong W-K, Graham B, Boyd BJ. Photoswitchable molecules in long-wavelength light-responsive drug delivery: from molecular design to applications. Chem Mater. 2018;30:2873–87.
27. Fredrich S, Goestl R, Herder M, Grubert L, Hecht S. Switching diarylethenes reliably in both directions with visible light. Angew Chem Intl Ed Engl. 2016;55:1208–12.
28. Irie M, Lifka T, Kobatake S, Kato N. Photochromism of 1,2-Bis(2-methyl-5-phenyl-3-thienyl) perfluorocyclopentene in a Single-Crystalline Phase. J Am Chem Soc. 2000;122:4871–6.
29. Takami S, Kawai T, Irie M. Photochromism of dithiazolylethenes having methoxy groups at the reaction centers. Eur J Org Chem. 2002;22:3796–800.
30. Yamaguchi T, Irie M. Photochromic and fluorescent properties of bisfurylethene derivatives. J Mater Chem. 2006;16:4690–4.
31. Kuroki L, Takami S, Shibata K, Irie M. Photochromism of single crystals composed of dioxazolylethene and dithiazolylethene. Chem Commun. 2005;6005–7.
32. Uchida K, Ishikawa T, Takeshita M, Irie M. Thermally irreversible photochromic systems. reversible photocyclization of 1,2-Bis(thiazolyl) perfluorocydopentenes. Tetrahedron. 1998;54:6627–38.
33. Hohlneicher G, Mueller M, Demmer M, Lex J, Penn JH, Gan L-X, et al. 1,2-diphenylcycloalkenes: electronic and geometric structures in the gas phase, solution, and solid state. J Am Chem Soc. 1988;110:4483–94.
34. Zhu W, Yang Y, Métivier R, Zhang Q, Guillot R, Xie Y, et al. Unprecedented stability of a photochromic bisthienylethene based on benzobisthiadiazole as an ethene bridge. Angew Chem Intl Ed. 2011;50:10986–90.
35. Wu Y, Guo Z, Zhu W-H, Wan W, Zhang J, Li W, et al. Photoswitching between black and colourless spectra exhibits resettable spatiotemporal logic. Mater Horiz. 2016;3:124–9.
36. Irie M, Lifka T, Uchida K, Kobatake S, Shindo Y. Fatigue resistant properties of photochromic dithienylethenes: by-product formation. Chem Commun. 1999;747–50.
37. Jean-Ruel H, Cooney RR, Gao M, Lu C, Kochman MA, Morrison CA, et al. Femtosecond dynamics of the ring closing process of diarylethene: a case study of electrocyclic reactions in photochromic single crystals. J Phys Chem A. 2011;115:13158–68.
38. Kellogg RM, Groen B, Wynberg H. Photochemically induced cyclization of some furyl- and thienylethenes. J Org Chem. 1967;32:3093–100.
39. Irie M. Chapter 5 Diarylethenes with heterocyclic aryl groups. In: Crano JC, Guglielmetti RJ editors, Organic photochromic and thermochromic compounds volume 1: main photochromic families, 207–22. New York: Plenum, 1999.
40. Matsuda K, Irie M. Diarylethene as a photoswitching unit. J Photochem Photobiol C Photochem Rev. 2004;5:169–82.
41. Warford CC, Lemieux V, Branda NR. Chapter 1. Multifunctional diarylethenes. In: Feringa BL, Browne WR, editors. Molecular switches. 2nd ed. Weinheim: Wiley-VCH, 2011:3–35.
42. Zhang J, Tian H. The endeavor of diarylethenes: new structures, high performance, and bright future. Adv Opt Mater. 2018;6:1701278.

43. Yokoyama Y, Nakatani K, eds. Photon-working switches. Tokyo: Springer, 2017.
44. Stellacci F, Bertarelli C, Toscano F, Gallazzi MC, Zerbi G. Diarylethene-based photochromic rewritable optical memories: on the possibility of reading in the mid-infrared. Chem Phys Lett. 1999;302:563–70.
45. Minkin VI. Bistable organic, organometallic, and coordination compounds for molecular electronics and spintronics. Russ Chem Bull Int Ed. 2008;57:687–717.
46. Bertarelli C, Bianco A, Castagna R Pariani G. Photochromism into optics: opportunities to develop light-triggered optical elements. J Photochem Photobiol C Photochem Rev. 2011;12:106–25.
47. Feringa BL, van Delden RA, Koumura N, Geertsema EM. Chiroptical molecular switches. Chem Rev. 2000;100:1789–816.
48. Nakagawa T, Ubukata T, Yokoyama Y. Chirality and stereoselectivity in photochromic reactions. J Photochem Photobiol C Photochem Rev. 2018;34:152–91.
49. Bisoyi HK, Li Q. Light-driven liquid crystalline materials: from photo-induced phase transitions and property modulations to application. Chem Rev. 2016;116:15089–166.
50. Göstl R, Senf A, Hecht S. Remote-controlling chemical reactions by light: Towards chemistry with high spatio-temporal resolution. Chem Soc Rev. 2014;43:1982–96.
51. Wang L, Li Q. Photochromism into nanosystems: towards lighting up the future nanoworld. Chem Soc Rev. 2018;47:1044–97.
52. Hüll K, Morstein J, Trauner D. In vivo photopharmacology. Chem Rev. 2018;118:10710–47.
53. Ihrig SP, Eisenreich F, Hecht S. Photoswitchable polymerization catalysis: state of the art, challenges, and perspectives. Chem Commun. 2019;55:4290–8.
54. Mostafavi SH, Tong F, Dugger TW, Kisailus D, Bardeen CJ. Noncovalent photochromic polymer adhesion. Macromol. 2018;51:2388–94.
55. Pu S-Z, Sun Q, Fan C-B, Wang R-J, Liu G. Recent advances in diarylethene-based multi-responsive molecular switches. J Mater Chem C. 2016;4:3075–93.
56. Nevskyi O, Sysoiev D, Dreier J, Stein SC, Oppermann A, Lemken F, et al. Fluorescent diarylethene photoswitches – A universal tool for super-resolution microscopy in nanostructured materials. Small. 2018;14:1703333.
57. Tsivgoulis GM, Lehn J-M. Photonic molecular devices: reversibly photoswitchable fluorophores for nondestructive readout for optical memory. Angew Chem Int Ed. 1995;34:1119–22.
58. Guo F, Guo Z. Inspired smart materials with external stimuli responsive wettability: a review. RSC Adv. 2016;6:36623–41.
59. Dunne A, Francis W, Delaney C, Florea L, Diamond D. Stimuli-controlled fluid control and microvehicle movement in microfluidic channels, reference module in materials science and materials engineering. Elsevier, 2017. http://doi.org/10.1016/B978-0-12-803581-8.04043-1 Accessed: 25 Feb 2020.
60. Towns A. Colorants: general survey. Phys Sci Rev. 2019;4. DOI: 10.1515/psr-2019-0008.
61. Norsten TB, Branda NR. Axially coordinated porphyrinic photochromes for non-destructive information processing. Adv Mater. 2001;13:347–9.
62. Higashiguchi K, Matsuda K, Tanifuji N, Irie M. Full-color photochromism of a fused dithienylethene trimer. J Am Chem Soc. 2005;127:8922–3.
63. Tsujioka T, Irie M. Electrical functions of photochromic molecules. J Photochem Photobio C Photochem Rev. 2010;11:1–14.
64. Gilat SL, Kawai SH, Lehn J-M. Light-triggered molecular devices: photochemical switching of optical and electrochemical properties in molecular wire type diarylethene species. Chem Eur J. 1995;1:275–84.

65. Kawai T, Kunitake T, Irie M. Novel photochromic conducting polymer having diarylethene derivative in the main chain. Chem Lett. 1999;28:905–6.
66. Gentili PL, Giubila MS, Germani R, Heron BM. Photochromic and luminescent compounds as artificial neuron models. Dyes Pigm. 2018;156:149–59.
67. Galbraith CA, Galbraith JA. Super-resolution microscopy at a glance. J Cell Sci. 2011;124:1607–11.
68. Hell SW. Nanoscopy with Focused Light (Nobel Lecture). Angew Chem Int Ed. 2015;54:8054–66.
69. Pujals S, Feiner-Gracia N, Albertazzi L. Unveiling complex structure and dynamics in supramolecular biomaterials using super-resolution microscopy. In: Azevedo HS, da Silva RM, editor(s). Self-assembling biomaterials: molecular design, characterization and application in biology and medicine. Kidlington: Woodhead. Chapter 12 2018:251–74.
70. Minoshima M, Kikuchi K. Photostable and photoswitching fluorescent dyes for super-resolution imaging. J Biol Inorg Chem. 2017;22:639–52.
71. Uno K, Bossi ML, Konen T, Belov VN, Irie M, Hell SW. Asymmetric diarylethenes with oxidized 2-Alkylbenzothiophen-3-yl units: Chemistry, fluorescence, and photoswitching. Adv Opt Mater. 2019;7:1801746.
72. Uno K, Bossi ML, Irie M, Belov VN, Hell SW. Reversibly photoswitchable fluorescent diarylethenes resistant against photobleaching in aqueous solutions. J Am Chem Soc. 2019;141:16471–8.
73. Kathan M, Hecht S. Photoswitchable molecules as key ingredients to drive systems away from the global thermodynamic minimum. Chem Soc Rev. 2017;46:5536–50.
74. Szymański W, Beierle JM, Kistemaker HA, Velema WA, Feringa BL. Reversible photocontrol of biological systems by the incorporation of molecular photoswitches. Chem Rev. 2013;113:6114–78.
75. Zhang J, Wang J, Tian H. Taking orders from light: progress in photochromic bio-materials. Mater Horiz. 2014;1:169–84.
76. Xiao C, Zhao W-Y, Zhou D-Y, Huang Y, Tao Y, Wu W-H, et al. Recent advance of photochromic diarylethenes-containing supramolecular systems. Chin Chem Lett. 2015;26:817–24.
77. Lerch MM, Hansen MJ, van Dam GM, Szymanski W, Feringa BL. Emerging targets in photopharmacology. Angew Chem Intl Edn. 2016;55:10978–99.
78. Reeßing F, Szymanski W. Beyond photodynamic therapy: light-activated cancer chemotherapy. Curr Med Chem. 2017;24:4905–50.
79. Velema WA, Szymanski W, Feringa BL. Photopharmacology: beyond proof of principle. J Am Chem Soc. 2014;136:2178–91.
80. Presa A, Brissos RF, Cabellero AB, Borilovic I, Korrodi-Gregório L, Pérez-Tomás R, et al. Photoswitching the cytotoxic properties of platinum(II) compounds. Angew Chem Intl Edn. 2015;54:4561–5.
81. Albert L, Vázquez O. Photoswitchable peptides for spatiotemporal control of biological functions. Chem Commun. 2019;55:10192–213.
82. Babii O, Afonin S, Garmanchuk LV, Nikulina VV, Nikolaienko TV, Storozhuk OV, et al. Direct photocontrol of peptidomimetics: an alternative to oxygen-dependent photodynamic cancer therapy. Angew Chem. 2016;128:5583–6.
83. Babii O, Afonin S, Ischenko AY, Schober T, Negelia AO, Tolstanova GM, et al. Structure-activity relationships of photoswitchable diarylethene-based β-hairpin peptides as membranolytic antimicrobial and anticancer agents. J Med Chem. 2018;61:10793–813.
84. Schober T, Wehl I, Afonin S, Babii O, Iampolska A, Schepers U, et al. Controlling the uptake of diarylethene-based cell-penetrating peptides into cells using light. ChemPhotoChem. 2019;3:384–91.

85. Fleming C, Remón P, Li S, Simeth NA, König B, Grøtli M, et al. On the use of diarylmaleimide derivatives in biological contexts: An investigation of the photochromic properties in aqueous solution. Dyes Pigm. 2017;137:410–20.
86. Komarov IV, Afonin S, Babii O, Schober T, Ulrich AS. Efficiently photocontrollable or not? Biological activity of photoisomerizable diarylethenes. Chem Eur J. 2018;24:11245–54.
87. Higashiguchi K, Taira G, Kitai J-I, Hirose T, Matsuda K. Photoinduced macroscopic morphological transformation of an amphiphilic diarylethene assembly: reversible dynamic motion. J Am Chem Soc. 2015;137:2722–9.
88. Kobatake S, Takami S, Muto H, Ishikawa T, Irie M. Rapid and reversible shape changes of molecular crystals on photoirradiation. Nature. 2007;446:778–81.
89. Nakagawa Y, Morimoto M, Yasuda N, Hyodo K, Yokojima S, Nakamura S, et al. Photosalient effect of diarylethene crystals of Thiazolyl and Thienyl derivatives. Chem Eur J. 2019;25:7874–80.
90. Mamiya J-I, Kuriyama A, Yokota N, Yamada M, Ikeda T. Photomobile polymer materials: photoresponsive behavior of cross-linked liquid-crystalline polymers with mesomorphic diarylethenes. Chem Eur J. 2015;21:3174–7.
91. Kuenstler AS, Hayward RC. Light-induced shape morphing of thin films. Curr Opinion Colloid Interface Sci. 2019;40:70–86.

Robert Christie and Adrian Abel

26 Diketopyrrolopyrrole (DPP) pigments

Abstract: Since their industrial introduction in the 1980s, DPP pigments now constitute a highly important group of high-performance carbonyl pigments. The DPP system was first discovered by accident in 1974, and was subsequently re-investigated by Ciba Geigy who recognized its potential to provide commercial organic pigments. DPP pigments exhibit strong similarities compared with quinacridone pigments, in terms of their molecular and crystal structures and their properties, including low solubility and excellent fastness properties. X-ray crystal structural analysis has demonstrated that their technical performance is the result of intermolecular hydrogen bonding and π–π stacking interactions in the crystal lattice structure. Based on a simple retrosynthetic analysis, an efficient synthetic process was developed by Ciba Geigy for their large-scale manufacture. DPP pigments currently provide orange through to reddish violet shades and have become of special importance in providing brilliant saturated red shades with the outstanding durability required for applications such as automotive paints.

Keywords: diketopyrrolopyrrole pigments, DPP, retrosynthetic analysis, intermolecular hydrogen bonding, polymorphism, high-performance pigments, parallel stacking arrangement

26.1 Fundamentals

The most significant development in organic pigments in the late twentieth century was the discovery of diketopyrrolopyrrole (DPP) molecular structures based on a previously unknown chromophoric system. Indeed, this was undoubtedly the most important development since the discovery of the phthalocyanine pigments around fifty years earlier. DPP pigments, by appropriate substituent variation, can provide orange through red to reddish violet shades and have become of special importance in providing brilliant saturated red shades with the outstanding durability required for applications such as automotive paints. Indeed, in this respect they have occasionally been referred to as "red phthalocyanines". Their excellent thermal stability and migration resistance means that they are also of considerable interest for the pigmentation of plastics. The DPP pigment range has

This article has previously been published in the journal Physical Sciences Reviews. Please cite as: R. Christie, A. Abel, Diketopyrrolopyrrole (DPP) Pigments *Physical Sciences Reviews* [Online] 2021, 6. DOI: 10.1515/psr-2020-0167

https://doi.org/10.1515/9783110587104-026

emerged as an outstanding example of how investigations that apply fundamental principles of synthetic and mechanistic organic chemistry can lead to a successful global commercial outcome.

26.2 History

The formation of a diketopyrrolopyrrole (DPP) was first reported by Farnum and coworkers in 1974 [1]. The researchers were attempting to use a Reformatsky reaction to synthesize a 4-membered ring heterocyclic compound from the reaction of benzonitrile with ethyl bromoacetate in toluene heated under reflux for several hours with zinc. They observed that the color of the reaction mixture changed progressively from yellow, to green, to brown and to red. Although unsuccessful in their original aim, they reported the isolation of low yield of a product described as "a highly insoluble, brilliant red, crystalline compound" with molecular formula $C_{18}H_{12}0_2N_2$, identified as a diphenyldiketopyrrolopyrrole. They also proposed a mechanism for its formation. Around 1980, researchers from Ciba-Geigy AG (now BASF) in Switzerland came across Farnum's article in a literature review [2] and successfully repeated the process to obtain small amounts of the material. They recognized structural similarities compared with other high-performance carbonyl pigments, for example the quinacridones, and recognized its potential as a high-performance pigment. At the same time, they realized that a much more efficient synthetic route would be required. Ciba Geigy published a series of patents on DPP pigments starting from the early 1980s onwards [3–5]. Investigations by research chemists at Ciba Geigy demonstrated that the original mechanism proposed for the formation of the molecule was incorrect. Using retrosynthetic design, an alternative, efficient, versatile, "one-pot" manufacturing procedure for DPP pigments from readily available starting materials was devised. In 1986 the first DPP pigment was introduced to the market as CI Pigment Red 254. This pigment is a bright, mid red shade with good opacity and excellent fastness properties. For most of the period when the pigment was under patent protection, the pigment commanded a high price. As patent expiry approached, the price drifted downwards. There is a view that this may have been intended to minimize antagonism from customers who may have felt exploited, and to deter competitors from entering the market as they could not be clear about where the price might settle. Additional pigments covering a range of colors from mid orange to blue shade red have been introduced. Many other companies from different parts of the world, including traditional manufacturers such as BASF (who acquired the former Ciba products), Clariant, DCL, Heubach, Toyo, and Chinese newcomers such as CINIC and Union Colour, now participate in this market. The lowering of the price as production has expanded is leading to the decline of several pigments, including some mid-priced azo pigments and high-priced carbonyl pigments, both on technical and commercial grounds. They are even competing in some traditional classical azo pigment application areas, such as those enjoyed by CI Pigment Red 112 and CI Pigment Red 170.

26.3 Structures and properties

The essential structural feature of this group of pigments is the 1,4-diketopyrrolo (3,4-c)pyrrole (DPP) heterocyclic system (**1**), consisting of two fused five-membered ketopyrrole rings, as illustrated in Figure 26.1 [6–9]. It has been considered as a lactam analogue of pentalene. The alternative nomenclature 2,5-dihydropyrrolo[4,3- c] pyrrolo-1,4-dione has also been used. The commercial products whose structures are disclosed, pigments (**1a-1f**), contain phenyl groups attached to the ketopyrrole rings at the available carbon atoms, and differ according to the nature and position of the single substituent in the phenyl rings, as listed in Table 26.1. Compound (**1b**) is the original product obtained by Farnum *et al* [1]. The pigments offer excellent fastness to light and weather, suitable for automotive finishes, and excellent thermal stability and migration resistance suitable for the pigmentation of plastics. They can also be used in high grade printing inks.

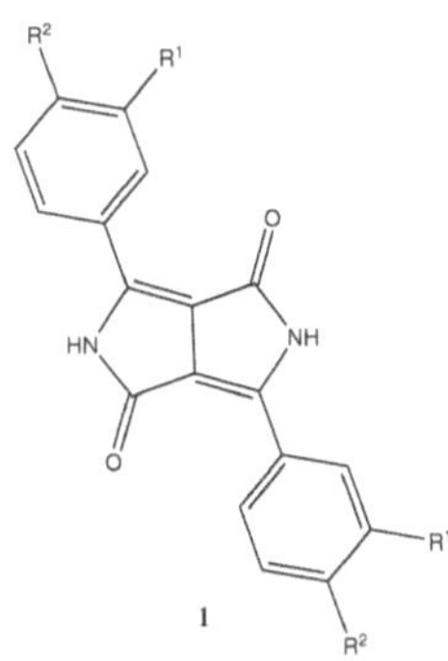

1

Figure 26.1: General structure (**1**) of diketopyrrolopyrrole (DPP) pigments.

Table 26.1: Substituents in diketopyrrolopyrrole (DPP) pigments.

Compound	CI Pigment orange or red	R^1	R^2
1a	R254	H	Cl
1b	R255	H	H
1c	R264	H	C_6H_5
1d	R272	H	CH_3
1e	O71	CN	H
1f	O73	H	$C(CH_3)_3$

DPP pigments exhibit strong similarities compared with quinacridone pigments. They are symmetrical molecules containing similar heterocyclic structural features, and the compounds are similarly remarkable in providing low solubility and an excellent range of fastness properties from such small molecules. X-ray structural analysis of the diphenyl DPP (**1b**) has demonstrated that this is due to strong intermolecular hydrogen bonding and π–π stacking interactions in the crystal lattice structure [10], as with the quinacridones. The central DPP unit is essentially planar and the dihedral angle between the planes of the phenyl substituent rings is only 7°. Since the DPP molecules contain two oppositely directed amide groups, they arrange into a pattern of ribbons due to intermolecular hydrogen bonding, as shown in Figure 26.2. The short distances between adjacent molecules in the stacks, arranged in columns, is evidence of strong π–π interactions. These structural features are responsible for the low solubility and high stability of the pigments. The effect of the crystal structures on the colors of a range of DPP pigments has been attributed to relative differences in the nature of the intermolecular forces, the head-to-tail intermolecular hydrogen bonding arrangement causing a bathochromic shift, while the parallel stacking arrangement, involving π–π interactions, contributes significantly to hypsochromic shifts [11]. The crystal structure of the most important DPP pigment, CI Pigment Red 254 (**1a**) has been obtained from X-ray powder diffraction data [12]. As with many classes of organic pigments, DPPs have been found to exhibit polymorphism. For example, CI Pigment Red 254 (**1a**) exists in two crystal modifications. The α-form, which is thermally more stable, is a mid to bluish red shade, while the β-form is more yellowish red. Since their successful introduction as pigments, the DPP molecular system has become one of the most extensively investigated structural systems because of their unique structures and their potential as functional colorants, for example as fluorescent probes [13], organic semiconductors [14], and in organic solar cells [15].

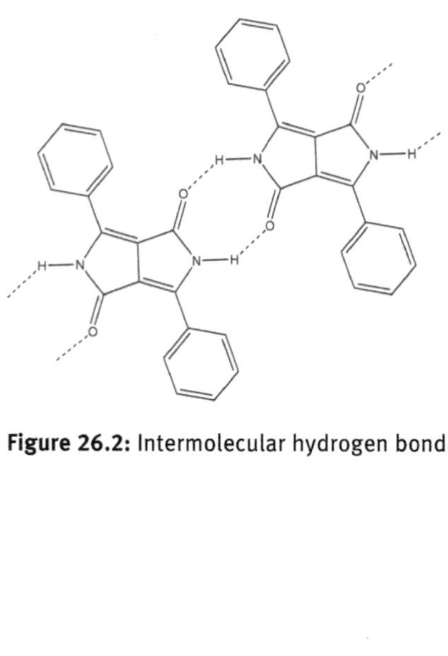

Figure 26.2: Intermolecular hydrogen bonding in the crystal structure of DPP pigment (**1b**).

26.4 Synthesis and manufacture

Studies carried out in the laboratories of Ciba Geigy confirmed that the original method could not form the basis of an efficient synthesis of the pigments. Subsequently, they carried out simple retrosynthetic analysis, illustrated schematically in Figure 26.3, which suggested that benzonitrile and diesters of succinic acid could be suitable starting materials for the synthesis. The synthetic process subsequently devised by Ciba Geigy involves the treatment of a succinic acid diester (1 mol) with an aromatic cyanide (2 mol) in an alcoholic solvent and in the presence of a strong base to provide the symmetrical derivatives which are the commercial products (Figure 26.4). Investigations carried out to optimize the reaction conditions demonstrated that the best results are obtained when succinate diesters of tertiary or secondary alcohols are used as starting materials, when the reactions are performed in the presence of tertiary alkoxides as bases, and in tertiary alcohols as solvents [7–9]. The reaction, although a one-pot process, proceeds through intermediate (**2**). Preparation and isolation of this intermediate under controlled conditions allows the synthesis of unsymmetrical DPP derivatives. The pigments may be isolated by simple filtration and washing. By appropriate optimization of particle size, DPP pigments may be manufactured in either transparent or opaque grades. Classical conditioning processes involving solvent treatments provide larger size particles and thus opaque grades, while smaller particles leading to transparent grades are obtained by various milling processes or by pH-dependent re-precipitation procedures. Several patents claim *in situ* formation of small particle size DPP pigments [8]. Clariant have been pioneering the manufacture of its DPP reds using the established route, but with the diesters obtained from succinic acid derived from renewable resources rather than from crude oil.

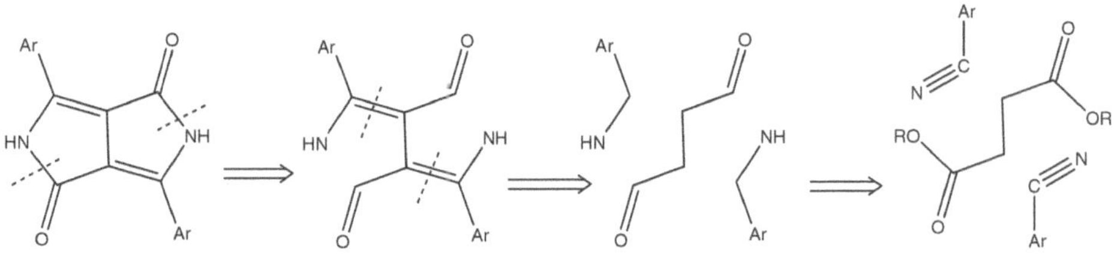

Figure 26.3: The retrosynthetic analysis that led to development of the commercial synthesis of DPP pigments.

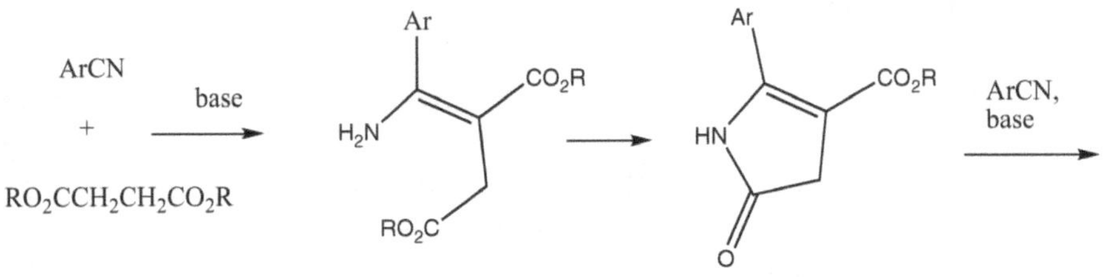

2

1

Figure 26.4: Synthesis of DPP pigments.

26.5 Applications

26.5.1 CI Pigment Red 254 (1a)

This pigment, the first in the DPP series, was introduced as Irgazin DPP Red BO and quickly developed an important position in the market, which it still occupies. It is a medium red shade, with a bluish undertone in reduction. The pigment exists in two different crystal modifications, a bluer shade α-modification and a yellower shade β-modification. Its main application is paints, where it is used in most sectors including the demanding automotive original finish market. Occasionally referred to as "Ferrari Red", the pigment was used to provide the characteristic clean bright red color of the Ferraris, and has also been used on Alfa Romeos, BMWs, and Chevrolet Corvettes. It is also used in decorative finishes, where its clean shade and excellent lightfastness lends itself to tinting systems. Fastness to weathering is somewhat dependent on the grade of the pigment, although most opaque grades with a larger size particle size show only a small degree of fading even after external exposure for two years. It has excellent fastness to most solvents, although rated just below excellent in butanol and ketones. It has excellent fastness to over-coating even after 30 minutes at 160°C. The commercial grades of this pigment do not show a marked difference in color, although the smaller particle versions are generally somewhat yellower than the more opaque grades. The pigment is also important in the coloration of plastics. Its mid-red shade with high saturation makes it equally useful in full shades and in pale reductions. In PVC it offers excellent fastness to light and weather. It is completely fast to migration in plasticized PVC, and its good dielectric properties mean that it can be used in sheathing for cables. In polyolefins it is stable to 300°C even in pale reductions, offering excellent fastness properties, including for the coloration of polypropylene fibers. Standard grades can affect the dimensional stability of injection moldings causing warpage, but grades are available which minimize this effect. In polystyrene, it is stable to 300°C and shows excellent fastness to light. In ABS, its heat stability reduces to 270°C and lightfastness is slightly less than excellent. It is not generally recommended for engineering polymers or for most fibers, such as nylon, PET, and acrylics. A reduction in the price of this pigment has brought it into contention for some printing ink applications, especially for packaging, laminates, and security inks, and it is proposed for use in inkjet printing. The pigment is acceptable in most applications where good toxicological properties are required, such as in toys and food packaging.

26.5.2 CI Pigment Red 255 (1b)

This pigment is yellower than CI Pigment Red 254 (**1a**), described as scarlet, with similar fastness to light and weather, and heat stability, but with slightly lower fastness to solvents. It is recommended for almost all paint applications including automotive original equipment. The pigment can also be used in both rigid and plasticized PVC where it offers very good to excellent lightfastness.

26.5.3 CI Pigment Red 264 (1c)

This pigment is much bluer than CI Pigment Red 254 (**1a**) but with similar fastness properties. It can be produced in either transparent or highly opaque grades, its high tinctorial strength making it an ideal tinting pigment. Both grades have excellent fastness to light and weather, and good solvent resistance. The transparent grade provides high saturation, durability and a flop tone that makes it attractive in metallic shades. Pigment grades prepared using novel surface treatment technologies can be used in an extended range of applications, for example to provide much improved rheology in solvent-based systems without adversely affecting its performance in water-based coatings. The treatments do not have any negative effects on the high level of durability of the original pigment. In plastics, it is suitable for both plasticized and rigid PVC offering excellent lightfastness. In polyolefins it shows heat stability to 300°C and offers very good to excellent lightfastness in both full shade and reduction, although its fastness to weathering is a little lower than that of CI Pigment Red 254. There are grades available that are non-warping in injection moldings. It can be used to color polypropylene and PET fibers, and can be used for nylon if the user carries out careful testing to ensure that it is suitable for the application. A recent addition to the pigment grades has further optimized transparency and higher saturation for use in PVC and polyolefins. Grades of this pigment are available for most applications in which there are regulations requiring relatively benign toxicological properties.

26.5.4 CI Pigment Red 272 (1d)

This pigment offers a yellowish red shade, high chroma, and good opacity, but is of lesser importance than other DPP reds. It can be used in PVC, polyolefins, and polystyrene.

26.5.5 CI Pigment Orange 71 (1e)

This transparent yellowish orange pigment, which offers excellent heat stability is used mainly in the coloration of plastics, and also in some specialist ink applications. It can be used in both plasticized and rigid PVC where it shows excellent resistance to migration, excellent fastness to light, and very good to excellent resistance to weathering. In polyolefins it is stable to 300°C and provides very good to excellent lightfastness, making it very suitable for polypropylene fibers. It has a low tendency to cause warping in injected molded articles. It is used in polystyrene and, with some reservations, in styrenic copolymers, but is not recommended for engineering polymers. In printing inks, it provides good gloss and transparency and is suitable for applications such as metal decorative finishes, laminates, and both solvent and water-based packaging inks, where high levels of fastness properties are required.

26.5.6 CI Pigment Orange 73 (1f)

This pigment provides a similar shade to CI Pigment Orange 71 (**1a**) and is generally preferred for paint applications. It has a cleaner shade, medium opacity, and good solvent fastness. It has excellent heat stability, with good to excellent lightfastness. Its fastness to overcoating is rated as good.

Solid solutions are products in which two or more components form a single crystal lattice arrangement. DPP pigments can form such arrangements either with other DPP pigments or pigments of another chemical species. The best-known example of the latter involves a quinacridone as the second component. One product is a very transparent, yellowish red pigment that is used for automotive metallic finishes, giving a unique flop tone. Another is bluish red, also very transparent, and is used for metallic finishes.

References

1. Farnum DG, Mehta G, Moore GG, Siegal FP. Attempted Reformatskii reaction of benzonitrile, 1,4-diketo-3,6-diphenylpyrrolopyrrole, a lactam analogue of pentalene. Tetrahedron Lett. 1974;29:2549–52.
2. Ranganathan D, Ranganathan S. Further challenging problems in organic reaction mechanisms. New York: Academic Press, 1980:206.
3. Iqbal A, Cassar L. Process for colouring highly molecular organic material and polycyclic pigments. EP0061426. 15 Mar 1982.
4. Rochat AC, Iqbal A, Cassar L. Preparation of pyrrolo-(3,4-c)-pyrroles. EP 0094911. 11 May 1983.
5. Rochat AC, Iqbal A, Cassar L. Preparation of 1,4-diketopyrrolo-(3,4-c)-pyrroles. EP 0098808. 18 Jan 1983.
6. Hunger K, Schmidt MU. Industrial organic pigments, 4th ed, Weinheim: Wiley-VCH Verlag GmbH, Ch 3 2019.
7. Wahlquist O, Lenz R. In: Faulkner EB, Schwartz RJ, editors. High performance pigments, 2nd ed. Weinheim: Wiley-VCH Verlag GmbH, 2009 Ch 11.
8. Zollinger H. Color chemistry: syntheses, properties and applications of organic dyes and pigments, 3rd ed, Weinheim: Wiley-VCH, Ch. 8 2003.
9. Grzybowski M, Gryko DT. Diketopyrrolopyrroles: synthesis, reactivity, and optical properties. Adv Opt Mater. 2015;3:280–320.
10. Iqbal A, Jost M, Kirchmayr R, Rochat AC, Pfenninger J, Wallquist O. The synthesis and properties of 1,4-diketo-pyrrolo[3,4-C]pyrroles. Bull Soc Chim Belg. 1988;97:615–43.
11. Mizuguchi J. Correlation between crystal and electronic structures in diketopyrrolopyrrole pigments as viewed from exciton coupling effects. J Phys Chem A. 2000;104:1817–21.
12. Ivashevskaya SN. Crystal structure of Pigment Red 254 from X-ray powder diffraction data. Acta Crystallogr E Crystallogr Commun. 2017;73:507–10.
13. Kaur M, Choi DH. Diketopyrrolopyrrole: brilliant red pigment dye-based fluorescent probes and their applications. Chem Soc Rev. 2015;44:58–77.
14. Glowacki ED, Coskun H, Blood-Forsythe MA, Monkowius U, Leonat L, Grzybowski M, et al. Hydrogen-bonded diketopyrrolopyrrole (DPP) pigments as organic semiconductors. Org Electron. 2014;15:3521–8.
15. Li W, Hendricks KH, Wienk MM, Janssen RA. Diketopyrrolopyrrole polymers for organic solar cells. Acc Chem Res. 2016;49:78–85.

Robert Christie and Adrian Abel

27 Dioxazine pigments

Abstract: This chapter provides an overview of the structural and synthetic chemistry, and the industrial applications, of dioxazine pigments, a small group of high performance organic pigments. The color violet (or purple) has frequently assumed a prominent position in history, on account of its rarity and cost. The natural colorant Tyrian purple and the first synthetic textile dye, Mauveine, are prime examples of this unique historical feature. CI Pigment Violet 23, also referred to as Dioxazine Violet or Carbazole Violet, is one of the most universally used organic pigments, by far the most important industrial pigment in the violet shade area. Dioxazine Violet is also unique as the dominant industrial violet pigment providing a brilliant, intense violet color and an excellent all-round set of fastness properties. The pigment has a polycyclic molecular structure, originally described wrongly as a linear arrangement, and later shown to adopt an S-shaped arrangement on the basis of X-ray structural analysis. Two other dioxazine pigments are of rather lesser importance. The synthesis and manufacturing route to CI Pigment Violet 23 is described in the review. Finally, a survey of the principal current applications of the individual dioxazine pigments is presented.

Keywords: dioxazine pigments, violet color, CI Pigment Violet 23, S-shaped structure, carbazole

27.1 Fundamentals

Alongside copper phthalocyanines, the dioxazine pigment, CI Pigment Violet 23, commonly known as Dioxazine Violet or Carbazole Violet, is one of the most universally used organic pigments, suitable for virtually all application areas [1–3]. It is commonly referred to in general terms as a polycyclic pigment. Its molecular structure consists of nine fused rings organised in an S-shaped arrangement. The pigment provides a brilliant, intense violet color and an excellent all-round set of fastness properties. In fact, the pigment is reported to be almost 50% stronger than copper phthalocyanine blue, although it is around five times more expensive. CI Pigment Violet 37 is also a dioxazine structure, with some specialist applications.

This article has previously been published in the journal Physical Sciences Reviews. Please cite as: R. Christie, A. Abel, Dioxazine Pigments *Physical Sciences Reviews* [Online]2021, 6. DOI: 10.1515/psr-2020-0168

https://doi.org/10.1515/9783110587104-027

27.2 History

Dioxazine pigments currently occupy a unique position as high-performance organic pigments, especially CI Pigment Violet 23, which dominate the violet shade area. In modern times, fashion trends for violet and purple colors may come and go. However, "the color purple" has played a unique role historically, as it was rare and in high demand. In view of this, it was selected as the color of emperors, with laws enacted to prevent the color from being worn by anyone else. The early demand was satisfied by Tyrian Purple, dating back to the Phoenicians in 1200 BC. Indeed, some scholars claim that the word Phoenician means "the color purple", but this is perhaps uncertain [4]. The legend of the discovery of Tyrian Purple dates from the second century AD. The Greek author Pollux described Hercules beachcombing with his dog on his way to meet a nymph. The dog had a liking for sea snails and, after eating them, its mouth turned bright purple, a scene depicted in "Hercules' Dog Discovers Purple Dye", a Rubens painting in the Musée Bonnat-Helleu, Bayonne, France. When Hercules met the nymph, she requested a garment of the same color [5]. The natural dye was produced by the hypobranchial glands of the Murex sea snail, and it is only when exposed to the elements, including sunlight, that the color is observed [6]. An industry grew up around its production, mainly around Tyre, Lebanon although other sites on the Mediterranean coastline are known to have produced the dye and, indeed, other locations well beyond the Mediterranean, including Brittany, Wales, Norway, the Caribbean and the Pacific. While it was possible to "milk" the mollusc, by irritating it, this was a slow process. Producers preferred a less ecologically sustainable method that involved crushing them in vats, steeping for 3 days, then boiling the liquor. Each mollusc produced only a tiny amount of the dye. It has been estimated that it required 10,000 snails to provide just 1 gram of the dye, which would only yield enough dye to color the trim of a garment! The dye was therefore rare and very expensive. Huge mounds of the shells can still be found on the shores of the Mediterranean around Tyre. Use of the dye is even mentioned in the New Testament [7]. In Acts 16, Paul describes his visit to Philippi, Macedonia. "We sat down and preached to the women, who had come to the meeting. One of these women was called Lydia, a devout woman from the town of Thyatira who was in the purple dye trade." Lydia is claimed to be the first convert to Christianity in Europe and is considered the patron saint of dyers by the Roman Catholic church. The color is due to 6,6′-dibromoindigo. The Color Index describes it as a natural (vat) dye and designates it as CI Natural Violet 1 [8].

In the modern age, the use of Dioxazine Violet is virtually universal, certainly no longer reserved for emperors! Mixing blue and red pigments to produce violet leads to a dullness of shade, often close to brown. Dioxazine Violet is considered the optimum pigment to produce clean bright violet colors, as well as to redden blue pigments and to create bluer shade red pigments without losing purity of shade. Dioxazine compounds were discovered by Kranzlein, Gruene and Thiele (Farbwerke Hoechst) in the late 1920s. Sulfonation of these compounds produced a series of direct dyes that could be used for the coloration of cotton. It was not until 1952 that Hoechst patented the structure of Dioxazine Violet as a pigment [9], introduced commercially a year later as Permanent Violet RL. Geigy introduced two "brilliant purple" pigments, CI Pigments Violet 34 and 35, into their Irgazin range, although they are no longer marketed. CI Pigment Violet 37 was developed by Ciba and still plays a role, mainly in printing inks.

27.3 Structures and properties

The structures of dioxazine pigments are based on triphenodioxazine, a system of 5 fused 6-membered rings (Figure 27.1). The parent compound is orange. By far the most important dioxazine pigment is CI Pigment Violet 23 (**1**) (Figure 27.2), commonly referred to as Dioxazine Violet, or Carbazole Violet based on the presence of two heterocyclic carbazole rings in its structure, the most significant violet pigment for high performance applications. The pigment is characterised by a brilliant intense violet color, excellent lightfastness and resistance to heat and solvents, although it is rather expensive. For many years, the pigment was reported as structure (**2**) (Figure 27.2), a linear arrangement of nine fused rings. For some time there had been a suspicion that this was erroneous, and ultimately the correct S-shaped structure (**1**) was confirmed by X-ray crystal analysis [3, 10]. Following publication of the evidence, the Color Index Editorial Board agreed to make the change in the internet version of the Index. As a consequence, many companies that had registered the pigment according to its Color Index name feared that the change could effectively de-register their product, as their registration implied the linear structure. Fortunately, authorities recognised that the change of structure was based on results from new analytical techniques and accepted that existing products were correctly represented by the S-shaped structure (**1**). Thus, a potentially serious problem was averted. In its crystal structure, the molecules are stacked with a high degree of π-overlap. Neighbouring stacks of molecules are arranged essentially perpendicular to adjacent stacks, efficiently occupying the cavity formed by the S-shape of the molecule. Although only π-π and van der Waals' intermolecular interactions are formed throughout the crystal, efficient packing ensuring that the pigment has high stability. CI Pigment

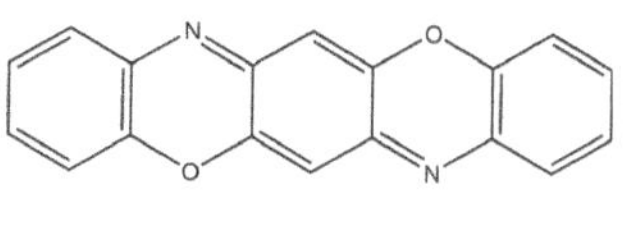

Figure 27.1: The triphenodioxazine ring system.

Violet 39 (**3**) is a reddish-violet dioxazine pigment which is used in speciality applications. CI Pigment Blue 80 (**4**) is a recently introduced dioxazine pigment containing benzimidazolone groups. It provides brilliant bluish-violet shades with excellent tinctorial strength and outstanding fastness properties.

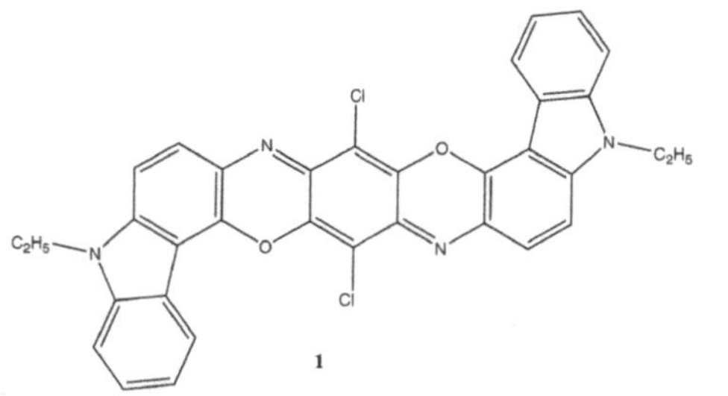

1

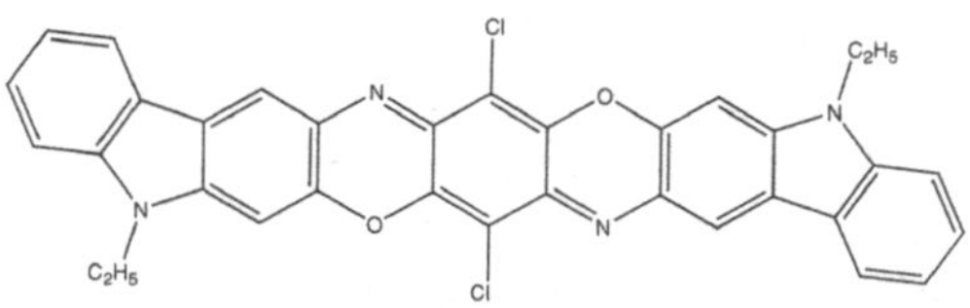

2

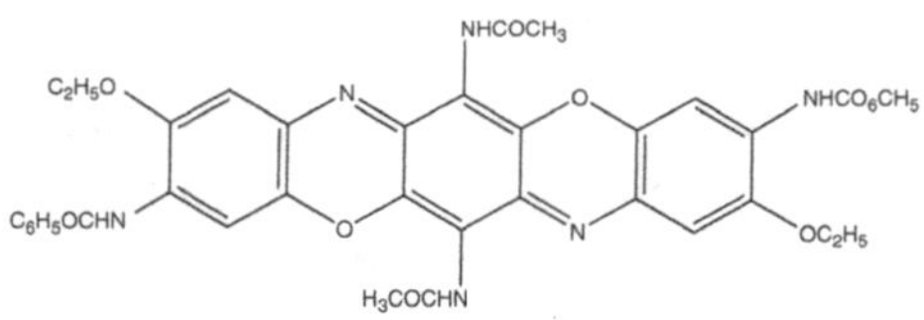

3

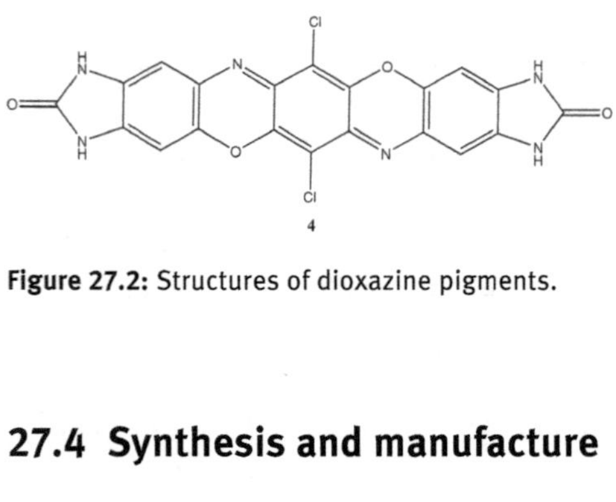

4

Figure 27.2: Structures of dioxazine pigments.

27.4 Synthesis and manufacture

The synthesis of the most important dioxazine pigment, CI Pigment Violet 23 (**1**) is illustrated in Figure 27.3. In this process, 3-amino-9-ethylcarbazole (**5**) (2 mol) is reacted with chloranil (**6**) (1 mol) in an organic solvent at an elevated temperature and in the presence of an acid scavenger such as dry sodium acetate to form intermediate (**7**). This intermediate is converted to dioxazine (**1**) by cyclization at elevated temperatures in the presence of a catalyst, for example, aluminum (III) chloride or benzenesulfonyl chloride. The crude pigment thus obtained is followed by a conditioning process to produce a commercial pigment [2, 11, 12]. The processes developed for this purpose were based on experience with copper phthalocyanine blue pigments. One method involves milling the crude product in a steel ball mill in the presence of salt and an organic solvent. Another process involves kneading the crude material in sulfuric acid. Over the years, the quality of the pigmentary products has significantly improved, especially in terms of color strength and brilliance of shade, mainly by developing finer particle size grades. These grades are susceptible to flocculation in certain applications. Lessons learned from experience with copper phthalocyanine pigments, which are also prone to flocculation, have facilitated development of additives used as surface treatments, dating as far back as 1954. The technique involves treating the crude pigment with dioxazine derivatives that have been modified with functional groups such as the sulfonate group, which may form salts of metals such as calcium, barium, strontium and aluminum, or polymeric quaternary ammonium salts [11–14]. The dioxazine part of the structure of these additives attaches to the surface of the pigment, while the functional group on the pigment particle surface offers improved compatibility with the application media and inhibits flocculation.

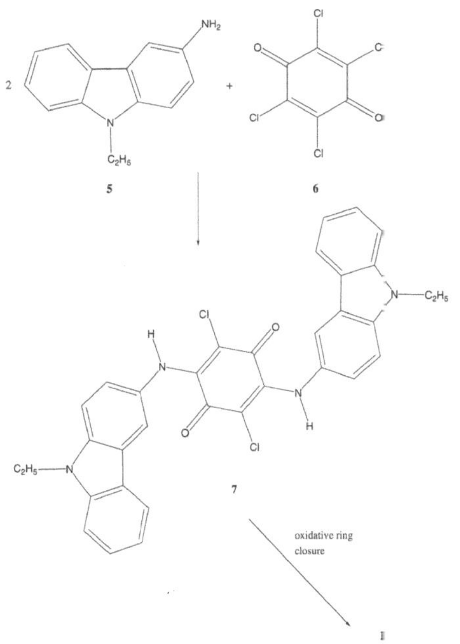

Figure 27.3: CI Pigment Violet 23.

27.5 Applications

CI Pigment Violet 23 (1). This pigment was originally produced with a blue violet shade. However, as more finely divided pigment grades with ever smaller particles were introduced, the tinctorial strength increased and the shade shifted to become redder and brighter. In fact, around 10% copper phthalocyanine blue would need to be added to current products to provide the same shade of violet that would have been achieved with pigments from fifty years ago. Most major pigment manufacturers now offer Dioxazine Violet in their product range. Since producers use different conditioning techniques, this inevitably means that different manufacturers offer different product standards, so that in the case of pigment selection it cannot be assumed that pigments from different sources are identical. In coatings, the lightfastness of CI Pigment Violet 23 is of the highest order, even when reduced to very pale shades with titanium dioxide. Thus, it is ideal for tinting systems used in architectural coatings and industrial finishes. Only the most demanding of modern automotive original finishes may limit its use, for example paler metallic and pearlescent finishes. An interesting use is in stoving paints for white goods such as refrigerators and washing machines. Rutile titanium dioxide, especially grades made by the sulfate process, may contain small amounts of iron contaminants which lead to a yellowish undertone. Tiny amounts of CI Pigment Violet 23 neutralise the yellowness and its violet undertone produces a more appealing white that is preferred by the public especially in Europe. The trend towards using titanium dioxide produced using the chloride process reduces the necessity for a blue or violet tint. Like many pigments with small particle size, problems with plate-out, which refers to an accumulation of additive on the surfaces of processing equipment, may be encountered when using certain hardeners, especially with the epoxy resins used in powder coatings. In inks, its extremely high color strength makes it ideal for packaging, especially when the requirement for lightfastness means that cationic dye complex pigments are unsuitable. It is transparent, and so gives attractive finishes printed on foil and for metal decorative inks. It is widely used for the coloration of polymers used in plastics, stable up to 280 °C in polyolefins and 260 °C in polystyrene. At higher temperatures the pigment starts to dissolve, leading to inferior migration stability in polymers other than those that are highly crystalline. Although it has slight solubility in plasticized PVC, it is still widely used. However, its solubility makes it less suitable in very pale shades. The fine particle size of Dioxazine Violet pigments may lead to warping in injection moulded articles. It is the standard violet pigment for textile printing, used alone or to tint blues towards red and reds towards blue. It can also be used for the mass coloration of polyester fibers and in the spin dyeing of viscose. As such small particles, CI Pigment Violet 23 is prone to flocculation, although flocculation-resistant grades are produced.

CI Pigment Violet 37 (3). This pigment was developed by Ciba (now incorporated into BASF) as an alternative violet pigment not covered by CI Pigment Violet

23 (**1**) patent restrictions. It is significantly redder in shade than CI Pigment Violet 23 and offers similar high levels of fastness properties. It can be used in most paint systems, although lower color strength limits its use. It finds use mainly in printing inks, particularly useful for nitrocellulose-based inks where it achieves high tinctorial strength. In oil-based inks, although somewhat weaker than CI Pigment Violet 23, it generally attains higher gloss and superior flow properties. In plastics, it can be used in similar applications to CI Pigment Violet 23, but its superior insolubility in plasticized PVC means it can be used at lower concentrations without loss of migration fastness. In polyolefins, its heat stability is similar to CI Pigment Violet 23, but this stability declines similarly when used at lower depths of shade. Information on its effect on shrinkage in polyolefins is contradictory. One source suggests it causes considerable shrinkage [2], but internal information from the product list of the manufacturer (BASF), claims it has only a small effect on warpage (as is also suggested for CI Pigment Violet 23) [15]. CI Pigment Violet 37 is recommended for the spin coloration of polypropylene.

CI Pigment Blue 80 (4). Pigment Blue R5R VP2548 is a relatively new product, based on a combination of dioxazine and benzimidazolone chemistry and technology, which was released early in the millennium by Clariant. Its shade lies somewhere between CI Pigment Violet 23 and CI Pigment Blue 60. Its fastness properties are of the highest order, even surpassing those of Indanthrone, CI Pigment Blue 60, and it also satisfies some demands that CI Pigment Violet 23 cannot quite achieve, such as in pale shade metallics. It is transparent and so ideal for metallic and pearlescent finishes. Another potential application is in coil coatings, where the color requires to remain unchanged even after many years of exposure [16].

References

1. Christie RM. Color chemistry. 2nd ed. London: RSC, Ch 9 2015.
2. Hunger K, Schmidt MU. Industrial organic pigments. 4th ed. Weinheim: Wiley-VCH Verlag GmbH, Ch 3 2019.
3. Chamberlain TR, In Faulkner EB, Schwartz RJ. High performance pigments. 2nd ed. Weinheim: Wiley-VCH Verlag GmbH, 2009 Ch 12.
4. https://www.etymonline.com/word/phoenician.
5. https://www.bbc.com/culture/article/20180801-tyrian-purple-the-regal-color-taken-from-mollusc-mucus.
6. Cooksey CJ. Tyrian purple: 6,6′-Dibromoindigo and related compounds. Molecules. 2001;6:736–69.
7. https://www.bibliacatolica.com.br/new-jerusalem-bible/acts/16/ (Verse 14).
8. https://color-index.com/fingerprint/view/4528.
9. Sakaguchi I, Hayashi Y. US 4,317,908. 7 Mar 1982.
10. Hunger K. The effect of crystal structure on color application properties of organic pigments. Rev Prog Color. 1999;29:71–84.
11. Spietschka E, Urban M, Paulus E. DE 2,742,575. (filed 5th July 2007 but withdrawn).

12. Sanders JD. Pigments for inkmakers. London: SITA Technology, 1989:141–9.
13. Struve WS. US 2,843,498. 15 Jul 1958.
14. Wooden G, Wallquist O. US 5,271,759. 21 Dec 1993.
15. https://www.basf.com/us/documents/en/general-business-topics/pigments/industries/plastics/Colors-and-Effects_PSG-Pigment-preparations-Polyolefins.pdf.
16. Wilker G, Kempter PW. What's blue stays blue: Dioxazine blue - a new, solvent-resistant pigment. Eur Coat J. 2002.

Robert Christie and Adrian Abel

28 Disazo (Bishydrazone) condensation pigments

Abstract: Disazo condensation pigments were developed in the mid-twentieth century as a range of high-performance azo pigments to meet the stringent demands placed on the technical properties required for applications in growing markets such as automotive paints, plastics, and fibers. The commercial products vary in shade from yellow through red to brown. Structurally, the pigments are related to the classical yellow azoacetoacetanilides or red azonaphthols by synthetically connecting two monoazo derivatives by a condensation reaction involving an aromatic diamine. The resulting pigment molecules are of extremely large molecular size with the presence of several amide groups, factors that determine their excellent set of fastness properties. The procedures used in their manufacture involve complex and demanding multistage processes, and this explains the higher cost of these pigments. The application performance attributes provided by the individual commercial products are discussed in detail in the final section.

Keywords: disazo condensation pigments, disazoacetoacetanilide pigments, disazonaphtharylamide, bishydrazone, condensation reaction, high-performance pigments, plastics

28.1 Fundamentals

The classical azo pigments provide bright intense colors at relatively low cost but, in general, they lack the exceptional fastness towards light, weathering, heat and solvents that are required by highly demanding applications. When faced with the demand for high-performance pigments as the twentieth century progressed, it is unsurprising that many industrial research laboratories chose to devote considerable resources to investigations into ways in which the technical performance of azo pigments might be upgraded. Consequently, a series of disazo condensation pigments were developed by Ciba to constitute a range of high-performance azo pigments varying in shade from yellow through red to brown [1,2,3]. This innovation was followed later by the introduction by Hoechst of the high-performance benzimidazolone azo pigments.

This article has previously been published in the journal Physical Sciences Reviews. Please cite as: R. Christie, A. Abel Disazo (Bishydrazone) Condensation Pigments *Physical Sciences Reviews* [Online] 2021, 6. DOI: 10.1515/psr- 2020-0169

https://doi.org/10.1515/9783110587104-028

28.2 History

Disazo condensation pigments were developed and introduced by Ciba (before the merger with Geigy in 1970) following a program of research and development initiated in the 1950s and continuing into the 1960s [4, 5]. The concept underlying the research was to select a chromophoric structure, based on knowledge and experience with the well-established classical azo pigments, to provide the color properties as well as contributing to technical performance, and subsequently to modify the structure by incorporating features that would raise the fastness properties to the level required of high-performance pigments. In the case of disazo condensation pigments, the strategy principally involved increasing the molecular size. Ciba developed a way of building up the molecular size using a two-stage process that involved azo coupling followed by a condensation reaction process involving an aromatic diamine to connect two monoazo structures formed in the first stage. It was also important that amide groups were formed in this process. This explains the origin of the term disazo condensation pigments, introduced under the trade name Cromophtal. By applying this concept to a multitude of permutations of potential diazo components, coupling components and diamines, a vast number of compounds could be synthesized covering the spectrum from greenish yellow through orange, red, bordeaux, and violet to brown. It has been estimated that well over 10,000 different potential pigments were synthesized at the time [5] and 550 of these compounds appear in the Chemical Abstracts Service (CAS) registry, although relatively few became commercial products. Several of the early products were discontinued when certain intermediates were withdrawn from the market. It is a tribute to the extensive research and development program conducted by Ciba that, for several years after their patents expired, only a couple of additional pigments were introduced to the market from other producers. Some of the early disazo condensation pigments have lost their technical/commercial advantage having been replaced in the yellow/orange sector by benzimidazolone azo pigments and in the orange/red sector by diketopyrrolopyrrole (DPP) pigments. However, probably due to developments in conditioning techniques used in their manufacture, a few later products have entered the market. There remain over a dozen disazo condensation pigments produced, many extensively used in the coloration of plastics, their main application.

28.3 Structures and properties

Disazo condensation pigments are among the most complex organic pigments structurally, some with a molecular weight of over 1,000. The yellow disazo condensation pigments are disazoacetoacetanilides of one of two structural types, represented by either structure (**1**) (Type I) (Figure 28.1) or structure (**2**) (Type II) (Figure 28.2). In Figure 28.2, the bridge A generally represents a phenylene or diphenylene group. The

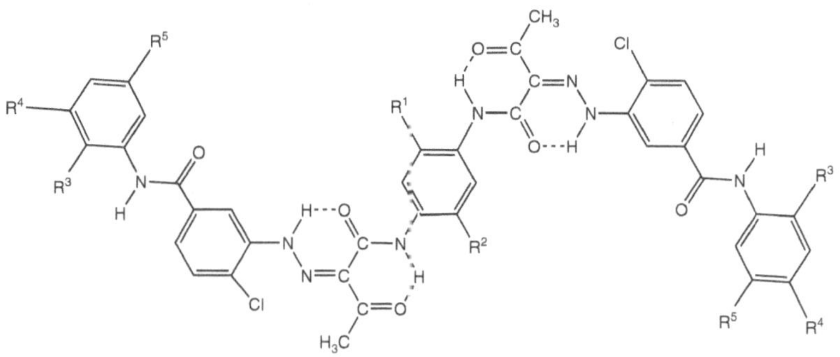

1

Figure 28.1: Structure of disazo condensation pigments based on azoacetoacetanilides (Type I).

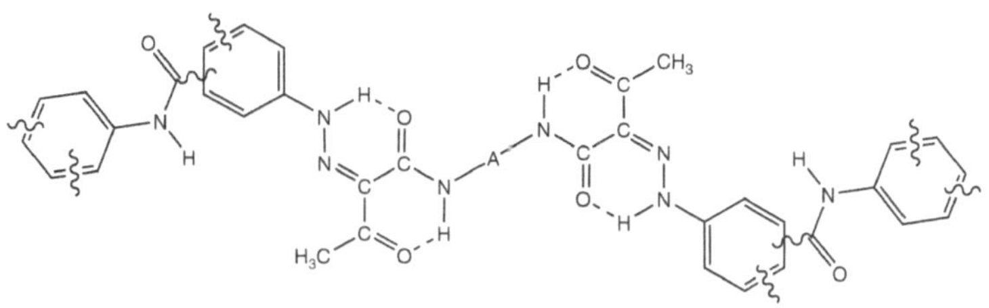

2

Figure 28.2: Structure of disazo condensation pigments based on azoacetoacetanilides (Type II).

commercial pigments that enjoy widespread use have Type I structures (**1**), identified in Table 28.1. A series of disazonaphtharylamides provide red and brown derivatives based on structure (**3**), as illustrated in Figure 28.3. The individual substitution patterns in this series of pigments are given in Table 28.2. The disazo condensation pigments may be considered as essentially "dimerized" monoazo (monohydrazone) pigments. The pigments owe their excellent technical performance, including lightfastness, solvent resistance, and thermal stability, mostly to their large molecular size and the presence of several amide groups.

Table 28.1: Substituent pattern in commercial disazo condensation pigments based on azoacetoacetanilides.

Compound	CI Pigment Yellow	R^1	R^2	R^3	R^4	R^5
1a	93	CH_3	Cl	CH_3	Cl	H
1b	94	Cl	Cl	CH_3	H	Cl
1c	95	CH_3	CH_3	CH_3	H	Cl
1d	128	CH_3	Cl	$OC_6H_4(p)Cl$	H	CF_3

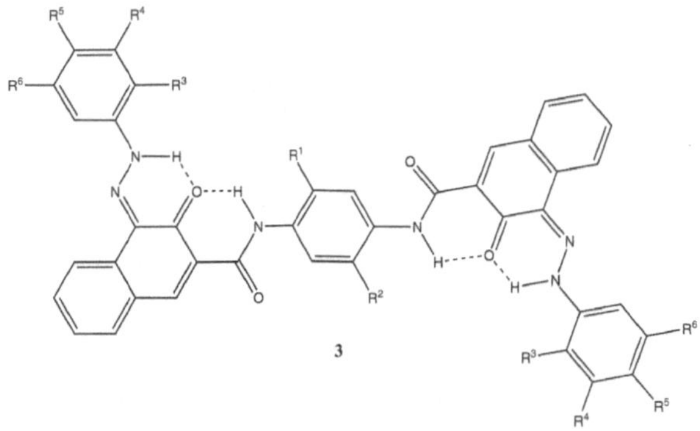

3

Figure 28.3: Structures of disazo condensation pigments based on azonaphtharylamides.

Table 28.2: Substituent pattern in disazo condensation pigments based on disazonaphtharylamides.

Compound	CI Pigment Red or Brown	R^1	R^2	R^3	R^4	R^5	R^6
3a	R144	Cl	H	Cl	H	H	Cl
3b	R166	H	H	Cl	H	H	Cl
3c	R214	Cl	Cl	Cl	H	H	Cl
3d	R220	CH_3	CH_3	CH_3	H	H	$CO_2(CH_2)_2Cl$
3e	R221	Cl	Cl	Cl	H	H	$CO_2CH(CH_3)_2$
3f	R242	Cl	Cl	Cl	H	H	CF_3
3g	R262	CN	H	Cl	H	H	CF_3
3h	Br23	Cl	H	NO_2	H	Cl	H
3i	Br41	*	*	Cl	Cl	H	H

* aromatic bridge is 1,8-naphthyl

28.4 Synthesis and manufacture

The procedures used in the synthesis, from which the pigments derive their name, is complex and demanding, and explains the higher cost of these pigments. The sequence used is illustrated in Figure 28.4 for the case of disazonaphtharylamide pigment (**3c**), CI Pigment Red 166, as an example. A monoazo compound containing a carboxylic acid group is prepared by a typical azo coupling reaction using, in the case illustrated in Figure 28.4, 3-hydroxy-2-naphthoic acid (βONA) (**4**), a coupling component also commonly used in the preparation of metal salt azo pigments. The acid intermediate (**5**) thus formed is dried azeotropically in an organic solvent, such as a dichlorobenzene, and is then converted into the acid chloride (**6**) using a phosphorus halide or thionyl chloride. This stage proceeds efficiently under mild conditions in the presence of N,N-dimethylformamide (DMF) as a catalyst. The final stage involves a condensation reaction between the acid chloride (**6**) (2 mol) and an aromatic diamine (1 mol), *p*-phenylenediamine (**7**) in this case, generally carried out in an organic solvent and in the presence of an acid scavenger such as sodium acetate or a tertiary amine. The conditioning processes used with disazo condensation pigments are often based on knowledge and experience of those used for copper phthalocyanine blue pigments. For example, a particular process used to enhance dispersibility, flow properties and gloss, mostly for paint applications, involves incorporation of soluble derivatives of the pigment. The disazoacetoacetanilide pigments are prepared by an analogous set of reactions, although the exact conditions used vary among different cases. A simpler and more cost-effective route to this type of pigment involving azo coupling of the diazonium salt (2 mol) with the appropriate bifunctional coupling component (1 mol) might be envisaged in principle. In practice, however, such an approach fails because the monoazo derivative formed initially is generally so insoluble that the second azo coupling reaction cannot take place efficiently.

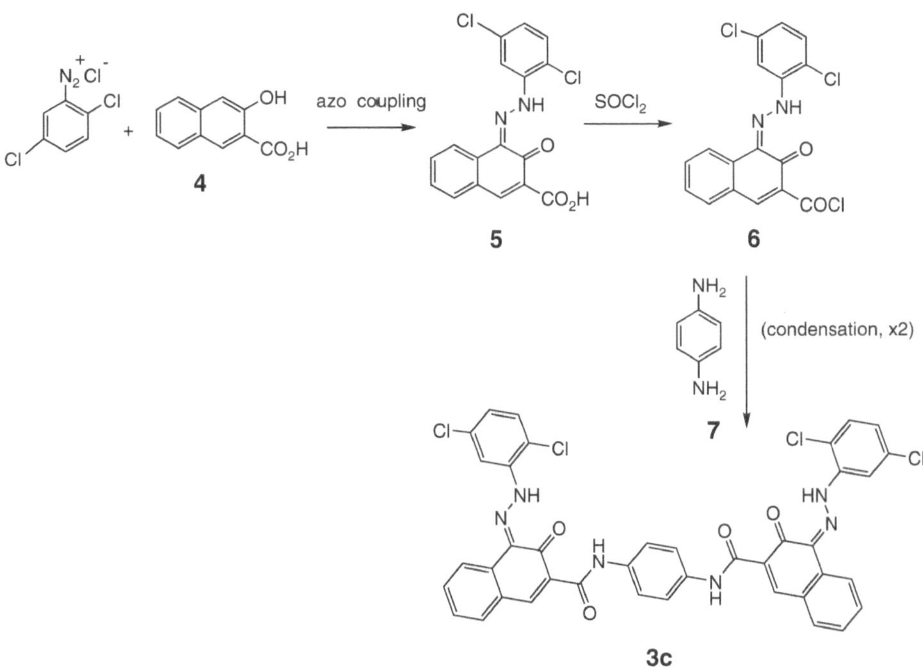

Figure 28.4: Synthesis of a disazo condensation pigment.

28.5 Applications

Because of the complexity of the processes involved in their manufacture, disazo condensation pigments are quite expensive, and this has inevitably had an influence on where they fit in the market. The products that have made it on to the market generally exhibit good tinctorial strength, very good solvent fastness, so that they are resistant to blooming and bleeding and, of course, they possess very good lightfastness. The fastness to solvents varies among the different pigments and even with the same pigment from different manufacturers. It has been suggested that it is not always the pigment that bleeds, but impurities in the pigment [5]. The lightfastness is not only dependent on the molecular and crystal structures but also on the particle size. Most disazo condensation pigments are produced in fairly large particle size grades, a feature that enhances lightfastness often to a rating of excellent, and most have very good fastness to weathering. However, it is their excellent heat stability that has made them such an ideal choice for their main market, the coloration of plastics. They are also recommended for specialist inks that demand very good fastness, such as floor-coverings, wallcoverings, and metal decorative inks, where their high cost can be justified. Their use in general industrial paints was never too important but at least one product was used for automotive finishes. Many of the early products were withdrawn when certain intermediates were removed from the market in the early 1970s.

28.5.1 CI Pigment Yellow 93 (1a)

This is a mid-greenish yellow pigment that produces moderate color strength and pure shades, first introduced under the designation Cromophtal Yellow 3G [6]. It is mainly used in PVC, where it offers very good resistance to plasticizers and provides lightfastness that is excellent in full shade and very good in white reductions. Fastness to weathering is very good in full shade, decreasing to a rating of good in paler reductions. It also finds application in polyolefins, in which it is stable to 280–290°C, depending on the dwell time, and its lightfastness is excellent in full shades but decreases to between good and very good in reductions. The pigment causes some dimensional instability in injection molded articles, but this can be minimized by injecting at higher temperatures. It has very good lightfastness in HDPE and can be used for the spin coloration of polypropylene. In inks, it can be used as a process color for special applications that require high fastness levels. The pigment can also be used for inkjet inks, for which a special grade is recommended, and is also specially recommended for use in toners for laser printers and copiers [7].

28.5.2 CI Pigment Yellow 94 (1b)

This pigment provides a very green shade. It was originally known as Cromophtal Yellow 6 G but is no longer on the current range of BASF pigments [6]. It provides excellent fastness to migration in plasticized PVC, where its lightfastness and fastness to weathering are excellent. Its disadvantage is its low tinctorial strength. In HDPE it is stable to 300°C and again offers very good to excellent lightfastness. The pigment does affect the dimensional stability of injectional moldings.

28.5.3 CI Pigment Yellow 95 (1c)

This pigment, introduced under the designation Cromophtal Yellow 3G, is a medium to red shade yellow, the reddest of the series [6]. It has high tinctorial strength, about 20% stronger than CI Pigment Yellow 93 (**1a**) but has slightly inferior lightfastness at equal depths of shade. It provides excellent resistance to migration and to plasticizers. The pigment has very high heat stability in polyolefins, in which it is stable to 280°C and provides excellent lightfastness in full shade, although only rated as good in white reductions. Its effect on dimensional stability of injection molded articles is slight. It can be used for the spin coloration of polypropylene but is of limited use in applications where the highest levels of lightfastness are demanded. It is recommended for use in polystyrene, but this does not extend to styrene copolymers such as ABS. It is especially recommended for water-based inks and can be used for metal decorative inks, inkjet inks and toners.

28.5.4 CI Pigment Yellow 128 (1c)

This pigment, designated as Cromophtal Yellow 8G provides transparency and the greenest shade in the yellow disazo condensation series, finding application beyond the plastics industry [8]. In plastics, it is mainly used in PVC, where it offers a medium level of tinctorial strength. It shows high resistance to migration and to plasticizers. Its lightfastness is rated as very good to excellent and can be used in the plastisols that are used for coil coatings. In polyolefins it is less stable than CI Pigment Yellow 93 (**1a**) and 95 (**1b**) but can still withstand temperatures up to 260°C. The pigment has only a slight effect on the dimensional stability of injection molded plastic articles. In inks, as it is fast to sterilization and calendering, and is used where high fastness levels are demanded, such as in metal decorative inks, or also where it is preferred on toxicological grounds. The pigment is specially recommended for water-based, solvent-based and UV cured inks and can be used in inkjet inks and toners. In the paint industry, it offers very good solvent fastness, but is no longer recommended for automotive paints in spite of its very good lightfastness and weathering properties. This area of the spectrum is well covered by other types of pigment, including the benzimidazolone azo pigments, which are usually more economical in use.

28.5.5 CI Pigment Red 144 (3a)

This pigment, introduced as Cromophtal Red BR, has become a market leader in the medium to bluish red part of the spectrum [9]. In PVC it shows high tinctorial strength with very good resistance to migration and lightfastness, but its weatherability in rigid PVC does not quite meet the highest standards. In polyolefins it has proved a popular choice, due to its high tinctorial strength, while stable to temperatures up to 300°C, and it provides very good to excellent lightfastness. It is suitable for polypropylene fibers but can affect the dimensional stability in injection moldings, causing warpage in articles such as crates. Its lightfastness in polyolefins is very good to excellent. It can also be used in polystyrene and ABS. The pigment satisfies most demands in specialist printing inks, as the finest particle size version offers reasonable transparency, while is also fast to sterilization and calendering. The pigment is suitable for metal decorative inks, and is specially recommended for water-based, solvent-based and UV cured inks.

28.5.6 CI Pigment Red 166 (3b)

Introduced as Cromophtal Scarlet R, the structure of this scarlet pigment differs from CI Pigment Red 144 (**3a**) only in the absence of substitution on the bridging aromatic ring [6]. In PVC It has moderate tinctorial strength, very good resistance to migration and plasticizers, very good to excellent lightfastness, and good to very good fastness to weathering. In polyolefins it is stable up to 300°C and has very good to excellent lightfastness. However, compared with CI Pigment Red 144 (3a) it has a more adverse effect on dimensional stability in injection moldings. It is suitable for polypropylene spin dyeing and is recommended for coloring polystyrene and ABS. Although there is not a large market for the pigment in the ink sector, it finds application where a high level of fastness properties is demanded and cost is a secondary factor, especially in packaging applications using flexographic and gravure printing. Its stability to calendering and sterilization mean that it is applicable to metal decorative inks. The pigment occasionally finds use in specialist industrial paints, where it exhibits good fastness to weathering.

28.5.7 CI Pigment Red 214 (3c)

This mid shade red pigment, introduced under the designation Cromophtal Red BN, is distinguished by its excellent fastness to light and over-coating [6]. It is manufactured widely and is especially important for the coloration of plastics. In PVC, it shows very good tinctorial strength and is cleaner than the yellower CI Pigment Red 144 (**3a**) and with similar tinctorial strength, but it is not totally fast to bleeding. It can be used in most common plastics, except for polyamides. In polyolefins, it is stable up to 300°C and is used in the spin coloration of polypropylene fibers. Unfortunately, it has a considerable adverse effect on the dimensional stability of items injection molded at all temperatures. In inks, it has specialist applications, such as in metal decorative inks, where it is stable to 200°C, and packaging gravure inks based on nitrocellulose and vinyl resins which use ketones and esters as solvents.

28.5.8 CI Pigment Red 220 (3d)

This pigment, introduced as Cromophtal Red G, is a yellowish shade red, intermediate between those of CI Pigment Red 144 (**3a**) and CI Pigment Red 166 (**3b**) but with lower tinctorial strength [10]. In PVC it has good, but not perfect, resistance to migration and plasticizers. Its lightfastness is very good, while also offering good weathering properties. In polyolefins, it is stable up to 300°C, and exhibits very good to excellent lightfastness, but in injection moldings it can cause warping in HDPE articles manufactured at lower temperatures. However, as the processing temperature is increased, the performance gradually improves until, by 260°C, they become free from distortion. It can be used for coloring polystyrene but is less suitable for other styrenic polymers.

28.5.9 CI Pigment Red 221 (3e)

This pigment, introduced as Cromophtal Red 2B, is just on the bluish side of mid red and is tinctorially the strongest product in the series [10]. It is recommended and widely used for coloring PVC, in which it is fast to migration and has good resistance to plasticizers. However, its lightfastness is a little lower than other red pigments in the series and, while in full shades it is rated as very good to excellent, in reductions it is rated as good to very good. Its weatherfastness, even in full shade, is rated as only moderate. It has limited suitability for polyolefins in which it is stable to 260°C and is unsuitable for polystyrene and engineering polymers. It also exists in a fine particle size form which is useful for printing on transparent foils.

28.5.10 CI Pigment Red 242 (3f)

This is one of the few disazo condensation pigments that was not first introduced by Ciba or Ciba Geigy (now BASF). It was introduced as Sandorin Scarlet 4RF by Sandoz (now Clariant) after the Ciba patents for the disazo condensation process had lapsed

[11]. It is a brilliant yellowish red, quite close to CI Pigment Red 166 (**3b**) in shade and in the level of fastness properties, including excellent to very good lightfastness for its main application areas in plastics. In PVC it has very good to excellent migration fastness and is highly insoluble in plasticizers, just slightly below the level provided by CI Pigment Red 166. It shows heat stability is up to 300°C, making it suitable for polyolefins, polystyrene, and ABS, and for polypropylene fibers. It has, unfortunately, a considerable adverse effect on the dimensional stability of injection moldings. In paints, it can be used for general industrial finishes, including vehicle refinishes, but there are other preferred pigments available with a similar shade. For example, in the Clariant sales range for paints, batches are not routinely tested in these media. In ink applications, its clean shade is an advantage. It is used where there are high demands for the level of fastness properties, such as in metal decorative inks and in systems based on powerful solvents such as ketones and esters, such as those used in nitrocellulose and vinyl inks. Its solvent resistance is less than excellent in aromatic hydrocarbons.

28.5.11 CI Pigment Red 262 (3g)

This pigment was introduced by Sandoz as Red 2BN. It had a bluish-red shade with high color strength and was used in applications where economy was important. In PVC it provided very good fastness to light and migration and was used for cable sheathing applications. In polyolefins, it is stable up to 300°C and does not adversely affect the dimensional stability of injection moldings. However, the pigment has recently been taken off the Clariant range and there are no other registrants in the Colour Index.

28.5.12 CI Pigment Brown 23 (3h)

This is a reddish-brown pigment introduced by Ciba as Cromophtal Brown 5R. In respect to its color, it competes with the benzimidazolone pigment CI Pigment Brown 25. It is specifically recommended for PVC where it has very good resistance to migration and is almost insoluble in plasticizers. Its lightfastness is good to excellent and, unusually, is slightly higher in white reduction than in full shade. The pigment is considered suitable for polyolefins but is not highly recommended, despite good to very good lightfastness, but its thermal stability of up to 260°C limits its suitability for such applications. Its tendency to produce warpage in injection moldings decreases as the processing temperature increases above 230°C. It can also be used for polystyrene, although only recommended after careful testing, and for various spin dyeing applications, including acrylics. Its use in paints is now less important than previously, but it is still designated as a pigment that can be used for applications where a good toxicologically profile is required. Its transparency provides interesting effects in metallic and pearlescent finishes. The paint industry uses this pigment in general industrial paints, in cases where its less than perfect solvent fastness can be tolerated. It starts to bleed at temperatures above 120°C.

28.5.13 CI Pigment Brown 41 (3i)

This is a very yellow shade brown pigment introduced by Clariant under the designation PV Fast Brown RL. In PVC it is fast to migration and plasticizers. Its lightfastness is excellent in full shade and in reductions, so that it is highly suitable for exterior applications. In polyolefins, it is stable to 300°C at concentrations higher than 0.03% and again has excellent fastness to light. The pigment affects the dimensional stability of injection moldings, and therefore is not considered suitable for plastic applications such as crates, but it is suitable for the spin coloration of polypropylene and acrylic fibers. It can also be used in polystyrene, but is not recommended for ABS.

The following pigments in this series have either no current manufacturer registered in the Colour Index or have little commercial importance: CI Pigment Orange 31, CI Pigment Red 248, and CI Pigment Brown 42. The technical recommendations regarding areas of use and the accompanying data have been obtained mainly from information published by BASF and Clariant [7, 12, 13].

References

1. Christie RM. Colour chemistry, 2nd ed. London: RSC, Ch 9 2015.
2. Hunger K, Schmidt MU. Industrial organic pigments, 4th ed. Weinheim: Wiley-VCH Verlag GmbH, Ch 2 2019.
3. Herren F, Faulkner EB, Schwartz RJ. High performance pigments, 2nd ed. Weinheim: Wiley-VCH Verlag GmbH, ch 13 2009.
4. Schmid M. Deutsche Farben Zeitschrift. 1955;9:252.
5. Gaertner HJ. Oil Colour Chem Assoc. 1963;46:36.
6. Schmid M, Mueller W. US 2,936,306. 8 May 1960. (Ciba).
7. BASF. Solutions for digital printing; EDC 4213e.
8. Mueller W. US 3,413,279. 26 Nov 1968. (Ciba).
9. Schmid M, Moser E, Danuser J, Mory R, Mueller W, Wuergler. US 2,741,656. 10 Apr 1956. (Ciba).
10. Schmid M, Streiff H. US 2,888,453. 26 May 1959. (Ciba).
11. Tcherinsky B, Wasern H. US 4,051,121. 26 May 1977. (Sandoz).
12. BASF. Organic and inorganic pigments for plastics. EVP 2013/04e.
14. Clariant. The coloration of plastics and rubber. PI 3002e/09.2019.

Robert Christie and Adrian Abel

29 Disazo (Bishydrazone) pigments based on acetoacetanilides

Abstract: Disazoacetoacetanilide pigments, more commonly known as diarylide yellows, are the most important group of yellow classical organic pigments. They were commercialized in the early 20th century many years after the introduction of the structurally related monazoacetoacetanilides (Hansa yellows). The molecules adopt the bis-ketohydrazone tautomeric form. X-ray single crystal structure investigations have provided an insight into the influence of the molecular geometry and crystal packing arrangements in the solid state on the properties of the pigments in application. The synthesis of diarylide pigments is relatively straightforward, the conditions essentially following those used for the corresponding monoazo pigments, so that the products are economically priced. In the case of these disazo pigments, suitable aromatic amines (1 mol) are bis-diazotized and the resulting bis-diazonium salts reacted with acetoacetanilide coupling components (2 mol), the two azo coupling reactions occurring at the same time. They are by far the dominant group of yellow pigments used in printing inks, well-suited for most standard process yellow inks. They were formerly important in the coloration of plastics but are no longer recommended for polymers processed above 200 °C, under which conditions toxic decomposition products are formed. Diarylide yellow pigments are characterized by high color strength, good to excellent solvent fastness, and good chemical stability, although they generally show inferior lightfastness.

Keywords: disazo pigments, disazoacetoacetanilide pigments, bishydrazone pigments, diarylide pigments, diarylide yellows, Hansa yellows, acetoacetanilides

29.1 Fundamentals

The most important classical yellow azo pigments are the azoacetoacetanilides, which are derived from azo coupling reactions with acetoacetanilide derivatives as the coupling component. A separate chapter is focused on monoazoacetanilide pigments. This chapter covers the structurally related, and industrially more important, disazoacetoacetanilides, commonly referred to as the diarylide yellows, with a couple of orange pigments of lesser importance. Today, disazoacetoacetanilide pigments represent over

This article has previously been published in the journal Physical Sciences Reviews. Please cite as: R. Christie, A. Abel, Disazo (Bishydrazone) Pigments Based on Acetoacetanilides *Physical Sciences Reviews* [Online] 2021, 6. DOI: 10.1515/psr-2020-0170

https://doi.org/10.1515/9783110587104-029

20% of all organic pigments manufactured, second in terms of tonnage only to copper phthalocyanines, with their commercial use mainly in coloration of printing inks.

29.2 History

The disazoacetoacetanilides can be considered as latecomers to the color palette of classical organic pigments, despite their discovery as long ago as 1911 by Griesheim Elektron (eventually forming part of Hoechst, now Clariant). In the early twentieth century, there was little demand for these pigments in the paint industry, then the largest user of pigments, as they were considered inferior to the monoazoacetoacetanilides (Hansa Yellows) on account of their lower lightfastness. In that period, there was relatively little color printing and plastics were virtually unknown. Bakelite, the first synthetic polymer had just been discovered. PVC only became useful when methods of making it flexible were developed in 1926. Polyethylene, although discovered in the nineteenth century, was not made commercially until 1944. IG Farben eventually introduced a range of diarylide yellows under the designation Vulcan Fast Yellow in 1935, initially for the rubber industry. The US ink industry recognized the potential for pigments of this type with high color strength and good resistance to heat and solvents in response to the rapid expansion seen in the color ink and plastics industries just before the outbreak of World War II. However, the pigments were not used in Europe until after the war. While they were used widely in plastics for many years, this use has declined due to toxicological concerns because of the recognition that toxic breakdown products are formed during high temperature processing. In the US, CI Pigment Yellow 12 is the dominant pigment with some use of CI Pigment Yellow 14 whereas CI Pigment Yellow 13 dominates in Europe, with CI Pigment Yellow 12 playing a lesser, but still important, role.

29.3 Structures and properties

Disazoacetoacetanilides (**1a**) – (**1l**) constitute a group of classical azo pigments that contain the most important of all yellow organic pigments, and a couple of oranges [1,2]. They were formerly known as benzidine yellows. However, this term has been replaced ubiquitously by the term diarylide yellows, as benzidine itself, a proven human carcinogen, is now banned for use in the manufacture of colorants in most parts of the world. 3,3′-dichlorobenzidine (DCB), a suspected carcinogen, is used in the synthesis of the more important diarylides. The German regulations that ban the use of benzidine also do not allow the use of the dichloro derivative in the manufacture of textile dyes, but its use is still allowed for pigments [3]. The diarylide yellows exhibit around twice the color strength and higher transparency compared with the corresponding monoazo pigments. Hence, they are ideally suited to

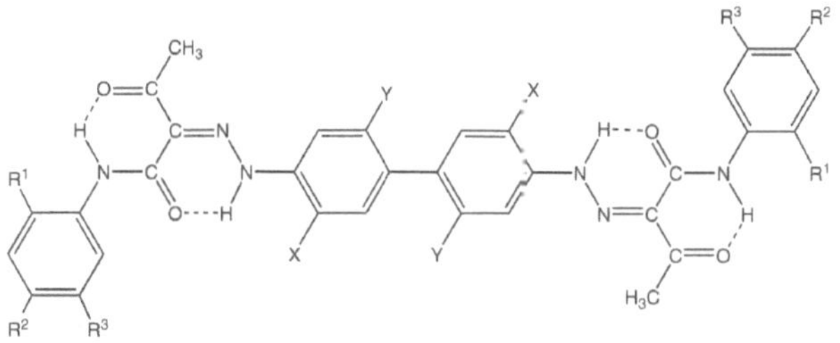

1

Figure 29.1: Structure of the disazoacetoacetanilides.

printing ink applications, in which CI Pigments Yellow 12, 13 and 14 (**1a-c**) are of prime importance. The products exist as bisketohydrazone forms (Figure 29.1), with significant intramolecular hydrogen bonding, structurally analogous to the monoazoacetoacetanildes (Hansa Yellows) [1,2,4]. The disazo pigments are symmetrical with respect to the biphenyl link in their structure, effectively "dimerized" monoazoacetoacetanilides. Extension into the orange part of the spectrum can be achieved by replacing 3,3′-dichlorobenzidine as the bis-diazo component with 3,3′-dimethylbenzidine, as is the case with pigment (**1k**), or 3,3′-dimethoxybenzidine (*o*-dianisidine) with pigment (**1l**). Alternatively, important orange disazopyrazolone pigments, structurally related to the disazoacetoacetanilides, are prepared using pyrazolone-based coupling components. This group of pigments is dealt with in a separate chapter.

Table 29.1: Substituent pattern in disazoacetoacetanilides.

Compound	CI Pigment Yellow or Orange	R^1	R^2	R^3	X	Y
1a	Y12	H	H	H	Cl	H
1b	Y13	CH_3	CH_3	H	Cl	H
1c	Y14	CH_3	H	H	Cl	H
1d	Y17	OCH_3	H	H	Cl	H
1e	Y55	H	CH_3	H	Cl	H
1f	Y63	Cl	H	H	Cl	H
1g	Y77	CH_3	H	Cl	Cl	H
1h	Y81	CH_3	CH_3	H	Cl	Cl
1i	Y83	OCH_3	Cl	OCH_3	Cl	H
1j	Y113	CH_3	Cl	H	Cl	Cl
1k	O15	H	H	H	CH_3	H
1l	O16	H	H	H	OCH_3	H

Crystal structures of CI Pigment Yellow 12 [5], 13 [6], 14 [6] and 83 [7] have been reported, determined by single crystal X-ray methods. In the case of CI Pigment Yellows 13 and 14, the molecules are symmetrical and approximately planar in their crystal lattice structures. In contrast, CI Pigment Yellow 12 is crystallographically non-centrosymmetric, with one half of the molecule essentially planar, while the other deviates significantly from planarity. The molecule is twisted by 27° about the biphenyl link, and the outer phenyl ring by a further 28°. CI Pigment Yellow 12 shows lower color strength than CI Pigment Yellows 13, 14 and 83, and the considerable deviation from planarity of the molecules in the solid state may well be a determining factor in this observation. The diarylide yellows show improved solvent fastness compared with the Hansa Yellows, a feature attributable to their larger molecular size, although the intermolecular interactions in the crystal structure are, as with most of the monoazo compounds, essentially due to van der Waals' forces. These disazo pigments generally show inferior lightfastness, inadequate for most paint applications. However, in a few cases, especially the reddish-yellow CI Pigment Yellow 83 (**1i**), the technical performance is significantly better so that it may be used in a wider range of applications, especially in coatings. In its crystal structure, the molecules are more nearly planar than is the case with CI Pigments Yellow 12, 13 and 14, and the compact molecular packing arrangement involves simple inclined stacks, a feature that no doubt contributes towards its enhanced photostability and insolubility.

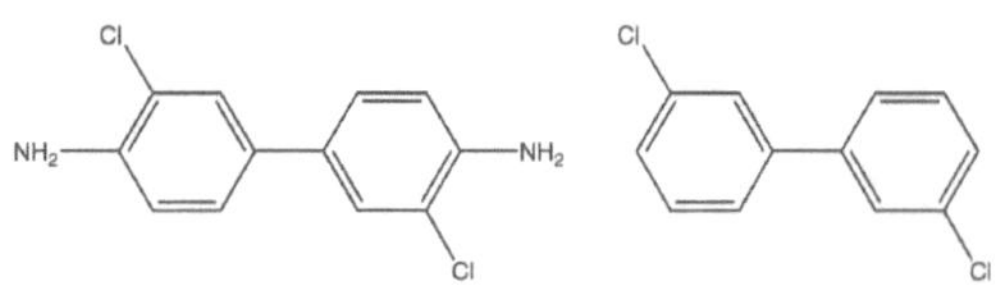

Figure 29.2: 3,3′-dichlorobenzidine (DCB) (**2**) and 3,3′-biphenyl, PCB 11 (**3**).

There are a few issues of toxicological and environmental concern associated with diarylide pigments. Most commercial pigments in this class are derived synthetically from 3,3′-dichlorobenzidine (DCB) (**2**), a suspected carcinogen. In view of this, when DCB is used in the manufacture of these pigments, special handling procedures are used to minimize human contact. Although the pigments themselves are generally considered as non-toxic, evidence has been reported that they may cleave thermally at temperatures above 200°C to give a monoazo compound and that prolonged heating above 240°C causes further decomposition leading to release of DCB [8] (**2**). As a result of these observations, the use of diarylide pigments is no longer recommended in applications where high temperatures are likely to be encountered, for example in thermoplastics, other than PVC which is processed below 200°C. In addition, it has been reported that traces of 3,3-dichlorobiphenyl (PCB 11) (**3**) may be formed during the manufacturing processes [9]. PCBs (polychlorinated biphenyls) are considered as PBTs, chemicals that are *persistent* (they do not readily break down in the environment), *bioaccumulative* (they accumulate in living tissues), and *toxic* (they cause adverse health effects) and their deliberate manufacture has long since been banned. When generated inadvertently, as in this case, the levels are required to be minimized to comply with regulatory requirements. The evidence for the distribution of PCB 11 in the environment arising from diarylide pigment manufacture and use has been reviewed, and mechanisms leading to its formation during pigment manufacture proposed [10].

CI Pigments Yellow 16 (**4a**) and 155 (**4b**) (Figure 29.3) are prominent examples of disazo yellow pigments (often referred to as bisazoacetoacetanilides) with a somewhat different structural arrangement compared with the diarylides. However, they are similar in that the molecules also exist as bisketohydrazones. CI Pigment Yellow 180 is a disazoacetoacetanilide pigment which offers significantly improved heat stability compared with the traditional diarylide yellows (**1**) and is used in plastics. Structurally, it contains benzimidazolone groups and is discussed in the separate chapter dealing specifically with benzimidazolone azo pigments.

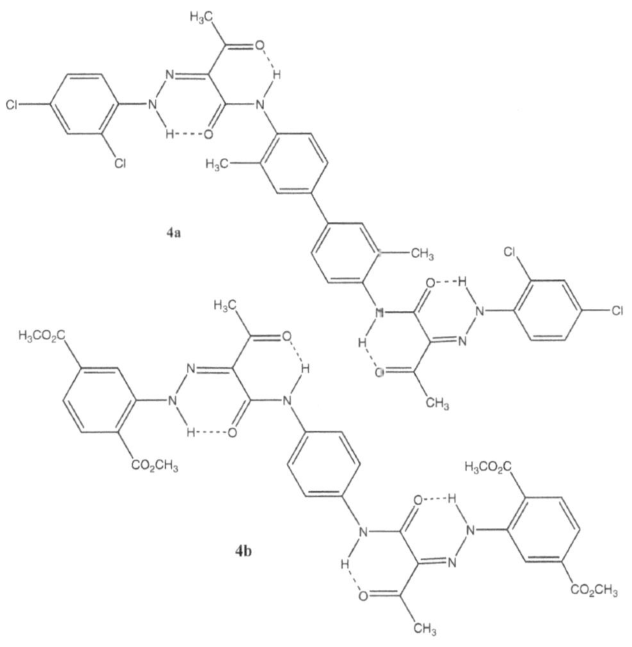

Figure 29.3: Structure of pigments (**4a**) and (**4b**).

29.4 Synthesis and manufacture

Diarylide pigments are prepared by bis-diazotizing (commonly, but incorrectly referred to as tetrazotizing) an aromatic diamine (1 mol) and reacting the resulting bis-diazonium salt with an acetoacetanilide coupling component (2 mol), as illustrated in Figure 29.4 for CI Pigment Yellow 12 (**1a**) as an example. In this case, and in most others in the series, the diamine is 3,3′dichlorobenzidine (DCB) (**2**) which forms a bis-diazonium salt (**5**). The coupling components are commonly referred to by key letters taken from their names. For example, the coupling component for CI Pigment Yellow 12 (**1a**), **a**ceto**a**ce**t**anilide, is known as AAA, for CI Pigments Yellow 13 (**1b**) and 81 (**1 h**) as AAMX, **a**ceto**a**cet-***m***-**x**ylidide, for CI Pigment Yellow 14 (**1c**), **a**ceto**a**cet-***o***-**t**oluidide as AAOT, for CI Pigment Yellow 17 (**1d**), **a**ceto**a**cet-***o***-**a**nisidide as AAOA, for CI Pigment Yellow 55 (**1e**), **a**ceto**a**cet-***p***-**t**oluidide as AAPT, and for CI Pigment Yellow 83 (**1i**), **a**ceto**a**cet**d**imethoxy**c**hlor**a**nilide, as AADMCA. Indeed, there is a convention, most often associated with the US, to distinguish the pigments themselves using these abbreviations. Because DCB is a suspected carcinogen, special careful industrial handling procedures are required in its use, ideally avoiding human contact. The pigments (**1**) are symmetrical disazo compounds, and this means that the two diazotization and coupling reactions effectively take place concurrently. In this way, the overall manufacturing procedures leading to the disazo pigments are essentially the same as those involved in the manufacture of the analogous monoazo pigments. Similar conditioning and finishing conditions are also used involving boiling the aqueous suspensions of the pigments with the addition of surfactants before filtration, drying and milling. Pigments (**4a**) and (**4b**) are also symmetrical molecules. However, they require an alternative synthetic strategy, involving the reaction of a diazotized aniline derivative (2 mol) with a symmetrical bis-acetoacetanilide that has two coupling sites (1 mol).

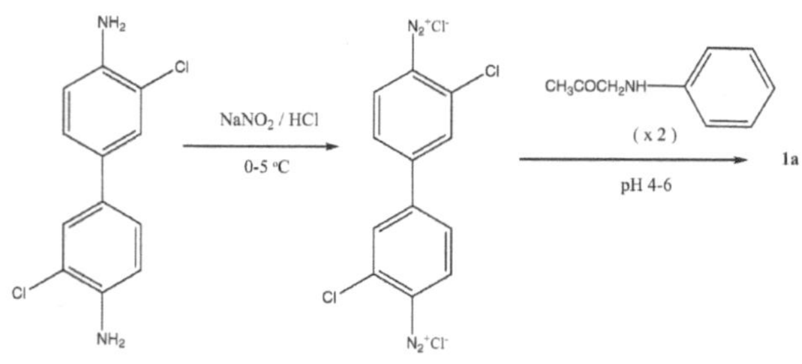

2 5

Figure 29.4: Synthesis of CI Pigment Yellow 12.

29.5 Applications

29.5.1 Diarylide yellow and orange pigments

Disazoacetoacetanilide pigments are by far the most important yellow pigments for printing inks, used in most of the standard process yellow inks. They are less important for paints as they generally lack the required lightfastness demands. They used to be important in plastics, but their use in this application has diminished as they are not recommended for any application that requires exposure to above 200°C, effectively limiting them to the coloration of PVC. Diarylide yellow and orange pigments are characterized by high color strength, good to excellent solvent fastness and good chemical stability although, with a few exceptions, they generally lack good lightfastness. The synthesis of most diarylide pigments is relatively straightforward so that they are economically priced in a competitive market. In contrast, building up the molecular size of other pigments of the azo chemical class, for example disazo condensation pigments, generally requires more complex and expensive manufacturing processes. The diarylide pigments are commonly resinated during manufacture resulting in pigments with high color strength and transparency, and good dispersibility Based on their properties, they are well-suited to printing inks, where high lightfastness is not usually demanded.

29.5.1.1 CI Pigment Yellow 12 (1a)

This pigment is structurally the simplest of the series. It shows lower levels of fastness properties compared other diarylide yellow pigments, especially to light, where it fades about four times faster than other members of the series. It is mid-yellow in shade and is used in letterpress and offset inks but, as it is greener than the European process standard yellow shade, it requires to be tinted with a redder pigment. Untreated, it is considered semi-transparent. However, as the particles are reduced in size the pigment becomes greener, cleaner, stronger, and more transparent, although there is a reduction in the lightfastness. It is often used in processes where yellow is printed first, such as newspaper printing in the US. Whereas Europe prefers CI Pigment Yellow 13 for offset inks, the US prefers CI Pigment Yellow 12, and it is almost exclusively the first choice. The US also prefers to use flush pastes as their starting point, a technique that they have brought to their subsidiaries in Europe with some success. CI Pigment Yellow 12 shows lower solvent fastness with a tendency to dissolve during the dispersion process, leading to re-crystallization during which the pigment becomes weaker and less transparent. As the pigment is often used at high concentrations in the ink, this effect increases the shear during the dispersion process and hence raises the temperature, exacerbating the effect. Special grades have been developed for publication gravure inks. In those grades, the pigment is treated with a primary amine (RNH_2), where R is a long aliphatic hydrocarbon chain that can react with the keto carbonyl group (C = O) derived from the coupling component, causing partial conversion of the pigment to a ketimine (C = NR), or Schiff's base, that has some solubility in the ink solvent, especially, but not only, in systems rich in aromatic solvents such as toluene. The treatment provides higher color strength and several other positive effects. The solubility of the ketamine derivative leads to a stabilized ink, compared with inks made from the untreated pigments, which are unstable due to the recrystallisation process that can occur. Also, the treatment enhances the ability of the ink to sit on the surface of, rather than becoming absorbed into, cheaper grades of paper. On drying, ketimine formation reverses and the pigment recrystallizes on the surface of the paper, returning it to its pigmentary state. Almost all publication gravure inks use amine-treated pigments. Different amines have different characteristics, some performing better in aromatic solvents while others are better in aliphatic solvents.

29.5.1.2 CI Pigment Yellow 13 (1b)

This pigment is preferred in Europe, especially for offset inks. It is a mid-shade yellow, slightly redder than CI Pigment Yellow 12 and thus close to the standard yellow for three- and four-color printing. The transparency of the pigment can be enhanced by resination, a major asset in 3 or 4-color printing when the yellow ink is printed last. The color strength of these pigments is approximately 25% higher than CI Pigment Yellow 12 and, while slightly more expensive, it is cost effective. The lightfastness and resistance to solvents are both superior, and the pigment has slightly greater stability to recrystallisation especially in offset inks produced in modern bead mills. However, this remains an area where pigment manufacturers and ink makers are constantly seeking improvements. In the US, the pigment is used only for some specialist applications, such as metal decorative inks. In liquid inks the American market prefers CI Pigment Yellow 14. There is some small use in paints, if there is no demand for good lightfastness or fastness to over-coating. The only polymer in which the pigment can be used is PVC in which it has some minor use. In this application it provides significantly better lightfastness than in prints. It does not migrate when used in rigid PVC and can meet standards for use in electrical cables, due to its satisfactory dielectric properties.

29.5.1.3 CI Pigment Yellow 14(1c)

This greenish yellow diarylide pigment is important in the US for liquid packaging inks, partly on account of the good flow properties that it provides in applications. It is more opaque and greener than CI Pigments 12 and 13. Its lightfastness is somewhat lower than CI Pigment Yellow 13, one of the reasons for its limited use in paste inks produced in Europe. It has slightly better solvent fastness than CI Pigment Yellow 12. It also finds application in textile printing. For use in publication gravure inks it is possible to treat the pigment with long chain amines and produce the effect described for CI Pigment Yellow 12. It is not used in the paint industry and is of limited interest in PVC as it blooms below certain concentrations. It is used for the mass coloration of viscose, provided that good lightfastness is not required.

29.5.1.4 CI Pigment Yellow 17 (1d)

This is the greenest of the original diarylide pigments and is quite transparent. It has been used to tint CI Pigment Yellow 13 to make it greener and more transparent, but its poor flow properties in application and low tinctorial strength limits its use. It tends to be more expensive than the other original pigments. Its lightfastness is variable depending on how it is tested. Accelerated fading tests using Xenon lamps degrade the pigment more than sunlight but in outdoor exposure it is found to be superior to CI Pigment Yellow 13. It can be resinated making it a useful pigment for packaging inks, based on nitrocellulose, vinyl chloride copolymers and polyamide resins. As it is stable up to 200°C it can be used in metal decorative inks. It was a favored pigment for plastics before the discovery of it degrading into toxic materials at temperatures above 200°C. It can be used for rigid PVC and PVC film, where it has similar lightfastness to CI Pigment Yellow 13. However, it blooms in plasticized PVC at low concentrations. It is rarely used in paints as it does not meet the lightfastness requirements and is not fast to over-coating. It can be used in textile printing as a cheaper alternative to certain more lightfast pigments. It is also recommended for toners used in laser printers and photocopiers on account of its good dielectric properties.

29.5.1.5 CI Pigment Yellow 55 (1e)

This is a less well-known diarylide yellow pigment, although it was discovered and patented as part of the original group of diarylide pigments. It is much redder than other members of the series. The pigment can be manufactured with reduced particle size to become stronger, less red, and transparent, but this results in inferior flow properties in application. Alternatively, larger particle grades of the pigment are weaker, redder and more opaque, with slightly enhanced lightfastness and flow. It is useful for liquid inks as a semi-opaque red-shade yellow, with reasonable lightfastness, although not sufficient for paint applications. It can be used to color rigid PVC and rubber but in plasticized PVC, while it has moderately good tinctorial strength, it tends to bleed. It has been used as an inexpensive alternative to the more popular CI Pigment Yellow 83 (**1 j**) in textile printing. Never a popular pigment in Europe, all current registered manufacturers are based in Asia.

29.5.1.6 CI Pigment Yellow 81 (1h)

This pigment has a very green shade (designated Yellow H10G), similar to the monoazoacetanilide CI Pigment Yellow 3 (Hansa Yellow 10 G), and in most applications it shows significantly higher color strength. At equal depth of shade, it has slightly lower lightfastness than the monoazoacetanilide CI Pigment Yellow 3, but its solvent fastness is much better, meeting most requirements, although it is more expensive. Consequently, it can be used in inks, paints, and plastics as well as in textile printing. In inks, it is used in metal decorative inks and can be used in inks based on vinyl copolymers, which use esters and ketones as solvents. Prints are fast to sterilization and calendering. In paint, it is fast to over-coating and is reasonably opaque. The pigment preceded the introduction of many of the benzimidazolone pigments, which became considered as alternatives, thus reducing its potential for use. In plastics, it can only be used in PVC, especially in rigid PVC as it may bloom in plasticized PVC at low concentrations but is highly resistant to bleeding. Its tendency to decompose at temperatures above 200° C forming unacceptable breakdown products makes it unsuitable for most other polymers.

29.5.1.7 CI Pigment Yellow 83 (1j)

This highly important pigment is at the opposite end of the yellow spectral range from CI Pigment Yellow 81 (**1 h**) as it offers a very reddish yellow shade. It is one of the most versatile pigments for use across the range of organic pigment applications, except for those involving exposure to temperatures above 200°C. In its original small particle size form, it is very strong tinctorially and transparent. In inks, the use of this stronger form predominates, often resinated and suitable for printing onto aluminum foil and in metal decorative inks. It has very good fastness to most solvents, is stable to recrystallization, and fast to sterilization and calendering. It is used widely in PVC, both rigid and plasticized, in which it is fast to migration. It has excellent lightfastness in deep shades, slightly less in pale shades. For the paint industry, the transparent grade can be used in many industrial finishes, but its lightfastness is inadequate for architectural finishes. It is normally used in reductions and in transparent and metallic finishes. In the 1970s, a highly opaque form of the pigment was introduced, not without some controversy surrounding its patent, which was challenged, but upheld, by a court in London. The treatment used to prepare this form involves growing the crystals to a defined size that gives maximum opacity. Although this form is tinctorially weaker, its high opacity and lower oil absorption allow it to be used at a higher concentration, ideal for replacing lead chromates. The lightfastness is also increased, which makes it suitable for some automotive original equipment and most refinish systems. The increase in lightfastness also allows it to be used in pale shades in both water-based and solvent-based architectural paints. The transparent grade is still one of the most popular pigments for printing on textiles, where it offers very good lightfastness and good dry clean fastness. It can also be used for the pulp coloration of paper and for decorative laminate papers.

Table 29.2 presents a summary of some of the properties of the diarylide yellow pigments that are used most frequently in inks. The data are compiled from several sources and refer to what is judged to be the basic ink pigment (except arguably for

Table 29.2: Summary of some properties of commonly used disazoacetoacetanilide (diarylide) yellow pigments in ink formulations.

Color	CI pigment yellow	Coupling component	Lightfastness		Solvent fastness			Heat stability /°C for 10 min.
			Full shade	Tint	Xylene	Ethanol	Methyl ethyl ketone	
Greenish yellow ↓	81	AAMX	VG	G	5	5	5	180
	17	AAOA	M-R	F-M	3	5	3–4	180
	14	AAOT	F-M	P	3	5	3–4	180
	12	AAA	M	P	3–4	3–4	4	180
	13	AAMX	M	M	4	4–5	4–5	180
Mid yellow ↓	55	AAPT	R	R	4	5	4	160
	83 transparent	AADMCA	R	R	3–4	4–5	3–4	180
Reddish yellow	83 opaque	AADMCA	VG	G	4	4–5	4	180

opaque CI Pigment Yellow 83). The data are presented to allow a broad comparison between the products, but only as a guide to selection as there are so many variables that can influence the properties in application. It is essential that potential users test the actual pigments they intend to use in their own system and following their own test procedures. In the table, qualitative lightfastness ratings are given: E- excellent, VG – very good, G – good, R – reasonable, M – moderate, F – fair, P – poor. These ratings are considered more reliable for the intended purpose than more precise quantitative data, for example based on Blue Wool Scale assessments. For solvent fastness, the ratings refer to the staining of the solvent, from 5 (no staining) to 1 (severe staining), when the print is immersed in the solvent, assessed against standard grey scales. The heat stability assessments refer to the maximum temperature to which the pigment is stable when subjected to that temperature for 10 minutes.

There are several pigments that are the products of mixed azo couplings where a diazo component is reacted simultaneously with two different coupling components in an appropriate ratio. These pigments are not simple mixtures of two pigments, but rather they contain molecules of the two symmetrical pigments together with a third molecule which is unsymmetrical, derived from each coupling component. Recognizing this fundamental difference, they were eventually given their own Colour Index designations.

29.5.1.8 CI Pigment Yellow 126

This mixed coupled product was introduced by Hoechst as Yellow DGR, manufactured using the coupling components for CI Pigments Yellows 12 and 13. Although it has been reported as being predominantly CI Pigment Yellow 12, its shade and strength are closer to CI Pigment Yellow 13. Although transparent versions were introduced, they never reached the transparency demanded for use in offset inks. It is, however, a useful pigment for water-based inks, where it is stronger and more lightfast than CI Pigment Yellow 12, making it more economical in use. There are products available, as both powder and dispersions, for water-based flexographic inks.

29.5.1.9 CI Pigment Yellow 127

Introduced by Hoechst, this pigment is manufactured using the coupling components for CI Pigments Yellow 13 and 17. It occupies the area normally satisfied by CI Pigment Yellow 13, resinated versions providing high transparency and tinctorial strength It is eminently suitable for use in packaging gravure inks where it offers excellent gloss, especially in nitrocellulose inks. In addition, it provides very good rheological behavior in application and generally good fastness properties, similar to CI Pigment Yellow 13, thus suitable for metal decorative inks. It also has much improved resistance to recrystallisation and is easier to disperse in bead mills.

29.5.1.10 CI Pigment Yellow 174

Manufactured using the coupling components for CI Pigments Yellow 13 and 14, this pigment is highly resinated to provide high color strength, transparency, and excellent gloss in application. The product was first introduced by Ciba-Geigy. Modern grades now offer excellent flow properties in offset process printing. In sheet-fed offset inks, particularly in VOC-free formulations, the pigment produces inks with excellent flow properties. In web offset printing, which often suffers from misting developing during the speedy running of the press, modern grades can significantly reduce this effect.

29.5.1.11 CI Pigment Yellow 176

Manufactured using the coupling components for CI Pigments Yellow 13 and 83, this pigment has largely superseded CI Pigment Yellow 127 for offset inks. It has high transparency and is suitable for 4-color process inks according to the European standards. It provides excellent rheology in application and it has been optimized to provide good ink-water balance. Its main application areas are in heat-set and sheet-fed offset printing inks and it has been found to perform well in VOC-free printing inks. Another grade of pigment has been developed specifically for water based flexographic inks.

29.5.1.12 CI Pigment Orange 15(1k)

This pigment provides a yellow shade orange, not too dissimilar to the disazopyrazolone, CI Pigment Orange 13, but with a lower level of fastness properties.

29.5.1.13 CI Pigment Orange 16 (1l)

This pigment is a yellow-shade orange but redder than CI Pigment Orange 13, known as dianisidine orange, particularly in the US. It has only modest fastness properties and so only finds use in the ink industry, where economy is more important than performance. The inferior rheology that it provides in application makes it difficult to use in deeper shades, but it is used to adjust the shade of diarylide yellow pigments towards redder shades. It is occasionally used for inexpensive textile prints. *O*-Dianisidine, the diamine used in its manufacture, is considered highly toxic and a possible carcinogen.

There are several other diarylide pigments that have been withdrawn either through lack of demand or in some cases non-availability of intermediates used in manufacture. Others are not well known internationally, having just a local market, frequently in Japan. These include CI Pigments Yellow 63, 87, 90, 121, 124, 136, 152, 170, 171, 172, and 198.

29.5.2 Bisacetoacetarylide pigments

The two pigments classified as bisacetoacetarylide pigments exhibit properties intermediate between those of the diarylide and disazo condensation pigments.

29.5.2.1 CI Pigment Yellow 16 (4a)

The oldest member of the group, this mid-yellow shade pigment was discovered by Laska and Zitscher of Griesheim Elektron in 1921. Its fastness to alcohols, ketones and esters is good but it is less fast to aromatic hydrocarbons. This can cause a problem with recrystallization following dispersion into hydrocarbon-based paints and packaging gravure inks, not only due to the pigment dissolving but also to a change to a crystal modification with a redder shade. Its main application is in general industrial paints where it shows good fastness to over-coating and does not bleed in stoving paints in full shade and achieves very good to excellent lightfastness, although significantly lower in paler shades. Its name Yellow NCG, recognizes that it was one of the first yellow organic pigments that could be used in nitrocellulose. It has been used in inks, but other pigments can now achieve the level of lightfastness and solvent fastness that made it useful to ink-makers. It is used to a limited extent in textile printing where it gives good fastness to dry cleaning. CI Pigment Yellow 16 was one of the first pigments to be offered in an alternative form for high opacity (the Hoechst 70 series) offering potential to meet the demand for heavy metal free paints. A solvent treatment under controlled conditions was used to grow the pigment crystals to provide much higher opacity but lower tinctorial strength. The smaller surface area of large particles produced lower viscosity in applications thus permitting use of a higher pigment concentration, thus further improving the covering power. Lightfastness also increased, roughly halving the rate of fading.

29.5.2.2 CI Pigment Yellow 155 (4b)

This is a clean and slightly greenish yellow pigment introduced as Sandorin Yellow 4 G by Sandoz, now Clariant. In its original form, it has high tinctorial strength, good resistance to solvents and very good lightfastness. It is used in a wide range of paint, ink, and plastics applications. In paints, it has excellent fastness to over-coating in alkyd-melamine systems with no bleeding when stoved for 30 minutes at 140°C. Its lightfastness is very good in strong shades and better than CI Pigment Yellow 16 (**4a**) in reductions with white pigment. It also has superior weatherability. A larger particle size version of the pigment offers higher opacity and better lightfastness, thus providing a good replacement for chrome yellow, and potential for application in vehicle refinishing, agricultural finishes, and commercial vehicles. In stoving paints, it is stable to 160°C. In plasticized PVC it is stable up to 180°C but is not free from bleeding. It has moderate tinctorial strength and very good to excellent lightfastness. It can be used to color polyolefins up to a temperature of 260°C. It can be used in all types of printing inks where it achieves high fastness properties. Special grades of both the standard and opaque versions have been developed to meet the needs of inkjet printing, including high purity and good stability in water-based, solvent-based, and UV-cured inkjet systems. Another growing area of use is as toners in laser printers and photocopiers. A special grade has been introduced that offers the pigment in a form that meets the demands for the electrostatically specified quality and neutral triboelectric behavior required for commonly used toner resins.

References

1. Christie RM. Colour chemistry. 2nd ed. London: RSC, Ch 9 2015.
2. Hunger K, Schmidt MU. Industrial organic pigments. 4th ed. Weinheim: Wiley-VCH Verlag GmbH, Ch 2 2019.
3. Second Amendment to the German Consumer Goods Ordinance. Bundesgesetzblatt, Part. 1. 1994:1670.
4. Christie RM, Standring, PN. Colour and constitution relationships in organic pigments. Part 2: Disazoacetoacetanilides. Dyes Pigm. 1989;11:109–121.
5. Barrow MJ, Christie RM, Lough AJ, Monteith JE, Standring PN. The crystal structure of CI Pigment Yellow 12. Dyes Pigm. 2000;45:153–160.
6. Christie RM, Monteith JE, Barrow M. The crystal and molecular structures of CI Pigments Yellow 13, 14 and 63. Dyes Pigm. 2002;55:79–89.

7. Christie RM, Barrow M, Badcock TD. The crystal and molecular structure of CI Pigment Yellow 83, a superior performance diarylide yellow pigment. Dyes Pigm. 2003;57:99–106.
8. Az A, Dewald B, Schnaitmann D. Pigment decomposition in polymers in applications at elevated temperatures. Dyes Pigm. 1991;15:1–14.
9. The Japanese Ministry of Economy Trade and Industry (METI). Compiled results or reanalysis of the presence of polychlorinated biphenyls (PCBs) as by-products in organic pigments. www.meti.go.jp/english/press/2013/0510_02.html. Accessed: 15 Nov 2014.
10. Rodenburg L, Guo J, Christie RM. Polychlorinated biphenyls in pigments: inadvertent production and environmental significance. Color Technol. 2015;131:353–369.

Robert Christie and Adrian Abel

30 Disazo (Bishydrazone) pigments based on pyrazolones

Abstract: The most important classical orange organic pigments are disazopyrazolones, also referred to as diarylide oranges. The first pigment in this series, CI Pigment Orange 13, was discovered in 1910 although it was a further 20 years before it was introduced as a commercial product. Currently, two orange disazopyrazolones are extremely important industrial organic pigments, while two red products are of lesser importance. The products are structurally analogous to the disazoacetoacetanilides (diarylide yellows), which are discussed in a separate chapter. For example, they are symmetrical compounds that exist in the bis-ketohydrazone tautomeric form. The pigments also exhibit similar technical and color properties compared with disazoacetoacetanilide pigments, for example providing high color strength and transparency, features that determine their importance as printing ink pigments. They are manufactured in a process that parallels those used for the disazoacetoacetanilide (diarylide) yellows, but with coupling components containing the pyrazolone heterocyclic system, in place of acetoacetanilides.

Keywords: disazopyrazolone pigments, disazopyrazolones, pyrazolones, pyrazolone oranges, bishydrazone, diarylide oranges, diarylide yellows, tartrazine, azo coupling, bis-diazotization

30.1 Fundamentals

There are a few classical orange pigments provided by the disazoacetoacetanilides and the naphthol-based pigment classes, as described in other chapters. However, the most important classical orange organic pigments are the disazopyrazolones, also known as diarylide or pyrazolone oranges [1, 2]. The structures of these pigments are derived from coupling components containing the pyrazolone heterocyclic system, based on the parent compound (**1**) (Figure 30.1), a five-membered heterocyclic ring containing two adjacent nitrogen atoms and a carbonyl group. Otherwise, the pigments are structurally similar to the disazoacetoacetanilides and exhibit similar performance properties.

This article has previously been published in the journal Physical Sciences Reviews. Please cite as: R. Christie, A. Abel, Disazo (Bishydrazone) Pigments Based on Pyrazolones *Physical Sciences Reviews* [Online] 2021, 6. DOI: 10.1515/psr-2020-0171

https://doi.org/10.1515/9783110587104-030

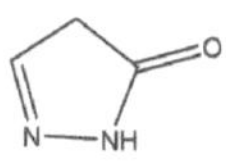

1

Figure 30.1: The pyrazolone ring system.

30.2 History

Disazopyrazolone pigments have a long history, which essentially parallels that of the disazoacetoacetanilides. Pyrazolones were discovered in 1883 by Knorr in Germany [3]. He subsequently prepared antipyrine, a drug based on the pyrazolone ring system, which proved effective as an analgesic and as a treatment for fever and was widely used until the discovery of aspirin. A pyrazolone was also used in the synthesis of the monoazo dye tartrazine (**2**) (Figure 30.2), discovered by Ziegler in 1884 and still, although sometimes controversially, used as a yellow food dye. Tartrazine can also be converted into an aluminum pigment lake, CI Pigment Yellow 100, mainly used in the US where it was used for metal decoration as a preferred pigment for food contact. It was also used for a few reddish yellow and orange monoazo pigments analogous to the monoazoacetoacetanilide (Hansa Yellow) series of pigments. In 1910, Laska of Griesheim Elektron synthesized CI Pigment Orange 13 (**3a**), although it was a further 20 years before the product was launched on the market. This pigment was originally used in the coloration of rubber, until it began to be used in printing inks in the mid to late 1940s. It was not until the early 1950s that CI Pigment Orange 34 (**3b**) also became a commercial pigment.

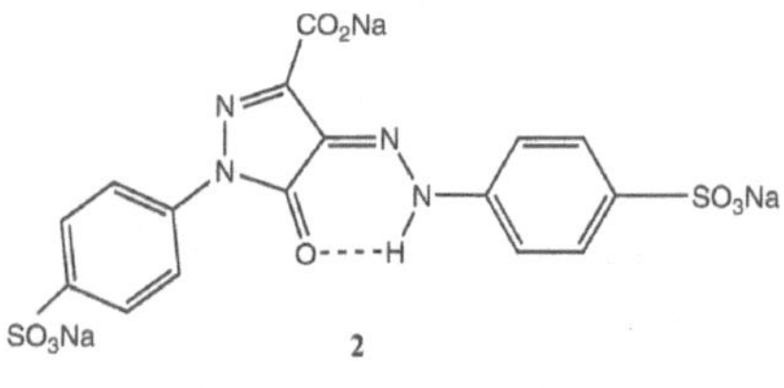

2

Figure 30.2: The structure of Tartrazine (**2**).

30.3 Structures and properties

The disazopyrazolones, notably compounds (**3a**) and (**3b**), are extremely important industrial orange organic pigments [1, 2]. The group also includes red products (**3c**) and (**3d**) which are of lesser importance. The molecular structures are illustrated in Figure 30.3, and the substituent patterns in Table 30.1. These products are structurally analogous to the disazoacetoacetanilide pigments. For example, they are symmetrical disazo compounds, which exist in the bis-ketohydrazone tautomeric form [4, 5]. These features explain why the two classes of diarylide pigments exhibit similar technical and color properties, for example providing high color strength and transparency, which in turn accounts for their commercial importance, especially as printing ink pigments.

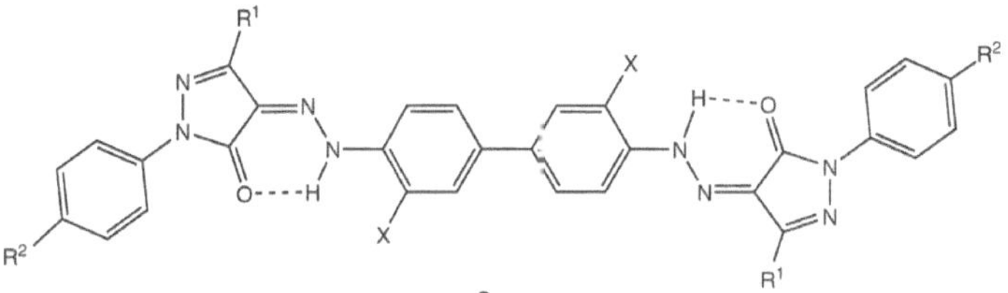

3

Figure 30.3: The structure of disazopyrazolones.

Table 30.1: The substituent pattern in disazopyrazolones.

Compound	CI Pigment	*X*	*R*¹	*R*²
3a	Orange 13	Cl	CH_3	H
3b	Orange 34	Cl	CH_3	CH_3
3c	Red 37	OCH_3	CH_3	CH_3
3d	Red 38	Cl	CO_2Et	H

30.4 Synthesis and manufacture

The diarylide orange pigments are prepared by bis-diazotization of an aromatic diamine (1 mol) and reacting the resulting bis-diazonium salt with a pyrazolone as the coupling component (2 mol), in a process that parallels that used for the diarylide yellows. The reaction sequence is illustrated in Figure 30.4 for CI Pigment Orange 13 (**3a**). In this case, the diamine is 3,3'dichlorobenzidine (DCB) (**4**), and the coupling component is 3-methyl-1-phenylpyrazolone (**5**). This coupling component has a reactive methylene group, which is the site at which azo coupling takes place when deprotonated to give the anion. The pyrazolone coupling components react in essentially the same way as acetoacetanilides and using essentially the same manufacturing conditions. Since DCB is a suspected carcinogen, special industrial handling procedures are required in its use, ideally avoiding human contact. As is the case with the diarylide yellows, the diarylide oranges (**1**) are symmetrical disazo compounds, and the two azo coupling reactions effectively take place concurrently.

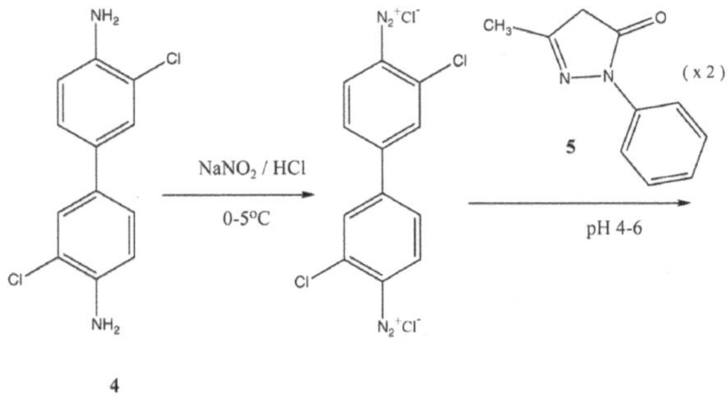

Figure 30.4: Synthesis of CI Pigment Orange 13 (**3a**).

30.5 Applications

The disazopyrazolone pigments cover the orange part of the spectrum through to orange-red and bluish-red. However, it is the two orange pigments that are by far the most important.

30.5.1 CI Pigment Orange 13 (3a)

The yellowest of these products, this pigment is tinctorially weaker than CI Pigment Orange 34 (**3b**). Its main use is in the ink industry, where its low fastness to light, similar to some of the disazoacetoacetanilide yellows, is tolerable. The pigment shows moderate to good resistance to solvents and can withstand temperatures up to 200°C, so that it is suitable for metal decorative inks. It is also stable to sterilization. As with diarylide yellow pigments, it starts to decompose at temperatures above 200 °C releasing toxic breakdown products, and thus effectively the only plastic polymer in which it can be used is PVC, and at concentrations above 0.1% [6, 7]. While satisfactory in rigid PVC, the pigment blooms in plasticized PVC and bleeds badly. It is occasionally used in general industrial paints, but it bleeds, especially into aromatic hydrocarbons, and has poor lightfastness. It can only withstand stoving temperatures of around 120 °C. Many pigment manufacturers offer this pigment in their range. The pigment has been used for polyester decorative laminate paper, where it offers a significantly higher lightfastness than in other application media.

30.5.2 CI Pigment Orange 34 (3b)

In its transparent form, this pigment offers a clean orange shade with high tinctorial strength, redder than CI Pigment Orange 13 (**3a**) and with moderate lightfastness. In some applications, it fades in light at about quarter the rate of CI Pigment Orange 13. Its solvent fastness is superior to that of CI Pigment Orange 13, notably in ketones and esters but, depending on processing conditions, it may tend to recrystallize following dispersion, partly due to its inferior heat stability, to which the finer particle size versions are only able to withstand temperatures up to a maximum of 140 °C, thus restricting its usefulness in metal decorative inks. In Europe, it is used in most printing processes, especially packaging inks based on nitrocellulose. However, it is rather less extensively used in the US. It is widely used for textile printing if there is no significant demand for lightfastness. It gives good dry-cleaning fastness. As with the diarylide yellows and CI Pigment Orange 13, it decomposes above 200 °C into hazardous breakdown products and so can only be used for coloring PVC. It blooms below concentrations of 0.1% in plasticized PVC, and also bleeds in this application. It is generally the preferred European choice for use in industrial paints, if good lightfastness is not required, and economy is critical. In the 1970s, Hoechst introduced a large particle version, which was more opaque, joining CI Pigment Yellow 16, a bisazoacetoacetanilide, as the second member of their so-called 70 series. Additionally, the opaque grade can be used at much higher concentrations than the traditional grade, while maintaining good flow and gloss properties in application. When used at the same concentrations, it is actually more opaque than molybdate orange, although molybdate orange can be used at much higher concentrations. The lightfastness of the opaque version is significantly better than the transparent pigment, fading at about half the rate in paler shades. To achieve even higher levels of fastness properties, more expensive orange pigments, such as benzimidazolone azo, disazo condensation, and DPP pigments, are required. Due to its good dielectric properties, CI Pigment Orange 34 is recommended for inkjet inks, and toners for photocopiers and laser printers, if the temperature in application does not exceed 200 °C. Many pigment manufacturers offer CI Pigment Orange 34 in their ranges, both as transparent and opaque grades.

30.5.3 CI Pigment Red 37 (3c)

This pigment has an orange-red shade that has never been used as much as CI Pigments Oranges 13 (**3a**) and 34 (**3b**). Its main uses are in PVC and rubber, but only when its poor lightfastness can be tolerated. It has very high tinctorial strength, and has good dielectric properties making it useful for the coloration of PVC cable sheathing. It now has no registrants in the Colour Index.

30.5.4 CI Pigment Red 38 (3d)

A bright medium red shade, this pigment provides technical properties that are much better than those of CI Pigment Red 37 (**3c**), including higher fastness to solvents and light. Thus, it is much more useful for the coloration of PVC in which it can withstand temperatures of up to 180 °C, and it also has good dielectric properties making it useful for cable sheathing. It is generally used for deep shades as it blooms at low concentrations. It was used extensively for the coloration of PVC and polyvinylidinechoride (PVDC) film.

Two further disazopyrazolones, CI Pigments Red 41 and 111 are no longer commercially important.

References

1. Christie RM. Colour chemistry, 2nd ed. London: RSC, Ch 9 2015.
2. Hunger K, Schmidt MU. Industrial organic pigments, 4th ed. Weinheim: Wiley-VCH Verlag GmbH, Ch 2 2019.
3. Knorr L. Einwirkung von Acetessigester auf Phenylhydrazin. Ber Dtsch Chem Ges. 1983;16:2597–9.
4. Whitaker A. Crystal structures of azo colorants derived from pyrazolone: a review. J Soc Dyers Colour. 1995;111:66–72.
5. Whitaker A. Acetoacetanilide colorants: intramolecular bonding and their relationship with pyrazolone colorants. J Soc Dyers Colour. 1996;112:17–18.
6. Second amendment to the German consumer goods ordinance (1994) Bundesgesetzblatt, part 1. 1670.
7. Az A, Dewald B, Schnaitmann D. Pigment decomposition in polymers in applications at elevated temperatures. Dyes Pigm. 1991;15:1–14.

Heinz Mustroph

31 Dyes: quantum chemical calculation of electronic spectra

Abstract: The basics of the quantum mechanical theory of the light absorption process, the simplifications of the theory in form of models and their application to dyes are reviewed. The factors governing the electronic transition energy, the intensity of the electronic transition and the vibrational fine structure of the absorption bands are examined.

Keywords: dyes, electronic spectroscopy, electronic transitions, Franck-Condon principle, vibronic transitions

31.1 The electronic spectrum

When electromagnetic radiation of intensity $I_0(\tilde{\nu})$ with the wavenumber $\tilde{\nu}$ [cm^{-1}] passes through a material, a portion of the radiation is absorbed and the remainder of intensity $I(\tilde{\nu})$ is transmitted. According to the Lambert-Beer law [eq. (31.1)] the absorbance A of a beam of monochromatic radiation at $\tilde{\nu}$ is proportional to a proportionality constant ε [$dm^3\ mol^{-1}\ cm^{-1}$], which is called the molar decadic absorption coefficient, to the concentration c [$mol\ dm^{-3}$], and to the absorption path length l [cm].

$$A = lg\, I_0(\tilde{\nu})/I(\tilde{\nu}) = \varepsilon c\, l \tag{31.1}$$

A graph of A versus $\tilde{\nu}$ or the wavelength λ [nm] or the frequency ν [s^{-1}] is called an absorption spectrum. The absorption spectra are classified into three types on the basis of the origin of the caused transitions between molecular states by absorption of a specific region of the electromagnetic spectrum.

In the microwave region the absorption of radiation causes transitions between rotational states, whereas in the infrared region vibrational states are excited. Since the energy difference of vibrational states of a molecule is greater than that of rotational states vibration spectra are associated with rotational transitions. Therefore, the vibration spectra are rotation-vibration spectra more exactly. In the ultraviolet (UV) and visible (VIS) region, respectively, electronic spectra arise from transitions between electronic states. Due to the large energy difference between electronic states, the electronic transitions are accompanied by rotational and vibrational transitions. Therefore,

This article has previously been published in the journal Physical Sciences Reviews. Please cite as: H. Mustroph, Dyes: Quantum Chemical Calculation of Electronic Spectra *Physical Sciences Reviews* [Online] 2019, 4. DOI: 10.1515/psr-2019-0040

https://doi.org/10.1515/9783110587104-031

the spectra should be called rotational-vibrational-electronic spectra. Nevertheless, the term "electronic spectra" is used in general for spectra which arise from transitions between electronic states induced by absorption of UV/VIS light.

With the development of the electronic spectroscopy of molecules in the 1920s it became possible to measure the absorption spectra of colorants. The color of molecules depends on their ability to absorb visible light (approximately 380 nm to 780 nm) [1] and is determined by the whole area of the absorption curve. The experimental absolute intensity of an absorption band I^{exp} is computed from the integrated area under the absorption curve:

$$I^{\text{exp}} = \int \varepsilon \, d\tilde{\nu} \tag{31.2}$$

However, given the complexity of determining I^{exp} single features of the absorption band which influence the color of dyes are used instead. The main used characteristics of the absorption bands are:

(1) the wavelength at the intensity maximum of the absorption band, the absorption maximum λ_{max}
(2) the intensity maximum of the absorption band ε_{max}
(3) the shape of the absorption band, which is semi-quantitatively described by the width of the band in cm^{-1} at $\varepsilon = ½\ \varepsilon_{max}$, the full width at half maximum (FWHM) $\Delta\tilde{\nu}_{1/2}$

With the introduction of electronic spectra the early structure-color theories tried to link molecular structure with λ_{max}. However, since theorists at the time did not possess a deep knowledge of electronic structure, the rules that they developed remained empirical. It was not possible to interpret the effect of molecular structure on ε_{max}, and $\Delta\tilde{\nu}_{1/2}$. This changed with the development of quantum mechanics. Nevertheless practical application of the theory to dyes remained a long way off.

31.2 Basics of quantum mechanics

Quantum mechanics was originated by Max Planck in the year 1900 with his idea that the energy in material oscillators E with the frequency ν is quantized to explain the emission spectrum of a black body ($E = h{\cdot}\nu$). Albert Einstein had hypothesized in 1905 that electromagnetic radiation itself consists of particles which he called energy quanta. Analogous to Planck's material oscillators he said the energy of the quanta is proportional to their frequency ($E = h{\cdot}\nu$).

The development of the quantum mechanics of atoms and molecules can be divided roughly into two periods: the "old" and the "new quantum mechanics of atoms and molecules" [2]. The discovery of the electron and atomic nucleus led to different models for the atomic structure. In 1913 Niels Bohr proposed a theory for the hydrogen atom. It is based on a mixing of quantum ideas and classical physics and Bohr introduced two principal assumptions:

(4) the *Bohr quantization condition* whereby the electrons move around the nucleus of an atom in definite orbits with an integer electronic quantum number n and a discrete electronic energy, E_a, E_b, E_c, … . He called these states the stationary electronic states

(5) the *Bohr resonance condition* which states that an atom in the initial electronic state with the Energy E_a can absorb or emit a photon of the energy $E = h{\cdot}\nu$ only if there exists a final stationary electronic state with the Energy E_b, whereby the energy difference between E_b and E_a ($\Delta E_{ab} = E_b - E_a = h{\cdot}\nu$) matches the energy of the absorbed or emitted photon

With Bohr's model of the hydrogen atom started the development of the quantum mechanics of atomic structure. Today the period during 1913–1925, based on a mixing of quantum mechanical approaches and classical physics, is called the "old quantum mechanics of atoms and molecules" [2].

The development of a "new quantum mechanics" started in 1925. In this year Louis de Broglie published his famous paper "Recherches sur la théorie des Quanta", where he depicted the particle and wave nature of electrons [3]. A year later this proposal led Erwin Schrödinger to describe electrons as standing wave surrounding the nucleus [4–7]. Therefore, this new quantum mechanics is also called wave mechanics.

$$H(r,R)\psi(r,R) = E\psi(r,R) \tag{31.3}$$

Equation (31.3) represents the Schrödinger equation (SE) with the total Hamiltonian operator $H(r, R)$ and the total wave-function of a molecular state $\psi(r, R)$, which depends on both the coordinates of all electrons r and the coordinates of all nuclei R.

The Hamiltonian operator consists of kinetic energy operators of the nuclei $T^{\mathrm{nu}}(R)$ and the electrons $T^{\mathrm{el}}(r)$ as well the potential energy operators of the electron-electron $V^{\mathrm{elel}}(r)$, nuclear-nuclear $V^{\mathrm{nunu}}(R)$ repulsion and electron-nuclear $V^{\mathrm{elnu}}(r, R)$ attraction, respectively.

$$H(r,R) = T^{\mathrm{nu}}(R) + T^{\mathrm{el}}(r) + V^{\mathrm{elel}}(r) + V^{\mathrm{nunu}}(R) + V^{\mathrm{elnu}}(r,R) \tag{31.4}$$

However, eq. (31.3) can be solved analytically only for a two-body system like the hydrogen atom, i. e. application of the approach to multi-atomic systems typical of colorants lay well out of reach. The initial stimulus for further developments stemmed primarily from electronic spectroscopy again.

Already on the basis of the old quantum mechanics, James Franck had presented in 1925 a theory of the photo-dissociation of diatomic molecules [8]. He assumed that the nuclei do not vibrate in the electronic ground state. Today the assumption of non-vibrating molecules in the electronic ground state is often called the *classical Franck approximation* and is widely used in the Computational Chemistry. However, non-vibrating molecules in the electronic ground state have nothing to do with reality. Most important was Franck's idea, if a light quantum is absorbed, the electronic energy of the molecule changes only and the electronic transition happens in a negligibly short time compared to the period of nuclear vibrations, so that the electronic transition does not change the momentum p and the coordinates of the nuclei. Based on Franck's idea, in 1926 Edward U. Condon developed a semi-classical theory combining classical mechanics with quantum mechanical vibrational states [9]. With it Condon included the concept of vibrating nuclei in the electronic ground state, different to Franck's assumption of non-vibrating molecules in the electronic ground state. He said the most favored vibrational-electronic (vibronic) transition will be at the semi-classical turning points of the nuclear vibration because the nuclei do not move here ($p = 0$) and, therefore, spent a larger proportion of the vibration period here [9–11]. With this theory it was possible to explain the intensity distribution of vibronic sub-bands in electronic spectra of diatomic molecules in the context of the old quantum mechanics.

Schrödinger's series of papers [4–7], Born's probability interpretation of $|\psi|^2$ [12] and Heisenberg's uncertainty principle [13] stimulated Condon to apply the new quantum mechanics to electronic spectroscopy of molecules. As mentioned already, eq. (31.3) can be solved analytically only for a two-body system. To approximate the total wave-function of a molecule Condon neglected rotation and translation and postulated that electronic and nuclear motions are separable. With it, the total wave-function of a molecular state can be expanded as a product of an electronic wave-function $\psi_n(r, R)$, which depends on the quantum number n of the electronic state and the coordinates of all electrons and all nuclei and a nuclear wave-function χ_{nv} (R), which depends on all quantum numbers and the coordinates of all nuclei [14].

$$\psi_{\mathrm{nv}}(r,R) = \psi_{\mathrm{n}}(r,R)\chi_{\mathrm{nv}}(R) \tag{31.5}$$

Referring to Condon's paper [14], Max Born and Robert Oppenheimer (BO) published their famous paper in 1927 [15]. They used perturbation theory to find approximate solutions for the total SE with the perturbation parameter κ, which is taken as $\kappa = (m/M)^{1/4}$, where m is the mass of an electron and M the average mass of the nuclei.

BO did not suggest the separation of the electronic and nuclear motions [eq. (31.5)]. Nevertheless, eq. (31.5) is often referred to as the Born-Oppenheimer Approximation (BOA). However, Condon first suggested this approximation. The merit of BO is their justification that Condon's postulate is usually fulfilled to a satisfactory approximation [15].

Referring to BO in a subsequent paper, Condon wrote the justification of his postulate is founded in the fact that nuclei (proton rest mass $M = 1.673 \cdot 10^{-27}$ kg) are so much heavier than electrons (electron rest mass $m = 9.109 \cdot 10^{-31}$ kg) that the electrons adjusts instantly to changes in the nuclear coordinates [16]. This is the most popular justification today.

Now a further problem remained. Due to $T^{nu}(R)$ even the SE for a many-body system with separated electronic and nuclear motions cannot be solved. To solve it, a second approximation is necessary. If the electrons in each nuclear position in an electronic state adjust instantly to changes in the nuclear coordinates the nuclei can be considered to be fixed in a special configuration. Then it is possible to consider the electrons as moving in the field of fixed nuclei. With this approximation the kinetic energy of the nuclei is zero [$T^{nu}(R) = 0$] and the complete Hamiltonian eq. (31.4) reduces to the electronic Hamiltonian $H^{el}(r, R)$:

$$H^{el}(r,R) = T^{el}(r) + V^{elel}(r) + V^{nunu}(R) + V^{elnu}(r,R) \tag{31.6}$$

This approximation is also often called BOA. However it was already used in the old quantum mechanics. Therefore, for greater clarity the term Fixed (Clamped) Nuclei Approximation should be used instead of BOA [10].

With this approximation the electronic energy E_n^{el} of an electronic state n is calculated by solving the electronic SE for fixed nuclei coordinates. If this is carried out for a large number of different nuclei coordinates R and E_n^{el} is plotted as a function of R [$E_n^{el}(R)$] an energy curve is obtained as illustrated in Figure 31.1 with $E_0^{el}(R)$ of the electronic ground state $n = 0$. It reaches its minimum at equilibrium coordinates of the nuclei R_e. In molecules $E_n^{el}(R)$ is the effective potential $U_n(R)$ in which the nuclei vibrate in an electronic state. Therefore, this function is called a potential energy surface (PES).

For simplicity vibrations of nuclei in large polyatomic molecules can be modeled by the harmonic oscillator approximation and the potential eq. (31.7) replaces $E_n^{el}(R)$.

$$U_n(R) = \tfrac{1}{2} k(R - R_e)^2 \tag{31.7}$$

With the harmonic oscillator approximation the energy of the vibrational states v in an electronic state n is

$$E_{nv}^{vi} = (v_n + \tfrac{1}{2}) h \cdot \nu_n^{vi} \tag{31.8}$$

in which $v_n = 0, 1, 2, 3, \ldots$ is the vibrational quantum number in the electronic state n and ν_n^{vi} is the frequency of the vibration in the electronic state n.

The value for $v_0 = 0$ in the ground electronic state $n = 0$ ($\tfrac{1}{2}\, h \cdot \nu_0^{vi}$) is called zero-point vibrational energy (ZPVE). Zero of potential energy can be chosen at any point as reference and the most convenient choice of where to set it to zero is at R_e. This is mentioned here, because many people believe “zero” refers to $T = 0$ K, which is not correct.

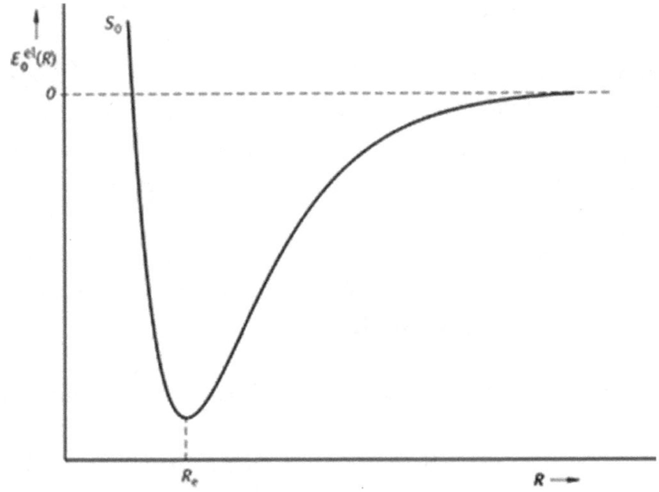

Figure 31.1: Curve of the electronic energy $E_0^{el}(R)$ of the electronic ground state $n = 0$ [potential energy surface (PES)] for a diatomic molecule as a function of different nuclei coordinates R with the minimum of $E_0^{el}(R)$ at equilibrium coordinates $R_{e.}$

In this chapter a short review of the basics could be given only. There are a lot of books dealing with the theory of electronic spectroscopy like e. g. [17–26], to which readers are referred for deeper insight.

31.3 Calculation of the electronic transition energy of molecules

As mentioned, electronic spectra arise from transition between electronic states and are accompanied by simultaneous transitions between electronic and vibrational states (neglecting rotational states as done by Condon). To consider all vibrational states would make the calculations very complicated.

At room temperature the ground vibrational state of the ground electronic state ($v_{OO} = 0$) is mostly populated, whereas the first excited vibrational state ($v_{01} = 1$) at 298 K is populated to $N_1 = N_0 \cdot 0.001$ only [11, 18]. Therefore, at room temperature it is a well-founded approximation to assume the main part of absorption bands arise from transitions between the lowest vibrational state of $n = 0$ and different vibrational states of $n = 1$ termed 0–0, 0–1, 0–2 . . . vibronic transitions.

With this approximation the transition energy of the vibronic transitions $\Delta E_{OO,\ 1v}$ is the difference between the minimum of the electronic energy surface $E_1^{el}(R_e)$ of the first electronic excited state $n = 1$ (S_1) plus the considered vibrational energy

$h{\cdot}\nu_1^{vi}(v_1 + ½)$ and the minimum of the electronic energy surface $E_0^{el}(R_e)$ of the electronic ground state $n = 0$ (S_0) plus the ZPVE ($½\, h{\cdot}\nu_0^{vi}$):

$$\Delta E_{00,\,1v} = E_1^{el}(R_e) + h \cdot \nu_1^{vi}(v_1 + ½) - [E_0^{el}(Re) + ½\, h \cdot \nu_0^{vi}] \tag{31.9}$$

These are the only calculated values, which are related to experimental values, the energies of the 0–v vibronic transitions [10, 11].

For the special case that the transition occurs between the two vibrational ground states, eq. (31.9) reduces to

$$\Delta E_{00,\,10} = E_1^{el}(R_e) + ½\, h \cdot \nu_1^{vi} - [E_0^{el}(R_e) + ½\, h \cdot \nu_0^{vi}] \tag{31.10}$$

Where, $\Delta E_{00,\,10}$ is the energy of the center line of the 0–0 transition (Figure 31.2).

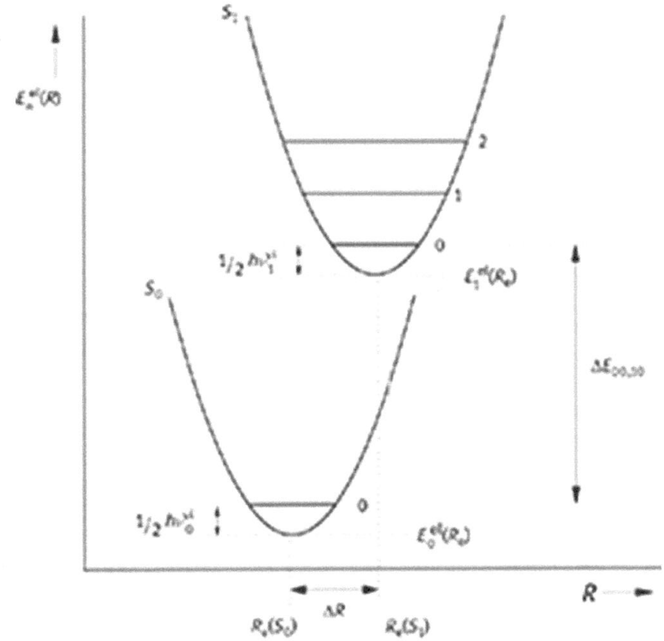

Figure 31.2: Based on the harmonic oscillator model schematic representation of the electronic energy $E_n^{el}(R)$ of the electronic ground state $n = 0$ (S_0) and the first excited state $n = 1$ (S_1) with the zero-point vibrational energy $½\, h{\cdot}\nu_0^{vi}$, the resulting 0–0 transition energy $\Delta E_{00,10}$ and the difference ΔR between the equilibrium nuclear coordinates in the excited and ground state $R_e(S_1)$ and $R_e(S_0)$, respectively.

It is worth noting that the 0–0 transition energy $\Delta E_{00,10}$ does not depend on the difference ΔR between the equilibrium nuclear coordinates in the ground and excited state $R_e(S_0)$ and $R_e(S_1)$, respectively (Figure 31.2). Often this point is not considered comparing experimental results and calculated transition energies.

So much for the theory: now to the practical applications. The first point is that with all these approximations it is not so easy to solve the SE for complex molecules to calculate the equilibrium nuclear coordinates in the ground and excited electronic state and the energies of the electronic as well the vibrational states.

In the early days of the quantum chemistry the most drastically simplifications to calculate transition energies were made by Erich Hückel to the so called Hückel molecular orbital theory (HMO) [27, 28]. Hückel ignored σ-electrons and considered π-electrons only, i. e., he constructed π-electron molecular orbitals (MO) as linear combination of $2p_z$ atomic orbitals (AO), introduced empirical parameters and solved the SE using variation method [27, 28]. Such simplified quantum mechanical methods using empirical parameters are called semi-empirical methods. With these approximations it was not possible to calculate the equilibrium nuclear coordinates in the ground and excited electronic states and the calculated transition energies were empirically correlated to λ_{max}, but there is no scientific basis to do so. The λ_{max} value is an important application parameter. However it is nothing other than the intensity maximum of an absorption band. In dependence on ΔR between $R_e(S_0)$ and $R_e(S_1)$ it can be caused by the 0–0 or 0–1 or another vibronic transition [11]. Therefore, calculation of λ_{max} is a risk, but within a series of structural similar dyes with similar ΔR values the numerical results of empirical correlations, may correspond well, sometimes with remarkable quantitative results.

Nevertheless, it was the first molecular orbital theory that could be applied to π-conjugated molecules. In addition, it has been the starting point for very successful theoretical interpretations of electronic spectra of conjugated and aromatic hydrocarbons as well as the basis for more advanced theories.

The next development in the field of semi-empirical quantum mechanical methods for the calculation of the properties of conjugated molecules was the Pariser-Parr-Pople (PPP) method [29]. The first substantial difference between PPP and HMO is that explicit the equilibrium nuclear coordinates in the electronic ground state $R_e(S_0)$ are introduced in the starting operations. Instead of the purely empirical parameters in HMO PPP employs atomic parameters derived from experimental data (valence state ionization potential and electron affinity) [29]. Both HMO and PPP ignore σ-electrons and consider only π-electrons. In addition, π-electron MO are constructed by linear combination of $2p_z$ AO. However, to get most suitable approximate MO and electronic energy the self-consistent field (SCF) method is used. Starting from trial MO and a rough Hamiltonian and applying the variation method new MO, new electronic energy and a new Hamiltonian are output, that is, the operator depends on the solution of the SE. The new Hamiltonian produces new MO and this new MO yields an updated operator. This process is continued until the MO no longer changes within defined limits. Essentially, the insertion of the MO reproduces itself. With the PPP method it becomes possible to calculate transition energies, absorption intensities, polarization of the bands as well as dipole moments of the ground and excited state of dyes. It was used for several decades [30–35].

However, one should pay attention to the underlying concept. Within the old quantum mechanics Franck had assumed that the nuclei do not vibrate in the electronic ground state. According to this over-simplification a vertical electronic transition starts from the minimum of the PES of S_0 ($p = 0$) to the same nuclear co-ordinates in S_1 as $R_e(S_0)$ with the same momentum of the nuclei ($p = 0$). That is, $p = 0$ in S_0 and in S_1 are defined at an arbitrary geometry without paying attention to the basics of classical mechanics. The point of intersection in S_1, where a vertical line from the minimum of the PES of S_0 cuts through the PES of S_1, is called the "Franck-Condon point", the resulting state a "Franck-Condon excited state" with the "Franck-Condon geometry", which is the same as R_e in S_0 (Figure 31.3). The energy difference between $E_0^{el}(R_e)$ in S_0 and the energy $E_1^{el}[R_e(S_0)]$ on the PES in S_1 is called vertical transition energy ΔE_{vert}:

$$\Delta E_{vert} = E_1^{el}[Re(S_0)] - E_0^{el}(Re) \tag{31.11}$$

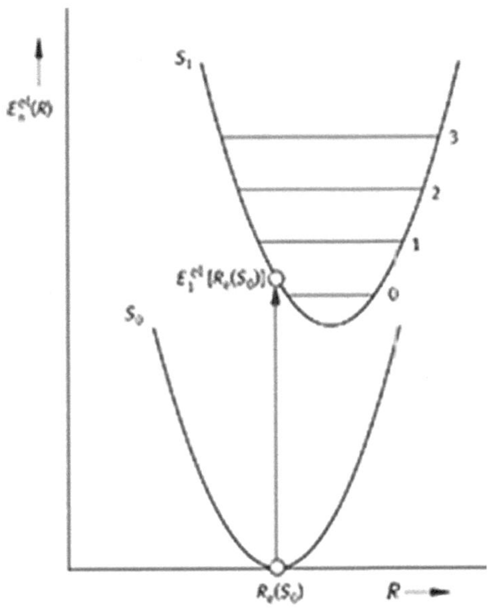

Figure 31.3: Potential energy surfaces (PES) of electronic ground S_0 and excited state S_1 with the harmonic oscillator as model for the electronic energy $E_n^{el}(R)$ and illustration of the vertical electronic transition from the minimum of the PES $R_e(S_0)$ in S_0 to the electronic energy $E_1^{el}[R_e(S_0)]$ of S_1 with the same nuclear co-ordinates as $R_e(S_0)$.

For simplicity the idea of non-vibrating nuclei from the old quantum mechanics is used in the PPP formalism, too. The fixed $R_e(S_0)$ are introduced in the starting operations for S_0 and the equilibrium nuclear coordinates for S_0 [$R_e(S_0)$] are used to calculate the electronic energy $E_1^{el}[R_e(S_0)]$ on the PES in S_1 at the equilibrium geometry of S_0, which is shown graphically in Figure 31.3. The calculated energy

$E_1^{el}[R_e(S_O)]$ on the PES in S_1 is an arbitrary determined value. Therefore, ΔE_{vert} does not correspond to any observable feature [10, 11] and thus this produced number does not reflect what happens in a dye molecule, i. e. transitions between vibronic levels (Figure 31.3). Due to its relative simplicity the calculation of ΔE_{vert} is a very popular choice and this value is correlated with λ_{max}. However λ_{max} is nothing other than the intensity maximum of an absorption band and can correspond to the 0–0, 0–1 or any other vibronic transition. Because the correlation of ΔE_{vert} and experimental λ_{max} values has no legitimate scientific basis it is sometimes good, sometimes poor and sometimes misleading.

These facts are often neglected when studies are reported in which calculated ΔE_{vert} and experimental λ_{max} values are correlated successfully for a series of similar dyes. If the difference between $R_e(S_O)$ and $R_e(S_1)$ of different dyes is nearly similar, so that the failure is a systematic one and the same vibronic transition is responsible for λ_{max} then the correlations will generally be reasonable. If the difference $R_e(S_O)$ and $R_e(S_1)$ is substantially different within the dye series and, therefore, λ_{max} values within the series originate from different vibronic transitions, than there will be little or no correlation as shown with unsymmetrical cyanine dyes [36].

The third step was the development of numerous all-valence-electron methods. Here all valence electrons, π and σ, are considered explicitly but only basis functions for them are used. All the semi-empirical methods use the SCF method and the Zero Differential Overlap (ZDO) approximation. This approach means that all the products of different basis functions are zero. The ZDO approximation greatly simplifies the computation of wave functions, but at the cost of reducing the quality of the calculation, too. Therefore, in addition to the strategy CNDO (Complete Neglect of Differential Overlap), other approaches like e. g. INDO (Intermediate Neglect of Differential Overlap), MINDO (Modified Intermediate Neglect of Differential Overlap) or NDDO (Neglect of Differential Diatomic Overlap) were created.

Especially for electronic spectra the CNDO/S method (Complete Neglect of Differential Overlap for Spectroscopy) [37], and the INDO/S method (Intermediate Neglect of Diatomic Differential Overlap for Spectroscopy), were developed [38]. In principle these approaches enable the calculation of n-π^* and π-π^* transitions. The λ_{max} values calculated by both methods for dyes lie typically at substantially shorter wavelengths than the corresponding experimental λ_{max} values. Neither method has the predictive reliability of PPP-based calculations [39]. In this respect, they do not represent progress in comparison with the PPP method, especially because all three only calculate ΔE_{vert}. As noted earlier, this value does not correspond to any observable feature.

Final goal of quantum chemistry is to find solutions to the electronic SE, based on a well-defined theoretical framework of approximations and when the only inputs are physical constants: no empirical parameters derived from experimental data are employed. This methodology is called *ab initio* ("from first principles").

For some decades the wave-function based approaches were the dominant methods to calculate electronic energies. However the computational complexity grows exponentially with the number of electrons and for it the practical dye structures were too large. The crucial breakthrough came with the seminal theorems of Hohenberg and Kohn [40] and the realization of this concept by Kohn and Sham to the Density Functional Theory (DFT), in which the ground state electronic energy is completely determined as a functional of the electron density [41]. DFT is nowadays one of the most popular methods for calculations of the ground state electronic structure. Compared to wave-function based ab initio and semi-empirical approaches DFT represents a good balance between accuracy and computational efficiency and a number of commercial programs is available. However, DFT is strictly limited to ground states, which excludes it from applications to electronic spectroscopy.

Nevertheless, several attempts have been made, to extend DFT to excited states [42]. The most popular method to treat excited states within the DFT method is the Time-Dependent Density Functional Theory (TDDFT) [42]. Many approximations used in ground state calculations, such as the Fixed (Clamped) Nuclei Approximation, are easily carried over to TDDFT.

In the following a lot of results of TDDFT calculations of conventional dyes were reported (for a review see e. g. [43]). However, in practice it is impossible to calculate the geometries of excited states of large dye molecules. Therefore, in a first step the ground state geometry is calculated by DFT and then, without calculating the geometries of excited states, ΔE_{vert} is calculated as in the case of the semi-empirical PPP and all-valence-electron methods [43, 44]. It is emphasized again that ΔE_{vert} does not correspond to any experimental value [11, 45].

One important point is that the calculation of the vibrational states and their energy is very demanding. The vibrational frequencies ν_1^{vi} in S_1 are usually smaller than ν_0^{vi} in S_0. The typical difference is about 200 cm^{-1}. Therefore, the error introduced by neglecting the ZPVE, which involves replacing the 0–0 transition energy by the purely electronic component, is relatively small. This simplifying assumption enables eq. (31.10) to be reduced to eq. (31.12).

$$\Delta E_{00,\,10} \approx E_1^{\mathrm{el}}(Re) - E_0^{\mathrm{el}}(Re) \equiv \Delta E_{\mathrm{adiab}} \tag{31.12}$$

The energy difference between the minima of the electronic energy in S_0 $[E_0^{\mathrm{el}}(R_{\mathrm{e}})]$ and in S_1 $[E_1^{\mathrm{el}}(R_{\mathrm{e}})]$ is called adiabatic transition energy $\Delta E_{\mathrm{adiab}}$. Because calculations involving vibrational states are so computationally demanding that the purely electronic transition energy $\Delta E_{\mathrm{adiab}}$ is often calculated rather than $\Delta E_{00,\,10}$, the vibronic transition energy.

For the sake of completeness, it must be mentioned that electronic absorption spectra in solution usually exhibit solvent dependence (solvatochromism). To include solvatochromic effects in quantum chemical calculations models are used that are based on classical electrostatic solvent models, e. g., the polarizable continuum model. Here the solvent is approximated by a polarizable continuum, while the solute molecule is placed in a cavity [46]. These models, however, do not account for changes in the intensity ratio of vibronic sub-bands brought about by solvent effects and, therefore, are of limited value.

In summary, electronic absorption spectroscopy is the oldest form of spectroscopy and nowadays electronic absorption spectra of molecules in solution can be easily measured. However, implementation of the corresponding theory is very demanding and the quantum chemical calculations of the vibronic transitions of a molecule in solution require very complex procedures. The advent of relatively cheap computing power in the early 1980s led to the proliferation of semi-empirical methods like PPP and all-valence electron procedures. However, all these methods calculate ΔE_{vert}. "First-principles" methods without any empirical parameters like e. g. TDDFT enabled further advances. But independent from these advances mostly ΔE_{vert} is calculated with TDDFT, too. So, despite considerable efforts in TDDFT, the accurate calculation of electronic absorption spectra remains a challenging task that has yet to be addressed by an approach rooted only in theory.

31.4 Calculation of the intensities of electronic transition in molecules

Usually electronic transitions in atoms lead to single narrow absorption lines. In molecules the electronic transitions are accompanied by simultaneous rotational and vibrational transitions. Therefore, the electronic spectra of molecules do not consist of a number of single lines. Due to rotation and vibration of the molecules the spectra are broad in general. So it is not so easy to calculate the absolute intensity of electronic transitions in molecular electronic spectra.

An absorption spectrum in solution is described by a function of ε in $dm^3\ mol^{-1}\ cm^{-1}$, in dependence on $\tilde{\nu}$ in cm^{-1}. However, there is no direct relationship with the intensity maximum of the absorption band ε_{max} and a theoretical term. This basic problem was solved by Robert S. Mulliken [47], who considered the experimental absolute intensity of an absorption band I^{exp} (eq. (31.2)).

In the theory of electronic transitions the determining term is the electric dipole contribution. When the Bohr resonance condition is satisfied, the probability of the transition from an initial molecular state $\psi_a(r, R)$ to a final state $\psi_b(r, R)$ depends on the electric transition dipole moment between the two states $\boldsymbol{M}_{ab}$:

$$\mathbf{M}_{ab} = e \int \psi_b(r,R) \left[-\sum_i r_i + \sum_I z_I R_I \right] \psi_a(r,R)\ \mathrm{d}r\mathrm{d}R \tag{31.13}$$

Here $-e\,\Sigma_i\, r_i = \mu(r)$ is the sum of dipole moments of each electron, r_i the coordinates of the ith electron and $e\,\Sigma_I\, z_I\, R_I = \mu(R)$ is the sum of dipole moments of each nucleus, z_I the nuclear charge and R_I the coordinates of the Ith nucleus.

Like Condon, Mulliken neglected rotation of molecules and used also Condon's approach that electronic and nuclear motions are separable (eq. (31.5)).

Considering an electronic transition from the vibrational state v of the ground electronic state S_O to the vth vibrational state of the first excited electronic state S_1 the electric transition dipole moment is given as

$$\boldsymbol{M}_{\mathrm{Ov}-1\mathrm{v}} = \iint \psi_1(r,R)\chi_{1\mathrm{v}}(R)[\mu(r)+\mu(R)]\psi_O(r,R)\chi_{O\mathrm{v}}(R)\,\mathrm{d}r\mathrm{d}R \tag{31.14}$$

where $\psi_O(r, R)$ is the electronic wave-function for S_O, $\psi_1(r, R)$ for S_1, $\chi_{Ov}(R)$ is the nuclear vibrational wave-function for S_O and $\chi_{1v}(R)$ for S_1. The electric transition dipole moment of a molecule is a transient dipole moment which connects two different molecular states. Note that it is not related to the permanent electric dipole moment of one molecular state (see Appendix)!

The integral of eq. (31.14) can be split off and factorized into a product of the electronic transition dipole moment $\boldsymbol{M}_{01} = \int\psi_1(r, R)\,[\mu(r)]\,\psi_O(r, R)\,\mathrm{d}r$ from S_O to S_1 and the overlap integral $\boldsymbol{S}_{OO,\,1v} = \int\chi_{1v}(R)\,\chi_{Ov}(R)\,\mathrm{d}R$ between the wavefunctions $\chi_{Ov}(R)$ and $\chi_{1v}(R)$ that are involved in the transition:

$$\boldsymbol{M}_{\mathrm{Ov}-1\mathrm{v}} = \int \psi_1(r,R)[\mu(r)]\psi_O(r,R)\mathrm{d}r \int \chi_{1\mathrm{v}}(R)\chi_{O\mathrm{v}}(R)\mathrm{d}R \tag{31.15}$$

The intensity of an electronic transition is proportional to the square of the electronic transition dipole moment $|\boldsymbol{M}_{01}|^2$. This theoretical value does not describe the absorption band area. Only Mulliken transformed this quantity to the dimensionless oscillator strength f

$$f_{01} = [(8\pi^2\cdot m_e\cdot c)/(3h\cdot e^2)]\cdot\tilde{\nu}_{01}|\boldsymbol{M}_{01}|^2 = 4.702\cdot 10^{-7}\cdot\tilde{\nu}_{01}|\boldsymbol{M}_{01}|^2. \tag{31.16}$$

where m_e is the electron rest mass and c the speed of the electromagnetic radiation in vacuum.

The oscillator strength depends on the electronic transition dipole moment and thus on the length of the conjugated system and the transition energy. The calculated oscillator strength is related to the absorption band area, however not in a direct way.

To compare it with experimental values the experimentally oscillator strength f^{exp} is obtained by integrating the area under the absorption band [11, 47].

$$f^{\mathrm{exp}} = \ln 10\,[(m_e\cdot c^2/\pi\cdot N_A\cdot e^2)]10^3\int\varepsilon\,\mathrm{d}\tilde{\nu} = 4.315\cdot 10^{-9}\int\varepsilon\,\mathrm{d}\tilde{\nu} \tag{31.17}$$

For absorption bands near a Gaussian function $f^{\exp}$ can be approximately calculated by the simple relationship:

$$f^{\exp} \approx 4.315 \cdot 10^{-9}\left[\mathrm{moldm^{-3}cm^{2}}\right] \cdot \varepsilon_{\max}\left[\mathrm{dm^{3}mol^{-1}cm^{-1}}\right] \cdot \Delta\tilde{\nu}_{1/2}\left[\mathrm{cm^{-1}}\right] \tag{31.18}$$

Vice versa, inversion of eq. (31.17) enables the determination of ε from calculated oscillator strengths. Usually Gaussian functions are used to calculate the individual band shapes. However, as mentioned earlier, an electronic transition between two electronic states of molecules is accompanied with simultaneous transitions between rotational and vibrational states. Therefore, the shape of an absorption band due to an electronic transition is determined by the distribution of sub-bands caused by rotational and vibrational transitions, which cannot be described by a symmetrical Gaussian function.

In summary, to simulate a single absorption band or an entire electronic spectrum the oscillator strength for an individual electronic transition is calculated and in a second step the shape is calculated by inversion of eq. (31.18) using Gaussian functions, which cannot describe the fine structure of an absorption band.

31.5 Calculation of the shape of absorption bands in molecular electronic spectra

As described above, the intensity of an electronic transition is proportional to the square of the transition dipole moment $|\boldsymbol{M}_{01}|^2$. However, in molecules the whole molecule rotates, the nuclei vibrate and the transitions between electronic states are accompanied by simultaneous transitions between rotational and vibrational states. Therefore, the shape of an experimental absorption band of a molecule is not a strict Gaussian function, because it includes rotational and vibrational sub-bands due to the transitions between rotational and vibrational states. Usually the rotational fine structure of spectra in solution cannot be resolved and the rotational sub-bands afford only $\Delta\tilde{\nu}_{1/2}$ of each vibrational sub-band. Therefore, the shape of an absorption band is determined by the spacing of the vibrational sub-bands and by their intensity distribution.

Due to these influences and to handle this complex matter, Condon neglected rotational states and considered only vibrational states. The intensity distribution of the vibronic transitions can be described by the so-called Frank–Condon principle (FCP), resulting in eq. (31.15) [9–12, 15].

The square of the overlap integral $\boldsymbol{S}_{00,\ 1v}$ between the wavefunctions χ_{0v} (R) and χ_{1v} (R) that are involved in the transitions in eq. (31.15), determines intensity distribution of the single vibronic transitions

$$(\mathbf{S}_{0\mathrm{v},\,1\mathrm{v}})^2 = \left| \int \chi_{1\mathrm{v}}(R)\chi_{0\mathrm{v}}(R)\mathrm{d}R \right|^2 = I_{0\mathrm{v}-1\mathrm{v}} \tag{31.19}$$

However, solving eq. (31.19) is not easy. Numerous suggestions have been proposed for the calculation of the overlap integrals. The first approach for the evaluation of Franck–Condon overlap integrals of diatomic molecules was given by Elmer Hutchisson [48]. Sharp and Rosenstocke extended Hutchisson's method of generating functions to polyatomic molecules [49]. Nowadays in commercial available quantum chemical program packages the coherent-state method of Doktorov et al. is intensively used [50, 51]. This method can be used only with the help of software packages.

Even then, Hutchisson realized that already the simple harmonic oscillator model leads to complex formalism [48]. To simplify it furthermore, Carl J. Ballhausen introduced the following rigorous approximations [52, 53].

(6) At typical room temperatures, i. e. ~ 290–300K, the vibrational ground state of S_0 is by far the most populated. Only a very small proportion of the molecules will be in anything other than their vibrational ground state. In this case of negligible population of vibrational excited states in S_0, the vibronic transitions are considered to take place only from v = 0 in S_0 to different vibrational states in S_1.

(7) Differences in ν in S_0 and S_1 are usually small so the vibrational frequencies in S_0 and S_1 are taken to be equal. A consequence of this assumption is that the shapes of both PES in S_0 and S_1 are the same [$k(S_0) = k(S_1)$; $k = 4\pi^2 m_r \nu^2$ – force constant of the harmonic oscillator; m_r – reduced mass; ν – frequency of the vibration of a diatomic molecule].

(8) The transition of an electron from S_0 to S_1 leads to an increase of the equilibrium bond length $R_e(S_1)$ in comparison with that of $R_e(S_0)$. Therefore, the two electronic states are described by PES of the same shape but whose minima are in different locations [$R_e(S_0) < R_e(S_1)$] – the *displaced harmonic oscillator approximation.*

Making these approximations Ballhausen introduced the so-called coupling strength S:

$$S = \tfrac{1}{2}\, k/h\nu\, [Re(S_1) - Re(S_0)]^2 \tag{31.20}$$

Due to approximation (7) ν is the vibrational frequency of the symmetric harmonic valence vibration of a diatomic molecule in S_0 and in S_1. With the approximations (6) to (8) it is necessary only to consider changes in equilibrium bond length [$R_e(S_1) - R_e(S_0)$] for both electronic states in order to predict the intensity distribution of the vibronic sub-bands in a given electronic transition. It enabled Ballhausen to reduce eq. (31.19) to eq. (31.21) for the calculation of the relative intensities of vibronic transitions between the vibrational ground state (v = 0) in S_0 and vibrational states v in S_1.

$$(S_{00,1v})^2 = I_{0-v} = e^{-S}S^{v}/v! = I_{0-0}S^{v}/v! \tag{31.21}$$

Equation (31.21) makes possible the calculation of the normalized intensity $I_{0\text{-}v}$ (transition probability) for a molecule from v = 0 in S_0 toward the vth vibrational state in S_1 if the coupling strength S is known. Conversely, eq. (31.21) permits determination of S from the experimental ratio of $I_{0\text{-}v}/I_{0\text{-}0}$, i. e. derived from intensities of sub-bands appearing in the fine structure of spectra where one assumes an assignment of sub-bands to particular vibronic transitions.

The calculation of overlap integrals in a diatomic molecule is a challenge and needs rigorous approximations. Compared with a diatomic molecule the calculation of multidimensional overlap integrals in polyatomic molecules is much more demanding. It is therefore worth examining the applicability of the diatomic FCP on polyatomic molecules.

The basic approach for polyatomic molecules was developed by Gerhard Herzberg and Edward Teller [54]. In a diatomic molecule, the only possible vibration is the Raman active symmetric valence vibration, which does not change the symmetry of the molecule. Herzberg and Teller deduced that in polyatomic molecules only a totally symmetric vibrational wave-function will have a non-zero overlap integral with the totally symmetric wave-function of the vibrational ground state (v = 0). The overlap integral with any non-totally symmetric vibrational wave-function must be zero. Due to this theory within an intense (symmetry allowed) electronic transition in a polyatomic molecule, vibrations can appear as a progression only if the symmetry is the same in the excited electronic state as in the ground state; i. e. it can couple only with totally symmetric vibrations, which change the size but not the symmetry of the molecule, just as is the case for a diatomic molecule. Therefore, only Raman-active vibrations can couple with a symmetry allowed electronic transition [53, 54] (see also Appendix).

So, the number of normal vibrations in a polyatomic molecule to be considered is drastically reduced. In aiming to apply the diatomic FCP to polyatomic molecules it is a prerequisite that the intensity distribution of the sub-bands must be largely determined by one single generalized parameter in the form of one dominant, totally symmetric vibration, which changes the bond length only. In cyanine dyes ($\tilde{\nu} \approx 1{,}200\,\text{cm}^{-1}$) [55] and in polyenes ($\tilde{\nu} \approx 1{,}450\,\text{cm}^{-1}$) [56, 57], this vibration is the symmetric carbon–carbon valence vibration of the polymethine chain [11, 53]. In this way, it is possible to apply the diatomic FCP to polyatomic molecules.

In addition to the rotational sub-bands caused by rotation of the whole molecule, in polyatomic molecules symmetric torsional vibrations around the carbon–carbon bonds of the polymethine chain afford $\Delta\tilde{\nu}_{1/2}$ of the dominant vibronic sub-bands but not their distribution.

Due to the simple model [eq. (31.21)], the distribution of the vibronic sub-bands depends only on the difference $\Delta R_e = R_e(S_1) - R_e(S_0)$. The S_0–S_1 transition can be ascribed roughly to a one-electron transition from the highest occupied MO (HOMO) to the lowest unoccupied MO (LUMO). The HOMO has bonding and the LUMO antibonding character. Therefore, an electron in the antibonding LUMO leads to an increase of $R_e(S_1)$ in comparison with $R_e(S_O)$. The values for ΔR_e depend on the electronic structure of the considered dye.

To illustrate the performance of this simple model and the influence of the electronic structure on the vibrational fine structure, the spectra of three dyes will be discussed. For it a value of $\frac{1}{2}\ k/hv = 0.004\ \text{pm}^{-2}$ is assumed and only different values for ΔR_e are used.

Symmetrical cyanine dyes show a low bond length alternation, that is, symmetric torsional vibrations around the carbon–carbon bonds of the polymethine chain are decreased. Therefore, the electronic spectra of symmetrical cyanine dyes exhibit a clear cut vibronic fine structure. Second point, in symmetrical cyanine dyes the change of the molecular geometry is relatively low [$R_e(S_1) \gtrapprox R_e(S_O)$] and the absorption intensity is largely concentrated in the 0–0 vibronic transition (Figure 31.4, Table 31.1).

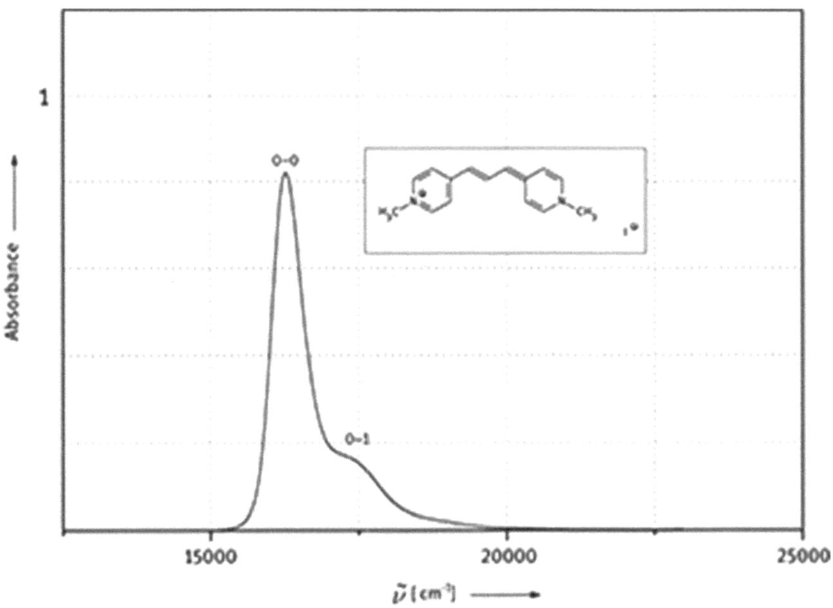

Figure 31.4: Structure of 1,1′-dimethyl-pyridinium trimethine cyanine and absorption spectrum in CH_2Cl_2 with the relative intensity $I_{0\text{-}0} = 0.82$ of the absorption maximum.

The chromophoric system of the perylenetetracarboxylic acid diimide is a rigidified polycyclic system and the possibilities of torsional vibrations are limited. Therefore, also the electronic spectrum in Figure 31.5 exhibits a clear cut fine structure. However, this system shows aromatic character and the change of the molecular geometry in S_1 is larger [$R_e(S_1) > R_e(S_O)$] in comparison with cyanine dyes and, consequently, the electronic transition intensity is spread over more and higher members of the vibrational

Table 31.1: Calculated relative intensities $I_{0\text{-}v}$ (cal) based on eq. (31.21) with the assumption $\frac{1}{2}\,k/hv = 0.004\ \mathrm{pm}^{-2}$ in eq. (31.20) and an assumed value $\Delta R_e = R_e(S_1) - R_e(S_0) = 7$ pm. The experimental values $I_{0\text{-}v}$ (exp) were deduced setting up $I_{0\text{-}v}$ (cal) of the absorption maximum as basis value.

v	$I_{0\text{-}v}$ (cal)	$I_{0\text{-}v}$ (exp)
0	0.82	0.82
1	0.16	0.17
2	0,02	0.02

progression. Here again, it remains λ_{max} still corresponds to the 0–0 vibronic transition (Figure 31.5, Table 31.2).

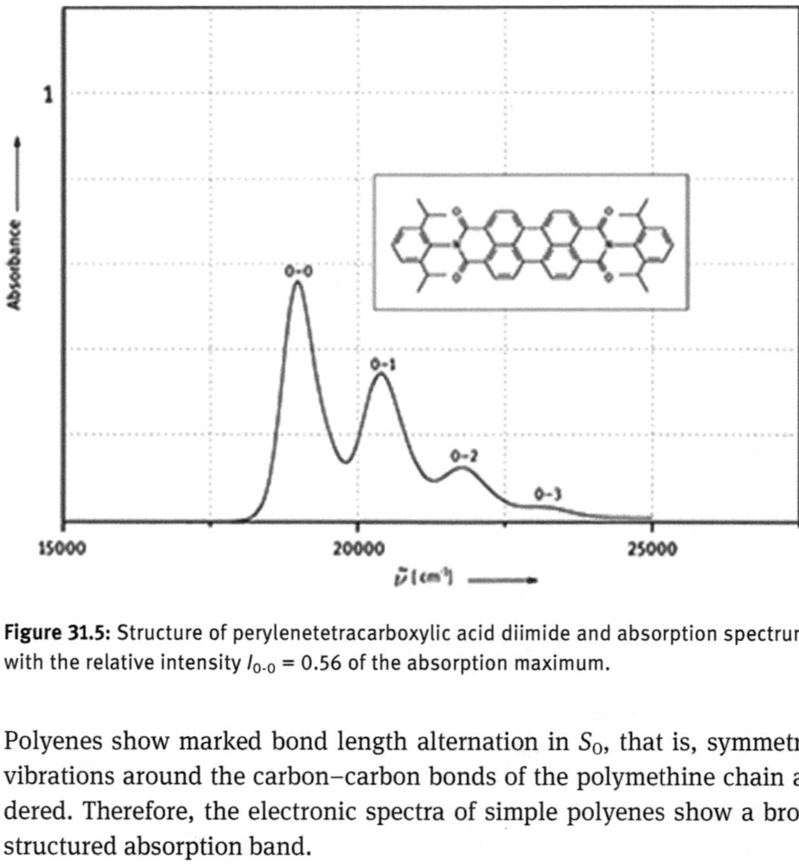

Figure 31.5: Structure of perylenetetracarboxylic acid diimide and absorption spectrum in CH_2Cl_2 with the relative intensity $I_{0\text{-}0} = 0.56$ of the absorption maximum.

Polyenes show marked bond length alternation in S_0, that is, symmetric torsional vibrations around the carbon–carbon bonds of the polymethine chain are less hindered. Therefore, the electronic spectra of simple polyenes show a broad and less structured absorption band.

In stilbene the vibrational structure is ascribed to the symmetric valence vibration of the central ethylene bond. However, the vibrational fine structure of the stilbene spectrum is blurred, which is caused by Ph–C torsional vibrations. Changing from a fluid solution to a glassy solution sharpens the sub-band due to the decrease of the

Table 31.2: Calculated relative intensities $I_{0\text{-}v}$ (cal) based on eq. (31.21) with the assumption $\frac{1}{2}\,k/hv = 0.004\ \text{pm}^{-2}$ in eq. (31.20) and an assumed value $\Delta R_e = R_e(S_1) - R_e(S_0) = 12$ pm. The experimental values $I_{0\text{-}v}$ (exp) were deduced setting up $I_{0\text{-}v}$ (cal) of the absorption maximum as basis value.

v	$I_{0\text{-}v}$ (cal)	$I_{0\text{-}v}$ (exp)
0	0.56	0.56
1	0.32	0.34
2	0.09	0.12
3	0.02	0.03

Ph–C torsional vibrations. In 5,6,11,12-tetrahydrochrysene the torsional vibrations around the Ph–C bonds are reduced due to the alkylene chains. Therefore, the electronic spectrum in Figure 31.6 exhibits a clear cut fine structure. It can be considered as a rigidified stilbene, and is an illustrative sample for polyenes.

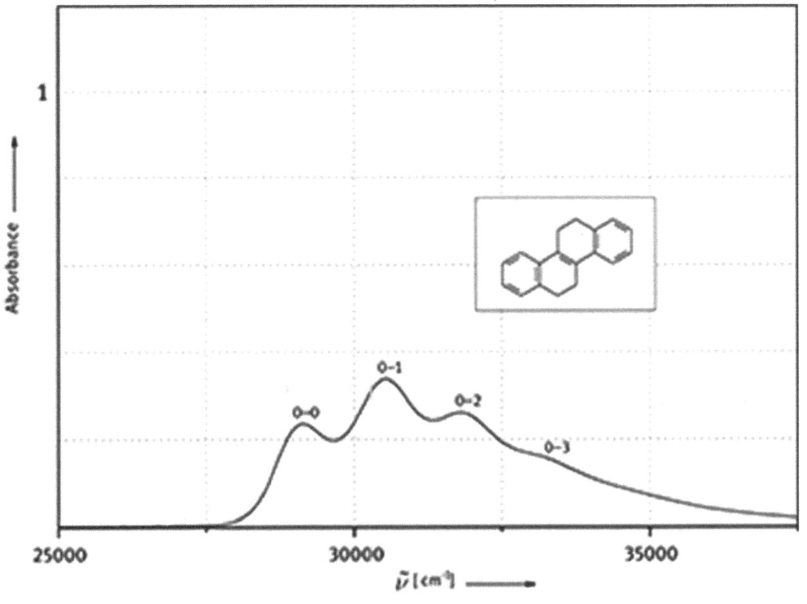

Figure 31.6: Structure of 5,6,11,12-tetrahydrochrysene and absorption spectrum in CH_2Cl_2 with the relative intensity $I_{0\text{-}1} = 0.34$ of the absorption maximum.

On transition of an electron from HOMO to LUMO the change of the molecular geometry in S_1 is much larger $[R_e(S_1) \gg R_e(S_0)]$ in comparison with cyanine dyes. Therefore, the electronic transition intensity is spread over more and higher members of the vibrational progression (Figure 31.6, Table 31.3) and λ_{max} is represented by the 0–1 vibronic transition.

Table 31.3: Calculated relative intensities $I_{0\text{-}v}$ (cal) based on eq. (31.21) with the assumption ½ k/hv = 0.004 pm^{-2} in eq. (31.20) and an assumed value $\Delta R_e = R_e(S_1) - R_e(S_0)$ = 19 pm. The experimental values $I_{0\text{-}v}$ (exp) were deduced setting up $I_{0\text{-}v}$ (cal) of the absorption maximum as basis value.

v	$I_{0\text{-}v}$ (cal)	$I_{0\text{-}v}$ (exp)
0	0.24	0.24
1	0.34	0.34
2	0.25	0.26
3	0.12	0.15
4	0.04	0.07
5	0.01	0.03

In summary, armed with the model of the harmonic oscillator and Ballhausen's rigorous approximations it is possible to get a good description of the influence of the difference of the equilibrium bond length $R_e(S_1)$ and $R_e(S_0)$ on the vibrational fine structure of electronic absorption bands.

References

1. Huges HK. Suggested nomenclature in applied spectroscopy. Anal Chem. 1952;24:1349.
2. Pauling L, Wilson EB. Introduction to quantum mechanics. New York: McGraw-Hill, 1935:25–49.
3. de Broglie L. Recherches sur la théorie des Quanta. Ann Phys (Paris). 1925;10:22.
4. Schrödinger E. Quantisierung als Eigenwertproblem. Ann Phys (Berlin). 1926;384:361.
5. Schrödinger E. Quantisierung als Eigenwertproblem. Ann Phys (Berlin). 1926;384:489.
6. Schrödinger E. Quantisierung als Eigenwertproblem. Ann Phys (Berlin). 1926;385:437.
7. Schrödinger E. Quantisierung als Eigenwertproblem. Ann Phys (Berlin). 1926;386:109.
8. Franck J. Elementary processes of photochemical reactions. Trans Faraday Soc. 1925;21:536.
9. Condon E. A theory of intensity distribution in band systems. Phys Rev. 1926;28:1182.
10. Mustroph H. Potential-energy surfaces, the born–oppenheimer approximations and the Franck–condon principle: Back to the roots. ChemPhysChem. 2016;17:2616.
11. Mustroph H, Ernst S, Senns B, Towns AD. Molecular electronic spectroscopy: from often neglected fundamental principles to limitations of state-of-the-art computational methods. Color Technol. 2015;131:9.
12. Born M. Zur Quantenmechanik der Stoßvorgänge. Z Phys. 1926;37:863.
13. Heisenberg W. Über den anschaulichen Inhalt der quantentheoretischen Kinematik und Mechanik. Z Phys. 1927;43:172.
14. Condon EU. Coupling of electronic and nuclear motions in diatomic molecules. Proc NAS. 1927;13:462.
15. Born M, Oppenheimer R. Oppenheimer. Zur Quantentheorie der Molekeln. Ann Phys (Berlin). 1927;389:457.

16. Condon EU. Nuclear motions associated with electron transitions in diatomic molecules. Phys Rev. 1928;32:858.
17. Jaffé HH, Orchin M. Theory and applications of ultraviolet spectroscopy. New York: Wiley, 1962.
18. Murrell JN. The theory of the electronic spectra of organic molecules. London: Metheuen, 1963.
19. Sandorfy C. Electronic spectra and quantum chemistry. Englewood Cliffs: Prencice-Hall, 1964.
20. Scott AI. Interpretation of the ultraviolet spectra of natural products. Oxford: Pergamon, 1964.
21. Suzuki H. Electronic absorption spectra and geometry of organic molecules. New York: Academic Press, 1967.
22. Orchin M, Jaffé HH. Symmetry, orbitals and spectra. New York: Wiley, 1971.
23. Griffiths J. Colour and constitution of organic molecules. London: Academic Press, 1976.
24. Fabian J, Hartmann H. Light absorption of organic colorants. Berlin: Springer, 1980.
25. Gordon PF, Gregory P. Organic chemistry in colour. Berlin: Springer, 1983.
26. Banwell CN. Fundamentals of molecular spectroscopy, 3rd ed. Maidenhead: McGraw-Hill, 1983.
27. Hückel E. Quantentheoretische Beiträge zum Benzolproblem. Z Physik. 1931;70:204.
28. Streitwieser A. Molecular orbital theory for organic chemists. New York: Wiley, 1961.
29. Parr RG. Quantum theory of molecular electronic structure. New York: Benjamin, 1963.
30. Klessinger M. Pariser-Parr-Pople-Rechnungen an Molekülen mit Aminogruppen. II. Cyanine, Merocyanine und Quadrupol-merocyanine. Theoret Chim Acta. 1966;5:251.
31. Mustroph H, Epperlein J. MO-Berechnungen der Absorptionsmaxima verschiedener Monoazofarbstoffe mit modifiziertem Parametersatz. Z Phys Chem (Leipzig). 1980;261:158.
32. Kogo Y, Kikuchi H, Matsuoka M, Kitao T. Colour and constitution of anthraquinonoid dyes. Part I – Modified PPP calculations and substituent effects. J Soc Dyers Colour. 1980;96:475.
33. Kogo Y, Kikuchi H, Matsuoka M, Kitao T. Colour and constitution of anthraquinonoid dyes. Part II – Modified PPP calculations and annelation effects. J Soc Dyers Colour. 1980;96:526.
34. Griffiths J. Practical aspects of colour prediction of organic dye molecules. Dyes & Pigm. 1982;3:211.
35. Matsuoka M, Takagi K, Obayashi H, Wakasugi K, Kitao T. Colour and constitution of 1,4-Naphthoquinonoid dyes – Modified PPP MO calculations and substituent effects. J Soc Dyers Colour. 1983;99:257.
36. Mustroph H, Reiner K, Senns B. Bond length alternation in unsymmetrical cyanine dyes and its influence on the vibrational structure of their electronic absorption spectra. Color Technol. 2017;133:469.
37. Del Bene J, Jaffé HH. Use of the CNDO method in spectroscopy. I. Benzene, Pyridine, and the Diazines. J Chem Phys. 1807;48:1968.
38. Ridley JE, Zerner MC. An intermediate neglect of differential overlap technique for spectroscopy: pyrrole and the azines. Theoret Chim Acta. 1973;32:111.
39. Adachi M, Nakamura S. Comparison of the INDO/S and the CNDO/S method for the absorption wavelength calculation of organic dyes. Dyes & Pigm. 1991;17:287.
40. Hohenberg P, Kohn W. Inhomogeneous electron gas. Phys Rev. 1964;136:B864.
41. Kohn W, Sham LJ. Self-consistent equations including exchange and correlation effects. Phys Rev. 1965;140:A1133.
42. Koch W, Holthausen MC. A Chemist's guide to density functional theory. 2nd ed. Weinheim: Wiley-VCH, 2001.

43. Fabian J. TDDFT-calculations of Vis/NIR absorbing compounds. Dyes & Pigm. 2010;84:36.
44. Laurent AD, Adamo C, Jacquemin D. Dye chemistry with time-dependent density functional theory. Phys Chem Chem Phys. 2014;16:14334.
45. Grimme S. Calculation of the electronic spectra of large molecules. Rev Comput Chem. 2004;20:153.
46. Mennucci B, Caprasecca S, Guido CA. Chapter four – computational studies of environmental effects and their interplay with experiment. Adv Phys Org Chem. 2016;50:203.
47. Mulliken RS. Intensities of electronic transitions in molecular spectra. J Chem Phys. 1939;7:14.
48. Hutchisson E. Band spectra intensities for symmetrical diatomic molecules. Phys Rev. 1930;36:410.
49. Sharp TE, Rosenstocke HM. Franck–condon factors for polyatomic molecules. J Chem Phys. 1964;41:3453.
50. Doktorov EV, Malkin IA, Man'ko VI. Dynamical symmetry of vibronic transitions in polyatomic molecules and the Franck–condon principle. J Mol Spectrosc. 1975;56:1.
51. Doktorov EV, Malkin IA, Man'ko VI. Dynamical symmetry of vibronic transitions in polyatomic molecules and the Franck–condon principle. J Mol Spectrosc. 1977;64:302.
52. Ballhausen CJ. A vibrational analysis of the MnO_4^- Bands. Theoret Chim Acta. 1963;1:285.
53. Mustroph H, Towns A. Fine structure in electronic spectra of cyanine dyes: are sub-bands largely determined by a dominant vibration or a collection of singly excited vibrations? ChemPhysChem. 2018;19:1016.
54. Herzberg G, Teller E. Schwingungsstruktur der Elektronenübergänge bei mehratomigen Molekülen. Z Phys Chem B. 1933;21:410.
55. Pouradier J. Remarque sur le Spectre D'Absorption des Cyanines en Solution. J Chim Phys. 1964;61:1107.
56. Sondheimer F, Ben-Efraim DA, Wolovsky R. Unsatured macrocyclic compounds. XVII. The prototropic rearrangement of linear 1,5-Enynes to conjugated polyenes. The synthesis of a series of Vinylogs of Butadien. J Am Chem Soc. 1675;83:1961.
57. Spangler CW, Little DA. Synthesis and characterization of representative Octa-1,3,5,7-tetraenes and Deca-1,3,5,7,9-pentaenes. J Chem Soc Perkin. 1982;1:2379.

A Appendix

In the chapter "Calculation of the Shape of Absorption Bands in Molecular Electronic Spectra" the applicability of the diatomic FCP to polyatomic molecules was discussed. In their basic approach for polyatomic molecules Herzberg and Teller have shown that vibrations can appear as a progression only if the symmetry is the same in the excited electronic state as in the ground state; i. e. an intense symmetry allowed electronic transition can couple only with symmetric vibrations, which change the size but not the symmetry of the molecule, just as is the case for the symmetric valence vibration in a diatomic molecule.

This lays the initial foundation anti-symmetric vibrations cannot couple with an electric dipole allowed electronic transition. In addition it is fundamental knowledge the anti-symmetric vibrations are IR active and not Raman active.

In the cases of homonuclear as well as in heteronuclear diatomic molecules the symmetric valence vibration changes the polarizability, so vibrations in both molecule types are Raman active. This point is also clear.

In many text books one can read the following or similar selection rules for IR active transitions:

"The condition for a normal vibration j to be IR active is a change in molecular dipole moment μ during vibration."[A1]

"The dipole moment μ is zero for a homonuclear diatomic molecule, resulting in $\boldsymbol{R}_\text{v} = 0$ and all vibrational transitions being forbidden. For a heteronuclear diatomic molecule μ is non-zero and varies with x."[A2]

"The gross selection rule for a change in vibrational state brought about by absorption or emission of radiation is that *the electric dipole moment of the molecule must change when the atoms are displaced relative to one another*. Such vibrations are said to be **infrared active**."[A3]

"In order for energy to be transferred from the IR photon to the molecule via absorption, the molecular vibration must cause a change in the dipole moment of the molecule. This is the familiar selection rule for IR spectroscopy, which requires a change in the dipole moment during the vibration to be IR active.

Homonuclear diatomic molecules such as H_2, N_2, and O_2 have no dipole moment and are IR inactive (but Raman active) while heteronuclear diatomic molecules such as HCl, NO, and CO do have dipole moments and have IR active vibrations."[A4]

Due to these rules the symmetric valence vibration in heteronuclear diatomic molecules shall be IR active because the change in molecular electric dipole moment during vibration Thus, often "seemingly suitable" IR vibration frequencies in polyatomic molecules are used to explain the vibrational fine structure in electronic spectra. However, is this justified?

Therefore, it is important to clarify, is the symmetric valence vibration in diatomic molecules IR and/or Raman active?

From the experimental point of view all these statements are invalid. Neither in the high-resolution rotation-vibration absorption spectra of HCl nor CO or other heteronuclear diatomic molecules is there an IR active normal vibration. This message is especially surprising, because the IR spectrum of HCl is acquired and analyzed already in undergraduate physical chemistry laboratories, yet in the experimental spectra there is no hint of a vibration absorption band [A5, A6].

Due to the cited vibrational selection rule, the transition $v = 0 \rightarrow v' = 1$ shall be IR active. But where is the absorption band of this vibrational transition?

In molecules there are transitions between separate rotational, vibrational and electronic states and the lower energy transitions can be associated with the excitation of higher energy transitions. The rotation-vibration absorption spectrum of HCl shows two branches, the *R* ($v = 0 \rightarrow v' = 1$; $\Delta J = +1$) and *P* ($v = 0 \rightarrow v' = 1$; $\Delta J = -1$) branches of the rotational transitions, whereas the absorption band for the vibrational transition $v = 0 \rightarrow v' = 1$ is absent (Table 4). It is expected that the frequency of the "missing vibrational absorption band" $v = 0 \rightarrow v' = 1$ is between the $R(0)$ ($v = 0, J = 0 \rightarrow v' = 1, J' = 1$) and $P(1)$ ($v = 0, J = 1 \rightarrow v' = 1, J' = 0$) rotational transitions (v - vibrational quantum number, *J* - rotational quantum number).

Table 4: Some transition wave-numbers of the rotation-vibration absorption spectrum of $H^{35}Cl$ [A5].

v	*J*	v′	*J*′	$\tilde{\nu}$ [cm^{-1}]
0	2	1	3	2,944.58
0	1	1	2	2,925.58
0	0	1	1	2,905.99
0	1	1	0	2,864.83
0	2	1	1	2,843.31
0	3	1	2	2,821.25

The spacing between rotational absorption bands varies slowly as a function of the vibrational states. Assuming constant spacing between the $R(0)$ and the $P(1)$ transitions (Table 4) the midpoint has a value of 2,885 cm^{-1}. This midpoint frequency corresponds with the frequency of the "missing vibrational absorption band" $v = 0 \rightarrow v' = 1$, which is obviously a non IR active vibration.

The rotation-vibration scattering spectrum (Raman spectrum) of HCl was first time measured by R. W. Wood [A7]. He determined the frequency of the "missing vibrational absorption band" to be 3.4645 μm (2,886 cm^{-1}) [A7].

So, since 1929 it has been known that the frequency of the "missing vibrational absorption band" in the rotation-vibration absorption spectrum of HCl can be observed in rotation-vibration scattering spectrum.

Also in the rotation-vibration absorption spectrum of CO the band of the fundamental CO vibrational transition is absent (Table 5) [A8].

Table 5: Some transition wave-numbers of the rotation-vibration absorption spectrum of $^{12}C^{16}O$ [A8].

v	J	v′	J′	$\tilde{v}$ [cm^{-1}]
0	2	1	3	2,154.44
0	1	1	2	2,150.83
0	0	1	1	2,147.05
0	1	1	0	2,139.32
0	2	1	1	2,135.48
0	3	1	2	2,131.49

The calculated midpoint between the $R(0)$ and the $P(1)$ transitions (Table 5) has a value of 2,143 cm^{-1}. It corresponds with the CO Raman frequency of 2,143 cm^{-1} [A9].

In summary, in the rotation-vibration absorption spectra of both heteronuclear diatomic molecules the v = 0 → v′= 1 vibrational transition cannot be detected, that is, the vibration is non IR active. However, the expected frequency of the "missing vibrational absorption band" (estimated by the midpoint between the $R(0)$ and the $P(1)$ transitions) can be observed in the Raman spectra.

So, back to the roots of the theory: To calculate the intensity of vibrational transitions one has to evaluate the vibrational transition dipole moment $\boldsymbol{M}_{v\text{-}v'}$ for the vibrational transition v → v′

$$\boldsymbol{M}_{v-v} = \int \chi_v(R)\ [\mu(R)]\chi_v\ (R)\ \mathrm{d}R \qquad (30.22)$$

where $\mu(R)$ is the molecular dipole moment operator, $\chi_v\ (R)$ and $\chi_{v'}\ (R)$ are the nuclear vibrational wave-functions in the vibrational states v and v′ and R are the nuclear coordinates.

The molecular dipole moment is a function of R and can be expanded using a Taylor series expansion about the equilibrium nuclear coordinates R_e of the initial vibrational state,

$$\mu(R) = \mu(Re) + (\partial\mu(R)/\partial R)_{R=Re} \cdot R \qquad (30.23)$$

where $\mu(R_e)$ is the permanent electric dipole moment. With it the vibrational transition dipole moment is given as

$$\boldsymbol{M}_{v-v'} = \mu(R_e) \int \chi_{v'}(R)\chi_v(R)\ \mathrm{d}R + (\partial\mu(R)/\partial R)_{R=Re} \int \chi_{v'}(R)[R]\chi_v(R)\ \mathrm{d}R \qquad (30.24)$$

Because of the orthogonality of the vibrational wave-functions for v = v′ the integral in the first term is 1, resulting in the permanent electric dipole moment of the molecule $\mu(R_e)$ in the vibrational state v.

The function of R is of *ungerade* (u) [odd] parity, whereas the nuclear vibrational wave-function of the vibrational ground state $\chi_v(R)$ is of *gerade* (g) [even] parity. Hence the second integral will be ≠ 0 only, if the parity of $\chi_{v'}(R)$ is u. In general, the parity of both vibrational wave-functions must be different that the integral will be ≠ 0. In the case v = v′ the parity of both vibrational wave-functions is either g or u and the second integral is zero.

For v ≠ v′ the integral in the first term is zero, because of the orthogonality of the vibrational wave-functions resulting in eq. (30.25).

$$\boldsymbol{M}_{v-v'} = (\partial\mu(R)/\partial R)_{R=Re} \int \chi_{v'}(R)[R]\chi_v(R)\,\mathrm{d}R \tag{30.25}$$

With the harmonic oscillator functions the *parity selection rule* Δv = ± 1, ± 3, . . . derives from the second integral.

Therefore, the intensity of a parity allowed vibrational transition is expressed through the first partial derivation of $\mu(R)$ and the correct selection rule for a normal vibration to be IR active reads as follows: "During a normal vibration IR radiation can be absorbed only if the condition for the first partial derivation of the molecular electric dipole moment $\mu(R)$ is fulfilled."

$$[(\partial\mu(R)/\partial(R)_{R=Re}]^2 > 0 \tag{30.26}$$

To calculate $\boldsymbol{M}_{v\text{-}v'}$ an accurate electric dipole moment function of symmetric valence bond vibrations in the ground electronic state is required. There are some quantum chemical program packages, but one cannot trust all their results, especially for the v = 0 → v′= 1 transitions.

A possible reason may be a wrong choice of the used electric dipole moment function. As an example of a proofed function for the description of diatomic vibration is the semi-empirical exponential Mecke-type function [A10],

$$\mu(R) = \mu(R_e)[R/R_e]^n \exp(-\alpha[R - R_e]) \tag{30.27}$$

where n and α are empirical parameters to fit the available experimental data best.

Already $\mu(R_e) = 0$ for homonuclear diatomic molecules leads to $\mu(R) = 0$. However, what happens for $\mu(R_e) > 0$?

The exponential Mecke-type function generally exhibits asymptotic behavior as $R \to 0$ and $R \to \infty$. Most important, the function reaches its maximum at $R = R_e$, that is,

$$(\partial\mu(R)/\partial R)_{R=Re} = 0$$

With this consideration the theory supports for all diatomic molecules, that the symmetric valence vibration is non IR active.

In summary, for the correlation with experimental vibration frequencies the key question was, is the symmetric valence vibration in diatomic molecules IR and/or

Raman active? As shown by experiment the symmetric valence vibration in diatomic molecules is non IR active irrespective of whether they possess a homonuclear or heteronuclear molecular structure. The exponential Mecke-type function – a nearly accurate electric dipole moment function for the vibration in diatomic molecules – reaches its maximum at R_e and therefore the first partial derivation of $\mu(R)$ at $R = R_e$ is zero. This result supports, the symmetric valence vibration in diatomic molecule can be only Raman active and non IR active.

References

A1. M. Hof in *Handbook of Spectroscopy*, Vol. 1, G. Gauglitz, T. Vo-Dinh (Eds.), Wiley-VCH, Weinheim, 2003, p. 39.

A2. J. M. Hollas, *Modern Spectroscopy*, 4th Ed., Wiley, Chichester, 2004, p. 138.

A3. P. Atkins, J. de Paula, *Physical Chemistry*, 9th Ed. University Press, Oxford, 2010, p. 812.

A4. P. J. Larkin, *IR and Raman Spectroscopy – Principles and Spectral Interpretation*, Elsevier, Amsterdam, 2011, p. 14.

A5. R. W. Schwenz, W. F. Polik. Analysis of the Infrared Spectra of Diatomic Molecules. *J. Chem. Educ.* **76** (1999) 1302.

A6. H. J. Park, N. M. Kurien, T. R. Rybolt. Simple Acid Vapor Method for Production of HCl and DCl Gas for IR Spectroscopy. *J. Chem. Educ.* **95** (2018) 2293.

A7. R. W. Wood. The Raman effect in gases. – Part I. HCl and NH_3. *Phil. Mag.* **7** (1929) 744.

A8. R. T. Lagemann, A. H. Nielsen, F. P. Dickey. The Infra-Red Spectrum and Molecular Constants of $C^{12}O^{16}$ and $C^{13}O^{16}$. *Phys. Rev.* **72** (1947) 284.

A9. T. Yoshino, H. J. Bernstein. Intensity in the Raman effect: VI. The photoelectrically recorded Raman spectra of some gases. J. *Mol. Struct.* **2** (1958) 213.

A10. R. Mecke. Dipolmoment und chemische Bindung. *Z. Elektrochem.* **54** (1950) 38.

Gerhard Pfaff

32 Effect pigments

Abstract: Metal effect pigments and special effect pigments are the two classes of effect pigments. Effect pigments, also termed luster pigments, consist of µm-sized thin platelets that show strong lustrous effects when oriented in parallel alignment in application systems, such as coatings, plastics, printing inks, and cosmetic formulations. Metal effect pigments show a metal-like luster after parallel orientation in the application medium, which has its origin in reflection of light at the surface of the platelet-like metal particles. Special effect pigments, on the other hand, consist of transparent flakes with a high refractive index. Oriented in a parallel way, most of them show a characteristic pearl luster generated by multiple reflections. Many of the special effect pigments are classified as interference pigments because their color generating mechanism is based completely or predominantly on the phenomenon of interference.

Keywords: interference pigments, metal effect pigments, pearl luster pigments, special effect pigments

Effect pigments are divided in the two classes metal effect pigments (metallic effect pigments) and special effect pigments including pearl luster pigments (pearlescent pigments, nacreous pigments) and interference pigments. All effect pigments consist of µm-sized thin platelets that show strong lustrous effects when oriented in parallel alignment in application systems, such as coatings, plastics, printing inks, and cosmetic formulations.

Effect pigments are a specialty among the colorants that provide additional color effects, such as angular color dependence (luster, iridesdence, color travel) or texture, when applied in an application medium [1, 2]. The term “luster pigments” is also used for effect pigments because almost all effect pigments are characterized by lustrous effects in their applications. Effect pigments consist of predominantly platelet-like particles. These particles align readily with a parallel orientation to the surface to which they are applied. This collective orientation of a plurality of plate- let-like particles leads to characteristic luster effects, which are generated by the reflection of incident light from the smooth surface of the pigment platelets.

In former times, the terms “metal effect pigments” and “pearl luster pigments” were mostly used for the classification of effect pigments respectively luster pigments. The term “metal effect pigments” describes the main effects of this pigment class still completely. The term “pearl luster pigments”, on the other hand, describes only in part the broad variety of luster pigments existing nowadays without the me-

https://doi.org/10.1515/9783110587104-032

tallic character of metal effect pigments. Therefore, the more comprising term "special effect pigments" has become more commonly used for this pigment class.

Metal effect pigments consist of metallic flakes. They show a metal-like luster after parallel orientation in the application system, which has its origin in reflection of light at the surface of the platelet-like metal particles. Special effect pigments, on the other hand, consist of transparent flakes with a high refractive index. Oriented in a parallel way, most of them show a characteristic pearl luster generated by mul- tiple reflection. Many pearl luster respectively nacreous pigments show interference colors. These types are also referred to as interference pigments. Interference pig- ments are special effect pigments whose color generating mechanism is based completely or predominantly on the phenomenon of interference [1, 2].

Several effect pigments consisting also of transparent platelets and following the optical phenomenon of multiple reflection by light interaction do not show the characteristic pearl luster. Other interference pigments are based on non-transpar- ent platelets. These types also do not exhibit pearlescent effects. All these varieties have broadened the range of effect pigments in the last decades and require a more complex classification. The implementation of the term "special effect pigments" is a result of this growing complexity. It comprises all platelet-like effect pigments, which do not belong to the metal effect pigments. Taking this into account, special effect pigments are pearl luster pigments as well as transparent and non-transpar- ent interference pigments, which lead in dependence on their composition and structure in the application medium to pearl lustrous or non-pearl lustrous effects in combination with interference phenomena [1, 2].

Special effect pigments are in the vast majority inorganic pigments, distinguish- ing themselves by high luster, brilliance and iridescent color phenomena based on optically thin films. Specialties among the special effect pigments are platelets of monocrystalline natural fish silver, some flaky organic pigments, such as diketo- diaryl-pyrrolopyrroles, dichloro-chinacridones, and metal phthalocyanines as well as pigments based on comminuted liquid crystal polymer films [1, 3, 4]. The appear- ance of special effect pigments in the application can be explained by reflection and deflection of the light at thin single and multiple layers. Similar effects are found in nature, for example in pearls, clamshells, birds, fishes, gemstones, miner- als, or insects. The investigation of the optical principles of the natural pearl luster shows that the brilliant colors can be attributed to structured biopolymers and lay- ered structures, which are formed by bio-mineralization [1–7].

Figure 32.1 shows a comparison of the optical principles for the interaction of light with special effect pigments (pearl luster pigments, interference pigments) and metal effect pigments as well as with absorption pigments (colored and black) and white pigments. The particles of white, colored and black pigments are mostly irregularly formed. White pigment particles interact with visible light mainly by diffuse reflection of the incident light in all directions (scattering) without absorption. Colored pigment particles reflect one part of the light diffusely and absorb wavelength-sensitive another part of the light (selective absorption). The particles of black pigments absorb all wavelengths of the visible spectral range (complete absorption).

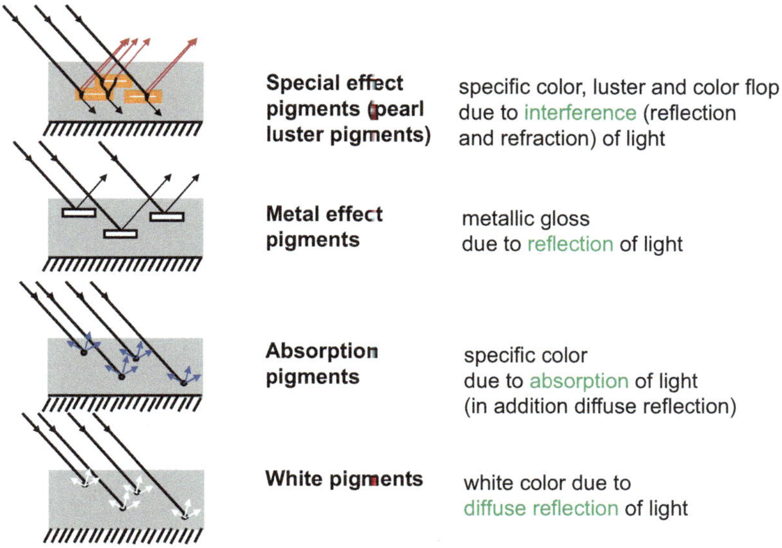

Figure 32.1: Optical interaction of visible light with particles of different pigment types in an application system, e. g., in a coating [2].

The typical particle sizes of white, colored and black pigments are in the range from 0.1 to 1 µm. Particle sizes of about 0.4 to 0.7 µm are in the range of the wavelengths of the visible light. In contrast, effect pigments have particle sizes predominantly in the range of 5 to 100 µm. Accordingly, effect pigments are with regard to their average platelet diameter significantly larger than the wavelengths of visible light. The thickness of the pigment platelets is, however, with values below 1 µm, in the order of the thickness of optical interference layers (exception: pigments based on liquid crystal polymers). The aspect ratio (ratio of diameter to thickness) of effect pigments is therefore with values of up to 200 very high in most cases.

The particles of metal effect pigments are nontransparent for light and reflect the entire incident light in one direction. They act similar to small mirrors and lead when orientated parallel in the application system to a reflecting metal luster (metallic effect). Special effect pigments, such as pearl luster pigments and transparent interference pigments, reflect only one part of the incident light, while a second part of the light enters into the transparent particles. This second part of the light is partly reflected at the interfaces in the inside of the pigments or at the bottom side of the platelets. A remaining third part of the light leaves the platelets on the underside and can interact with pigment particles located underneath to create further reflections (multiple reflection). The reflected parts of the light generate interference phenomena due to optical superimposition. This leads together with the multiple reflection to the appearance of a luster coming from the depths, comparable with pearl luster. The difference in the refractive indices between the high-refractive materials of special effect pigments (typically 1.8 to 2.9) and the application medium (typically 1.5 to 1.6) is decisive for the optical interactions and consequently for the appearance of the pigmented system.

Table 32.1 presents an overview of effect pigments. The pigment types are divided in metallic platelets, oxide coated metallic platelets, coated mica platelets (instead of mica other platelets such as silica, alumina, or borosilicate can be used), plate- let-like single crystals, comminuted thin PVD-films, and comminuted films of liquid crystal polymers.

Table 32.1: Overview of effect pigments.

Pigment type	**Examples**
Metallic platelets	Al, Zn/Cu, Cu, Ni, Au, Ag, Fe (steel), C (graphite)
Oxide coated metallic platelets	Surface oxidized Cu-, Zn/Cu-platelets, Fe_2O_3-coated Al-platelets
Coated mica plateletsa	non-absorbing coating: TiO_2 (rutile), TiO_2 (anatase), ZrO_2, SnO_2, SiO_2 selectively absorbing coating: FeOOH, Fe_2O_3, Cr_2O_3, TiO_{2-x}, TiO_xN_y, $CrPO_4$, KFe[Fe (CN)$_6$], organic colorants totally absorbing coating: Fe_3O_4, TiO, TiN, $FeTiO_3$, C, Ag, Au, Fe, Mo, Cr, W
Platelet-like single crystals	BiOCl, $Pb(OH)_2 \cdot 2\ PbCO_3$, α-Fe_2O_3, α-Fe_2O_3 x $nSiO_2$, $Al_xFe_{2-x}O_3$, $MnyFe_{2-y}O_3$, $Al_xMn_yFe_{2-x-y}O_3$, Fe_3O_4, Cu(II) phthalocyanine
Comminuted thin PVD-films Comminuted films of liquid crystal polymers	Al, Cr/MgF_2/Al/MgF_2/Cr polysiloxanes

[a]Instead of mica other platelets such as silica, alumina, or borosilicate can be used

Special effect pigments and metal effect pigments are described in detail in own chapters.

References

1. Pfaff G. In special effect pigments. Pfaff G, editor. 2nd ed. Hannover: Vincentz Verlag, 2008:16.
2. Pfaff G. Inorganic pigments. Berlin/Boston: Walter de Gruyter GmbH, 2017:205.
3. Pfaff G, Reynders P. Angle-dependent optical effects deriving from submicron structures of films and pigments. Chem Rev. 1999;99:1963.
4. Maile FJ, Pfaff G, Reynders P. Effect pigments – past, present and future. Progr Org Coat. 2005;54:150.
5. Pfaff G. Perlglanzpigmente. Chem Unserer Zeit. 1997;31:6.
6. Snow MR, Pring A, Self P, Losic D, Shapter J. The origin of the color of pearls in iridescence from nano-composite structures of the nacre. Amer Miner. 2004;89:1353.
7. Katti KS, Katti DR. Why is nacre so tough and strong? Mater Sci Eng C. 2006;26:1317.

Heinz Langhals

33 Fluorescence and fluorescent dyes

Abstract: The handling and control of light is becoming more and more attractive in science and technology such as data processing and requires functional chromophores. As a consequence, fluorescent materials are of special importance because they allow the processing of light energy. Thus, basics of fluorescence are reported as prerequisites for planning complex functional structures. Various fluorescent systems are presented beginning with historic observations followed by a detailed discussion of light absorption and emission indicating fluorescent chromophores as molecular resonators; molecular dynamics and intermolecular interactions are leading to complex functional materials.

Keywords: fluorescence, chromophores, optical functional materials, light emission, molecular dynamics

33.1 Introduction

Fluorescence, the spontaneous light emission of irradiated materials was empirically found. Linen weavers in the seventeenth century noticed brilliant purely white shades of garments after treatment with extracts of horse chestnuts where fluorescence caused optical bleaching. The background of this effect was not clear at that time. The pharmacist W. Raab [1] described special optical effects of such extracts where the active compound Aesculin (RN 531-75-9) was further investigated by F. Rochleder and R. Schwarz [2] and reported by Berzelius [3]. However, there was still minor interest in such organic materials. More attractive for chemists were shiny minerals such as uranium ores. Further progress brought about the development of new and systematic methods in organic chemistry in the middle of the nineteenth century by chemists such as Adolf von Baeyer. He developed Fluorescein (see below) as a synthetic organic material forming impressively fluorescent aqueous solutions in daylight. There were only rare applications of fluorescent materials in the beginning [4] such as tracer experiment for the detection of streams of ground-water; the photostabiliy of the first optical whitening agents was low restricting their general application. Moreover, the majority of organic compounds are not fluorescent so that a search for such special classes of compounds delayed the further developments where the mostly limited photostability was an important obstacle. Meanwhile, the chemistry and physics of fluorescent dyes are well developed with numerous applications in an increasing field such as optical whitening agents for paper,

This article has previously been published in the journal Physical Sciences Reviews. Please cite as H. Langhals Fluorescence and Fluorescent Dyes *Physical Sciences Reviews* [Online] 2020, 5. DOI: 10.1515/psr-2019-0100

https://doi.org/10.1515/9783110587104-033

textiles and various other white or bright materials where the impressive optical effect of fluorescence is not only of interest for marking inks, but also for safety wear and signage, but also for security printing and leak proofing. The temporary storage of light energy forms the basis for dye lasers and the spontaneous light emission is important for OLEDs. The green fluorescent protein brought about an appreciable progress in biochemistry. Applications and further developments of fluorescent materials are favored by the knowledge of the interaction of light with matter. Thus, the interaction of light with matter obtains an increasing importance both in science and technology. Visible light is electromagnetic radiation with frequencies in the range between 0.4 and 0.8 PHz; see Figure 33.1. Thus, the molecular handling of light may be named petahertz technology [5].

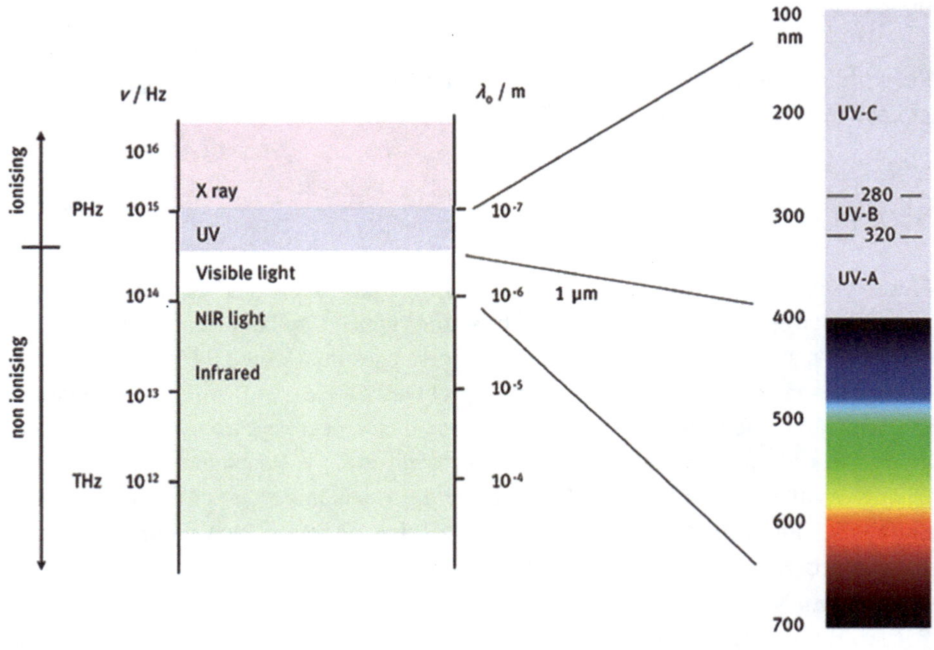

Figure 33.1: Significant regions of electromagnetic radiation for fluorescence. Left scale: Frequency, right scale vacuum wavelengths.

Visible light is usually characterized by the vacuum wavelength λ_o in nm with the velocity of light c according to eq. (33.1) (the velocity of light in vacuo c_o for λ_o) because of historic measurements by means of optical grids; see Figure 33.1. A further common measure is the wavenumber $\tilde{\nu}$.

$$\nu = c/\lambda = c \cdot \tilde{\nu} \tag{33.1}$$

The energy E of radiation is not continuously absorbed by matter, but in small quanta according to Einstein's Formula (2) where h is Planck's constant, c is the velocity of light, ν is the frequency and λ is the wavelength.

$$E = h \cdot \nu = h \cdot c/\lambda \quad (33.2)$$

The absorbed energy E may induce various chemical processes; thus, the frequency ν in cps or the proportional wavenumber $\tilde{\nu}$ (in cm^{-1} known as Kayser K or the multiple unit kK) are more clear in chemistry where energetic processes are dominant. However, wavelengths λ are mostly reported where one has to keep in mind that they are the inverse of energy.

33.2 Elastic and inelastic interaction of light with matter

$$\mathbf{n} = n - i \cdot \kappa \quad (33.3)$$

A continuous model of the electromagnetic field describes sufficiently precisely the interaction of propagating electromagnetic waves with matter in macroscopic dimensions and comparably low frequencies such as for radio transmitters [6]. A propagating electromagnetic wave induces more or less damped oscillations where a complex index of refraction **n** in eq. (33.3) is an appropriate description. The real component n represents the contribution to the optical index of refraction and the imaginary κ a linear measure of the absorptivity.

Thus, n represent elastic interactions with matter without loss of energy and κ inelastic interactions with the absorption of the radiation. The dependence of the parameters n and κ on the frequency ν can be calculated by means of classical electrodynamics and results in Formula (4) where ν_o means a resonant frequency of matter and γ characterizes the aptitude for light absorption; a and b combine mathematical and natural constants and a proportionality factor concerning the dimension of ν.

$$\mathbf{n} = a \cdot \frac{\nu_o^2 - \nu^2}{\left(\nu_o^2 - \nu^2\right) + \gamma^2\nu^2} - i \cdot b \cdot \frac{\gamma\nu}{\left(\nu_o^2 - \nu^2\right) + \gamma^2\nu^2} \quad (33.4)$$

Multiple resonances ν_o require a sum of individual terms according to eq. (33.4).

The wavelengths of light of about 500 nm extends to macroscopic dimensions just below 1 µm [7]; the molecular components of matter are appreciably smaller and thus, can be considered as homogeneous media where eq. (33.4) describes the interaction with sufficient precision. The calculated spectral dependence of the contribution to the index of refraction n characterizing the elastic light scattering is reported in Figure 33.2 (dashed curves), top for a scale linear in energy and in Figure 33.2, bottom more familiar for linear in wavelengths λ. The index n exhibits maxima and minima in the region of resonance, however, changes monotonously off-resonance. Colorless materials absorb light in the UV below 400 nm, but are transparent in the visible. As a consequence, the index of refraction is high for short wavelengths such as for blue light (compare the right dashed branch in Figure 33.2, bottom) and decreased for longer wavelengths such as for red light; this is known as normal dispersion and is applied in dispersing optical components such as prisms. A completely different spectral dependence is observed if light is absorbed in the visible shown in Figure 33.2 (dashed curves) where there is a reverse of the index of refraction known as anomalous refraction; the gem tourmaline is the most prominent example for such an optical behavior. Such anomalous dispersion in the visible causes an unusual sequence of color for the refraction of white light. The appendant absorption spectrum calculated from κ resembles a Lorentz curve (Agnesi curve) and is shown in Figure 33.2, dotted curves. A comparably slim maximum of κ and appreciable tailing to shorter and longer wavelengths can be seen; the points of inflection of κ are close to the maxima and minima of the index of refraction *n* and the maximum close to the zero pass.

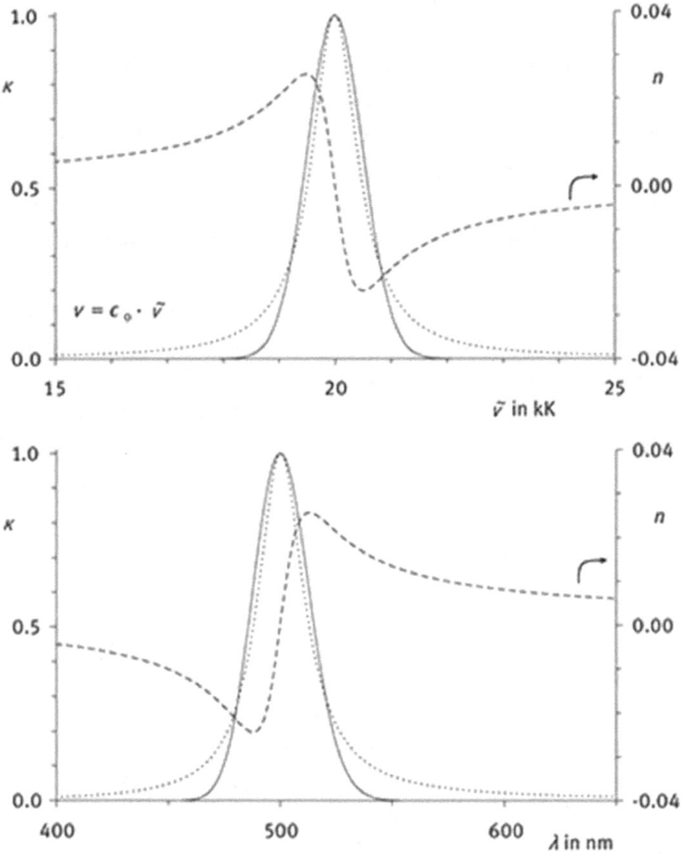

Figure 33.2: Normalized spectral dependence of the calculated absorptivity κ (dotted curves) and the contribution *n* to the index of refraction (dashed curves) compared with a Gaussian curve for absorption (solid curves). Top: Linear in wavenumber $\tilde{\nu}$ (proportional to the frequency ν) bottom: Linear in the wavelengths λ.

A homogeneous and isotropic medium would be required for the realization of a suitable continuous process; however, the resonators for visible light are local microscopic in their molecular dimensions and individually stochastically operating. The absorption (κ) is controlled by the transition probability being maximal at the resonance frequency and more and more lowered off resonance. This means a scattering in energy around the maximum depending on various other degrees of freedom such as by coupling with translation and libration vibrations in condensed phases, respectively. Such a distribution of multiple contribution of individually influenced stochastic processes results typically in a Gaussian dependence in energy and thus, in the electronic transition probability is best represented [8] in eq. (33.5), middle for multiple resonances n where $E_{(\nu,\lambda)}$ is the absorptivity proportional to κ depending of the frequency ν. $E_{\max(i)}$ are the maximal absorptivities of the individual Gaussian bands i, $\nu_{\max(i)}$ the positions and $\sigma_{(i)}$ the half widths. The inverse of the frequency ν must be applied for the wavelength λ, Formula (5), right [9], where the factor 100 simplifies the interconversion between the wavelengths in nm and the wavenumbers in cm^{-1}. (Other functions such as the Lorentzian [10] or log-normal function are less appropriate for the description of experimental spectra).

$$E_{(\nu,\lambda)} = \sum_{i=0}^{n} E_{\max(i)} e^{-\frac{\left(\nu-\nu_{\max(i)}\right)^2}{2\sigma_{(i)}^2}} = \sum_{i=0}^{n} E_{\max(i)} e^{-100\frac{\left(\frac{1}{\lambda}-\frac{1}{\lambda_{\max(i)}}\right)^2}{2\sigma_{(i)}^2}} \tag{33.5}$$

The Gaussian band of the absorptivity is symmetric if recorded linearly in energy such as the frequency ν and the wavenumber $\tilde{\nu}$, respectively; see Figure 33.2, top. It becomes asymmetric for a linear scale in wavelengths λ with a steeper edge to short and a less steep edge to long wavelengths; Figure 33.2, bottom. Gaussian curves (Figure 33.2, solid curves) are much faster, exponentially damped than Lorentzian curves (Figure 33.2, dotted curves) and cause for normal dispersion a spectral region where the index of refraction n is still comparably high, whereas the absorptivity κ is already damped to insignificantly small values; this is of special advantage for the generation of materials with high index of refraction for optical dispersing devices or light scattering. The latter is important for white pigment because strongly light scattering crystallites with micrometer dimensions (elastic light scattering) are required such as titanium dioxide (rutile) and zinc sulfide as well as organic pigments [11] such as the pteridines in wings of butterflies [12] or technical [13] organic white pigments. Their strong light absorption close to the visible region makes these substances also interesting as UV sun protectors [14].

The theoretical treatment of the interaction of light with matter with the resulting eq. (33.2) visualized in Figure 33.2 is experimentally verified with aqueous solutions of fluorescein (RN 2321-07-5, see below **2**) where the optical properties are reported in Figure 33.3. The absorption spectrum (solid curve) is exponentially damped at the bathochromic edge (highest energy) where a Gaussian analysis according to eq. (33.5) indicates several individual bands (bars). The composed spectrum on the basis of this analysis (dotted dashed curve) is nearly completely covered by the experimental spectrum and indicates the quality of the fit (actual to theoretical comparison).

$$n = a \cdot \frac{\nu_o^2 - \nu^2}{(\nu_o^2 - \nu^2) + \gamma^2\nu^2} + n_o = a \cdot \frac{1/\lambda_o^2 - 1/\lambda^2}{(1/\lambda_o^2 - 1/\lambda^2) + \gamma^2/\lambda^2} + n_o = a \cdot \frac{\lambda^2 - \lambda_o^2}{(\lambda^2 - \lambda_o^2) + \gamma^2\lambda_o^2} + n_o \tag{33.6}$$

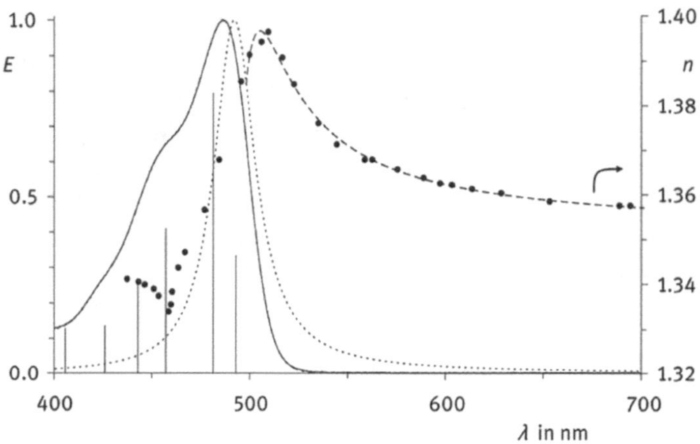

Figure 33.3: Normalized absorption spectrum of Fluorescein (RN 2321-07-5) in aqueous solution (solid curve, left scale *E*) and Gaussian analysis for the individual absorption bands (λ_{max} = 492.8 nm, 2 σ^2 = 0.239 kK, E_{max} = 0.333 and λ_{max} = 481.3 nm, 2 σ^2 = 0.669 kK, E_{max} = 0.794 for the most bathochromic) with bars for the positions and intensities and the simulated spectrum on the basis of the analysis: Dashed dotted curve, nearly completely covered by the experimental spectrum. Index of refraction *n* with the right scale; points: Experimental data according to ref [15].; dashed curve: Least square fit according to eq. (33.4), left (ν_o = 23.53 kK corresponds to 492 nm, γ = 2.076, a = 2.089, n_o = 1.347). Dotted curve: Calculated κ values (correspond to *E*) by means of the data obtained from *n*.

The experimental index of refraction [15] is shown in Figure 33.3 by points. The bathochromic branch above 500 nm is fitted to the parameters a, λ_o, γ and n_o of eq. (33.6) where n_o is the index of refraction of the medium and the other term the real component of eq. (33.4). n_o (1.347) is close to the applied medium water (1.332). The anomalous dispersion below 500 nm is complicated because of the overlay of several bands and was not further considered. The best fit for more than 500 nm is indicated by the dashed curve where the limited precision of the measurements have to be taken into account (some aggregation of the dye may have further influenced the measurements). The normal dispersion can be clearly seen where a monotonous decrease of the index of refraction is found above 510 nm. The obtained parameters λ_o and γ allow the construction of a hypothetical absorption band (dotted curve), if there would be a homogeneous macroscopic light absorption. This band is close to the first Gaussian band, however, tails much more into the bathochromic region than the experimental exponentially damped and is a further indicator for the individually stochastically operating molecular resonators. This is of importance for highly refractive materials with low residual light absorption.

33.3 Molecular resonators

Inelastic light scattering requires the uptake of light energy and thus, appropriate energetic eigenvalues of the involved matter. A partial absorption of light energy proceeds in Raman spectroscopy with some energy transfer to molecular vibrations where the residual spectrum is characteristic for the chemical structure. A complete absorption of light energy causes electronic excitation where a characteristic impression of color is generated for absorption in the visible.

A typical energetic sequence of discrete electronic energetic eigenvalues is schematically indicated in Figure 33.4, left. These can be filled with electrons according to the aufbau principle where each distinguishable level can be filled with maximal two electrons with antiparallel spins according to Hund's rule and the Pauli principle, respectively. There is an even number of electrons for the very most organic compounds and one energetically upmost occupied orbital (HOMO). As a consequence, all electron spins are paired and such molecules form a singlet state named S_0. The index 0 indicates the electronic ground state; see Figure 33.4, left. Accordingly, the corresponding first electronically exited state is S_1 and so on; see Figure 33.4, left. The restrictions by the Pauli principle are still no more given for the S_1 and higher states because different orbitals are involved in the energetic upmost electrons. As a consequence, a second electronically excited state can be realized with two parallel spins in the two upmost orbitals forming a triplet state because the added electron spins; accordingly, this is named the T_1 state. The antiparallel arranging of spin is unfavorable where the required energy of this paring depends on the chemical structure. As a consequence, the T_1 state is generally located more or less below the S_1 state. The electronic excitation is interlinked with vibronic eigenvalues and with molecular rotation in the gas phase with some line broadening by the Doppler effect. As a consequence, a comparably complicated system results. Thus, the electronic energies of individual occupied orbitals were summarized for simplification in Jablonski's diagram in Figure 33.4, right, resulting in the basic electronic levels S_0, S_1 and so on. The vibronic and rotational levels are indicated by larger and smaller lines in Figure 33.4, right. Rotation is hindered in the condensed phase resulting in libration vibrations. The latter are not as well defined as the energetic levels of rotation in the gas phase because of fluctuating interactions with nearest neighbors and coupling with translation motions. As a consequence, a further line broadening and a quasicontinuum of energetic levels results for the latter in the spectral region of rotation spectroscopy. Finally, the comparably strong coupling of vibration and libration vibrations with the levels of electronic excitation further increases the complexity; however, density maxima still remain around the initial levels of vibration. As a consequence, there is a densely packed ladder of energetic levels for vibration and libration vibrations where S_0 ground state forms the lower energetic limit. A similar sequence of energetic levels will be found for the S_1 state, preferently for complex molecules, because one chemical bond only is half loosen in S_1 compared with S_0 and this concerns comparably weak π-bonds for most dyes molecules (see below). Such sequences of energetic levels are also found for S_2 and higher levels and for T_1 and higher triplet states where density maxima and minima are less pronounced for the latter. Light absorption proceeds between energetic levels of these ladders as a fast process in about 10^{-15} s where a movement of the heavy nuclei is of minor importance (however, see ref [16, 17].) and the molecular geometry remains essentially unaltered ("vertical" processes); a subsequent relaxation into a local energetic minimum may proceed.

$$\frac{n_+}{n_-} = \frac{g_+}{g_-} \cdot e^{-\frac{\Delta E}{RT}} \tag{33.7}$$

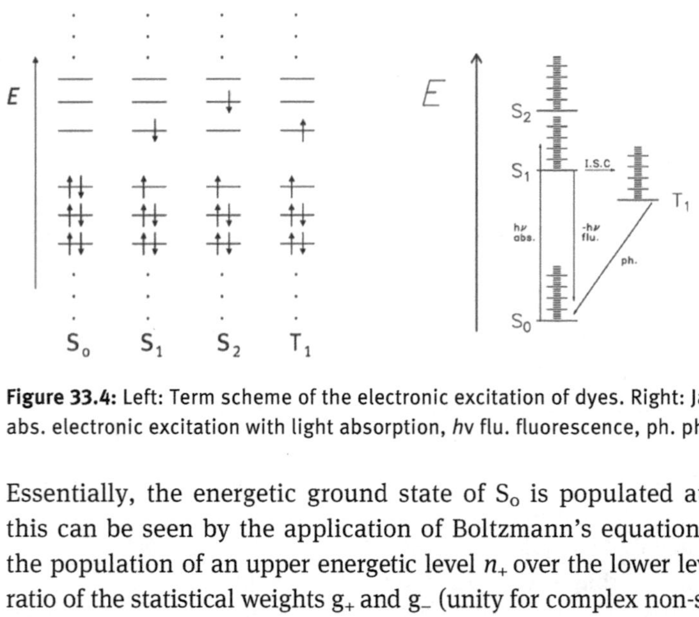

Figure 33.4: Left: Term scheme of the electronic excitation of dyes. Right: Jablonski's diagram: *hv* abs. electronic excitation with light absorption, *hv* flu. fluorescence, ph. phosphorescence.

Essentially, the energetic ground state of S_0 is populated at room temperature; this can be seen by the application of Boltzmann's equation (7) where generally the population of an upper energetic level n_+ over the lower level n_- is given by the ratio of the statistical weights g_+ and g_- (unity for complex non-symmetric molecules) multiplied with the exponential ratio of the energetic difference ΔE of the involved levels and the thermal energy RT. Thus, the thermal energy at room temperature (2.4 kJ·mol^{-1} at 20 °C) is not sufficient for a significant population of upper vibronic levels (about 18 kJ·mol^{-1} at 1500 cm^{-1}). As a consequence, the S_0 ground state can be taken as the initial state for electronic excitation at room temperature. The molecular multiplicity is not affected by the process of light absorption proceeding into electronically excited S states because of spin conservation. An excitation into T states would require a simultaneous inversion of an electron spin, is indicated as a spin forbidden process and thus, proceeds with a very low probability; some paramagnetic substances, for example molecular oxygen, may favor such transitions (compare the very weakly brownish color of highly pure air-saturated benzene). The light-induced electronic transition from the S_0 ground state may proceeds not only into the S_1 ground state (0-0-transition), but also into vibronically excited states; sharp individual electronic transitions were found in the gas phase (with the mentioned some line broadening by the Doppler effect), whereas strong line-broadening proceeds in the condensed phase with increasing molecular interactions where there are still maxima of transition probability close to the initial molecular vibronic levels. As a consequence, the ladder of states is projected to the UV/Vis absorption spectrum [18] where a bathochromic limit is found and a more or less structured continuation to shorter wavelengths (compare that the wavelength is the inverse to the energy of absorption; see eq. (33.1) above). The efficiency of molecular light absorption is characterized by the wavelength-dependent molar absorptivity interrelated with the oscillator strengths; the latter is important for quantum chemical calculations. The absorption spectrum linearly recorded in energy such as the frequency or wave number can be composed by a sum of individual Gaussian bands according to eq. (33.5) where the branch at longest wavelengths is exponentially damped [19] with the square of wavenumber; accordingly, the damping proceeds with the inverse square for the wavelength. Very weak absorptions at longer wavelengths sometimes observed with practical samples proved to be caused by low residual contents of by-products and could be removed by means of progressed chemical purification.

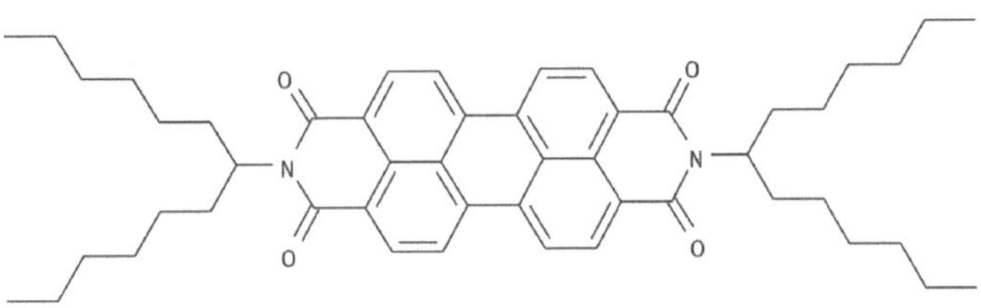

S-13 (**1**; CAS registry number RN 110590-84-6)

S-13 (**1**; CAS registry number RN 110590-84-6)

Optical excitation with even higher energy at shorter wavelengths allows reaching higher electronically states such as the S_2 and S_3 state. As a consequence, the structured absorption continues to shorter wavelengths where there may be intense absorption bands in the UV.

33.4 Basics of fluorescence

The vibronically excited S_1 state looses the vibronic excitation in a very short time in the order of 0.1 ps (10^{-13} s) to become thermal energy. The strong coupling of the vibronic modes causes the equilibration of the vibronic energy within one or a few periods so that the S_1 ground state is reached exhibiting a longer lifetime of some nanoseconds (10^{-9} s). A similarly fast relaxation of higher electronically excited states to the S_1 ground state is observed for nearly all organic molecules (azulene is a prominent exception [20]) so that this state is generally the origin for the further light-induced processes. A simple interconversion (I.C.) from the S_1 state to a highly vibronically excited S_o state may proceed and is followed by a fast relaxation to the ground state and the dissipation of the energy, finally to produce heat. This describes the function of the majority of textile dyes and technical pigments where the selective light absorption causes an impression of color. Some dyes are able to re-emit the absorbed energy with a process named fluorescence [21]. Starting with the S_1 ground state the ladder of energetic levels of the S_o state will be reached where the maximal energy will be obtained by the transitions between the vibronic ground states. As a consequence, there is a limit of the spectrum of light emission at high energy and short wavelengths, respectively (hypsochromic limit). Moreover, the energetic ladder of the S_o state is very similar to the ladder of the S_1 state in most cases for complex molecules because only one chemical bond is half loosened and this influences unimportantly the vibronic force constants (see above). As a consequence, the wavelengths-dependence of the fluorescence spectrum looks about like mirror-type of the absorption spectrum (known as the Stokes' mirror type spectra); this is demonstrated with the typically structured spectrum in Figure 33.5 of the for the perylene derivative [22] S-13 (dye **1**).

The process of fluorescence seems to be simply the reverse of the absorption at the first glance (Stokes' mirror image of absorption and fluorescence spectra). However, there are important differences. Firstly, the intensities of the sequence of vibronic bands of fluorescence (Figure 33.5, right spectrum with a blue curve) obviously become faster damped with increasing wavelengths than in the absorption spectrum to shorter wavelength (Figure 33.5, left spectrum with a magenta curve). This can be attributed to the mismatch of the wavelength of radiation with the dimension of the antenna. For example, the mean wavelengths of absorption and emission of **1** in Figure 33.4 is about 500 nm where suitable dimensions of antennae are $\lambda/2$ and thus, 250 nm (the index of refraction has to be considered for fine tuning). However, the chromophore of **1** extends over no more than 1 nm; this is the effective length between the two nitrogen atoms as the termini of the conjugated system. Thus, the mismatch is important for absorption and becomes slightly minor for the higher vibronic bands at shorter wavelengths. However, the mismatch becomes worse for higher vibronic bands in fluorescence because of longer wavelengths. This effect is described by Rüdenberg's formula [23] for classical treatment of electromagnetic radiation with radio antennae and in the same manner with the equivalent Ross' equation [24] obtained by quantum

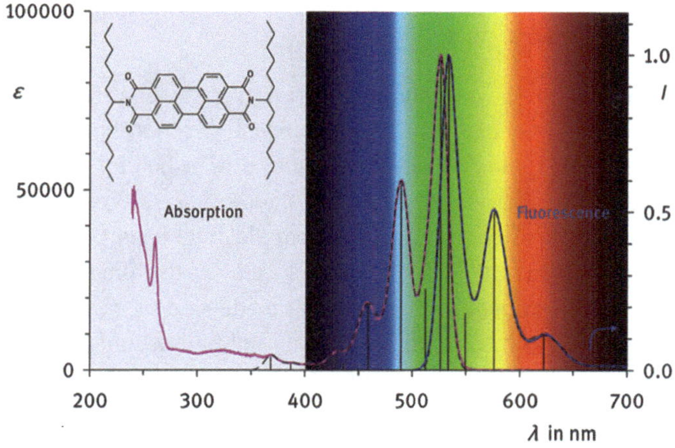

Figure 33.5: Absorption (left magenta curve) and fluorescence (right blue curve) spectra of the perylene dye **1** (S-13) in chloroform. Bars: Positions and intensities of the calculated individual Gaussian vibronic bands. Dotted curves: Simulated spectra on the basis of the Gaussian analysis (mainly covered by the experimental spectra). The fluorescence quantum yield of **1** is close to unity.

mechanical treatment. The 0–0 transition should be identical in absorption and fluorescence; however, one generally finds the fluorescence maximum slightly or even more pronounced situated at longer wavelengths known as the Stokes' shift. Relaxation processes of the excited state into a new geometry are therefore responsible because the optical excitation changes slightly the force constants of chemical bonds; this shift is small in the spectra of firm, rigid compounds such as for **1** in Figure 33.4. Larger changes in molecular geometry by relaxation may cause appreciable Stokes' shifts. Finally, the ratio of emitted light quanta over the number of absorbed is known as the fluorescence quantum yield Φ and can extend from zero to unity. There are few examples of dyes with fluorescence quantum yields larger than 0.7.

33.5 Optical excitation of fluorescent molecular structures

A light wave passing a diluted solution of well-separated independently operating dye molecules is not continuously damped such as radio waves from absorbing materials, but induces individual molecular electronic transitions. These individual processes are not determined; however, the transition probability remains constant. The very large number of dye molecules in practical samples averages the individual processes to a constant damping of the wave. The light-absorbing [25] is well described by Lambert Beer's law of eq. (33.8) where I_o is the intensity of the incoming monochromatic light and I the intensity after the absorbing light pass where the decadic logarithm (*lg*) is applied for historic convenience to obtain the absorptivity E.

$$lg\frac{I_0}{I}=E=\varepsilon\cdot c\cdot d \tag{33.8}$$

The dimensionless E is proportional to the concentration c and the path length d where the factor ε is known as the coefficient of extinction (the coefficient of absorption A may be more characteristic for the chromophore because ε becomes diminished for strongly fluorescent samples; however, this effect is very small for standard spectrometers because of the small efficient solid angle of the applied nearly parallel light beam). For convenience, d is measured in cm and c in mol/L and thus $L\cdot mol^{-1}\cdot cm^{-1}$ is the dimension of ε as the decadic molar coefficient of extinction. The wavelengths-dependent ε can range from small value to more the 100,000 in maxima of strongly light-absorbing chromophores. Thus, even weakly light-absorbing materials would efficiently absorb light if the path length is long enough; for example, even obviously clear, transparent natural water weakly absorbs light so that the grounds of lakes become dark for more than 20 m depths. The coefficient ε depends from the wavelength λ and is generally recorded as UV/Vis spectra. Maxima of ε are commonly applied for the characterization of dyes where the integral of the absorption band is a more appropriate measure for the oscillator strengths of the chromophore because it can be interrelated with quantum chemical methods; however, the determination is more difficult because of partial overlap with bands of higher excitation. On the other hand, the half widths of most absorption bands are similar. Thus, the absorptivity at the maximum is well suitable for a rough, global characterization of chromophores. Lambert-Beer's law is frequently applied for the optical determination (E) of chemical concentrations c with comparably high precision and comparably low experimental effort; however, one has to take care (i) to allow to operate the spectrometer in the appropriate range (about $0.3 < E < 1.0$), (ii) that there is no interaction of dye molecules such as aggregation (sufficient dilution), and (iii) to apply monochromatic radiation because eq. (33.8) is valid only for one wavelength, but not for a spectral range. Such errors become minimal for measurement in the maxima because of minimal change of ε in a small spectral rage around the maxima.

33.6 Fluorescence

The application of Jablonski's diagram (Figure 33.4, right) may transmit the impression that the absorption of a light pulse would cause an echo of fluorescence similar to the RADAR for radio waves, possibly with some delay caused by molecular processes; *however, this is definitively not the case.* The macroscopic electromagnetic light wave passes many individual chromophore with barriers for electronic excitation with constant transition probabilities where the individual processes are not predictable. In a very rough calculation a concentration $3.01\cdot10^{-6}$ mol/L of a strongly light-absorbing dye with ε = 100 000 is required for the absorption of 50% of the incoming light (50% transition probability) within 1 cm path of an optical standard cuvette. In a cylinder with the diameter of the wavelength of about 500 nm and 1 cm path about 4 million dye molecules are involved until a 50% transition probability for absorption is reached. Here, one can see that absorption will proceed if only the optical path is long enough.

The conditions for fluorescence are absolutely different because the energy of electronic excitation is localized to an individual molecular structure where a barrier to the evolution of a free electromagnetic wave has to be passed. This proceeds by tunneling [26] this barrier such as is also known for the radioactive decay. Indeed the fluorescence decay generally proceeds exponentially according to the same first-order-law with the time constant τ of some nanoseconds where the probability of the process is independent from the already elapsed time (compare the well-known radioactive decay; multi exponential decays are observed if more than one process is involved). The fluorescence of optically excited molecules has to compete with other processes of deactivation such as the internal conversion I.C. Competing processes for deactivation lower the fluorescence quantum yield Φ and the experimental time constant where the natural time constant τ_0 can be calculated by means of eq. (33.9).

$$\tau = \Phi \tau_0 \tag{33.9}$$

The two time constants τ and τ_0 become essentially equal for highly fluorescent chromophores ($\Phi \approx 1$) such as **1** where 4 ns were observed in chloroform. The time constant τ_0 depends on the transition probability from the S_1 ground state to the S_0 levels and is roughly inversely proportional to the molar absorptivity. This is quantitatively indicated by the Strickler–Berg-Equation (10) [27] initially established by Förster [28] where the inverse of the fluorescence lifetime τ_0 is equal to Einstein's transition probability $A_{u \to l}$ from the upper (u) to the lower (l) state. This is equal an expression of some constants such as Avogadro's number N_A, ln (10) for the adaption of the decadic absorptivity ε to SI units, the velocity of light c, the wavenumber $\tilde{\nu}_{u \to l}$ for the transition from u to l, n the index of refraction, g_l and g_u the statistical weight and the integral over the absorption band linear in wavenumber.

$$\frac{1}{\tau_0} = A_{u \to l} = 1000 \frac{8\pi \ln(10) c \tilde{\nu}_{u \to l}^2 n^2}{N_A} \frac{g_l}{g_u} \int \varepsilon d\tilde{\nu} \tag{33.10}$$

Equation (33.10) can be applied in many cases, preferment for series of similar compounds, however, there are exceptions. Short fluorescent lifetimes τ_0 are expected for strongly light-absorbing dyes with comparably broad absorption bands such as **1** where 4 ns were observed; however, the weaker light absorbing diketoterthiophenes unexpectedly exhibit an appreciably shorter fluorescence lifetime [29] of only 0.4 ns.

Moreover, the fluorescence lifetime τ_0 according to eq. (33.10) is estimated to be a molecular property and thus, independent from the chemical concentration of highly diluted solutions of isolated molecules. In contrast to the theory, a fluorescence lifetime of 5 ns is found for an already diluted $2 \cdot 10^{-5}$ molar solution where the lifetime is diminished to 3.8 ns by further dilution to $1 \cdot 10^{-7}$ molar [30]; see Figure 33.6. This unexpected result may be interpreted in terms of evanescent waves [31] and the tunnelling where the evanescent waves were reflected by the other identical not electronically excited chromophores thus, retarding the tunnelling. This new far-reaching effect over more than 100 nm basically allows molecular addressing by macroscopic structures. Finally, fluorescence lifetime and the influencing by the molecular surrounding are obtaining increasing importance for imaging [32] in biology and medicine (FLIM) because of detailed microscopic information.

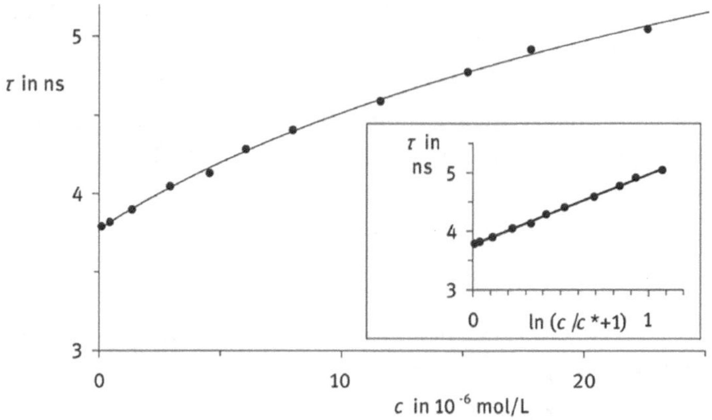

Figure 33.6: The fluorescence lifetime τ of **1** in chloroform as a function of concentration *c* in mol/L. Inset: Linearization of the curved plot with $c^* = 1.17 \cdot 10^{-6}$ mol/L.

33.7 Prerequisites for strong fluorescence

To obtain highly fluorescent materials, natural light emission characterized by the natural lifetime τ_0 has to compete with other processes of deactivation such as internal conversion (I.C.) to highly vibronically excited electronic ground states. Such competing processes are efficient for the majority of organic compounds, whereas low molar absorptivities diminish the transition probability for spontaneous light emission; as a consequence, organic materials, preferentially dyes, are generally non fluorescent ($\Phi < 0.001$) and comparably few basic chemical structure with high fluorescence quantum yields are known. Special chemical and photo physical requirements can favor fluorescence such as high molar absorptivities increasing the transition probabilities, and rigid chemical structures for lowering possibilities for radiationless decays by means of internal conversion according to the loose bolt mechanism defined by Lewis and Calvin [33]. There are only few strongly fluorescent materials. Fluorescein [34] (**2**, RN 2321-07-5), developed by Adolf von Baeyer, is one of the most prominent examples, is intensely green fluorescent in aqueous solution; uranine is a further name for **2** and its di-sodium salt (RN 518-47-8), respectively, because the yellowish color in absorption of aqueous solutions and the yellowish green fluorescence resemble the color of solutions of uranylium compounds and uranium glasses, respectively, the most prominent fluorescent materials at the time of the discovery of **2**.

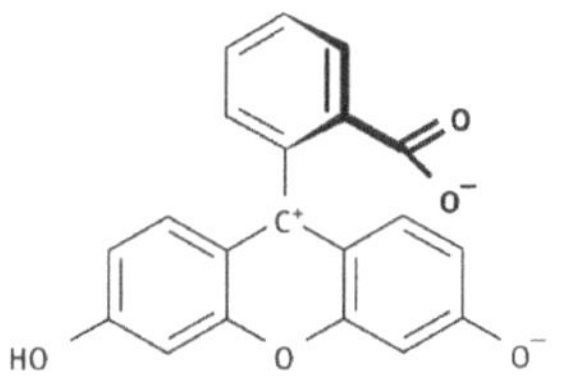

2

Basic structural prerequisites for strong fluorescence can be rationalized with **2**. The arrangement of two donor-substituted (–OH and $-O^-$) phenyl groups with a formal carbenium ion in between as an acceptor allows a high molar absorptivity and a comparably bathochromic absorption in the middle of the visible region according to the principle of König and Ismialski [35]. The linking of the two phenyl units by means of an oxygen atom forming the rigid heterocycle xanthene restricts the intramolecular mobility of the chromophore; degrees of freedom coupling to the system of electronic excitation are most important concerning fluorescence quenching, whereas peripheral alkyl groups such as rotatable methyl groups are of minor influence. Further restriction of internal rotation in **2** causes the carboxylate group because of steric interactions with the neighboring hydrogen atoms in the *peri*-positiones of the xanthene. The limitation to light chemical elements in **2** (carbon, hydrogen and oxygen) is favorable for intense fluorescence because heavy elements favor the inter system crossing (I.S.C.) to the triplet as a competing and deactivating process: A molecule of light elements form essentially isolated spin and orbital systems with weak coupling, whereas heavy elements support spin orbital coupling (Russell-Sounders coupling) [36, 37] and favor I.S.C. Moreover, the orthogonal arrangement of the carboxyphenyl group disfavors aggregation as a possible further source of fluorescence quenching. Finally, the negative charge lowers the tendency of aggregation in the aqueous phase by electrostatic repulsion. As a consequence, a fluorescence quantum yield of more than 90% is observed aqueous solutions of **2**.

The UV/Vis spectra of **2** are recorded in Figure 33.7 with the left magenta curve for the light absorption. This mainly extends to the spectral range of blue light and generates a yellow impression because of the complementary color. The emitted green fluorescent light (right blue curve) extends around the sensitivity maximum of the human eye and generates an impressive color effect with a brilliant shade because of the small half widths of the emission spectrum.

The photostability of **2** is comparably low. The donor groups –OH and $-O^-$ favor oxidative degradation. Their exchange to electrically neutral nitrogen-based

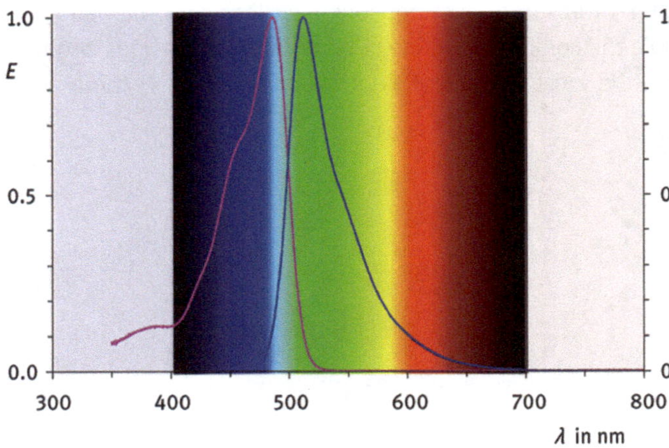

Figure 33.7: Absorption (left, magenta curve, left scale) and fluorescence (right, blue curve, right scale) of Fluorescein (**2**, RN 2321-07-5) in water.

donor groups increase the photostability appreciably where alkylation to form tertiary amines prevent degradation processes initiated by deprotonation. The donor effect of a dialkylamino group concerning chromophores is roughly about the same effect of a deprotonated phenolic oxygen atom. As a consequence, the replacement of the donor groups in **2** by means of diethylamino groups in **3** (Rhodamine B as the chloride) causes an appreciable bathochromic shift in absorption and fluorescence and a significant improvement of the photostability; see Figure 33.8. This class of compounds is known as the rhodamine dyes where **3** is one of the most prominent.

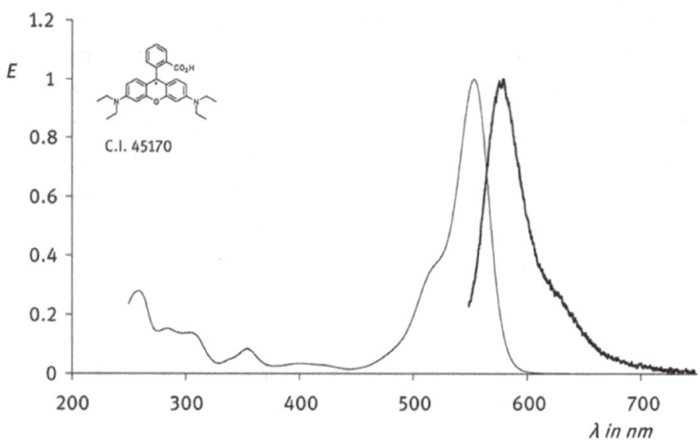

Figure 33.8: UV/Vis spectra of Rhodamine B chloride (**3**, RN 81-88-9, C.I. 45,170) in water. Left: Absorption spectrum, right: Fluorescence spectrum.

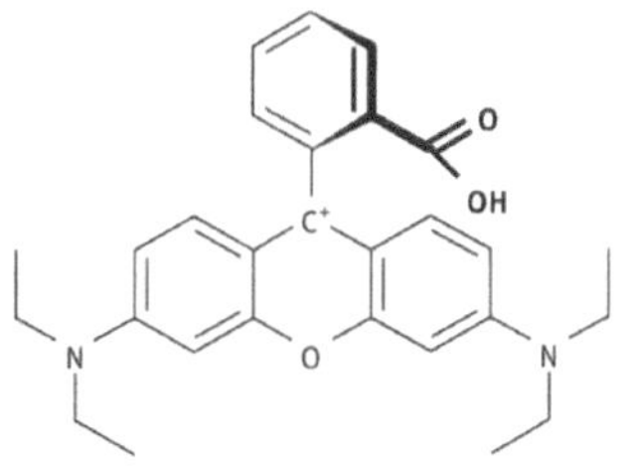

3

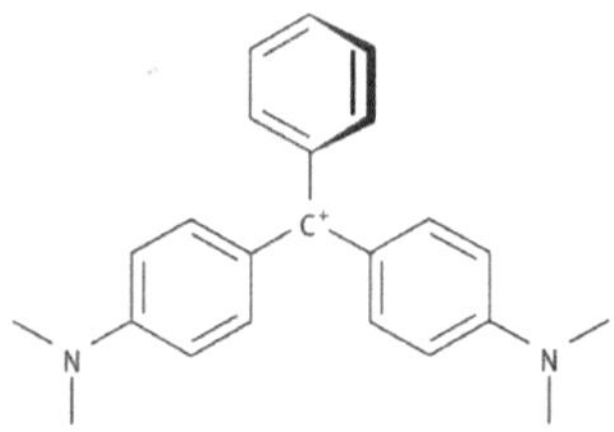

4

The effect of intra molecular mobility on the fluorescence can be clearly seen with the comparison of the strongly fluorescent Rhodamine B (**3**) with the non-fluorescent Malachite Green (**4**, RN 569-64-2). The fluorescence deactivation of the latter can be attributed to the freedom of rotation around the single bonds of the dimethylaminophenyl groups to the formal carbenium ion. These degrees of freedom are significant concerning the electronic excitation, whereas freedoms of rotation in the alkyl side-chains such as with **3** are of unimportant influence concerning the chromophoric system. The photostability of **3** is appreciably improved compared with **2** because of the more suitable donor groups; however the photostability is only just acceptable for many applications. Some inter system crossing (I.S.C.) to the triplet state and subsequent sensitizing of the triplet oxygen to form singlet oxygen is an important path for the photochemical degradation of dyes because the latter exhibits reactions of an electron depleted reactive olefin such as Diels–Alder reactions and degrades chromophores. The I.S.C. becomes even more pronounced by the introduction of heavy elements such as the four bromine atoms in Eosin (**5**, RN 17372-87-1). The strong electron withdrawing effect in **5** gently lowers the tendency for oxidation and increases the acidity of both phenolic groups so that an appreciable bathochromic shift in the absorption is reached by two strong donor groups and a full deprotonation proceeds in aqueous solution; as a consequence, Eosin (**5**) is applied as red ink where the fluorescence is much weaker than in **2**. The replacement of the bromine atoms in **5** by the even heavier iodine atoms (Erythrosin, RN 568-63-8) favors further the I.S.C. Finally, a chlorination for diminishing the tendency of oxidative degradation makes the similar Rose Bengale (**6**, RN 632-69-9) useful as a singlet oxygen sensitizer in preparative chemistry ($\Phi_{I.S.C.} > 0.9$ [38] where $\Phi_{flu.} \approx 0.1$ remained). Chlorination is generally frequently applied for the stabilization of dyes.

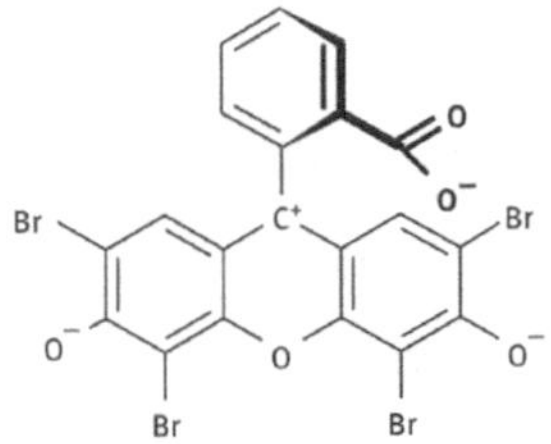

5

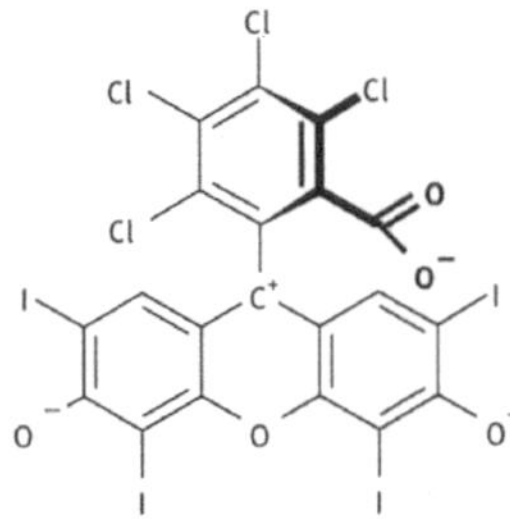

6

The limited photostability of xanthene derivatives prompted the development of many other fluorescent chromophores [39, 40] with high fluorescent quantum yields [41, 42]. Examples [43] are the coumarines **7**, where the disubstituted amino group as a donor induces bathochromic spectral shifts of fluorescence until the visible. The fluorescent natural product Aesculine (see above) is a similar coumarine, however less stable than **7** because of OH groups as a donors instead of the amino group in **7**. Various aryl and heteroaryl substituents may be applied in **7** forming a formal stilben substructure. Indeed, stilbenes substituted with amino groups such as **8** where the sulphonate groups are introduced for water solubility are applied as fluorescent whitening agents. Finally naphthalimides such as **9** are highly photo stable. **9** absorbs mainly in the UV and exhibits a blue fluorescence. The exchange of the methoxy group as a weak donor by an aminogroup as a stronger one causes a bathochromic shift both in absorption and fluorescence. Numerous further fluorescent materials are known with many applications such as the fluorescent natural product quinine (RN 130-95-0).

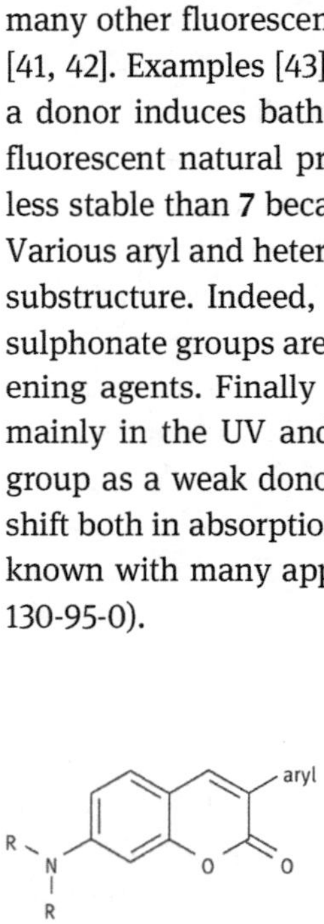

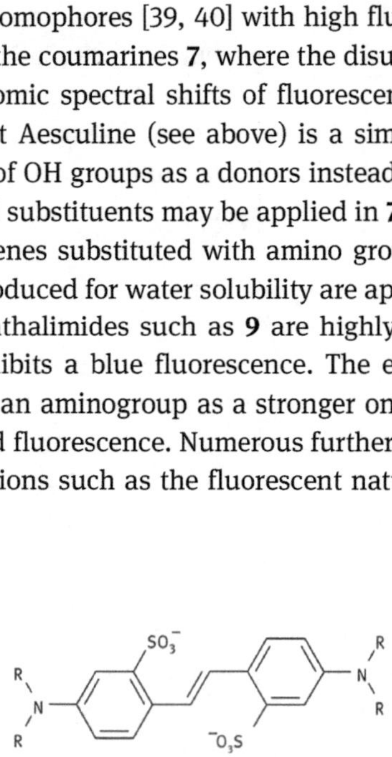

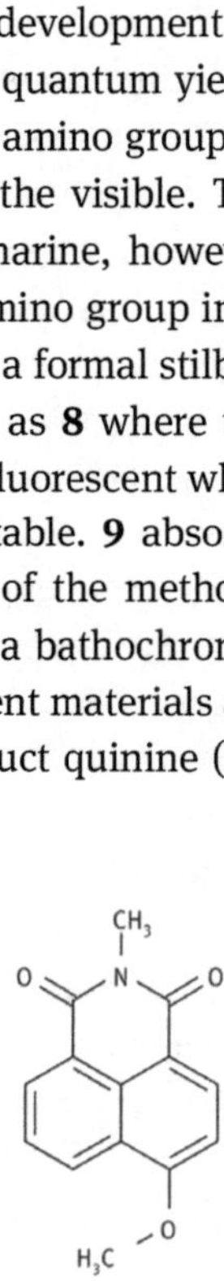

7 8 9

Unusually high photostabilities and fluorescence quantum yields exhibit the *peri*-arylenes **10** [5] known for $n = 1$ to $n = 4$ and $n = 6$. Moreover, there is only one electronic transition in the visible parallel to the *N*-*N*-connection line making the analyses of interactions of chromophores clear. As a consequence, here we concentrate further discussions mainly to **10**.

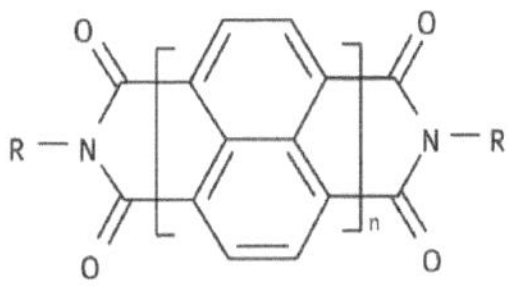

10

The naphthalenecarboximides (**10**, $n = 1$) absorb in the UV [44] and may be applied as white pigment [45] or for sun protection [46]; see Figure 33.9. The fluorescence is comparably weak because the S_1 state is close to a further, non-fluorescent state [47] and looses energy of excitation by their interaction. The perylenebiscarboximides (**10**, $n = 2$ where **1** is a special derivative because of the solubilizing *sec*-alkyl groups) as the next higher homologues generally exhibit a very low solubility and are applied as photostable pigments [48] such as the methyl derivative (R=CH_3) C.I. Pigment Red 179 (RN 5521-31-3). The low solubility of the perylenebiscarboximides [49] and their higher homologous could be resolved by solubility increasing groups R such as 2,5-di-*tert*-butylphenyl group (RN 83054-80-2) [50], 2,6-di–*iso*-propylphenyl group [51] or the even more efficient long-chain secondary alkyl groups [52] such as the here reported 7-tridecyl group [53]. The latter is a good compromise between efficiency and peripheral ballast to the chromophore where the two identical long-chain hexyl groups prevent the introduction of stereogenic centers to keep the purification of the dyes simple and efficient. The soluble perylene dyes are highly fluorescent where quantum yields close to unity were observed so that **1** may be used [54] for the calibration of fluorescence spectrometers because of the extraordinarily high photostability [55] as a prerequisite for long-term stable signals.

The terrylenecarboximide (**10**, $n = 3$, RN 1403692-73-8) [56] is the next higher homologous and forms strongly red fluorescent blue solutions where a quantum yield close to unity allows the application for the calibration in the more bathochromic spectral region. Finally, the quaterrylenecarboximide (**10**, $n = 4$, RN 168101-28-8 P) [57, 58] absorbs partially in the NIR. The fluorescence spectrum extends in the NIR above 800 nm; fluorescence quantum yields were not yet determined. The sexterrylenecarboximide (**10**, $n = 6$, RN 168101-28-8 P) [59] fully absorbs in the NIR with a maximum at 945 nm; it is still not investigated if there is fluorescence at very long wavelengths.

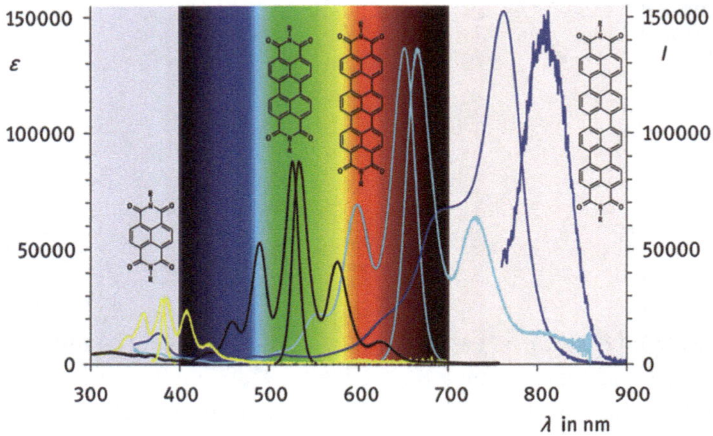

Figure 33.9: UV/Vis absorption (left curves and left scale, in L·mol·cm^{-1}) and to the absorption normalized fluorescence spectra (right curves and right scale) of *peri*-arylenes **10** in chloroform; R means the long-chain secondary alkyl substituent 7-tridecyl for solubilisation, and 1,5-dichloro-3-pentyl for *n* = 1. From left to right naphthalenebiscarboximides (**10**, *n* = 1, yellow curves), perylenebiscarboximides (**10**, *n* = 2, black curves), terrylenebiscarboximides (**10**, *n* = 3, turquoise curves), and quaterrylenebiscarboximides (**10**, *n* = 4, blue curves with limited accuracy in the bathochromic branch of fluorescence).

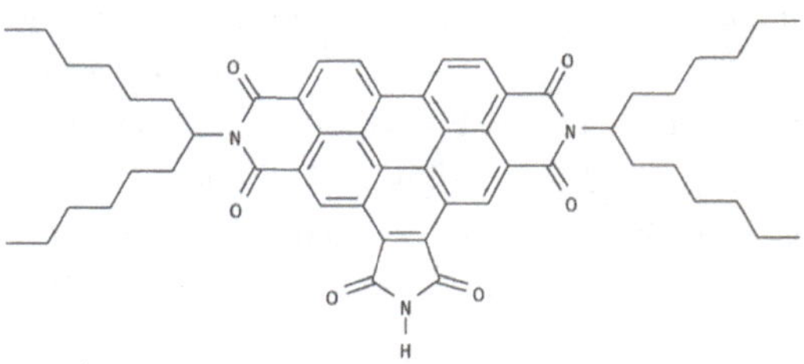

11

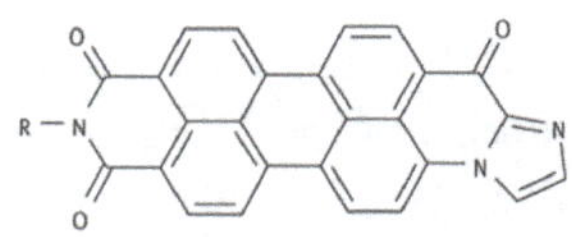

12

Derivatives of (**10**, *n* = 2) such as **1** are mostly applied for fluorescence applications because absorption and fluorescence are in the middle of the visible region and the perylenetetracarboxylicbisanhydride (RN 128-69-8, C.I. Pigment Red 224) as the starting material for syntheses is a technical mass product [60]. The position of the absorption and fluorescence spectra can be fine-tuned by means of an extension of the core and by alterations of the peripheral carboximide structure. A core substitution with donor groups cause bathochromic shifts [61–63], whereas hypsochromic shifts can be obtained by the lateral extension of the core such as in **11** [64]. The introduction of an imidazole ring into the periphery in **12** shifts the absorption and fluorescence spectra [65] exactly into spectral region of for spectroscopic investigations interesting hydrocarbon terrylene; however, **12** is appreciably more chemically and photo chemically stable than the labile terrylene.

33.8 Dynamic processes

The Stokes' shift of **1** and similar fluorescent dyes is comparably small causing an appreciable overlap between the absorption and fluorescence spectra. A larger spectral separation would bring about an appreciable advantage for some applications of fluorescent dyes such as for fluorescent solar collectors [66] and dye lasers [67] because of diminished re-absorption of the fluorescent light.

The Stokes' shift can be increased by means of dynamic processes M and M' in the timespan between the optical excitation of about 10^{-15} s with subsequent thermal relaxation to the S_1 ground state of about 10^{-13} s and the lifetime of the exited S_1 state of some nanoseconds; see Figure 33.10. The movement of electrons is in the timescale of the electronic excitation and does not cause effects of relaxation. However, the movement and relaxation of the heavier nuclei proceeds more slowly, but is still fast enough to be finished within the lifetime of the excited state. Fast chemical processes M are suitable to reach the S_1' state with some energy loss before deactivation; the remaining energy of excitation becomes diminished and thus, the fluorescence is bathochromically shifted [66]. Finally the energetic position of the S_o' state must be above the S_o state to reach the latter by M' in a cyclic process to avoid a photochemical consumption of the fluorescent dye. The stronger the molecular interactions for M and M' are the larger is the effect, but the more are the molecules stressed such as with excimer formation and this is frequently unfavorable for the photo stability. As consequence, one has to find compromises.

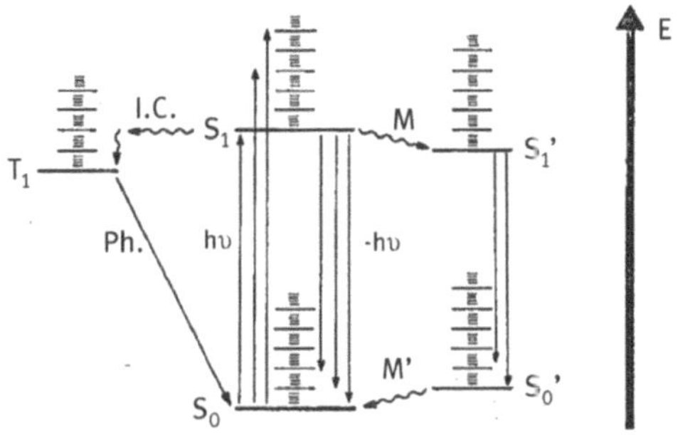

Figure 33.10: Jablonski-diagram for dynamic processes [66]. *hv*: Light absorption and -*hv*. Fluorescence, M and M': Relaxation processes.

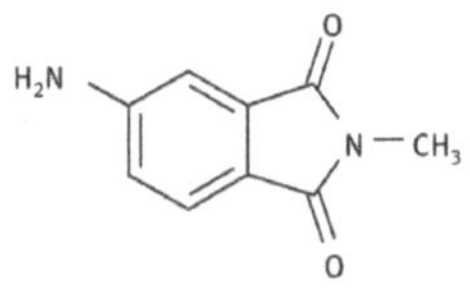

13

Solvent relaxation is a very gentile influencing where dye **13** [68, 69] is a strongly fluorescent suitable candidate for studies because the S_0 state is only weakly polar with a small dipole moment, whereas the S_1 state exhibits a large dipole moment because of charge shift from the amino to the carbonyl group; as a consequence, the S_0 ground state is only weakly solvated in polar media and this solvent shell is conserved in the S_1 state after electronic excitation; however, the geometry of this shell is unfavorable because the induced large dipole moment. As a consequence, the S_1' state is reached by solvent relaxation. Bathochromically shifted fluorescence proceeds to the S_0' state with a low dipole moment and again an unfavorable solvent shell. Finally, the S_0 state is reached back again by the relaxation M'. The spectral shift of the solvatochromism in fluorescence is so pronounced that dye **13** is suggested as an empirical fluorescent solvent polarity probe [70]; for more details see ref [71]. The required polar media for an increase of the Stokes' shift can be realized with polar solvents and is more difficult attainable with polymers where a copolymerization with a polar monomer was finally successful [72]. The limitations concerning the polarity of the mediums are not given for molecules that change their geometry as the optically induced relaxation. This was demonstrated with DPP fluorescent dyes [73] (initially named diketopyrrolopyrrole, presently pyrrolopyrroledione) according to Figure 33.11. The phenyl groups were twisted out of the chromophore plain by steric interactions of the substituents R and thus, electronically decoupled causing a hypsochromic shift in the absorption. The electronic excitation increases the double bond character of this bond and turns the phenyl groups against the sterical stress in plane if the size of R is appropriate and thus, couples the phenyl groups to the chromophore, induces a bathochromic shift of fluorescence and an increase of the Stokes' shift.

The dynamic process around a single bond is combined with a solvatochromic solvent sensitivity in fluorescence in dye **14** (RN 2253747-09-8) [74]; Figure 33.12. Optical excitation induces a charge shift from the methoxynaphthalene in **14** to the perylene being a driving force M for a planarization (iPLICT mechanism [75]; incomplete planarized intra molecular charge transfer). The resulting increase of the dipole moment with the formation of the excited state allows a fine tuning of the energy by means of solvent effects and thus the color of fluorescence (-$h\nu$'). The concept of steering the optical properties of fluorescent dyes by means of molecular dynamics looks attractive; however, this means a balancing act because flexibility may cause fluorescence quenching by the loose bolt mechanism of Lewis and Calvin [76, 77], moreover, a strong tilting of the ground state structure may end-up in a TICT mechanism [78] (twisted

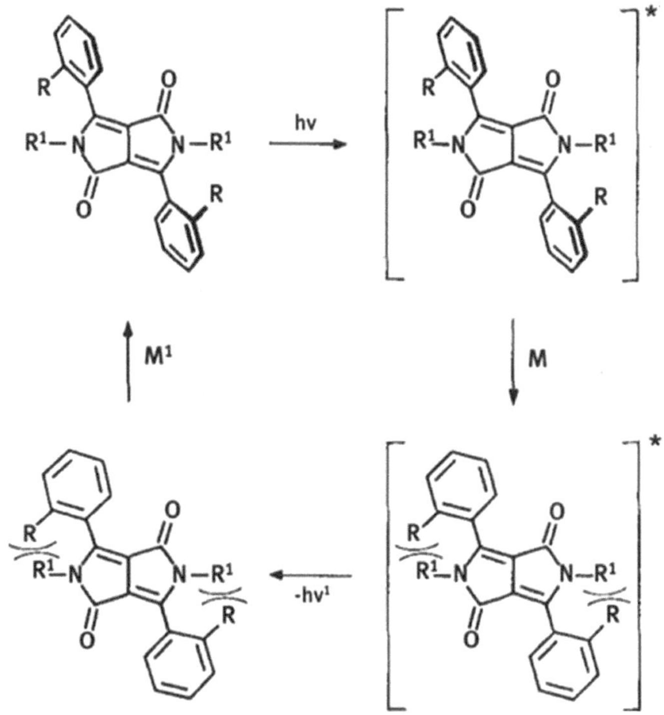

Figure 33.11: Dynamics in optically exci:ed DPP fluorescent dyes by the mobility of a single bond; R = CH_3, relaxation M and M'.

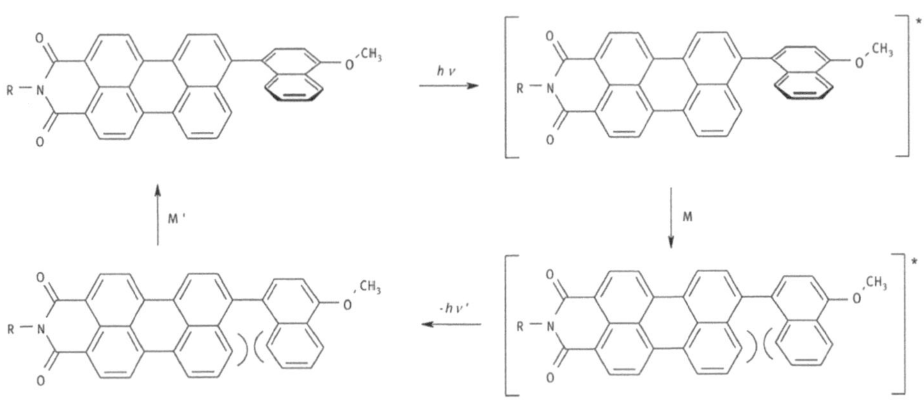

Figure 33.12: Dynamic process in perylene derivatives combined with a charge shift; dye **14**, R = 7-tridecyl. *h*ν: Optical excitation inducing a charge shift from the methoxynaphthalene to the perylene. M and M': Relaxation. - *h*ν': Fluorescence.

intra molecular charge transfer) with orthogonal substructures in the excited state; however, this seems to be less favorable for high fluorescence quantum yields [79].

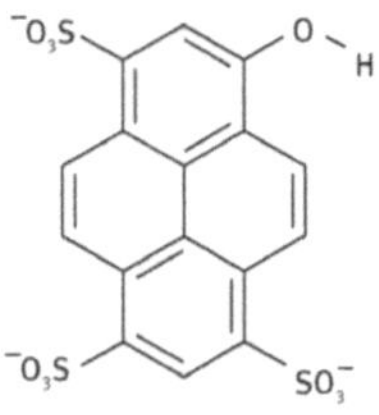

15

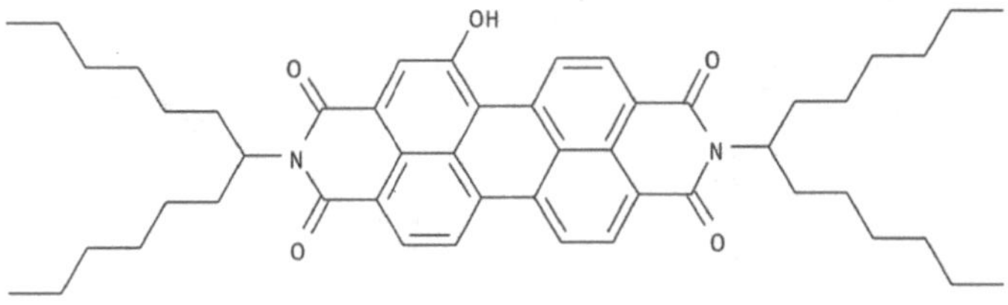

16

An even more pronounced influencing of electronically excited states is given by the breaking and forming chemical main valences. The breaking and forming of O–H and N–H bonds is fast enough to compete with the fluorescence decay. The OH group of the hydroxypyrenetrisulphonate **15** (RN 6358-69-6) [80] in neutral to slightly acidic aqueous solution is not dissociated. Light absorption with this weak donor group is found at comparably short wavelengths. Optical excitation shifts charge from the OH group into the aromatic nucleus and increases the acidity considerably. As a consequence, the OH group is deprotonated in a fast reaction forming the stronger donor group O^- and a considerable bathochromic shift in the fluorescence where a large Stoke' shift is observed with an intense green fluorescence and quantum yields close to unity by this ESPT (exited state proton transfer) mechanism [81]. The necessary aqueous medium with controlled acidity and the limited photostability of **15** mean restrictions for many applications. The ESPT mechanism can be also realized with the more stable perylene dyes such as **16** (RN 1018856-45-5) [82] even in more lipophilic media where amines may be applied as proton acceptors. The ESPT mechanism allows a separation between absorption and fluorescence spectra for **16**; see Figure 33.13.

The ESPT mechanism requires a proton acceptor; This restriction can be avoided by means of an intramolecular proton acceptor where the light-induced process is named ESIPT (excited state intra molecular proton transfer). 5,5'-Dihydroxybipyridyl (RN 36145-03-6) [83, 84] exhibits suitable structure elements for such a process where a fluorescence quantum yield of only 0.31 is found [85]. An appreciably higher fluorescence quantum yield of about 50% is found for the dimethyl derivative **17** (RN 34237-07-5) [83, 86] where a baseline-separation of the absorption and fluorescence spectrum is obtained; see Figure 33.14. A double proton transfer is discussed as the reason for the spectral separation; however, details of the mechanism are still under consideration [85, 87–90].

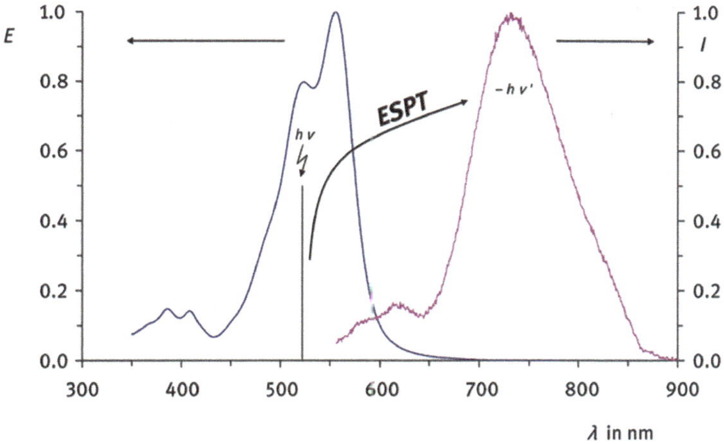

Figure 33.13: Absorption (left blue curve and left scale) and fluorescence spectrum (right magenta curve and right scale) of **16** in *N,N*-dimethylaniline as the proton acceptor. Bar: Wavelength for the optical excitation for fluorescence.

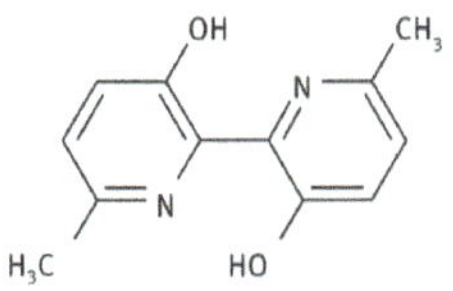

17

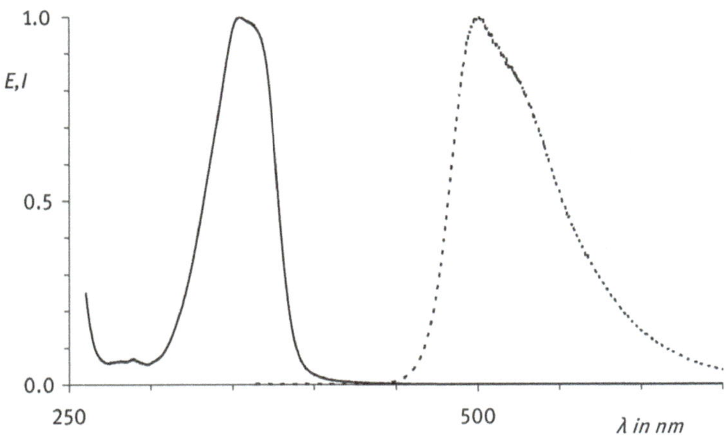

Figure 33.14: Absorption (solid curve left) and fluorescence spectra (dotted curve right) of the dimethylbipyridinediole **17** in chloroform.

Bipyridinedioles such as **17** exhibit chelating properties, for example for copper ions, forming non fluorescent complexes and can be applied for the analytical determinations of the ions with high sensitivity [91]; the large Stokes' shift avoiding re-absorption of the fluorescent light is of special advantage for such applications.

33.9 Energy transfer

Fluorescent materials conserve the energy of excitation for several nanoseconds and allow the processing of this energy by fast processes and thus useful for applications such as for molecular electronics. The energy transfer [92] from one chromophore to some other is an important process where the collection an transport of light energy in the photosynthesis reaction center [93] is the most prominent example. Essentially two mechanisms for the energy transfer are generally accepted (i) the Dexter mechanism [94] requiring a direct contact of orbitals and (ii) the Förster resonant energy transfer (FRET) [95] as a consequence of resonantly interacting dipoles of the hypsochromically absorbing energy donor and the more bathochromically absorbing and fluorescent acceptor. This type of interaction is farer reaching and may cover distances until more than 5 nm. Förster [28, 95] and Perrin [96, 97] established eq. (33.11) for the quantitative description of FRET where the rate constant of energy transfer (k_T) corresponds to the inverse sixth power of the distance R between the centers of the transition dipoles.

$$k_T = \frac{1000 \cdot (\text{In } 10) \cdot \kappa^2 \cdot J \cdot \Phi_D}{128 \cdot \pi^5 \cdot N_A \cdot \tau_D \cdot n^4 \cdot R^6} \tag{33.11}$$

J in eq. (33.11) is the overlap integral between the fluorescence spectrum of the energy donor and the absorption spectrum of the energy acceptor, Φ_D is the fluorescence quantum yield of the donor, N_A Avogadro's number, τ_D the fluorescence lifetime of the energy donor, and n the index of refraction.

$$\kappa = \cos(\theta_T) - 3 \cos(\theta_D) \cos(\theta_A) \tag{33.12}$$

The geometry factor κ is a sum of scalar products and can be calculated with eq. (33.12) where θ_T is the angle between the electronic transition moments μ of D and A. θ_D is the angel between the transition moment of the donor and θ_A of the acceptor, respectively, and the interconnecting vector. The R^{-6} dependence of the rate of energy transfer is widely applied for the determination of inter and intra molecular distances such as a molecular ruler and obtained a special importance for biochemical investigation. The orientation factor κ^2 becomes 2/3 for mobile structures with tumbling chromophores as a consequence of integration over all orientations. Equation (33.11) is generally accepted; however, its practical application is still subject of considerations [98, 99]. As a consequence, we tested the applicability of eq. (33.11) where κ becomes zero for orthogonal transition moments between the donor and the acceptor if one of the transition moments is orthogonal to the interconnecting vector. As a consequence, both chromophores should be completely decoupled and operate independently such as with dual fluorescence.

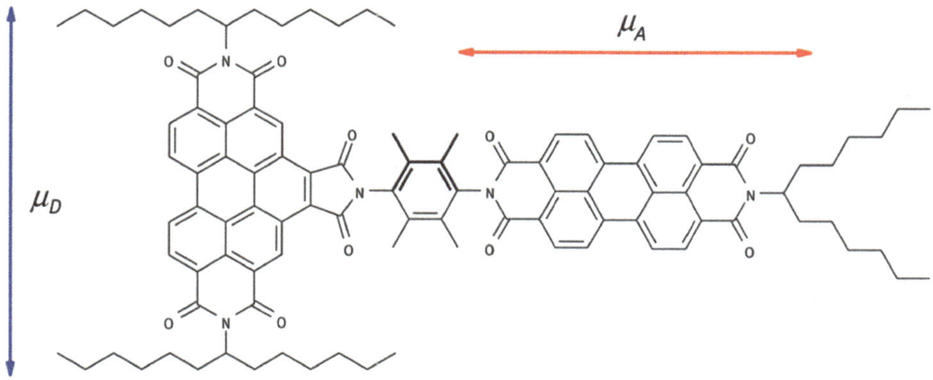

18

We prepared the dyes **18** for a test of eq. (33.11) [100] where the condition κ = 0 is fulfilled because the transition moment of the hypsochromically absorbing benzoperylene as the energy donor (blue vertical vector on the left) is orthogonal to the bathochromically absorbing perylene as the energy acceptor (red horizontal vector on the right); the interconnecting vector is orthogonal to the transition moment of the donor. As a consequence of κ = 0, energy transfer should be suppressed and independent fluorescence of both chromophores should be observed. In contrast to the theory, a fast and efficient energy transfer within 9.4 ps proceeded [101] with close to 100% fluorescence quantum yield for optical excitation of the fluorescence donor. The distance between the two orthogonal chromophores could be successively prolonged until 5 nm without inhibition of the energy transfer. An analysis of the rate of energy transfer as a function of the inter chromophore distance indicated the R^{-6} dependence as an only rough approximation [102]. Moreover, the overlap integral J in eq. (33.11) was successively diminished by means of the selection of suitable chromophores; however, no significant influence on the efficiency of energy transfer was found [103]. The unusual energy transfer was attributed to noise-induced processes with the contribution of molecular vibrations where this mechanism was supported by quantum chemical calculations [104, 105]. As a consequence, the well-established dipole–dipole interaction as the mechanism of FRET seems to be not the only way for a longer-distance energy transfer, but other ways with the contribution of molecular dynamics have to be considered, in particular for molecular geometric conditions with low valued of κ.

33.10 Charge transfer

Neighboring electron rich or electron depleted systems can induce electron transfer reaction of fluorescent chromophores where distances up to 5 nm may be spun for suitable chemical structures. The transfer of the energy rich electron means a photo-induced reduction where the water splitting and subsequent reduction of carbon dioxide in photosynthesis [93] is the most prominent process. A photochemical oxidation is favored in less electron rich chromophores such as the here discussed perylenebiscarboximides with four electron withdrawing carbonyl groups and means a uptake of an electron by the photo-induced half vacant HOMO; see Figure 33.15. The transition probability of the electron transferred to the LUMO back to the substituent is so low that generally radiationless processes dominate and no fluorescence is observed. On the other hand, the direct optical excitation from the substituent into the LUMO can be observed with comparably low probability such as with a sulfur-containing heterocycle attached to the chromophore [64].

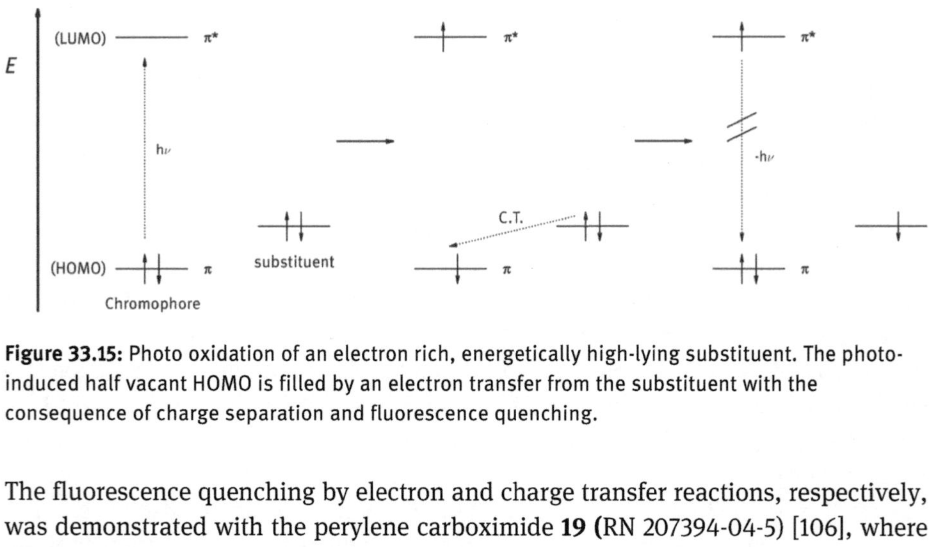

Figure 33.15: Photo oxidation of an electron rich, energetically high-lying substituent. The photo-induced half vacant HOMO is filled by an electron transfer from the substituent with the consequence of charge separation and fluorescence quenching.

The fluorescence quenching by electron and charge transfer reactions, respectively, was demonstrated with the perylene carboximide **19** (RN 207394-04-5) [106], where the electron rich amino group is attached to the chromophore.

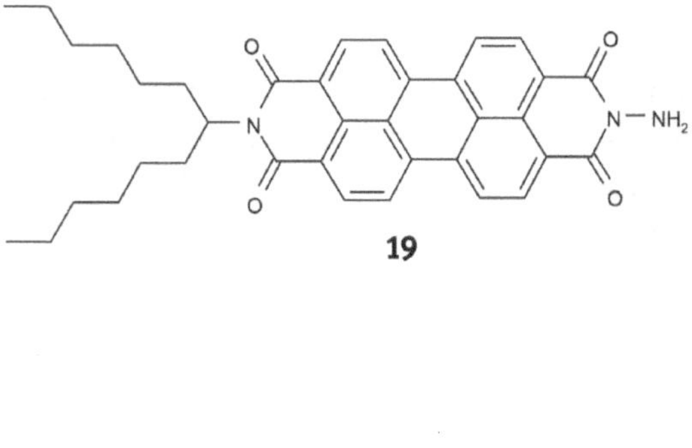

19

As a consequence, the fluorescence becomes quenched by an electron transfer from the amino group to the perylene chromophore [107]; the amino group is electronically decoupled from the chromophore because of orbital nodes in HOMO and LUMO at the linking nitrogen atom of the carboximide [108]. Some re-orientation in the electronically excited state of **19** seems to be important for this process [109]. This fluorescence quenching may be applied for analytical applications such as for the determination of aldehydes [110]. On the other hand, the electron transfer in Figure 33.15 causes a charge separation and may initiate photochemically induced reductions. The charge separation may proceed to other chromophores such as corroles [111, 112] or even by more distant colorless electron rich groups [113] where colorless local electron rich structures may be realized by the α-effect [114, 115]; this is given in isoxazolidines where a charge separation of 1.6 nm cold be induced [116].

33.11 Chemiluminescence

The S_1 state of dyes exhibits fluorescence in suitable chromophores where it is unimportant how the state was reached. The main topic of this contribution was the light-induced fluorescence; however, α-, β- or γ-radiation may cause the electronic excitation of dye molecules as well. Even chemical reactions may end-up in electronically excited states where the enforcing of symmetry forbidden reactions [117] by plenty of reaction enthalpy is an attractive way. This chemiluminescence is widely used in biochemistry [118, 119] because of sensitivity. The oxidation of Luminol (3-hydroxyphthalazine, RN 521-31-3) catalyzed by iron ions is a prominent example. Nature realized chemiluminescence, for example, in the firefly (*Photonius·Pyr*alis) with the mechanism of Figure 33.16.

The starting carboxylic acid is transferred in an energy consummating process enzymatically into the mixed anhydride with a derivative of diphosphoric acid. This increases the acidity of the *α*-proton of the carboxylic acid allowing a deprotonation with a subsequent autoxidation to the hydroperoxide anion. The strong *α*-effect nucleophile forms a dioxetanone derivative of the chromophore by intramolecular ring closure and loss of AMP. The symmetry forbidden ring scission leaves a neutral CO_2 and an optically excited chromophore that exhibits the nice green light of chemiluminescence. The staring carboxylic acid is recovered by enzymatic carboxylation and reduction as well as ATP from AMP.

Chemiluminescent emergency sticks operate with a similar mechanism where oxalic chloropheylesters are the staring material; Figure 33.17. The reaction with hydrogenperoxide and base give partial hydrolysis and subsequent ring closure with the anion of this strong α-effect nucleophile forming the dioxetanedione. The ring scission of the latter leaves a neutral CO_2 and one in the excited state. The fluorescence of CO_2 is weak; however, a sensitising of suitable fluorescent dyes results in bright chemiluminescent light [120, 121].

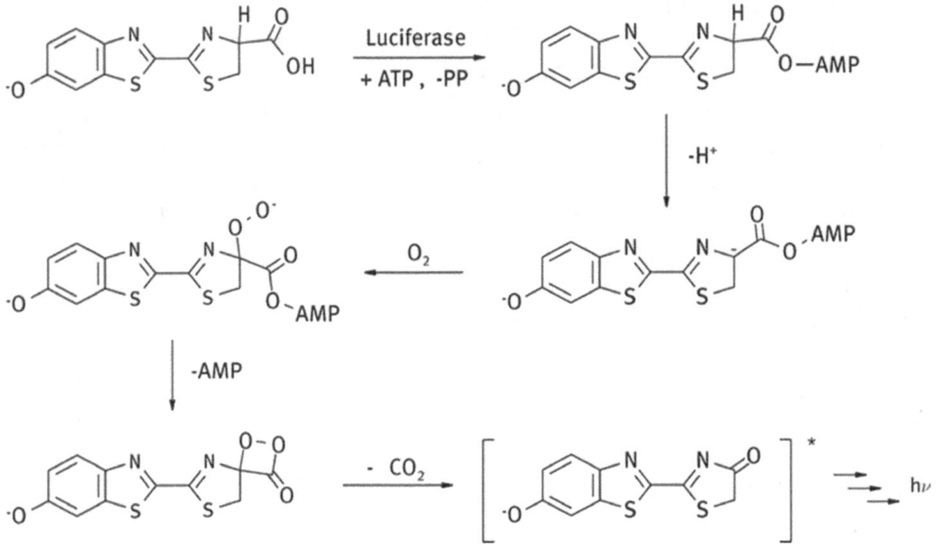

Figure 33.16: Chemical pathway for chemiluminescence in the firefly of Photonius Pyralis (η = 88%).

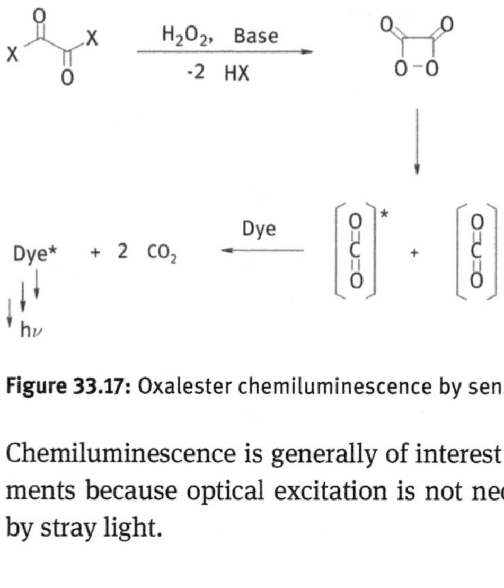

Figure 33.17: Oxalester chemiluminescence by sensitising fluorescent dyes.

Chemiluminescence is generally of interest in analytics for highly sensitive measurements because optical excitation is not necessary and thus, there is no interference by stray light.

33.12 Aggregation and fluorescence

The extended double bond systems in dyes, in particular large aromatic or hetero aromatic systems, exhibit strong dispersion interactions and the tendency of juxtaposing to form aggregates. This is favored in stronger polar media such as water. There are consequences for light absorption by such interactions; this is discussed by means of perylene dyes **1** and **10** ($n = 2$), respectively, because there is only one electronic transition moment in the visible parallel to the long axis of the chromophore; thus, there are clear conditions. One can distinguish between two significant basic types of geometric arrangements: (i) a co-planar sandwich-like orientation of the chromophores and the transition moments, respectively. Scheibe named this orientation *H*-aggregate because of hypsochromic shifts of the absorption [122]; see Figure 33.18, left blue curve and schematic indication of the transition moments down, within the curve.

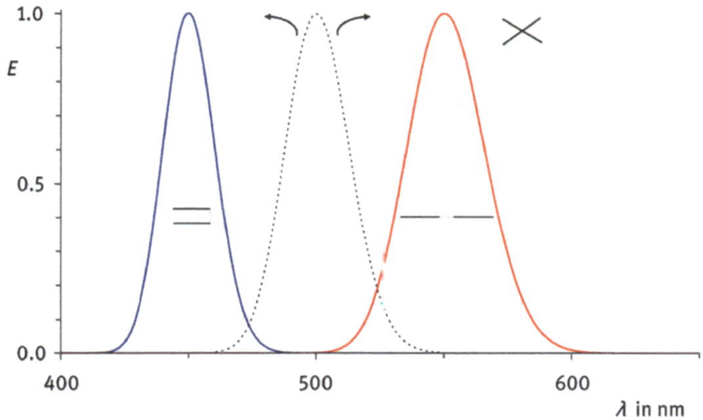

Figure 33.18: Schematic light absorption of a dye with the absorption maximum at 500 nm (black, dotted curve). Blue curve left: Hypsochromically shifted absorption by *H* aggregation and schematic indication of the transition moments down, within the curve. Red curve right: Bathochromically shifted absorption by *J* aggregation and schematic indication of the transition moments down, within the curve. Right top: Schematic skew arrangement of the transition moments with hypsochromic and bathochromic absorption bands: Davydov splitting.

The movement of the significant electrons for electronic transitions becomes antiparallely synchronized in coplanar arrangements of aggregated **1** such as in polar media because of Coulomb interactions. As a consequence, charges proximate averaged more than in isolated chromophores causing a higher energy of electronic excitation and a more hypsochromic absorption, respectively (exciton interactions). This compact arrangement of molecular antennae diminish their effective lengths and decrease the molar absorptivity ε [123]. The coupled transition moment exhibits point symmetry for fluorescence where competing radiationless processes become dominant so that fluorescence will be quenched according Förster's analysis [124]. Such aggregation is favored with increasing of the chemical concentration of dyes in many cases. The consequences concerning suppressing fluorescence are known as concentration quenching. A linear arrangement (ii) causes a synchronous, parallel movement of the involved electrons and thus, a larger distance with even proximate opposite charges lowers the required energy of excitation and a bathochromic shift of the absorption, respectively; see Figure 33.18, right red curve and schematic arrangement of the transition moments down, within the curve. This type of aggregate is known as *J*- aggregates according to the discoverer Jelley [125]. An effective prolongation of the antenna causes an increase of the absorptivity ε [126]. There is also an increased transition moment for light emission where fluorescence of *J*-aggregates is observed; however, fluorescence quantum yields reach 20% only for few examples because of the lability of the arrangement. Such optical behavior is both observed for dimers and higher aggregates.

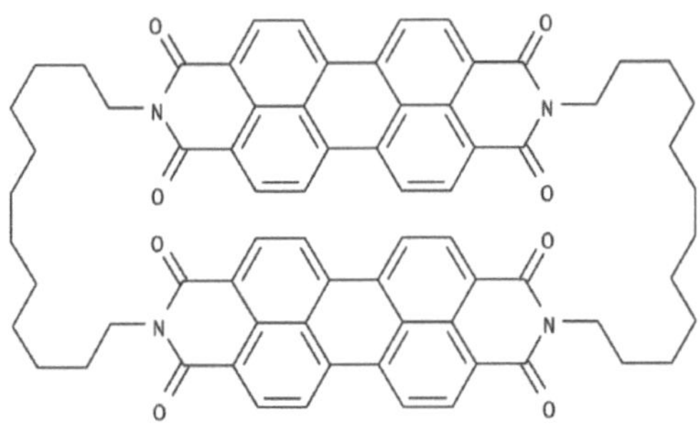

20

A skew arrangement of transition moments is schematically indicated in Figure 33.18, right top and allows both transitions. These are known as Davydov splitting [127]. A dominant *H* component can absorb light where the minor *J* component may be responsible for fluorescence (inner energy transfer). This could be realized for **1** in micells [128] or intramolecular with the more stable cyclophanes [129, 130].

The hypsochromically shifted absorption of the cyclophane **20** (RN 214078-84-9) [129] compared with **1** is shown in Figure 33.19 where the bathochromically shifted fluorescence causes an increased Stokes' shift. Cyclophanes of dyes may be efficiently synthesized by means of olefin metathesis [130].

33.13 Optical impression of fluorescence

Fluorescent materials are of interest for special optical effects. The extinction *E* of a Gaussian curve of the schematic light absorption in Figure 33.2 was transformed to the residual light intensity *I* by passing of white light ($I_o = 1$ for any wavelengths) a sample with $E = 1$ at 500 nm is shown in Figure 33.20, left. There one can see the broad loss of intensity in the important region of high intensities because of the broad socket of the absorption. This makes the generation of clear colors difficult by means of absorbing dyes and means a problem for color print; for exact color metric see ref [131, 132].

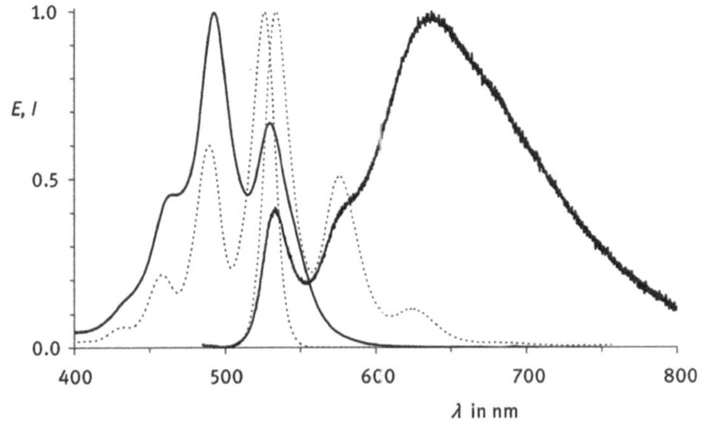

Figure 33.19: UV/Vis absorption (left) and fluorescence spectra (right) in chloroform. **20**: Thick solid curves compared with **1** (thin, dotted curves).

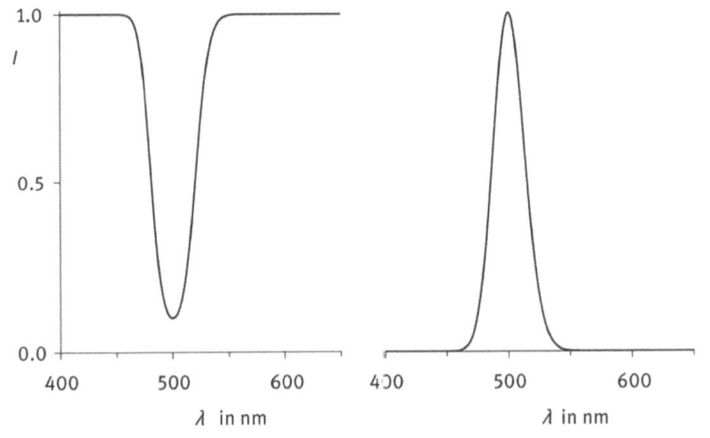

Figure 33.20: Schematic comparison of a typical spectral dependence of light intensity I. Left: Passing of white light with the Intensity $I_0 = 1$ through a sample with the absorptivity $E = 1$ at 500 nm. Right: Same spectral dependence of fluorescence light with maximal intensity $I = 1$ at 500 nm.

The spectral intensity of a fluorescence band with the same parameters is opposite and results in a comparably slim maximum; Figure 33.20, right. As a consequence, very clear and deep colors, nearly like spectral colors, can be obtained. This is demonstrated by means of color metrics with the comparison of absorption and emission colors of a perylene derivative [79].

33.14 Closing remarks

Generally, fluorescence can be detected with high sensitivity and is of special interest for uncomplicated analytical methods. This is of special interest in biochemistry by coding the Green Fluorescent Protein (GFP). There is only a limitation to high concentration because of the tendency for aggregation. A linear dependence of the fluorescence intensity is given until very low concentrations, for less than picomolar [133] and even further until single molecules [134]. Absorbed light quanta should be re-emitted for such applications as efficiently as possible. However, they might also initiate various processes in suitable chemical structures. The temporary stored energy by the dyes in the order of some nanoseconds enables many possibilities because of possible induction of photo chemical processes, energy transfer even for molecular electronics, for light collection and for single electron transfer where charge separation allows organic photovoltaics [135]. This can be controlled by means of molecular architecture [136] of suitable chemical structures. Structures for the transport of electrical charge are reaching resolutions as small as 5 nm in conventional electronics causing many technological problems because of the small size. The next steps of diminution of the active structure means molecular electronics where dyes such as **1** with a size of about 1 nm are attractive and would appreciably increase both the density of integration and frequency of cycle. The knowledge about the processes of fluorescence forms the basis for such developments.

References

1. Raab W. Vorläufige Anzeige von der Entdeckung des Schillerstoffes, einer neuen vegetabilischen Substanz. Archiv f d ges Naturlehre. 1827;10:121–3.
2. Rochleder F, Schwarz R. Über einige Bitterstoffe. Sitzungsberichte der k Akademie d Wiss Naturw Classe. 1852;9:70–9.
3. Berzelius J. Jahres-Bericht über die Fortschritte der physischen Wissenschaften. Tübingen: Heinrich Laub Verlag, 1837:283–7.
4. Nichols EL, Merritt E. Studies in luminescence, III. Phys Rev. 1904;19:18. DOI: 10.1103/PhysRevSeriesI.19.18. Chem Abstr. 1906:7432.
5. Langhals H. Chromophores for picoscale optical computers. In: Sattler K editor, Fundamentals of picoscience, 705–27.Bosa Roca/US: Taylor & Francis Inc. CRC Press Inc, 2013. ISBN 13: 9781466505094, ISBN 10: 1466505095
6. Bergmann L, Schaefer C. Lehrbuch der Experimentalphysik, Band 2, Elektromagnetismus: Elektromagnetismus. (W.Rath, co-author) Berlin: De Gruyter, 2015; ISBN: 9783110188981.
7. Bergmann L, Schaefer C. Lehrbuch der Experimentalphysik / Optik. vol. 3. (H. Niedrig, ed.; W. Eichler, co-author) Berlin: De Gruyter, 2004; ISBN 978-3-11-017081-8.
8. Langhals H. A re-examination of the line-shape of the electronic spectra of complex molecules in solution. Log-Normal Function Versus Gaussian. Spectrochim Acta Part A. 2000;56:2207–10.

9. Langhals H. The rapid identification of organic colorants by UV/Vis-spectroscopy. Anal Bioanal Chem. 2002;374:573–8.
10. Langhals H. UV-Visible spectroscopy and the potential of fluorescent probes. In: Frimmel FH, Refractory organic substances in the environment. Weinheim: Wiley-VCH, 2002:200–14. ISBN 3-527-30173-9.
11. Langhals H, Ritter U. γ-hydroxyalkyl naphthalene-tetracarboxdiimides: Organic white pigments. Eur J Org Chem. 2008;3912–15.
12. (a) Daubner SC, Fitzpatrick PF. Pteridines. Encyclopedia Biol Chem. (Eds.: W. J. Lennarz, M. D. Lane) 2004;3:556–60; Chem Abstr. 2005;143:55015; (b) Oliphant LW, Hudon J. Pteridines as reflecting pigments and components of reflecting organelles in vertebrates. Pigm Cell Res. 1993;6:205–8; Chem Abstr. 1994;120:159184; (c) Ziegler I. Pterins: pigments, cofactors and signal connection in cellular interaction. Naturwissenschaften. 1987;74:563–72; (d) Pfleiderer W, Natural pteridines – a chemical hobby. Adv Experim Med Biol. 1993;338:1–16; ISSN 0065–2598; (e) Landymore AF, Antia NJ, White-light promoted degradation of leucopterin and related pteridines dissolved in seawater, with evidence for involvement of complexation from major divalent cations of seawater. Marine Chem. 1978;6:309–25; Chem Abstr. 1978;89:220590.
13. Bugnon P, Karrer K, Hahn M, Sieber W. (Ciba Specialty Chemicals Holding Inc., Switzerland). PCT Int Appl. WO 2006003093 (30.6.2004); Chem Abstr. 2006;144:109847.
14. Langhals H, Eberspächer M. Water nanomicellar solutions naphthalenetetracarboxilic acid bisimides used as sunscreens. Ger Offen. DE 102013014353.5 27 Aug 2013, Ger Offen DE 102014012594 26 Mar 2015; Chem Abstr. 2015;162:454180.
15. Rohn W. Anomale Dispersion einiger organischer Farbstoffe. Ann Phys. 1912;38:987–1013. DOI: https://doi.org/10.1002/andp.19123431007.
16. Kuhn W. The physical significance of optical rotary power. Trans Faraday Soc. 1930;26: 293–308. DOI: 10.1039/TF9302600293.
17. Kirkwood JG. On the theory of optical rotatory power. J Chem Phys. 1937;5:479–421.
18. Roduner E, Krüger T, Forbes P, Kress K. Optical spectroscopy. Fundamentals and advanced applications. London: World Scientific, Publishing Europe Ltd, 2019: ISBN 978-1-78634-610-0.
19. Langhals H The determination of overlap between UV/Vis absorption and fluorescence spectra' Ber Bunsenges Phys Chem 1979;83:730–2.
20. Langhals H, Eberspächer M, Hofer A. Learning about structural and optical properties of organic compounds through preparation of functional nano micelles while avoiding hazardous chemicals or complicated apparatus. J Chem Educ. 2015;92:1725–9. DOI: http://dx.doi.org/10.1021/acs.jchemed 5b00145.
21. Wolfbeis OS. ed. Fluorescence Spectroscopy. New methods and applications. Berlin: Springer Verlag, 1993:ISBN 3-540-55281-2.
22. http://www.cup.lmu.de/oc/langhals/S-13/S-13data.html.
23. (a) Rothammel K. Antennenbuch. 5th edn. Stuttgart: Telekosmos Verlag, 1976. (b) Rüdenberg R. Der Empfang elektrischer Wellen in der drahtlosen Telegraphie. Ann d Phys. 1908;330:446–6. (c) Rüdenberg R. Der Empfang elektrischer Wellen in der drahtlosen Telegraphie. Ann d Phys Leipzig. 1908;25:466–500.
24. McCoy EF, Ross IG. Electronic states of aromatic hydrocarbons: The Franck-Condon principle and geometries in excited states. Aust J Chem. 1962;15:573–90.
25. Schmidt W. Optische Spektroskopie. 2nd ed. Weinheim: Wiley-VCH, 2000:ISBN 3-527-29828-2.
26. Hund F. Zur Deutung der Molekelspektren III. Zeitschrift für Physik. 1927;43:805–26.
27. Strickler SJ, Berg RA. Relationship between absorption intensity and fluorescence lifetime of molecules. J Chem Phys. 1962;37:814–22.

28. Förster T. Fluoreszenz organischer Verbindungen. Göttingen: Vandenhoeck & Ruprecht, 1951:158.
29. Schlücker T, Dhayalan V, Langhals H, Sämann C, Knochel P. Soluble adamantyl-substituted oligothiophenes with short fluorescence decay: An approach for ultra fast optical signal processing. Asian J Org Chem. 2015;4:763–9. DOI: 10.1002/ajoc.201500150.
30. Langhals H. A concept for molecular addressing by means of far-reaching electromagnetic interactions in the visible. J Electr Electron Syst. 2017;215. DOI: http://dx.doi.org/10.4172/2332-0796.1000215.
31. Hecht E. Optik. München: Oldenbourg, 2009. ISBN: 978-3-486-58861-3.
32. Periasamy A, Clegg RM. Flim microscopy in biology and medicine. Boca Raton, Fla.: CRC Press, 2010. ISBN 978-1-4200-7890-9.
33. Lewis GN, Calvin M. The color of organic substances. Chem Rev. 1939;25:273–328.
34. Baeyer A. Ueber eine neue Klasse von Farbstoffen. Ber Dtsch Chem Ges. 1871;4:555–8.
35. König W. Ueber den Begriff der "Polymethinfarbstoff" und eine davon ableitbare allgemeine Farbstoff -Formel als Grundlage einer neuen Systematik der Farbenchemie'. J Prakt Chem. 1926;112:1–36.
36. Wiese WL. Atomic oscillator strengths for light elements - progress and problems. J Korean Phys Soc. 1998;33:207–13.
37. Liu Y. A new method for obtaining Russell-Saunders terms'. J Chem Educ. 2011;88:295–8.
38. Ludvíková L, Friš P, Heger D, Šebej P, Wirza J, Klán P. Photochemistry of rose bengal in water and acetonitrile: A comprehensive kinetic analysis'. Phys Chem Chem Phys. 2016;18:16266–73.
39. Berlman IB. Handbook of fluorescence spectra of aromatic molecules, New York: Academic Press, 1971. LCCC-Nr. 78–154388.
40. Sauer M, Hofkens J, Enderlein J. Handbook of fluorescence spectroscopy and imaging. Weinheim: Wiley VCH, 2011: ISBN 978-3-527-31669-4.
41. Krasowitzkii BM, Bolotin BM. Organic luminescent materials, Weinheim: Wiley-VCH, 1988: ISBN 3-527-26728-X.
42. https://www.aatbio.com/resources/quantum-yield/. 15 Jan 2020.
43. Christie RM. Fluorescent dyes. Rev Prog Coloration. 1993;23:1–18.
44. Langhals H, Jaschke H. Naphthalene amidine imide dyes by transamination of naphthalene bisimides. Chem Eur J. 2006;12:2815–24.
45. Langhals H, Ritter U. γ-hydroxyalkyl naphthalene-tetracarboxdiimides: Organic white pigments. Eur J Org Chem. 2008;3912–15.
46. Langhals H, Eberspächer M. Water nanomicellar solutions naphthalenetetracarboxilic acid bisimides used as sunscreens. Ger Offen. DE 102014012594.7 27 Aug 2014. Chem Abstr. 2015;162:454180.
47. Adachi M, Murata Y, Nakamura S. Spectral similarity and difference of Naphthalenetetracarboxylic dianhydride, perylenetetracarboxylic dianhydride, and their derivatives. J Phys Chem. 1995;99:14240–6.
48. Zollinger H. Color chemistry: Syntheses, properties, and applications of organic dyes and pigments. Zürich: Verlag Helvetica Chimica Acta, 2003:ISBN 10: 3906390233 ISBN 13: 9783906390239.
49. Hunger K, Schmidt MU, Heber T, Reisinger F, Wannemacher S. Industrial organic pigments: production, crystal structures, properties, applications. 4th ed. Weinheim: Wiley-VCH, 2018: ISBN 978-3-527-32608-2.
50. Langhals H. Primary methods of generating solar power by using the targeted modification of fluorescent systems. Habilitationsschrift, Albert-Ludwigs-Universität Freiburg 1981, translation 9 Aug 2019. DOI: https://doi.org/10.5282/ubm/epub.68484.

51. Graser F, Feichtmayr F. Perylenetetracarboxylic acid diimide dyes. Eur Pat Appl. 1982;97, 7 July 1982. EP 55363 A1 19820707. Chem Abstr.218033.
52. https://www.cup.lmu.de/oc/langhals/Perylenebiscarboximides/index.html. 15 Jan 2020.
53. Langhals H, Demmig S, Potrawa T. The relation between packing effects and solid state fluorescence of dyes. J Prakt Chem. 1991;333:733–48.
54. Langhals H, Karolin J, Johansson LB. Spectroscopic properties of new and convenient standards for measuring fluorescence quantum yields. J Chem Soc Faraday Trans. 1998;94:2919–22.
55. Langhals H, El-Shishtawy R, von Unold P, Rauscher M. Methoxyperylene bisimides and perylene lactame imides: novel, red fluorescent dyes. Chem Eur J. 2006;12:4642–5. supporting information.
56. Langhals H, Walter A, Rosenbaum E, Johansson LB. A versatile standard for bathochromic fluorescence based on intramolecular FRET. Phys Chem Chem Phys. 2011;13:11055–9.
57. Langhals H, Schönmann G, Feiler L. A two-step synthesis of quaterrylenetetracarboxylic bisimides - novel NIR fluorescent dyes. Tetrahedron Lett. 1995;36:6423–4.
58. Langhals H, Büttner J, Blanke P. A two-step synthesis for quarterrylene bisimides. Application of the "green route" method. Synthesis. 2005;364–6.
59. Langhals H, Zgela D, Lüling R. Sexterrylenetetracarboxylic bisimides: NIR dyes. J Org Chem. 2015;80:12146–50. DOI: 10.1021/acs.joc.5b02092.
60. Schweizer HR. Künstliche organische Farbstoffe und ihre Zwischenprodukte, Berlin: Springer-Verlag, 1964:LCCC Nr. 63–23133.
61. Stauble M, Weber K. *peri*-Dicarboxylic acid imide dyes. CIBA Ltd., US patent US 2914531 19591124 Chem Abstr. 1960;54:47263.
62. Syebold G, Wagenblast G. New perylene and violanthrone dyestuffs for fluorescent collectors. Dyes Pigm. 1989;11:303–17.
63. Langhals H, Christian S, Hofer A. The substitution of aromatics by amines at room temperature with negative energy of activation: Amino peri-arylenes as metal-free components for dye-sensitized solar cells. J Org Chem. 2013;78:9883–91.
64. Langhals H, Kirner S. Novel fluorescent dyes by the extension of the core of perylene-tetracarboxylic bisimides. Eur J Org Chem. 2000;365–80.
65. Langhals H, Jaschke H, Ring U, von Unold P. Imidazolo perylene imides: A highly fluorescent and stable replacement of terrylene. Angew Chem. 1999;111:143–5. Angew. Chem. Int. Ed. Engl. 1999;38:201–3.
66. Langhals H. Primary methods of generating solar power by using the targeted modification of fluorescent systems, Habilitationsschrift, Albert-Ludwigs-Universität Freiburg. 1981. Open Access LMU UB München 15 Jan 2020. DOI: https://doi.org/10.5282/ubm/epub.68484.
67. Schäfer FP, Drexhage KH. Dye lasers. Berlin: Springer-Verlag, 1977: ISBN 978-3-540-51558-6.
68. Bogert MT, Renshaw RR. 4-Amino-*o*-phthalic acid and some of its derivatives. J Am Chem Soc. 1908;30:1135–44.
69. Fritzsche K, Langhals H. Elektronenreiche Heterocyclen als Donorgruppen in Fluoreszenzfarbstoffen. Chem Ber. 1984;117:2275–86.
70. Zhmyreva IA, Zelinskii VV, Kolobkov VP, Krasnitskaya ND. Universal scale for the action of solvents on the electron spectra of organic compounds. Dokl Akad Nauk SSSR. 1959;129:1089–92. Chem Abstr. 1961; 55:141336.
71. Langhals H. Heterocyclic structures applied as efficient molecular probes for the investigation of chemically important interactions in the liquid phase. Chem Heterocycl Compd. 2017;53:2–10. DOI: 10.1007/s10593-017-2014-z.
72. Langhals H. 'Polarität von organischen Gläsern. Angew Chem. 1982;94:452–3. Angew Chem Int Ed Engl. 1982;21:432–433.

73. Potrawa T, Langhals H. Fluoreszenzfarbstoffe mit großen Stokes-Shifts - lösliche Dihydropyrrolopyrroldione. Chem Ber. 1987;120:1075–8.
74. Langhals H, Greiner R, Schlücker T, Jakowetz A. Light-driven molecular dynamics in perylenes with medium-controlled emission. J Org Chem. 2019; 84: 5425–30. https://pubs.acs.org/doi/pdf/10.1021/acs.joc.9b00409.
75. Haberhauer G, Gleiter R, Burkhart C. Planarized intramolecular charge transfer: A concept for fluorophores with both large stokes shifts and high fluorescence quantum yields. Chem Eur J. 2016;22:971–8.
76. Lewis GN, Calvin M. The color of organic substances. Chem Rev. 1939;25:273–328.
77. Langhals H, Unold PV. Tetracarboxylic Bisimide-lactame-ring-contractions: A novel type of rearrangement. Angew Chem. 1995;107:2436–9. Angew Chem Int Ed. 1995;34:2234–2236.
78. Rettig W. Charge separation in excited states of decoupled systems: Twisted intramolecular charge transfer (TICT) compounds and implications for the development of new laser dyes and for the primary process of vision and photosynthesis. Angew Chem. 1986;98:969–86. Angew Chem Int Ed. 1986;25:971–988.
79. Greiner R, Schlücker T, Zgela D, Langhals H. Fluorescent aryl naphthalene dicarboximides with large Stokes' shifts and strong solvatochromism controlled by dynamics and molecular geometry. J Mater Chem C. 2016;4:11244–52. DOI: 10.1039/C6TC04453K.
80. Tietze E, Bayer O. Sulfonic acids of pyrene and their derivatives. Justus Liebigs Ann Chem. 1939;540:189–210.
81. Förster T, Völker S. Laser-excited absorption spectroscopy of rapid proton transfer processes. II. Reactions with weak acids. Zeitschr Phys Chem (München, Germany). 1975;97:275–84. Chem Abstr 1976; 84:104801:.
82. Langhals H, Rauscher M. NIR absorption of perylene dyes and fluorescence with large Stokes shift by simple deprotonation. Z Naturforsch. 2013;68b:683–6.
83. Vogt LH, Jr., Wirth JG. Crystal and molecular structure of 2,2'-bis(6-methyl-3-pyridinol). J Am Chem Soc. 1971;93:5402–5.
84. Langhals H, Pust S. Fluoreszenzfarbstoffe mit großen Stokes-Shifts - eine einfache Synthese von (2,2'-Bipyridin)-3,3'-diol. Chem Ber. 1985;118:4674–81.
85. Bulska H. Intramolecular cooperative double proton transfer in [2,2'-bipyridyl]-3,3'-diol. Chem Phys Lett. 1983;98:398–402.
86. Naumann C, Langhals H. A simple synthesis of Dihydroxybipyridyls. Synthesis. 1990;279–81.
87. Johansson LB, Persson L, Langhals H. Conspicuous absorption and fluorescence spectroscopic properties of 3,3'-dihydroxy-2,2'-bipyridines in solution. J Chem Soc Faraday Trans. 1996;92:4909–11.
88. Stock K, Schriever C, Lochbrunner S, Riedle E. Reaction path dependent coherent wavepacket dynamics in excited state intramolecular double proton transfer. Chem Phys. 2008; 349:197–203.
89. Plasser F, Barbatti M, Aquino AJ, Lischka H. Excited-state diproton transfer in [2,2'-bipyridyl]-3,3'-diol: The mechanism Is sequential, not concerted. J Phys Chem A. 2009;113:8490–9.
90. Zhao J, Liu X, Zheng Y. Controlling excited state single versus double proton transfer for 2,2'-bipyridyl-3,3'-diol: Solvent effect. J Phys Chem A. 2017;121:4002–8.
91. Reynal A, Etxebarria J, Nieto N, Serres S, Palomares E, Vidal-Ferran A. A bipyridine-based "naked-eye" fluorimetric Cu^{2+} chemosensor. Europ J Inorg Chem. 2010;1360–5.
92. Valeur B, Berberan-Santos M. Excitation energy transfer. Molecular fluorescence: Principles and applications. 2nd ed. Weinheim: Wiley-VCH, 2012. DOI: 10.1002/9783527650002.ch8. ISBN 9783527328376.
93. Golbeck JH. Structure and function of photosystem I. Annual Rev Plant Physiol Plant Mol Biol. 1992;43:293–324.

94. Dexter DL. A theory of sensitized luminescence in solids. J Chem Phys. 1953;21:836–50.
95. Förster T. Zwischenmolekulare Energiewanderung und Fluoreszenz. Ann Phys. 1948;437: 55–75. https://doi.org/10.1002/andp.19484370105.
96. Perrin F. The fluorescence of solutions: Molecular induction, polarization and duration of emission and photochemistry. Ann Phys. [10]. 1929;12:169–275. Chem Abstr 1930;24:0.25377.
97. Perrin F. Activation by light and by collisions in thermal equilibrium. Chem Rev. 1930; 7:231–7. Chem Abstr 1930; 24:34772.
98. VanBeek DB, Zwier MC, Shorb JM, Krueger BP. Fretting about FRET: Correlation between κ and *R*. Biophy J. 2007;92:4168–78.
99. Muñoz-Losa A, Curutchet C, Krueger BP, Hartsell LR, Mennucci B. Fretting about FRET: Failure of the ideal dipole approximation. Biophys J. 2009;96:4779–88.
100. Langhals H, Poxleitner S, Krotz O, Pust T, Walter A. FRET in orthogonally arranged chromophores. Eur J Org Chem. 2008;4559–62.
101. Langhals H, Esterbauer AJ, Walter A, Riedle E, Pugliesi I. Förster resonant energy transfer in orthogonally arranged chromophores. J Am Chem Soc. 2010;132:16777–82.
102. Nalbach P, Pugliesi I, Langhals H, Thorwart M. Noise-induced Förster resonant energy transfer between orthogonal dipoles in photoexcited molecules. Phys Rev Lett. 2012;108:218302–1–218302-5.
103. Langhals H, Walter A. FRET in dyads with orthogonal ohromophores and minimal spectral overlap. 2019. submitted.
104. Renger T, Dankl M, Klinger A, Schlücker T, Langhals H, Müh F. Structure-based theory of fluctuation-induced energy transfer in a molecular dyad. J Phys Chem Lett. 2018;9:5940–7. DOI: 10.1021/acs.jpclett.8b02403.
105. Václav P, Sláma V, Lincoln CN, Langhals H, Riedle E, Mančal T, et al. Geometric dependencies of vibronically mediated excitation transfer in rylene dyads. arXiv.org, e-Print Archive, Physics. 2018:1–8. https://arxiv.org/pdf/1808.02467.pdf.
106. Langhals H, Jona W. Intense dyes through chromophore-chromophore Interactions: Bi- and trichromophoric perylene-3,4:9,10-bis(dicarboximide)s. Angew Chem. 1998;110:998–1001. Angew Chem Int Ed. 1998;37:952–955.
107. Langhals H, Jona W. Identification of carbonyl compounds by fluorescence: A novel carbonyl derivating reagent. Chem Eur J. 1998;4:2110–16.
108. Langhals H, Demmig S, Huber H. Rotational barriers in perylene fluorescent dyes. Spectrochim Acta. 1988;44A:1189–93.
109. Langhals H, Riedle E, de Vivie-riedle R. unpublished results. 2011.
110. Mohr GJ, Spichiger UE, Jona W, Langhals H. Using *N*-aminoperylene-3,4:9,10-tetracarboxyl-bisimide as a fluorogenic reactand in the optical sensing of aqueous propionaldehyde. Anal Chem. 2000;72:1084–7.
111. Tasior M, Gryko DT, Shen J, Kadish KM, Becherer T, Langhals H, et al. Energy- and electron-transfer processes in corrole-perylenebisimide-diphenylacetylene array. J Phys Chem C. 2008;112:19699–709.
112. Flamigni L, Ciuciu AI, Langhals H, Böck B, Gryko DT. Improving the photoinduced charge separation parameters in corrole-perylene carboximide dyads by tuning the redox and spectroscopic properties of the components. Chem an Asian J. 2012;7:582–92.
113. Flamigni L, Ventura B, Barbieri A, Langhals H, Wetzel F, Fuchs K, et al. On/Off switching of the perylene tetracarboxylic bisimide luminescence by means of substitution at the N position by electron rich mono-, di- and tri-methoxybenzenes. Chem Eur J. 2010; 16:13406–16.

114. Dessolin M. Reactivity of α-effect nucleophiles toward aryl acetates. Effect of substituent in leaving groups. Tetrahedron Lett. 1972;45:4585–8.
115. Liebman JF, Pollack RM. Aromatic transition states and the α effect. J Org Chem. 1973;38:3444–545.
116. Langhals H, Obermeier A, Floredo Y, Zanelli A, Flamigni L. Light-driven charge separation in isoxazolidine-perylene bisimide dyads. Chem Eur J. 2009;15:12733–44.
117. Woodward RB, Hoffmann R. The conservation of orbital symmetry. Angew Chem Int Ed. 1969;8:781–853.
118. Roda A. Chemiluminescence and Bioluminescence. Cambridge: Royal Society of Chemistry, 2011. ISBN 978-1-84755-812-1.
119. Brolin S, Wettermark G. Bioluminescence analysis, Weinheim: Wiley-VCH, 1992. ISBN 3-527-28194-0.
120. van Moer A, Ladyjensky J. Chemiluminescent solution based on substituted perylene. Eur Pat Appl. 1990;EP 403809 A2 19901227. Chem Abstr. 1991;114:256680.
121. Nowak B, Ladyjensky J. Multi-color chemiluminescent lighting device and method of making same. *US Patent* U.S. (1996), US 5508893 A 19960416. Chem Abstr. 1996;125:21964.
122. Scheibe G. Variability of the absorption spectra of some sensitizing dyes and its cause. Angew Chem. 1936;49:563.
123. Langhals H. Handling electromagnetic radiation beyond terahertz using chromophores to transition from visible light to petahertz technology. J Electric Electron Syst. 2014;3:125.
124. Förster T. Energiewanderung und Fluoreszenz. Naturwissenschaften. 1946;33:166–75.
125. Jelley EE. Spectral absorption and fluorescence of dyes in the molecular state. Nature. 1936;138:1009–10.
126. Langhals H, Jona W. Intense dyes through chromophore - chromophore Interactions: Bi- and trichromophoric Perylene-3,4:9,10-bis(dicarboximide)s. Angew Chem. 1998;110:998–1001. Angew Chem Int Ed Engl. 1998;37:952–955.
127. Davydow AS. Theory of molecular excitations, Transl. H. Kasha und M. Oppenheimer, Jr., New York: McGraw-Hill, 1962.
128. Langhals H, Pust T. Lipophilic optical supramolecular nano devices in the aqueous phase. Green Sustainable Chem. 2011;1:1–6.
129. Langhals H, Ismael R. Cyclophanes as model compounds for permanent, dynamic aggregates - induced chirality with strong CD effects. Eur J Org Chem. 1998;1915–17.
130. Langhals H, Rauscher M, Mayer P. A sustainable preparation of functional perylenophanes by domino metathesis. Green Sustainable Chem. 2019;9:38–77.
131. Chamberlin GJ, Chamberlin DG. Colour, Its measurement, computation and application, London: Heyden & Sons Ltd, 1980. ISBN 0-85501-222-6.
132. Richter M. Einführung in die Farbmetrik, 2nd ed. Berlin: de Gruyter, 1981:ISBN 3-11-008209-8.
133. Aubert C, Fünfschilling J, Zschokke-Gränacher I, Langhals H. Hochempfindliches Nachweisverfahren auf der Basis der Fluorescenz durch Laser-Anregung. Zeitschr Analyt Chem. 1985;320:361–4.
134. Renge I, Hubner CG, Renn A, Langhals H, Wild UP. Slow photochemical transformations of single dye molecules in polymer environment at room temperature. J Lumines. 2002;98:91–6.
135. Günes S, Neugebauer H, Sariciftci NS. Conjugated polymer-based organic solar cells. Chem Rev. 2007;107:1324–38.
136. Langhals H, Saulich S. Bichromophoric perylene derivatives: Energy transfer from non fluorescent chromophores. Chem Eur J. 2002;8:5630–43.

Andrew Towns

34 Fulgide dyes

Abstract: This article outlines the general structures and photochromism characteristic of fulgide dyes and their most important related analogues. It provides an overview of synthetic routes to such derivatives in addition to exemplifying how typical structural variations influence photochromic behavior. A brief survey then follows, giving a flavor of the applications that have been – and continue to be – sought for them in the capacity of functional dyes.

Keywords: photochromism,photochromic, colorant, dye, functional, fulgide, fulgimide

The photosensitivity of fulgide derivatives was first reported a century ago [1]. This class of dye formed the subject of many investigations with great strides being made in accessing a wide range of photochromic properties during the past four decades. While early types furnished relatively weak T-type photochromism, dye chemists have accrued much knowledge concerning the impact of structural permutations on photochromic and chemical properties. They now design colorants that exhibit intense P-type behavior across the visible spectrum and beyond. Consequently, while fulgide and related dyes have only enjoyed niche commercial use to date, this class remains of interest as a source of photoactivated switches in many disparate fields of research. The remainder of the article introduces the most important kinds of such dyes, outlining their general structural and photochromic characteristics. Overviews will be given of routes to their synthesis in addition to past and potential applications as functional colorants.

34.1 Fulgide dye characteristics and molecular structure

Although this article is titled "Fulgide dyes", it also covers their close analogues shown in Figure 34.1, all of which are derivatives of 1,3-butadiene-2,3-dicarboxylic acid. Colorants **1** that are anhydrides of this compound, i.e. possess a bismethylene-succinic anhydride skeleton, go by the name 'fulgide'. (The inspiration behind the term, derived from the Latin 'fulgere' in the sense of 'to shine' or 'to gleam', came from the glistening appearance of crystals of structurally simple early examples [2].) The most useful cousins of these dyes are the succinimide-based 'fulgimide' derivatives **2**. Other related types that exhibit pronounced photochromism include the less

This article has previously been published in the journal Physical Sciences Reviews. Please cite as A. Towns Fulgide Dyes *Physical Sciences Reviews* [Online] 2021, 6. DOI: 10.1515/psr-2020-0173

https://doi.org/10.1515/9783110587104-034

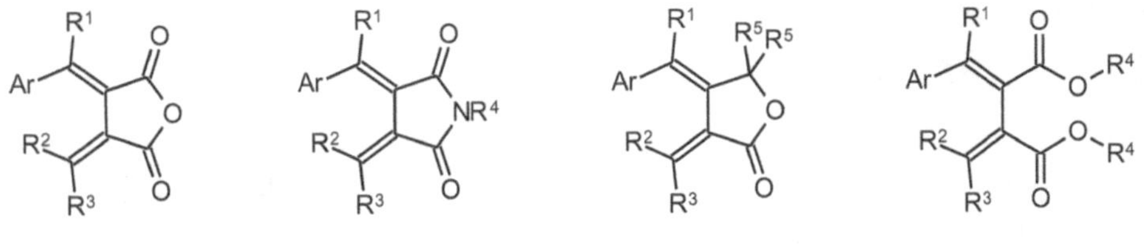

1 2 3 4

Ar = phenyl, hetaryl; $R^1 = R^2 = R^3$ = alkyl, cycloalkyl; R^4 = (substituted) alkyl, etc.; R^5 = H, alkyl

Figure 34.1: Generic structures of fulgide and related dyes.

commonly encountered lactones **3** and diesters **4**, named 'L-fulgenolides' and 'fulgenates', respectively [3]. Many examples of these and other sub-classes are known [4,5,6,7].

Derivatives **1**–**4** share the following features:

- at least one aryl ring attached to at least one of the methylene functions;
- substituents which orient the three double bonds at the core of their structures into a *s*–*cis*–*cis*–*cis*-like configuration as illustrated for the furylfulgide **5** in Scheme 34.1 where these bonds are highlighted in orange.

These criteria prime the dye for a light-driven photocyclization reaction to create intensely colored species, usually through absorption of ultraviolet (UV) and/or short-wavelength visible (blue) light. This conformation is exemplified by **5E** in Scheme 34.1. Use of the letter 'E' for this structure ties in with its designation as an *E*-isomer and distinguishes it from the non-photochromic *Z*-isomer **5Z.** The near-cyclic 1,3,5-hexatriene motif of **5E,** absent in **5Z,** undergoes 6π-electrocyclization, indicated by the white arrows in Scheme 34.1.

The process of ring-closure occurs in a matter of picoseconds. The shift in electron density creates a more bathochromic ring-closed 7,7*a*-dihydrobenzofuran **5C,** with a 1,3-cyclohexadiene fragment at its heart. (An alternative to the suffix 'C' often seen in connection with fulgides to denote this cyclized colored form is 'P'.) This isomeric photoproduct is intensely red because it comprises a polymethine chromophore [8], highlighted as pink in Scheme 34.1, that absorbs energies corresponding to wavelengths of light which lie in the visible region.

The substituents on the fulgide skeleton determine whether reversion by ring-opening can occur thermally (T-type) or photochemically (P-type). In the latter instance, only longer wavelength light triggers the switch back to the open form: in the absence of white light, or the specific visible wavelengths required, the colored photoisomer persists. In contrast, T-type fulgides revert spontaneously, but gradually, to their colorless or weakly colored ring-open isomers in the dark. They typically display an element of P-type character, whereby the thermal reversion is accelerated by irradiation with white light.

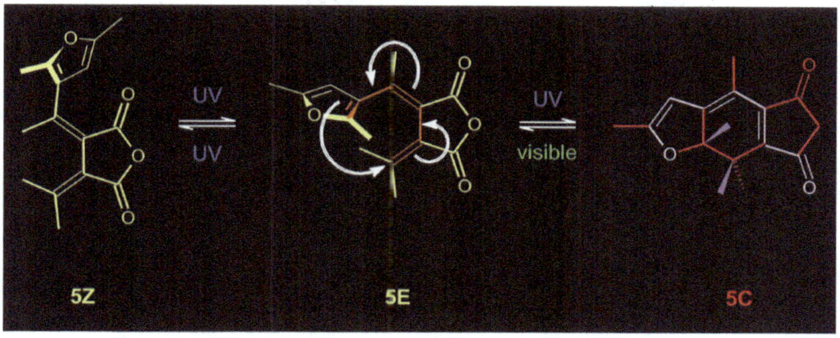

Scheme 34.1: Photoisomerization of a furylfulgide **5**.

It was not until the 1970s and 1980s that the pool of T-type fulgides became supplemented with the first examples of P-type analogues. This broadening of behavior and its refinement proved crucial in opening up the class to potential applications that depend on substances whose two-way switching of properties can be controlled in each direction by light. The introduction of two key structural modifications that block thermal reversibility provided the breakthrough:

- use of a heterocycle as the pendant substituent on one of the methylene functions;
- replacement of hydrogen atoms with alkyl groups on the carbon atoms at either end of the bond formed upon ring closure.

Colorants designed in this way obviate photochemical side-reactions associated with shift or loss of hydrogen atoms leading to irreversible rearrangements that compete with ring-opening. In addition, their ring-closed forms do not absorb strongly in the UV regions which efficiently produce reversible photocyclization, enabling high percentages of conversion. Most importantly, steric hindrance by the alkyl groups prevents ring-opening from occurring thermally, but not photochemically. Consequently, fulgides bearing these substituents can essentially be switched in a controlled manner between ring-closed and -opened states through irradiation with light alone. The furan ring and methyl groups (the latter shown in purple in Scheme 34.1) of fulgide **5** leads to efficient conversion of **5E** to **5C** upon exposure to UV, eventually becoming almost completely converted [9]; the colorant remains locked into the ring-closed colored form until irradiated with visible light.

How does the hindrance inhibit T-type behavior rather than P-type photochromism in this instance? The answer lies in the stereochemistry of the ring-opening reaction. The obstruction of the thermal pathway from **5C** relates to the feasibility of rotation of substituents on the single bond when it is ruptured. Ring-opening of fulgides is a pericyclic reaction with a concerted shift of six π-electrons. The interaction of orbitals involved in this kind of transition dictates that, for it to take place from the ground state (i.e. as a thermal process at typical ambient temperatures), the groups at each end of the breaking bond must rotate in opposite directions ('disrotatory' mode). Rotation in the same direction ('conrotatory mode') is thermally forbidden. Conversely, conrotation from a photochemically excited state of **5C** is allowed, while the disrotatory mode is not. In principle, both thermal disrotation and photochemical conrotation may occur to generate weakly colored ring-open species. However, movement of alkyl groups in a disrotatory fashion would lead to a steric clash and is thus energetically unfavorable. Conrotatory motion, on the other hand, avoids close approach of the methyl substituents and may take place following absorption of visible light energy. Scheme 34.2 captures the essence of what is stated above. It constitutes the specific application to **5C** of the general Woodward–Hoffmann rules for rationalizing pericyclic reactions. A thermal disrotatory mode during ring-opening would lead to a steric clash between the *syn*-methyl groups at the 7- and 7*a*-positions, which Scheme 34.2 shows in purple. No such clash occurs between the methyl substituents upon conrotation, so photochemical transformation to the ring-opened isomers readily occurs. Fulgide **5** thus exhibits P-type photochromism. Further mechanistic detail on photoisomerization as it relates to fulgides can be found in [4] and [6].

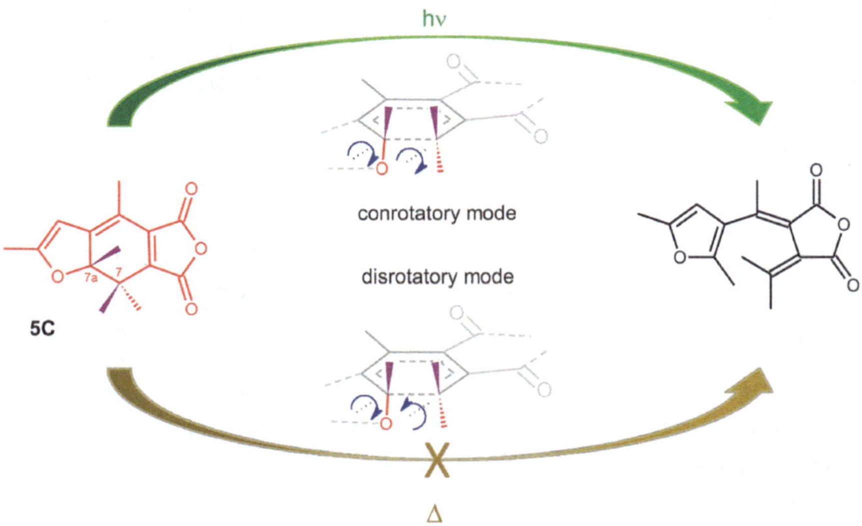

Scheme 34.2: Modes of ring-opening of fulgide **5**.

The brief summary outlined above gives a glimpse of the complex steric and electronic factors governing the photochromism of fulgide and related dyes. Molecular structure very heavily influences measures such as efficiency of reversible photo-coloration and -bleaching, more specifically quantum yields of ring-closure and -opening. It has a bearing on the occurrence of undesirable irreversible side-reactions, as touched upon above, and reversible competing pathways, both of which detract from photochromic performance. For example, fulgide dyes like **5** exist in ring-open form as one of two geometric isomers: in this instance, **5Z** and **5E**. Only the latter directly cyclizes to **5C** following absorption of a UV photon. The alternative pathway, isomerization of **5E** to **5Z** (see Scheme 34.1) merely renders photocoloration less efficient. The *Z*-isomer must be photoswitched back to the *E*-form before ring-closure can occur. Section 34.3 will outline some successful design strategies exist to minimize formation of unproductive isomers and to encourage the efficiency of photocoloration reactions from *E*-isomers. For much greater detail about the photochemical pathways involved as well as the complex electronic and steric influences on photoisomerization, please refer to the excellent review [10].

Fulgides and their related dyes manifest photocoloration when dissolved in solvents, polymers and glassy amorphous materials. Dye crystals may also exhibit photochromism when irradiated with UV. Such an effect tends to be a surface phenomenon see, for example, [11] – although the proportion of photoisomerization throughout the crystal lattice can be increased substantially by creating more free volume within it by incorporating a bulky, flexible group into the fulgide's structure [12]. (Another means is irradiation with a pulsed laser emitting in the long wavelength visible region, which produces coloration throughout crystals as a consequence of greater penetration into the solid and triggering of isomerization by means of two-photon absorption [13].)

Some members of the class exhibit color changes in response to stimuli other than light. Mechanical grinding induces striking irreversible [14, 15] and reversible [16] transformations of hue in certain fulgide-related derivatives. Although the labels 'tribochromism' and 'piezochromism' were applied, respectively, to these phenomena, their use in these contexts is not strictly correct since the former change was not reversible, while the latter required irradiation with white light to be reversed. The development of fulgides that are heliochromic, i.e. possessing a T-type sensitivity to strong sunshine [17], proved of more industrial interest owing to their potential utility in production of sunlight-responsive ophthalmic lenses. However, despite the endeavors in this direction (see Section 34.4.2), fulgide-based systems are not employed in the sector. P-type dyes thus dominate the minor existing uses to which the fulgide class is put and the academic work attempting to develop new technologies with them. The following section on dye chemistry therefore focuses mostly on fulgide colorants that can be switched controllably between states with light alone.

34.2 Synthesis of fulgides

The principal route of preparation of fulgides, dating from the late nineteenth century [18], requires multiple synthetic steps from commercially available raw materials as shown in Scheme 34.3. The first step is Stobbe condensation of a dialkylsuccinate **6** and an aliphatic ketone in the presence of strong base, typically an alkoxide (with *t*-butoxide in *t*-butanol often being effective) or hydride. The reaction mixture is quenched with water and the aqueous alkaline layer in which the crude half-ester Stobbe product **7** is dissolved then extracted with water-immiscible non-polar solvent to remove succinate dimer by-product. Acidification of the aqueous layer then leads to separation of the half-ester that is taken up in a non-polar water-immiscible solvent for isolation. Alternatively, the reaction mass may simply be drowned to aqueous acid and solvent-extracted. Stobbe condensate **7** is then esterified to a dialkylfulgenate **8** employing acid catalysis. A second condensation reaction is performed with an arylketone, again in the presence of an alkoxide. Other strong bases may work better depending on the extent of crowding in the succinate ester derivative and carbonyl compound. Hydrolysis of the half-ester product **9** with alkali furnishes the corresponding fulgenic acid derivative **10** that may then be dehydrated to afford the fulgide. Amongst the reagents used for this purpose, the cheapest are acetic anhydride and acetyl chloride. Purification by recrystallization, preceded by column chromatography as a last resort, furnishes the photochromic dye typically as a pale yellow, bright yellow or orange solid.

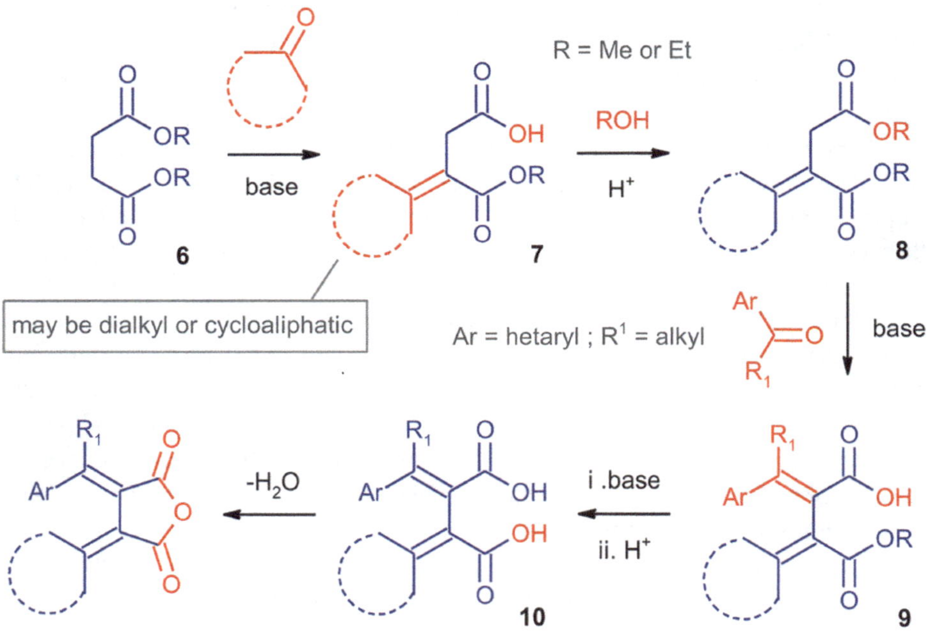

Scheme 34.3: Conventional synthetic route to fulgide dyes from dialkylsuccinates.

Only a few fulgide dyes are available off-the-shelf at gram scale. Their high cost is reflective of low production volumes and multiple synthetic steps. Pricing of fulgide **5** and its adamantylidene analogue **11**, respectively, known as Aberchrome 540 and Aberchrome 670 (see Figure 34.2), lies in the region of hundreds of dollars per gram – see, for example, [19] and [20]. Another dye, Aberchrome 999 P (**12**), was commercially available in its ring-closed, i.e. colored, form [21].

The first published syntheses of imide analogues **2** of fulgides appeared half a century later than the anhydrides [22]. The sub-class received its 'fulgimide' label in the

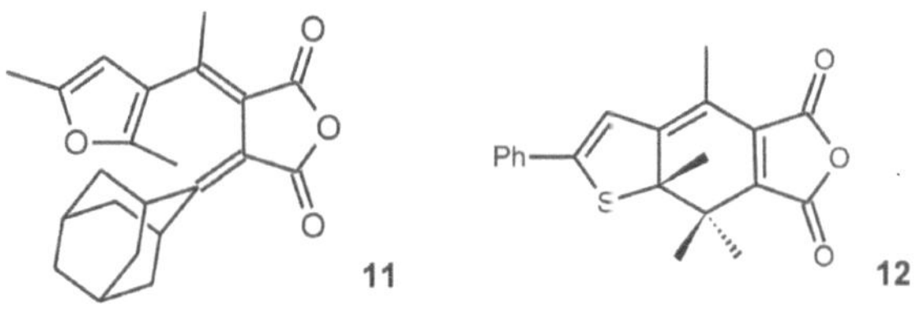

Figure 34.2: Molecular structures of two fulgide dyes that were commercialized.

1960s [23]. Several synthetic routes are available [5, 24]; Scheme 34.4 shows the most commonly used pathways. The parent fulgide acts as a synthon in one variant, involving a two-step conversion *via* a succinamic acid **13**. Heating of the anhydride with ammonia or primary amine (sometimes accompanied by a catalyst, e. g. zinc chloride [25]) produces this amidic acid precursor, which is then cyclized to the imide with a dehydrating agent tolerated by the other functional groups present. Mild cyclization conditions may allow the corresponding isofulgimide to be intercepted, whereas more forcing and/or acidic conditions ultimately generate fulgimides [24]. Another strategy, which is arguably more elegant since it cuts out some synthetic steps shown in Scheme 34.3, relies on reaction between the succinic acid half ester precursor **14** with a Grignard salt of a primary amine. Finally, in the case of the unsubstituted imide, the NH fragment may serve as an anchor point through further functionalization to make photochromic monomers or link the unit to polymers as generalized by **15** [5].

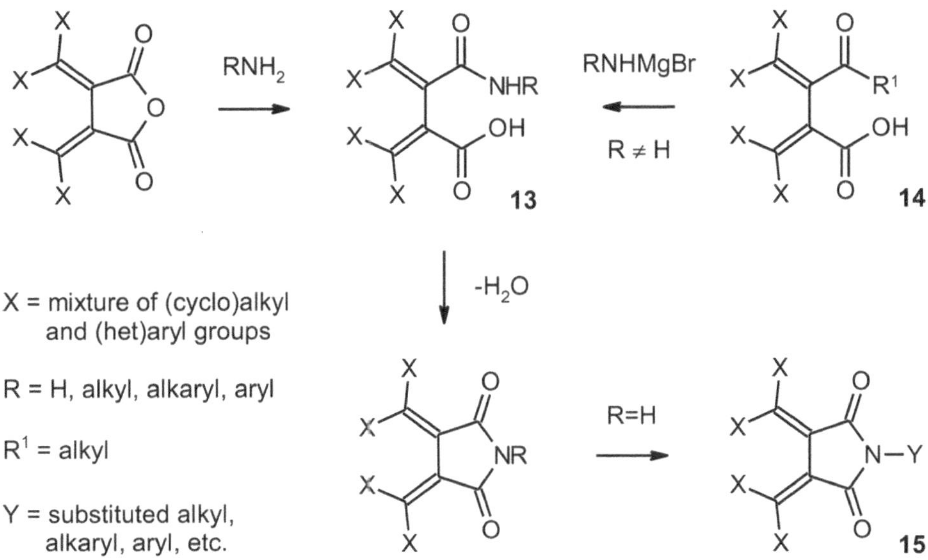

Scheme 34.4: Typical synthetic routes to fulgimide dyes.

Replacement of the central oxygen atom of the succinic anhydride ring with sulfur is possible through direct reaction of fulgides with sodium hydrosulfide in toluene, giving more bathochromic photocoloration while somewhat sacrificing photocoloration quantum yield [26]. However, as will be discussed in the next section, functionalization of fulgides with active methylene compounds yields analogues that cyclize to colorants with much larger red shifts, pushing absorption well towards the red end of the visible spectrum and even into the near infra-red. Knoevenagel condensation of fulgides with active methylene compounds, such as malononitrile [27] or methyl cyanoacetate [28], catalysed by aliphatic amines, produces fulgenate amine salts that dehydrate to fulgenolide analogues. The site of reaction depends on whether the *Z*- or *E*-isomer acts as synthon because of steric hindrance differing at each carbonyl group. For example, *E*-isomer **16** and *Z*-isomer **17** give salts **18** and **19**, respectively, which cyclize to the corresponding *E*- and *Z*-isomers **20** and **21** (see Scheme 34.5).

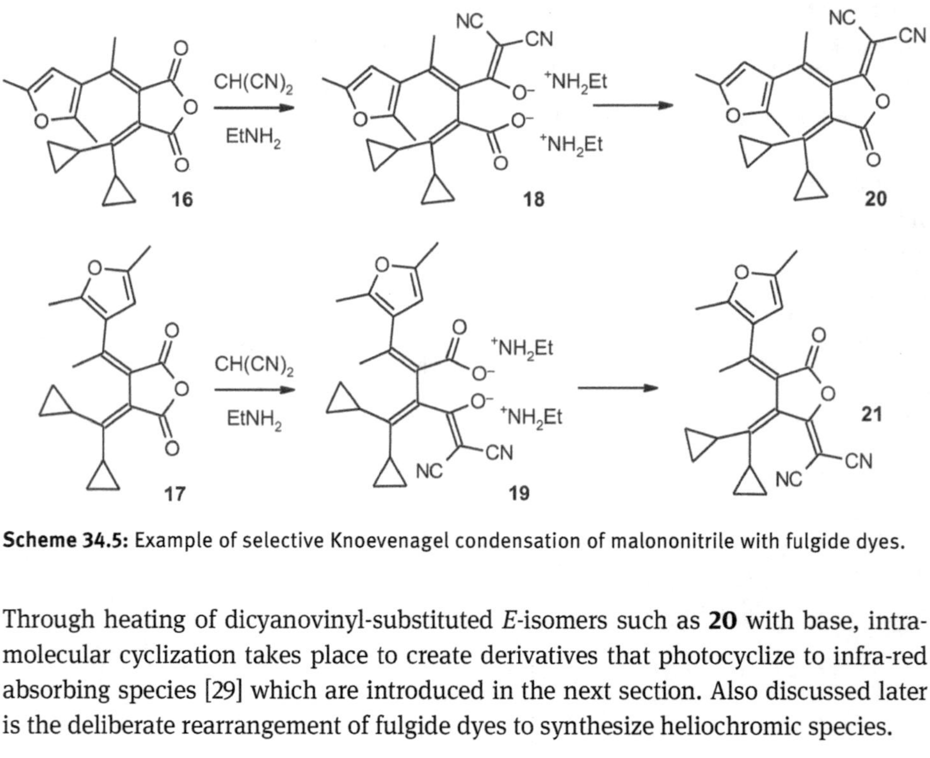

Scheme 34.5: Example of selective Knoevenagel condensation of malononitrile with fulgide dyes.

Through heating of dicyanovinyl-substituted *E*-isomers such as **20** with base, intramolecular cyclization takes place to create derivatives that photocyclize to infra-red absorbing species [29] which are introduced in the next section. Also discussed later is the deliberate rearrangement of fulgide dyes to synthesize heliochromic species.

34.3 Photocoloration and constitution of fulgide colorants

Much has changed since the following statement was published half a century ago: "Typically the fulgides are yellow to orange and change to a deeper orange, red, or reddish-brown on irradiation." [30]. Not only does the gamut of the class's photocoloration now span the visible spectrum and stretch into the near infra-red, but the range of behaviors exhibited by its members has been broadened from T-type to P-type switching. This section will illustrate how these innovations have been achieved through careful selection of substituents and draw attention to the key structural features of photochromic fulgides.

As was mentioned in Section 34.1, efficient P-type photochromism becomes possible in fulgides bearing a single aromatic ring substituent (i.e. of structure **1**) when it is a heterocycle whose 2-position is substituted with an alkyl group. The *E*- and photocyclized-isomers possess absorption spectra that do not overlap excessively, thereby encouraging intense photocoloration and effective photobleaching through high conversion to ring-closed and -opened states. The 2-substituents prevent hydrogen-shifting side reactions and assist in preventing thermal fading. Table 34.1 shows how the heteroatom strongly influences hue of photocoloration without significantly affecting the UV absorption of the *E*-isomer [31]. The deep red coloration of the furylfulgide example (Aberchrome 540, **5**) becomes increasingly bathochromic on switching to thienyl and pyrazolyl systems, albeit at the cost of lower conversion upon irradiation with UV.

Table 34.1: Influence of heterocycle on photostationary state properties of fulgides **1** (R^1–R^3 = Me) in toluene solution [31].

Ar	λ_{max} (nm) E-isomer	λ_{max} (nm) Ring-closed Isomer	Photoconversion
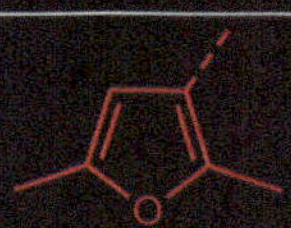	345	494	↑
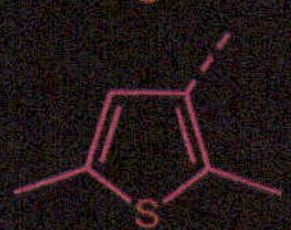	333	520	
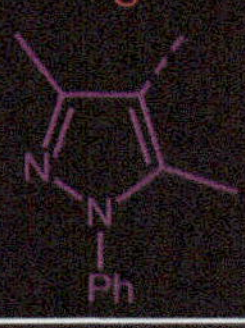	335	550	

The importance of steric influences in fulgide photochromism is evident upon comparison of analogues of **5** that are increasingly hindered as shown in Table 34.2. Increasing bulk of R (e. g. R = *i*-Pr or R = *t*-Bu compared to R = Me) suppresses E→Z photoisomerization owing to more pronounced crowding in the transition state [32]. The net effect is growth in the quantum yield for ring-closure. Rate of photocoloration thus increases in spite of the greater hindrance lowering UV absorption intensity as indicated by reduction in extinction coefficients when R possess greater bulk [33, 34]

Table 34.2: Influence of steric hindrance on absorption maxima (in nm), extinction coefficients (in $M^{-1} cm^{-1}$) and quantum yields Φ of photocoloration (*E*→*C*, 366 nm), photoisomerization (*E*→*Z*, 366 nm) and photobleaching (*C*→*E*, 492 nm) in 0.1 mmol chloroform solution at ~ 20°C unless marked otherwise [33, 34].

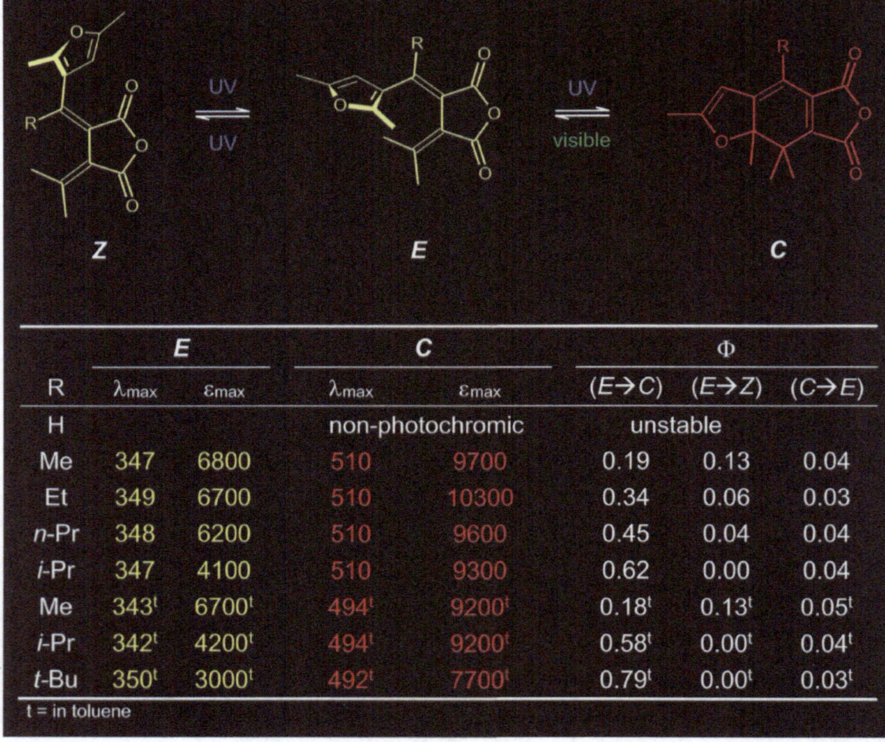

R	*E*		*C*		Φ		
	λ_{max}	ε_{max}	λ_{max}	ε_{max}	(*E*→*C*)	(*E*→*Z*)	(*C*→*E*)
H			non-photochromic		unstable		
Me	347	6800	510	9700	0.19	0.13	0.04
Et	349	6700	510	10300	0.34	0.06	0.03
n-Pr	348	6200	510	9600	0.45	0.04	0.04
i-Pr	347	4100	510	9300	0.62	0.00	0.04
Me	343^t	6700^t	494^t	9200^t	0.18^t	0.13^t	0.05^t
i-Pr	342^t	4200^t	494^t	9200^t	0.58^t	0.00^t	0.04^t
t-Bu	350^t	3000^t	492^t	7700^t	0.79^t	0.00^t	0.03^t

t = in toluene

Table 34.2 also reveals that, generally, solutions of *E*-forms and ring-closed photoisomers exhibit positive solvatochromism. In other words, as solvent polarity increases (e. g. on switching from toluene to chloroform) absorption maxima shift to longer wavelengths.

Greatly increasing steric crowding on the fulgide skeleton with a bulky and rigid group is also a valuable tool for boosting the efficiency of photochemical ring-opening, i.e. raising Φ(*C*→*E*). Replacing the 7,7-dimethyl pattern with a 2-adamantyl group increases strain on the 7,7*a*-bond, leading to much higher quantum yields of photobleaching [35] (see Table 34.3). Photocoloration is not much affected, i.e. Φ (E→C) remains high and Φ(E→Z) stays low, aside from a moderate bathochromic shift. A norbonyl group on the other hand is insufficiently bulky to make a big difference to photobleaching efficiency.

Table 34.3: Influence of steric hindrance on absorption maxima (in nm) and quantum yields Φ of photocoloration (*E*→*C*, 366 nm), photoisomerization (*E*→*Z*, 366 nm) and photobleaching (*C*→*E*, 492 nm) in toluene solution at ~ 20°C [32, 33, 35].

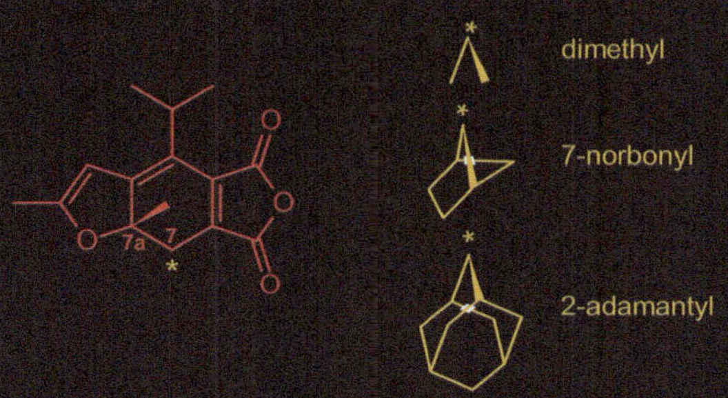

*	λ_{max}		Φ		
	E	*C*	(*E*→*C*)	(*E*→*Z*)	(*C*→*E*)
isopropylidene	342	494	0.58	0.00	0.04
7-norbonylidene	343	515	0.56	0.01	0.05
2-adamantylidene	337	520	0.51	0.02	0.26

The gamut of photocoloration exhibited by P-type fulgide-related dyes stretches across the visible spectrum and spills over into the infra-red (i.e. > 780 nm) as illustrated by Table 34.4.

Switching both 7-methyl groups (R) of furylfulgide **5C** for cyclopropyl rings, or replacing the heterocycle with a thiophene ring (X = O → S), produces a modest bathochromic shift in each case. Increasing the electron donor character of substituents on the heterocycle, or extending the conjugation (e. g. Y = Me → Ph), leads to sizeable red-shifting of absorption. The combination of these modifications transforms the deep red photocoloration of **5** to violet. However, bathochromism achieved in this manner is often accompanied by a reduction in photostability and a decline in the efficiency of photobleaching [27].

As was briefly mentioned in the previous section, (i) replacement of one of the carbonyl groups by a dicyanovinyl function furnishes derivatives such as **20C** and **21C** which are blue to bluish-green in toluene solution, and (ii) intramolecular cyclization of dye **20** produces a colorant **22** capable of furnishing very bathochromic P-type coloration [29] (see Scheme 34.6). The six-membered ring of its ring-closed form

Table 34.4: Absorption maxima (in nm) of cyclized dyes in toluene solution [29].

	X = O		X = S	
	Y = Me		Y = Me	Y = Ph
	R = Me	R = CP	R = Me	R = CP
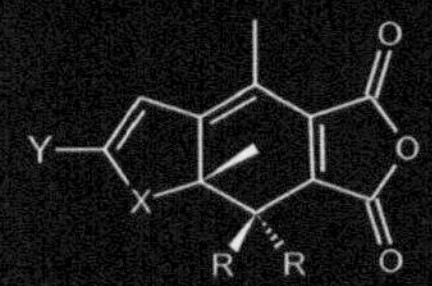 **5C**	496	514	520	566
 **20C**	601	594	598	653
 **21C**	610	634	642	669
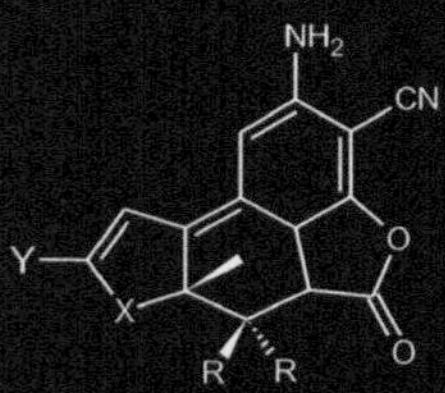 **22C**	698	720	681	776

CP = cyclopropyl

22C acquires aromatic character in its excited state (**22C***), narrowing the energy difference with the ground state, which translates to a substantial bathochromic shift. The increase in polarity upon excitation confers positive solvatochromism on **22C** so that absorption shifts firmly into the near infra-red region upon irradiation of the dye in polar solvents with UV or blue light (see Table 34.4).

The photochromism of fulgimides does not differ greatly from fulgide analogues [5, 6]. For example, *N*-phenylsuccinimide derivatives possess similar absorption maxima in toluene solution to corresponding fulgides (±10 nm) in addition to exhibiting comparable photo-responses [31]. However, they are more chemically stable to moisture and protic solvents, whilst also offering a point of attachment on their nitrogen atoms to other small molecules that does not significantly perturb photochromic characteristics. Fulgimide units may also be grafted onto macromolecules in this way.

This section provides only a very brief tour of the complex influences on photochromism of this dye class. Many more examples can be found in [5, 6, 7]. Although

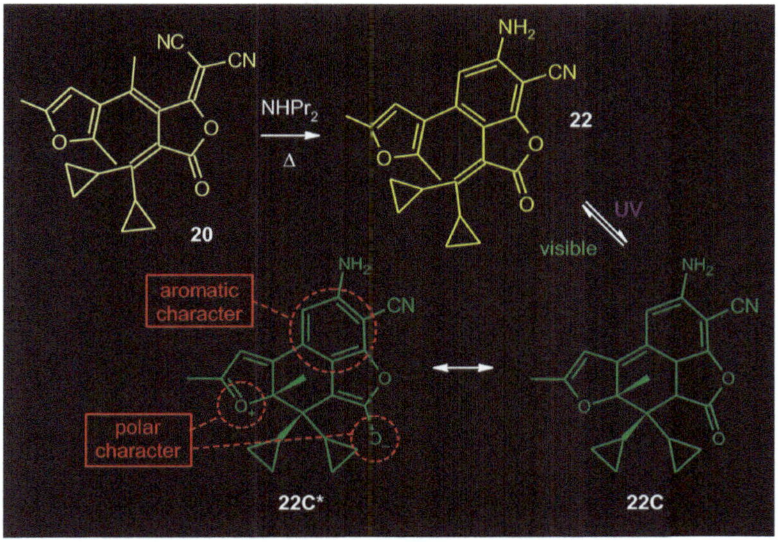

Scheme 34.6: Example of P-type infra-red absorbing fulgide-derived dye.

modifications to structure tend to simultaneously affect absorption properties of *E*-, *Z*- and *C*-isomers as well as the quantum efficiencies of interconversions between them, general design rules can be drawn up. Figure 34.3 highlights some of the most important ones for fulg(im)ides.

34.4 Applications of fulgide dyes

The discovery of the light sensitivity of some of the first examples of fulgides synthesized a few years into the twentieth century makes this class one of the earliest known groups of photochromic substances. The mechanism of its photochromism was not elucidated until over a half a century later [36]. Understanding of their behavior aided the design, as outlined in Section 34.1, of better performing dyes, which in turn stimulated interest in them. The report in 1981 [37] of the first fulgide, **5**, with useful P-type photochromism represented a major breakthrough: it heralded the dawn of renewed interest in the class with further development of robust P-type examples being a goal for their potential uses in memory elements of ultra-high density optical data storage or as switching components in photonic devices [6, 38]. (A close analogue of **5** that narrowly but crucially lacked features needed for useful P-type photochromism, e. g. a 2-methylfur-3-yl motif, was known as long ago as 1905 [39], yet its behavior was only rationalized decades later [40].) However, despite the advances made in dye design, these technologies have not progressed beyond prototypical systems. Research and development efforts during the 1980s and 1990s directed at rolling out fulgide-based photochromic ophthalmic lenses did not

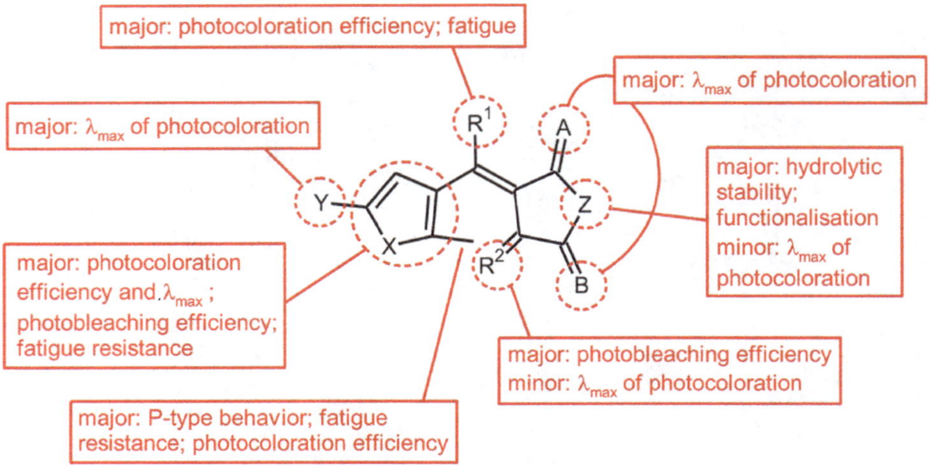

Figure 34.3: Some key structural influences on photochromism in the fulg(im)ide class.

ultimately meet with success (see Section 34.4.2). Commercial use of fulgide-derived functional photochromic materials remains restricted to small and specialized markets, e. g. actinometry (see Section 34.4.1), although attempts to exploit them more widely continue (see Section 34.4.3).

34.4.1 Actinometry

Fulgide dyes have long been sold as chemical actinometers to determine the exposure of systems to UV or visible light [5, 7]. For use in solvent solution or in dosimeter films, they were marketed by the now-defunct company Aberchromics Ltd, spun out from what at the time was University College Wales, Aberystwyth. Dyes **5** and **11** have long been used to quantify near UV exposure. Their quantum yields of photocoloration are known as a function of wavelength and temperature, enabling calculation of photon flux through measurement of absorbance of fulgide photocoloration. Conversely, **12** was advocated as a measure of visible light dosage through monitoring of photobleaching [21]. An indolylfulgide, originally of note as a relatively fatigue-resistant colorant [41], was recently re-visited and promoted as a convenient actinometer for a broad range of wavelengths, covering the near UV and much of the visible spectrum (340–700 nm) [42].

34.4.2 Ophthalmic lenses

The fulgide class was also the focus of an effort in the early 1980s to develop sunlight-responsive plastic lenses whose performance rivalled that of commercial photochromic glass. In this instance, T-type behavior was required to furnish rapid intense coloration triggered by near UV solar radiation that rapidly thermally bleached in its absence, i.e. heliochromism whereby (i) quantum yields of photocoloration from UV are high; (ii) quantum yields of photobleaching by white light are low; (iii) rates of thermal fade are rapid over the range of outdoor ambient temperatures. The T-type fulgides known before the discovery of their P-type counterparts were unsuitable; however the latter ironically offered a route to candidate 'heliochromes'. A collaboration between Pilkington plc, The Plessey Company plc, and Aberchromics Ltd in the UK led to the development of heliochromic purple to blue fulgides and yellow fulgimides to create neutral photocoloration [43]. In the former case, the heliochrome was produced by UV irradiation of a hot solution of specific P-type fulgides, inducing an irreversible isomerization to a T-type dye [4, 7, 44, 45]. This approach is exemplified by Scheme 34.7. Irradiation of **23E** leads to cyclization, forming purple **23C**, which then experiences a thermal 1,5-*H* shift of its 7*a*-proton, affording a ring-closed near colorless isomer **24**. This compound responds to UV by ring-opening to a colored (blue) species that thermally fades in a reversible manner. Note the emphasis by underlining to highlight the contrast between this type of photo-transformation and that of the fulgides previously described in this article which is the converse! It was even realized that the synthetic process could be shortened by heating and irradiating the dye within the lens matrix to effect this last step in the conversion. This clever approach ultimately proved unsuccessful. The fulgimides were tricky to make and/or insufficiently stable for commercial lenses, and so were replaced by yellow/orange naphthopyrans [46]. Unfortunately, the difference in temperature dependence of photochromism between the latter and the blue/purple fulgides led to undesirable variations in lens hue during darkening and fading owing to mismatches in photocoloration and thermal bleaching rates between dyes. In addition, the fulgides could not match the resilience of the naphthopyrans. Pilkington's attention thus turned in the mid-80s turned to lens formulations based upon blue/purple spironaphthoxazines instead. For a fascinating account of the development work with fulgides in regard to ophthalmic lens use (and what came after), see [47].

Tokuyama Corp. in Japan kept the torch burning for fulgide-based heliochromes when its R&D effort pertaining to photochromic lenses began in 1985 and extended into the 1990s [48]. Combinations of fulgimide with either spiroindolinonaphthoxazine or spiroindolinopyridobenzoxazine [49] colorants were explored [50,51,52]. Limited commercial success was achieved through inclusion of low levels of glycidyl methacrylate in lens monomer formulations. Usage alongside photochromic naphthopyrans enabled the creation of sunlight-responsive lenses with neutral shades. However, the fulgimides' relative lack of photostability [53] compared to spirooxazine-

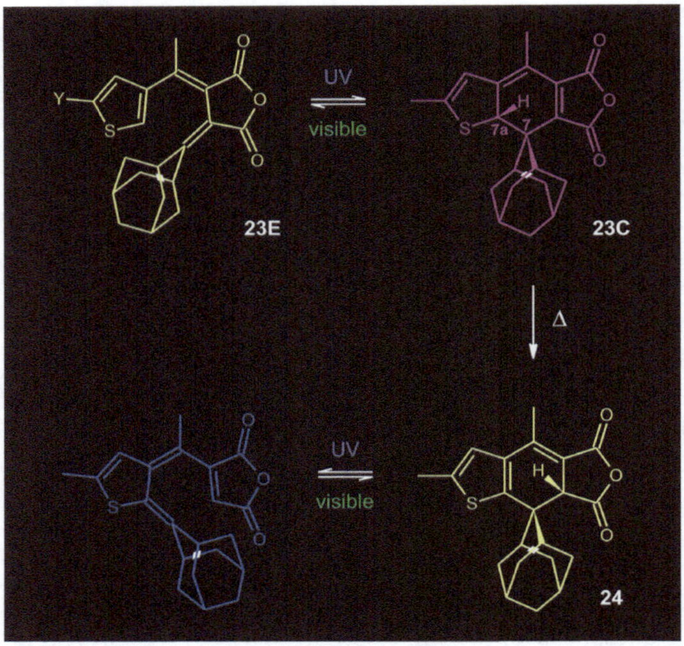

Scheme 34.7: Synthesis and photochromism of heliochromic fulgides exemplified by a thienyl derivative **24.**

based dyes [45] as well as the now-dominant family of naphthopyrans [54] ultimately contributed to them being squeezed out of use.

34.4.3 Other uses

Industry, academia and the military have expended much time and money trying to harness photochromic dyes as photoswitchable functional colorants in a plethora of disciplines from security printing to photonics to photopharmacology [17]. In terms of receiving such attention, the fulgide class has been no exception.

The long-held goal of adapting photochromic colorants to functional applications in digital optical data storage and processing continues to attract attention, some of it involving members of the fulgide class. Both fulgides and fulgimides remain of interest for various forms of image creation [55] as well as rewritable data storage [56–58] utilizing different techniques of non-destructive read-out. These include determination of fulgide dye state by monitoring properties such as refractive index [59] and fluorescence [60, 61]. Economic considerations aside, susceptibility to fatigue remains an obstacle to commercialization of fulgide-related systems. Proof of concept has been demonstrated of photoswitching between four different colors in mixed crystals produced from two fulgides; different irradiation wavelengths effect changes in one or both types of fulgide constituent. Such a system is aimed at applications like optical memory device fabrication, but they are too prone to fatigue [62].

34.5 Summary

The class of fulgide dyes and related derivatives is relatively old. For many years the photochromism of its members remained mysterious and of no practical use. Understanding of the phenomenon's origin gained around half a century ago, accompanied by a better appreciation of the influence of molecular structural features on photochromic properties, led to the envelope of these dyes' behavior being greatly expanded. The breakthrough discovery of efficient P-type switching in furylfulgides excited considerable interest in the class as a whole since it opened the door to their use in applications calling for two-way photocontrolled switching. In addition, the tuning of T-type photochromism to produce sunlight-responsive photocoloration formed the basis for industrial efforts to exploit them in ophthalmic lenses. Despite their promise in both these regards, cheaper and more robust technologies have relegated this class to the sidelines for now. Ongoing and very active research within numerous disciplines that relies on photochromic substances continues to make use of fulgide-based units. While dyes of this kind will remain only of commercial use in specialist applications for the short term, it is possible that further into the future they might enjoy another renaissance as functional photoresponsive materials.

References

1. Stobbe H. Die farbigen Anhydride der Butadiën-β,γ-dicarbonsäuren; ihr Verhalten gegen Licht und Warme (II. Abhandlung über Butadiënverbindungen.). Ber Deutschen Chemischen Ges. 1904;37:2236–40.
2. Stobbe H. Die Farbe der »Fulgensäuren« und »Fulgide« (7. Abhandlung über Butandiënverbindungen.). Ber Deutschen Chemischen Ges. 1905;38:3673–82.
3. Yokoyama Y, Kurita Y. Photochromism of fulgides and related compounds. Mol Cryst Liq Cryst Sci Technol Sect A Mol Cryst Liq Cryst. 1994;246:87–94.
4. Heller HG. Electronic materials: from silicon to organics. Miller LS, Mullin JB, editors. New York: Springer, 1991 Chapter 31 ("Photochromics for the future"):471–83.
5. Fan MG, Yu L, Zhao W. In: Crano JC, Guglielmetti RJ, editors. Organic photochromic and thermochromic compounds volume 1: main photochromic families. New York: Plenum, 1999 Chapter 4 ("Fulgide family compounds: synthesis, photochromism and applications"):141–206.
6. Yokoyama Y. Fulgides for memories and switches. Chem Rev. 100:1717–39.
7. Whittal J. In: Dürr H, Bouas-Laurent H editors. Photochromism molecules and systems . Amsterdam: Elsevier, 2003 Chapter 9 ("4n+2 Systems: Fulgides"):467–92.
8. Mustroph H. Polymethine dyes. Phys Sci Rev. 2020;5. DOI:doi.org/10.1515/psr-2019-0084.

9. Uhlmann E, Gauglitz G. New aspects in the photokinetics of Aberchrome 540. J Photochem Photobiol A: Chem. 1996;98:45–9.
10. Renth F, Siewertsen R, Temps F. Enhanced photoswitching and ultrafast dynamics in structurally modified photochromic fulgides. Int Rev Phys Chem. 2013;32:1–38.
11. Kaftory M. Photochromic and thermochromic compounds. I. Structures of (*E*) and (*Z*) isomers of 2-isopropylidene-3-[1-(2-methyl-5-phenyl-3-thienyl)ethylidene]succinic anhydride, C20H18O3S, and the photoproduct 7,7a-dihydro-4,7,7,7a-tetramethyl-2-phenylbenzo[*b*] thiophene-5,6-dicarboxylic anhydride (*P*), C20H18O3S. Acta Cryst. 1984;C40:1015–19.
12. Hettiarachchi CV, Weerasekara RK, Uekusa H. Crystalline state photochromism of 3-furylfulgides: impact of size and bond flexibility of the nonaromatic alkylidene group. Acta Cryst. 2015;B71:535–42.
13. Harada J, Nakajima R, Ogawa K. X-ray diffraction analysis of photochromic reaction of fulgides: crystalline state reaction induced by two-photon excitation. J Am Chem Soc. 2008;130:7085–91.
14. Asiri AM, Heller HG, Hursthouse MB, Karalulov A. Tribochromic compounds, exemplified by 3-dicyclopropylmethylene-5-dicyanomethylene-4-diphenylmethylenetetrahydrofuran-2-one. Chem Commun. 2000;799–800.
15. Asiri AMA, Heller HG, Hughes DS, Hursthouse MB, Kendrick J, Leusen FJJ, et al. A mechanophysical phase transition provides a dramatic example of colour polymorphism: the tribochromism of a substituted tri(methylene) tetrahydrofuran-2-one. Chem Cent J. 2014;8:70.
16. Asiri AMA, Cleeves A, Heller HG. Studies on piezochromic photochromic (*E*)-5-dicyanomethylene-3-[1-(2,5-dimethyl-3-furyl)ethylidene]-4-diphenylmethylenetetrahydrofuran-2-one and related photochromic compounds. J Chem Soc Perkin Trans.I 2000;1:2741–4.
17. Towns A. Photochromic Dyes. Phys. Sci. Rev. DOI: 10.1515/psr-2020-0191.
18. Stobbe H. II. Condensation des Acetophenons mit Bernsteinsäureester. Justus Liebigs Ann Chem. 1899;308:114–55.
19. Extrasynthese, (E)-Aberchrome 540 / 8200. www.extrasynthese.com/organics/12-e-aber chrome-540.html.
20. TCI America. Aberchrome 670 94856-25-4. www.tcichemicals.com/US/en/p/A2869.
21. International Union of Pure and Applied Chemistry Organic Chemistry Division Commission on Photochemistry. Chemical actinometry (IUPAC Technical Report). Pure Appl Chem. 2004;76:2105–46.
22. Goldschmidt S, Riedle R, Reichardt A. Über die Bisdiphenylenfulgide und die Spaltung der Bisdiphenylenefulgensäure in optisch active Komponenten. Justus Liebigs Ann Chem. 1957;604:121–32.
23. Hart RJ, Heller HG, Salisbury K. The photochemical rearrangements of some photochromic fulgimides. Chem Commun. 1968;1627–8.
24. Heller HG, Koh K, Elliot C, Whittall J. Fulgides and fulgimides for practical applications. Mol Cryst Liq Cryst Sci Technol Sect A Mol Cryst Liq Cryst. 1994;246:79–86.
25. Xiao J-P, Han Y, Chen Y, Li W, Fan M-G. Studies on synthesis and photochromism of a novel class of fulgimide: (Z)-4-Oxazolylfulgimide. Chin J Chem. 2004;22:100–2.
26. Badland M, Cleeves A, Heller HG, Hughes DS, Hursthouse MB. Photochromic heteroaromatic thiofulgides and dimethoxybutanoic acid lactones. Chem Commun. 2000;1567–8.
27. Heller HG, Koh K, Kose M, Rowles N. The design and development of new thermally stable infra-red active photochromic compounds. Mol Cryst Liq Cryst Sci Technol Sect A Mol Cryst Liq Cryst. 1997;297:73–80.

28. Kose M, Orhan E, Büyükgüngör O. Synthesis of novel photochromic methyl cyanoacetate-condensed fulgide derivatives. J Photochem Photobiol A: Chem. 2007;188:358–63.
29. Heller HG, Hughes DS, Hursthouse MB, Rowles NG. Thermally stable fatigue resistant near infrared active photochromic compounds, exemplified by 6-amino-7-cyano-3-(dicyclopropylmethylene)-4-(2,5-dimethyl-3-furyl)-benzofuran-2(3*H*)-one. Chem Commun. 2000;1397–8.
30. Margerum JD, Miller LJ. In: Brown GH, editor. Photochromism. New York: John Wiley & Sons, 1971 Chapter 6 ("Photochromic Processes by Tautomerism"):557–632.
31. Matsushima R, Sakaguchi H. Comparison of the photochromic properties of fulgides and fulgimides. J Photochem Photobiol A: Chem. 1997;108:239–45.
32. Yokoyama Y, Inoue T, Yokoyama M, Goto T, Iwai T, Kera N, et al. Effects of steric bulkiness of substituents on quantum yields of photochromic reactions of furylfulgides. Bull Chem Soc Jpn. 1994;67:3297–303.
33. Yokoyama Y, Goto T, Inoue T, Yokoyama M, Kurita Y. Fulgides as efficient photochromic compounds. Role of the substituent on furylalkylidene moiety of furylfulgides in the photoreaction. Chem Lett. 1988;17:1049–52.
34. Kiji J, Okano T, Kitamura H, Yokoyama Y, Kubota S, Kurita Y. Synthesis and photochromic properties of fulgides with a *t*-butyl substituent on the furyl- or thienylmethylidene moiety. Bull Chem Soc Jpn. 1995;68:616–9.
35. Yokoyama Y, Iwai T, Kera N, Hitomi I, Kurita Y. Steric effect of alkylidene groups of furylfulgides on the photochromic behavior. Chem Lett. 1990;263–4.
36. Santiago A, Becker RS. Photochromic fulgides. Spectroscopy and mechanism of photoreactions. J Am Chem Soc. 1968;90:3654–8.
37. Darcy PJ, Heller HG, Strydom PJ, Whittall J. Photochromic heterocyclic fulgides. Part 2. Electrocyclic reactions of (*E*)-α-2,5-Dimethyl-3-furylethylidene(alkyl-substituted methylene)-succinic anhydrides. J Chem Soc Perkin Trans I. 1981;202–5.
38. Glaze AP, Heller HG, Whittall J. Photochromic heterocyclic fulgides. Part 7. (*E*)-Adamantylidene-[1-(2,5-dimethyl-3-furyl)ethylidene]succinic anhydride and derivatives: model photochromic compounds for optical recording media. J Chem Soc Perkin Trans 2. 1992;2:591–4.
39. Stobbe H, Eckert R. Eine Parallele farbiger Furyl- und Phenyl-Fulgide. (11. Abhandlung über Butadiënverbindungen). Chem Ber. 1905;38:4075–81.
40. Heller HG, Oliver S. Photochromic Heterocyclic Fulgides. Part 1. Rearrangement reactions of (*E*)-α-3-Furylethylidene(isopropylidene)succinic anhydride. J Chem Soc Perkin Trans I. 1981;197–201.
41. Yokoyama Y, Takahashi K. Trifluoromethyl-substituted photochromic Indolylfulgide. A remarkably durable fulgide towards photochemical and thermal treatments. Chem Lett. 1996;1037–8.
42. Reinfelds M, Hermanns V, Halbritter T, Wachtveitl J, Braun M, Slanina T, et al. Broadly absorbing fulgide derivative as a universal chemical actinometer for the UV to NIR region. ChemPhotoChem. 2019;3:441–9.
43. Heller HG, Oliver SN, Whittal J, Johncock W, Darcy PJ, Trundle C. (The Plessey Co plc). Photochromic compounds and their use in photoreactive lenses. UK Patent GB2146327B. 1986.
44. Yokoyama Y, Nakata H, Sugama K, Yokoyama Y. Photochromism and kinetics of heliochromic Benzothienylfulgides. Mol Cryst Liq Cryst Sci Technol Sect A Mol Cryst Liq Cryst. 2000;344:253–8.
45. Heller HG. Encyclopedia of materials: science and technology, 2nd ed. Elsevier, 2001 ("Photochromic Materials (Organic)"):6904–9.

46. Heller HG, Oliver SN, Whittal J, Tomlinson I. (The Plessey Co plc). Photochromic spiropyran compounds. UK Patent GB2190379B. 1990.
47. Sothman B. Making the mold and breaking the mold (The rise and fall and rise of SOLA Optical). http://www.solahistory.com/
48. Kobayakawa T. Photochromism of spiro-oxazines. In: Third International Symposium on Photochromism, Fukuoka, 1999.
49. Towns A. Spirooxazine dyes. Phys Sci Rev. 2020;5. DOI:doi.org/10.1515/psr-2020-0013.
50. Imura T, Kida Y, Tanaka T. (Tokuyama Soda KK). Fulgimide compounds and production thereof. Japanese Patent JPS6438063(A). 1989.
51. Imura S, Tanizawa T, Kobayakawa T. (Tokuyama Corp.). Cyclopropyl-substituted spiro-photochromic compounds. European Patent Application EP0629626A2. 1994.
52. Kawabata Y, Momoda J, Nagoh H. (Tokuyama Corp.). Photochromic composition. European Patent Application EP0965628A1. 2000.
53. Mann C, Melzig M, Weigand U. (Optische Werke G. Rodenstock). World Patent Application WO 02/22594. 2002.
54. Towns A. Naphthopyran dyes. Phys Sci Rev. 2020;5. DOI:doi/10.1515/psr-2019-0085.
55. Kohno Y, Tamura Y, Matsushima R. Simple full-color rewritable film with photochromic fulgide derivatives. J Photochem Photobiol A: Chem. 2009;201:98–101.
56. Yao B, Wang Y, Menke N, Lei M, Zheng Y, Ren L, et al. Optical properties and applications of photochromic fulgides. Mol Cryst Liq Cryst. 2005;430:211–9.
57. Zong Z, Menke N, Yao B, Wang Y, Chen Y. Polarization multiplexing, angle multiplexing and circumrotation multiplexing holographic recording experiments with 3-indolylbenzylfulgimide/PMMA film. Appl Mech Mater. 2012;130-134:2035–41.
58. Chao L, Menke N, Yao B, Lei M, Wang Y, Sun X, et al. Optical storage properties of different fulgides. Adv Mater Res. 2013;660:24–9.
59. Menke N, Yao B, Wang Y, Dong W, Lei M, Chen Y, et al. Spectral relationship of photoinduced refractive index and absorption changes in fulgide films. J Mod Opt. 2008;55:1003–11.
60. Rybalkin VP, Pluzhnikov SY, Popova LL, Revinskii YV, Tikhomirova KS, Komissarova OA, et al. A novel approach to fluorescent photochromic fulgides. Mendeleev Commun. 2016;26:21–3.
61. Jiao Y, Yang R, Luo Y, Liu L, Xu B, Tian W. Fulgide derivative based solid-state reversible fluorescent switch for advanced optical memory. CCS Chem. 2021;3. DOI:doi.org/10.31635/ccschem.021.202000673.
62. Weerasekara RK, Uekusa H, Hettiarachchi CV. Multicolor photochromism of fulgide mixed crystals with enhanced fatigue resistance. Cryst Growth Des. 2017;17:3040–7.

Heinz Mustroph

35 Hemicyanine dyes

Abstract: Hemicyanine dyes are a subclass of polymethine colorants. At one end of their polymethine chain is an unsaturated heterocyclic ring possessing a nitrogen atom as would be found in a cyanine dye. However, the other end is terminated by a nitrogen atom that does not form part of an unsaturated heterocycle. The name alludes to their half-cyanine substitution pattern. Later the scope of the term "hemicyanine" was extended to the phenylogous dyes where there is a phenyl group between the two terminal nitrogen atoms. The first main technical application of hemicyanine dyes was in textile coloration. Nowadays hemicyanine dyes are used extensively as optical probes of cell membrane potential.

Keywords: cell membrane potential, cyanine dyes, Franck–Condon principle, hemicyanine dyes, optical probes, silver halide photographic materials, stilbazolium dyes, styryl dyes

35.1 Fundamentals

Cyanine dyes comprise two terminal nitrogen atoms which form part of separate unsaturated heterocyclic rings that are in conjugation with a polymethine chain [1-5]. Dyes where one terminal component is an unsaturated heterocyclic ring with a nitrogen atom in conjugation with the polymethine chain, as it is typically in *cyanine dyes*, and the second is obtained from a terminal nitrogen atom which is not part of a heterocyclic ring are named *hemicyanine dyes* (*ἥμι* = half) due to a suggestion of Frank L. White and Grafton H. Keyes [6]. The name alludes to their *half-cyanine* substitution pattern.

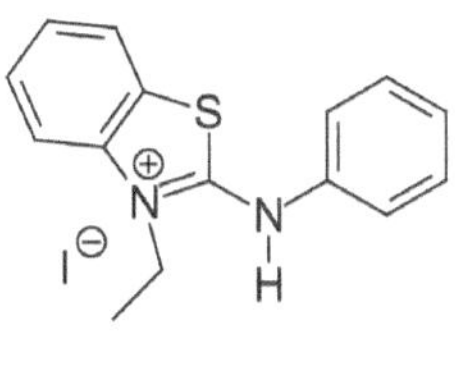

1

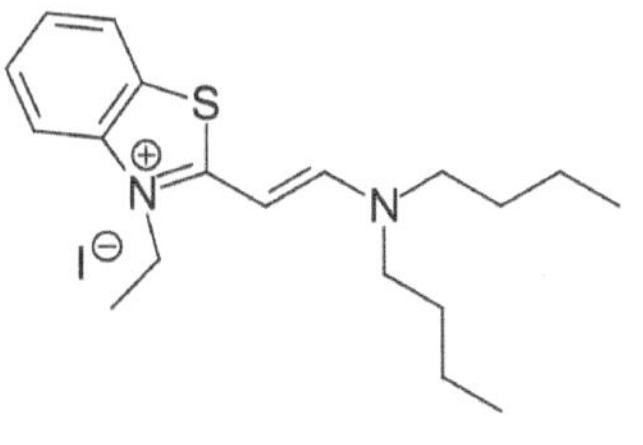

2

This article has previously been published in the journal Physical Sciences Reviews. Please cite as H. Mustroph Hemicyanine Dyes *Physical Sciences* Reviews [Online] 2021, 6. DOI: 10.1515/psr-2020-0174

https://doi.org/10.1515/9783110587104-035

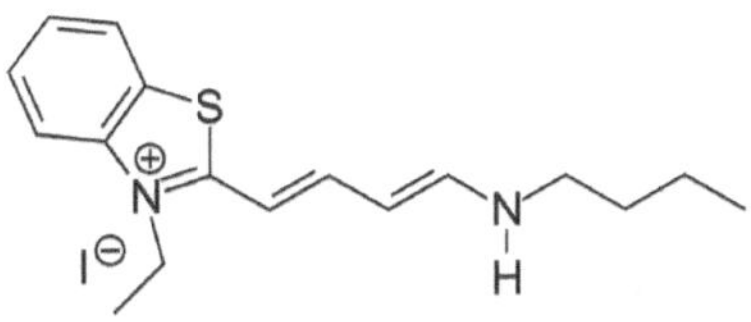

3

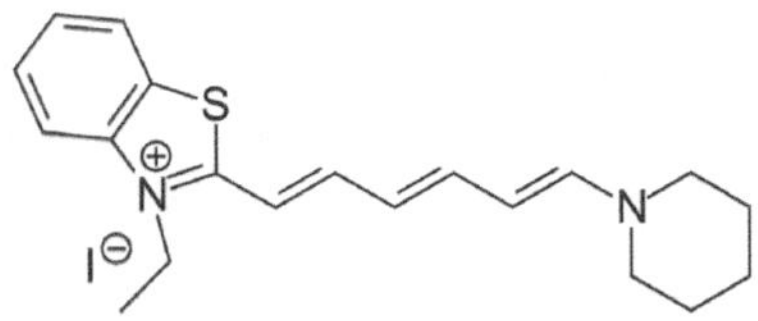

4

Hemicyanine dyes which do not contain a polymethine chain are termed zeromethine hemicyanine dyes **1**. Depending on the number of vinylene groups in their polymethine chains, others are referred to as dimethine **2**, tetramethine **3** and hexamethine hemicyanine dyes **4**. The amino group can be e. g. an arylamine **1**, a dialkylamine **2**, an alkylamine **3** or a cyclic dialkylamine **4** [6].

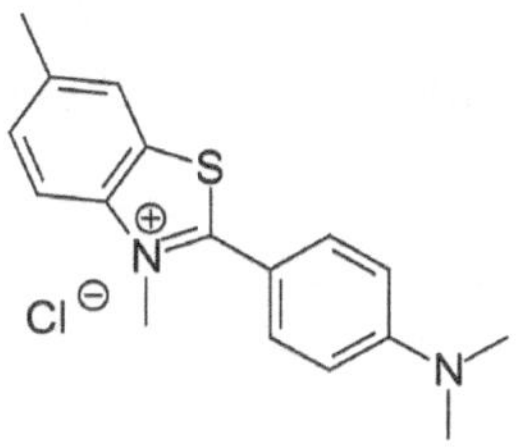

5

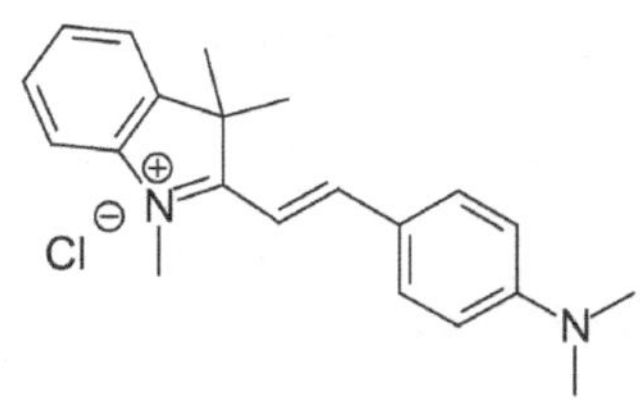

6

Norman Picus and Paul E. Spoerri extended the scope of the term "hemicyanine" to the phenylogous dyes where there is a phenyl group between the two terminal nitrogen atoms [7]. Doing so means that the class encompasses zeromethine **5** (thioflavine T), dimethine **6**, **7** and tetramethine hemicyanine dyes [8].

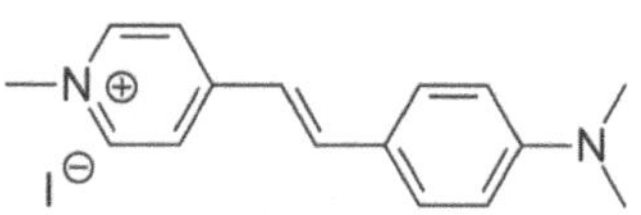

7

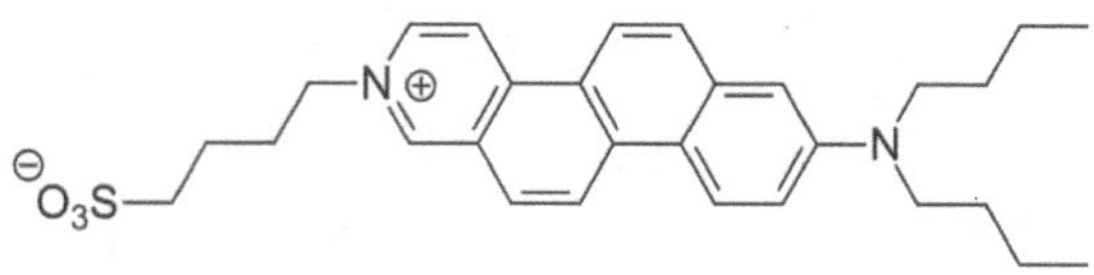

8

In the "stiff" hemicyanine dye **8** twisting vibration around the C–C single bonds is restricted and photoisomerism around the C=C double bonds is hindered by annelated benzene rings to improve the photophysical properties [9].

The examples **1–8** illustrate how this dye class covers a wide range of different chemical structures. So, special terms are used for the individual substructures. For example, dyes where the terminal nitrogen atom is directly coupled to the polymethine chain like e. g. in **2–4** are called *enamines*. The phenylogous zeromethine hemicyanine dyes like e. g. **5** are often referred with the wrong term *apocyanine dyes*. The term *styryl dyes* is often used to refer to those hemicyanine dyes that contain a styryl group like e. g. **6** or in the case **7** the dyes are often termed *stilbazolium dyes*.

All these dyes belong to the hemicyanine dye class. As cyanine dyes they are characterized by an odd number $2n + 3$ of π-centers and $2n + 4$ π-electrons (where n is the number of vinylene groups (–CH = CH–) and due to their electronic structure are part of the large group of polymethine dyes [10–13]. So, one would expect the hemicyanine dyes to have the characteristic features of the polymethine dyes like narrow absorption bands in the electronic spectrum, increasing molar absorption coefficient with increasing number of vinylene groups and each additional vinylene group in the polymethine chain giving a bathochromic shift of the 0–0 vibronic transition of about 100 nm (the so-called vinylene shift) [10–13].

Unfortunately, the electronic absorption spectra of hemicyanine dyes have not been systematically studied much regarding their vibrational fine structure and the underlying vibronic transitions, because most of the spectra are broad and do not exhibit a fine structure. Therefore, the experimental absorption maxima λ_{max} are used to discuss the structural influences on the spectra. However, the λ_{max} value is nothing more than the intensity maximum in an electronic spectrum. It corresponds to the vibronic (vibrational–electronic) transition with the highest Franck–Condon factor [14–16]. This is often not considered during the theoretical interpretation of the spectra of hemicyanine dyes and leads to wrong models and results in misinterpretations.

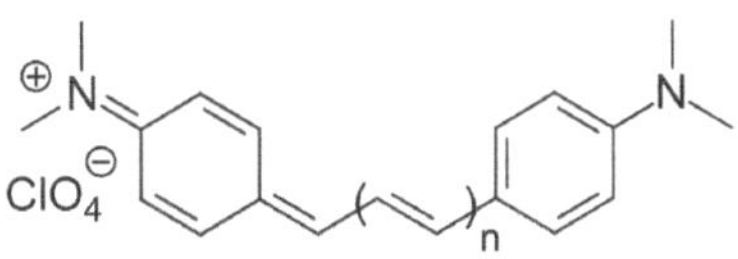

9

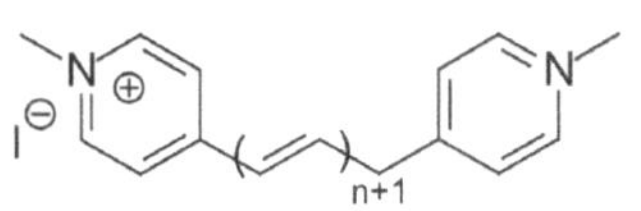

10

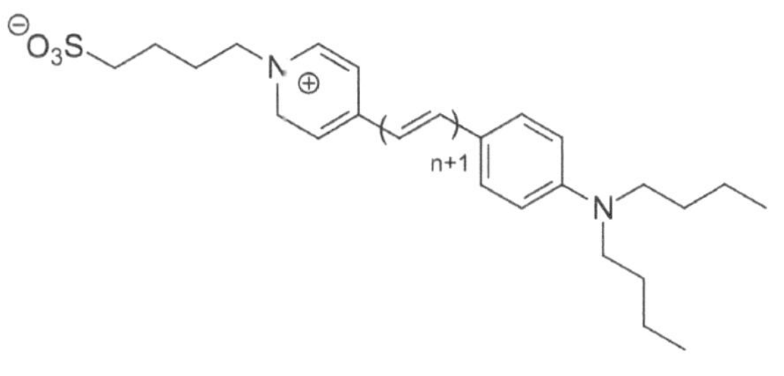

11

Table 35.1: Absorption maxima λ_{max} (nm) of the diarylmethine dyes **9** in acetic acid [17], the iso-π-electronic cyanine dyes **10** in DMSO [18] and the hemicyanine dyes **11** in ethanol as well the fluorescence maxima λ_{max}(Fl) (nm) of **11** [19] in relation to the number of vinylene groups *n*.

n	9	10	11	
	λ_{max}	λ_{max}	λ_{max}	λ_{max}(Fl)
0	610	606	500	615
1	693	707	526	701
2	790		544	789
3	883	920		

As mentioned, in polymethine dyes the energy of the 0–0 vibronic transition is mainly determined by the length of the conjugated system between the terminal nitrogen atoms. Dyes in the series **9–11** which have the same length of conjugated system between their terminal nitrogen atoms are iso-π-electronic. If the λ_{max} values represent the 0–0 vibronic transition energy all dyes **9–11** with the same length of conjugated system should show similar λ_{max} values and in all three series the vinylene shift. As shown in Table 35.1 only the λ_{max} values of both symmetrical dye series **9** and **10** meet the expected behaviour. In addition, the spectra of these dyes exhibit a clear fine structure and one can rightly assume that λ_{max} values correspond to the 0–0 vibronic transition.

The λ_{max} values of the hemicyanine dyes **11** do not meet both predictions. In contrast to dye series **9** and **10** the absorption spectra of series **11** show a broad and blurred shape. Therefore, one cannot say, which vibronic transition is represented by λ_{max}.

But there are some indications regarding the 0–0 vibronic transition energy. The fluorescence spectra of **11** show a high Stokes shift into the region of the absorption maxima of the both symmetrical dye series **9** and **10.** In addition, the fluorescence maxima λ_{max} (Fl) (nm) of **11** exhibit a clear vinylene shift (Table 35.1).

It is well known from the spectra of *merocyanine* [20, 21] and *unsymmetrical cyanine dyes* [22, 23] that an increasing difference between electron acceptor/donor abilities of the terminal end groups leads to a broad absorption band connected with a blurring of the fine structure.

The most popular model to explain the electronic spectra of polymethine dyes is based on the simple *Valence Bond theory* which describes the electronic structure by a linear combination of two *contributing structures* **12 L** (with the positive charge on the left hand side) and **12 R** (with the positive charge on the right hand side) to the *resonance hybrid* [20, 22]. It has already pointed out that the model with two contributing structures is too simple to describe the electronic structure of polymethine dyes fully [5, 21, 23]. Nevertheless, for some purposes it is sufficient to describe roughly the electronic structure of polymethines as a resonance hybrid between two contributing structures.

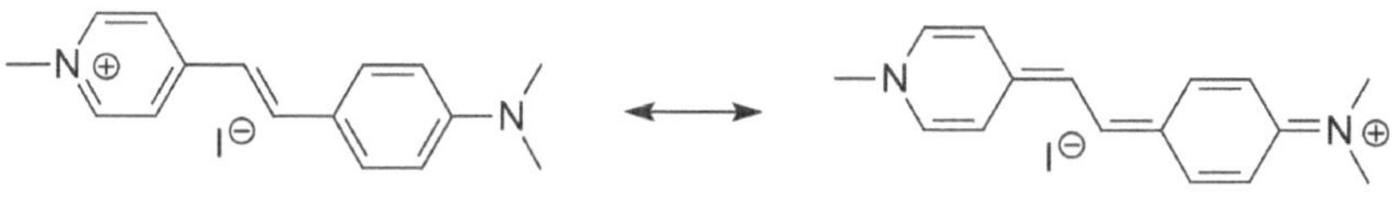

12L **12R**

The Eigen functions of the electronic ground state $\psi(S_0)$ and first electronic excited state $\psi(S_1)$ can be approximated as a linear combination of the wavefunctions of the two contributing structures ψ_L and ψ_R. The degree of mixing of **L** and **R** is determined by c^2 in eqs. (1) and (2), which reflects the difference of the electron acceptor/donor ability of the terminal end groups.

$$\psi(S_0) = \left(1 - c^2\right)^{1/2}\psi_L + c\psi_R \tag{35.1}$$

$$\psi(S_1) = c\psi_L - \left(1 - c^2\right)^{1/2}\psi_R \tag{35.2}$$

The electronic structure determines the equilibrium geometry R_e of the ground electronic state S_0 $[R_e(S_0)]$ and the first excited electronic state S_1 $[R_e(S_1)]$. From this model it follows that with increasing unsymmetrical electronic structure in S_0 [increasing deviation from $c^2 = 0.5$ in eq. (35.1)] the difference $R_e(S_1) - R_e(S_0)$ increases. However, using this model it is very important not to consider the pure electronic states and electronic transitions only. The effects of the changes in $R_e(S_0)$ to $R_e(S_1)$ have influences on the distribution of the spacing and intensity of the vibrational sub-bands, which is often not considered! In a simple and efficient way, it can be explained with the Franck–Condon principle for diatomic molecules. The differences of the equilibrium geometry in S_0 and in S_1 $[R_e(S_1) - R_e(S_0)]$ determines which vibronic transition will have the highest Franck–Condon factor [14–16]. If the difference is relatively low $[R_e(S_1) \gtrapprox R_e(S_0)]$ as in symmetric cyanine dyes then the absorption intensity is largely concentrated in the 0–0 vibronic transition. Already different terminal heterocyclic rings in unsymmetrical cyanine dyes cause an unsymmetrical electronic structure in S_1 and S_0. This leads to increasing differences $R_e(S_1) > R_e(S_0)$ which cause an intensification of higher 0-v vibronic transitions in parallel with unsymmetrical cyanine [23] and merocyanine dyes [21]. Therefore, only in combination with the Franck–Condon principle for diatomic molecules [14–16]. the model with the linear combination of the two contributing structures should be used.

To transfer the structural influence in polyatomic molecules to the model of diatomic molecules the bond-length alternation (BLA) in the electronic ground state was introduced. It is estimated as the average of the absolute difference between adjacent carbon–carbon equilibrium bond lengths in a polymethine chain. In symmetrical cyanine dyes the differences between carbon–carbon equilibrium bond lengths in a polymethine chain are small and, therefore, BLA [5]. It follows that the difference $R_e(S_1) - R_e(S_0)$ is small and λ_{max} values represent the 0–0 vibronic transition energy. With increasing unsymmetrical electronic structure in S_0 [increasing deviation from $c^2 = 0.5$ in eq. (35.1)] the difference $R_e(S_1) - R_e(S_0)$ increases and with it the difference between the absorption maxima of the hemicyanine dyes and the energy of the 0–0 vibronic transition of the iso-π-electronic symmetric cyanine dyes with the same length of the conjugated system (Table 35.1).

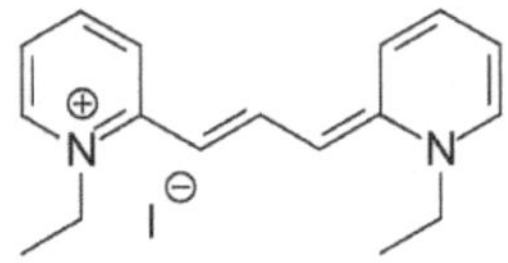

13

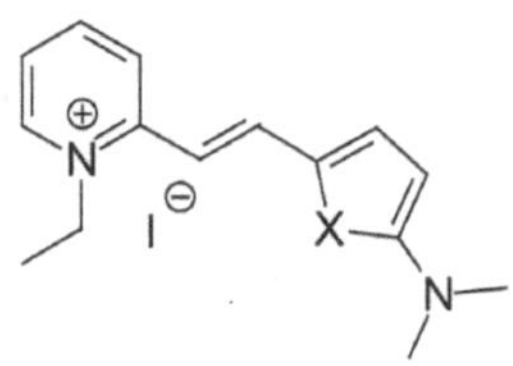

14a, b

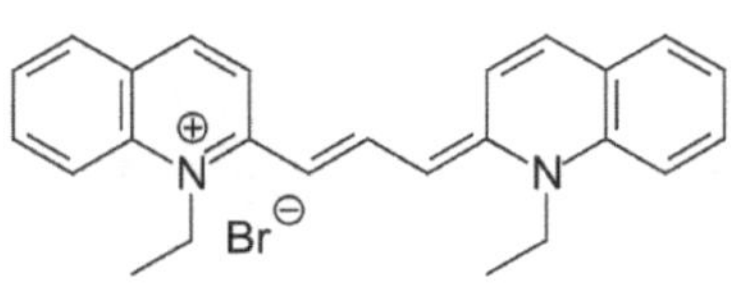

15

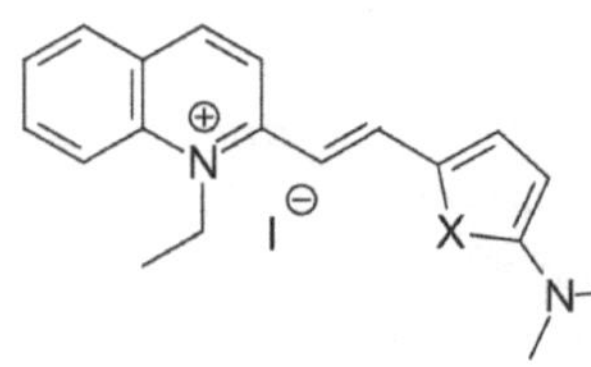

16a, b

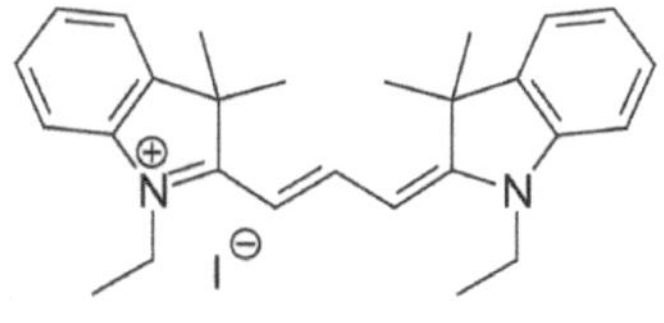

17

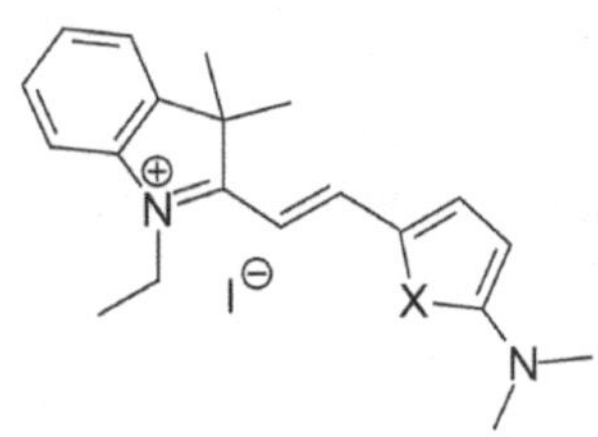

18a, b

a = –CH=CH–; **b** = O

For additional illustration of this effect some examples are given in Table 35.2. The largest asymmetry in the electronic structures is in **14a,** with the consequence that the greatest difference in λ_{max} values is between the symmetrical cyanine **13** and the hemicyanine dye **14a.** Replacement of *N,N*-dimethylaniline by 2-*N,N*-dimethylaminofuran leads to a less unsymmetrical electronic structure and, therefore, less deviation between the λ_{max} values of **13** and **14b** in comparison with **14a.** All spectra of **17, 18a** and **18b** show a fine structure and the λ_{max} values are nearly equal and represent the 0–0 transition energy.

Table 35.2: Absorption maxima λ_{max} (nm) of the cyanine dyes **13, 15, 17** [22] and the iso-π-electronic hemicyanine dyes **14a, 16a, 18a** [22] and **14b, 16b, 18b** [24] in nitromethane.

λ_{max}	λ_{max}	λ_{max}
13	**14a**	**14b**
562	459	536
15	**16a**	**16b**
608	527	587
17	**18a**	**18b**
548	550	555

In summary, hemicyanine dyes belong to the great group of polymethine dyes. When interpreting the spectra of hemicyanine dyes with respect to theory, one should always look at the fine structure of the absorption band in order to avoid wrong models and make misinterpretations.

35.2 History of the hemicyanine dyes

The first two hemicyanine dyes **19** and **20** were synthesized by Walter König to investigate these dyes as apparently analogous to compare with the corresponding cyanine dyes [25].

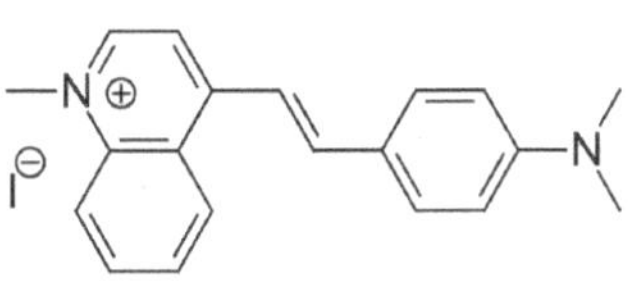

19

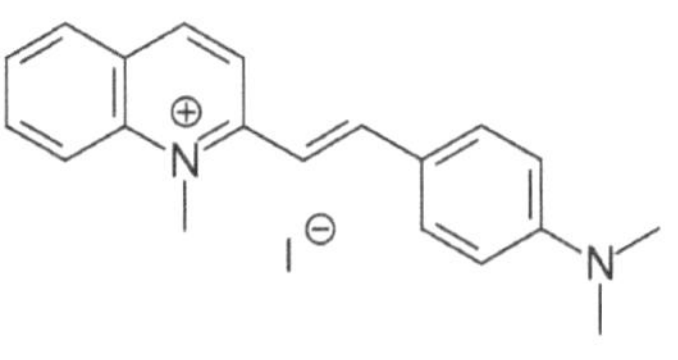

20

As mentioned above White and Keyes suggested the term hemicyanine for dyes where one terminal component is a heterocyclic ring with a nitrogen atom, as is typical in cyanine dyes, and the second is a terminal nitrogen atom which is not part of a heterocyclic ring like e. g. **1–4** [6]. In 1945 Brooker et al. used the term styryl dyes for the condensation products of 4-*N,N*-dimethylaminobenzaldehyde with a wide range of 2-methyl heterocyclic quaternary salts to obtain dyes like e. g. **6, 7, 14–20** [22]. Two years later Picus and Spoerri suggested in a footnote to the term "p-dimethylaminostyryl":

> Although the term, hemicyanine, was not originated to include the p-dimethylaminostyryl type of dye, C. E. K. Mees in 'The Theory of the Photographic Process,' The Mac Millan Co., New York, N. Y., 1946, p. 1034, indicates that this term may well be used to include this general type of dyestuffs. [7]

This suggestion has been adopted in part, but other terms for subclasses like e. g. styryl or stilbazolium dyes are often used. Especially the term "styryl dye" is used for different dyes classes as e. g. the streptomerocyanine dyes and is therefore not a happy choice for the general name of these dyes.

After the first synthesis in 1912, it took about ten years before the first technical use of hemicyanine dyes. Almost simultaneously William H. Mills and William J. Pope (manuscript received April 10, 1922) and Ernst König claimed the use of hemicyanine **21** as spectral green sensitizers for silver halide photographic materials [26, 27].

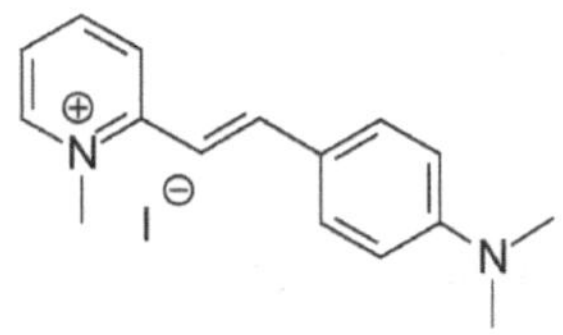

21

This discovery triggered extensive development work in the photographic industry [1, 3, 11, 22–30]. Later hemicyanine dyes were also developed for textile dyeing and Hi-Tec applications.

35.3 General synthetic routes to hemicyanine dyes

The first two hemicyanine dyes **19** and **20** were synthesized by condensation of 4-*N*,*N*-dimethylaminobenzaldehyde with lepidinium and quinaldinium quaternary salts with the aid of piperidine [25]. The condensation of 4-*N*,*N*-dialkylaminobenzaldehyde with 2-methyl heterocyclic quaternary salts with the aid of piperidine is the general synthetic route to "styryl dyes" like **6, 7, 14, 16, 18–21.** Reaction of 4-*N*,*N*-dialkylamino-cinnamaldehyde and aldehydes possessing longer polymethine chains with 2-methyl heterocyclic quaternary salts leads to the corresponding vinylogous dyes like e. g. **11.**

White and Keyes used the half dyes **24,** which are known from the synthesis of merocyanine and unsymmetrical cyanine dyes. They are synthesized from a 2-methyl heterocyclic quaternary salt **22** and diphenylformamidine hydrochloride **23** (n = 0), malondialdehyde dianil hydrochloride **23** (n = 1) or glutacondialdehyde dianil hydrochloride **23** (n = 2) followed by the reaction with acetic anhydride to the intermediate **24** with β-acetanilido vinyl groups. Next the intermediate **24** is reacted with the primary or secondary amines **25** in excess to the hemicyanine dyes **26** (Figure 35.1) [6].

In addition to these general synthetic routes there are some special procedures [1, 3, 8, 11]. For instance, the hemicyanine **5** is synthesized by heating *p*-toluidine with sulfur and subsequent alkylation with methyl sulfate.

The most important method for preparing zeromethine hemicyanine dyes is the reaction of *N*-heterocycles containing a carbonyl group in 2- or 4-position with secondary or tertiary aromatic amines in the presence of dehydrating agents. So, by condensation of *N*-ethyl-4-bromo-1,8-naphtholactam with *N*-methyl-4-ethoxy-diphenylamine the blue dye **27** is obtained.

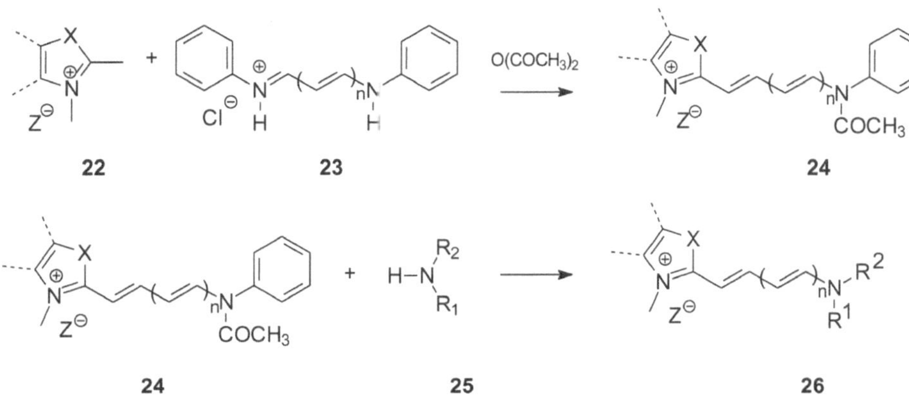

Figure 35.1: Reaction scheme.

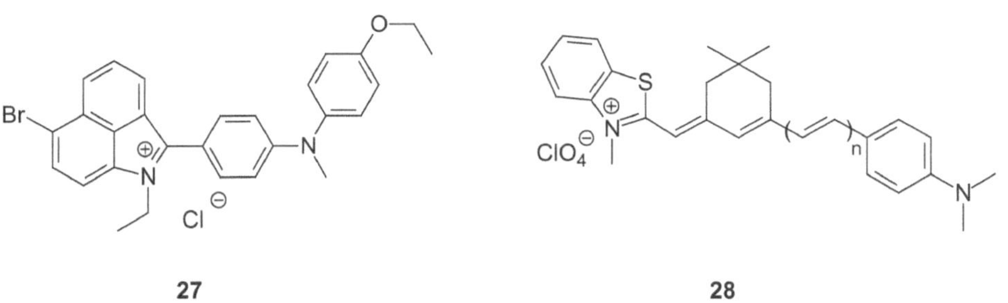

Another interesting synthesis variation is the condensation of a 2-methyl heterocyclic quaternary salt with isophorone and following condensation with 4-*N*,*N*-dimethylaminobenzaldehyd to **28a** (n = 1; Styryl 9 M) or with 4-*N*,*N*-dialkylaminocinnamaldehyde to **28b** (n = 2; Styryl 14).

35.4 Commercial uses of hemicyanine dyes

Despite the enthusiastic evaluation of hemicyanine dyes as spectral sensitizers [6, 26, 27], their technical importance remained limited, since at the time of their discovery there were already a large number of cyanine and merocyanine dyes with valuable sensitizing properties [28–30]. In addition, the absorption and thus the sensitization area is less influenced by the terminal heterocyclic rings than with the cyanines. Thus, the hemicyanine dyes were not of significant importance as spectral sensitizers in silver halide photography.

The first main technical application of hemicyanine dyes was in textile printing. So, e. g. **5** dyes silk and tanned cotton in greenish-yellow shades, **6** dyes polyacrylonitrile in brilliant red shades, **27** dyes polyacrylonitrile in very lightfast blue shades and **29** dyes polyacrylonitrile in neutral yellow shades [8, 31].

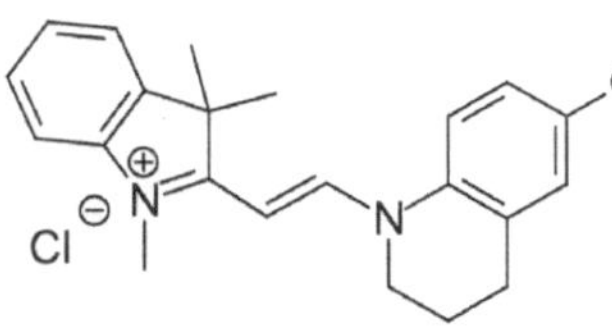

29

A lot of synthetic work in hemicyanine dye chemistry was done for the development of fast-response potential-sensitive dyes. In 1979 Alan S. Waggoner reviewed the current state of the use of dye molecules as optical probes of cell membrane potential [32]. Across a cell membrane there is an electrostatic potential. This membrane potential is an important physiological parameter of the living cell.

In these times it was the state-of-the-art to use cyanine, merocyanine and oxonol dyes as potential-sensitive probes. Waggoner divided the membrane potential probes into two classes based on the speed, size, and mechanism of the potential-dependent optical change – slow-response dyes and fast-response dyes. The first dyes respond to membrane potential changes in times of seconds, the second respond to membrane potential changes in less than milliseconds. Most cyanine, merocyanine and oxonol dyes are slow-response dyes.

From theoretical considerations Leslie M. Loew derived that the 4(*p*-AminoStyryl) Pyridinium (ASP) dyes should give fast-response dyes [33]. To investigate the influence of the dialkylamino substituents a series di-*n*-ASP **30** and (di-*n*-ASPPS) [4(*p*-AminoStyryl) Pyridinium PropylSulfonate] **31** were synthesized and tested [34].

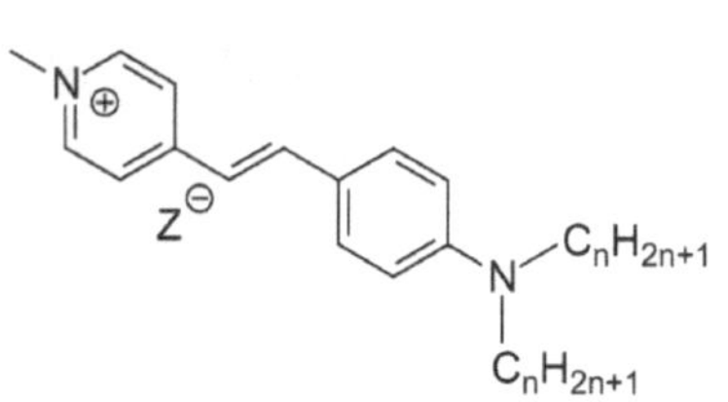

30

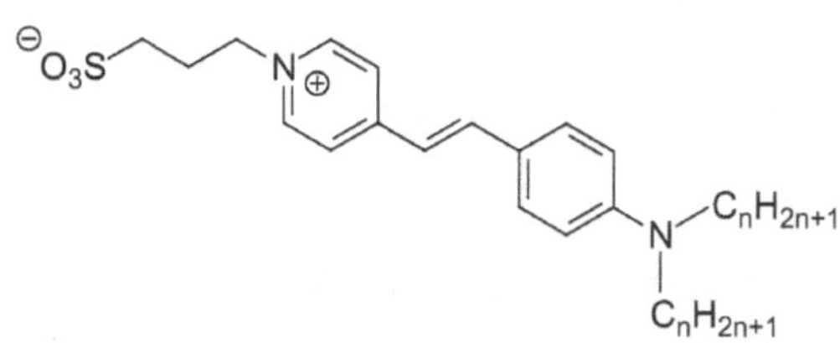

31

The next development to the ANEP (Amino Naphthyl Ethenyl Pyridinium) dyes with propylsulfonate substituent (di-nANEPPS) **32** is among the most sensitive of the fast-response probes [35].

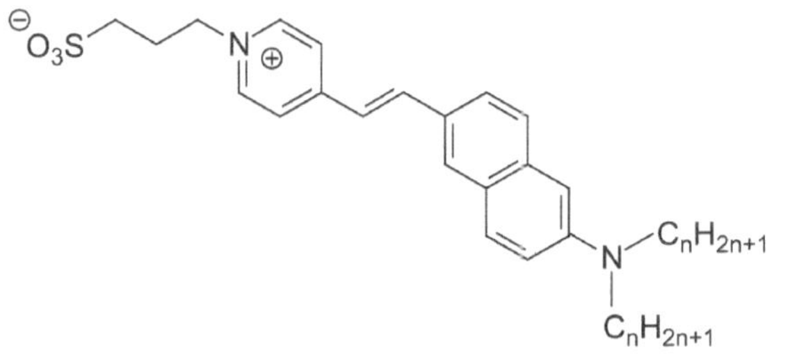

32

Rina Hildesheim (RH dyes) synthesized an extensive series of hemicyanine dyes with increasing number of vinylene groups, e. g. **11**, and different lengths of alkyl groups within their dialkylamino substituents [19].

The background for the wide range of dyes in practical use is that no single dye provides the optimal response under all experimental conditions.

So, a multitude of other hemicyanine dyes were synthesized and tested as fast-response probes. It is not possible to review all of them here. Only one interesting approach is reported here in brief [9]. In comparison with the RH dye series [19] in the ANNINE-*n* (ANellated hemicyaNINE) dyes, twisting vibration around the C–C single bonds is restricted and photoisomerism around the C = C double bonds is hindered. Here *n* describes the number of annelated rings, for instance, **8** is named ANNINE-4 and corresponds to **11** ($n = 0$; RH 364) and **33** is termed ANNINE-6 and corresponds to **11** ($n = 2$; RH 237).

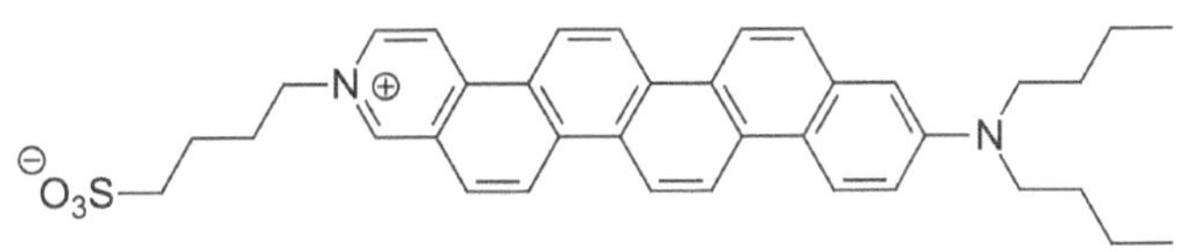

33

ANNINE-6 **33** exhibits a very high voltage-sensitivity. However, due to the number of annelated rings **33** has extremely low solubility in water and causes intracellular staining. To overcome the problem of staining, instead of the inner salt **33** the corresponding dye with two positive charges (ANNINE-6plus) **34** was synthesized which increases the water solubility substantially [36].

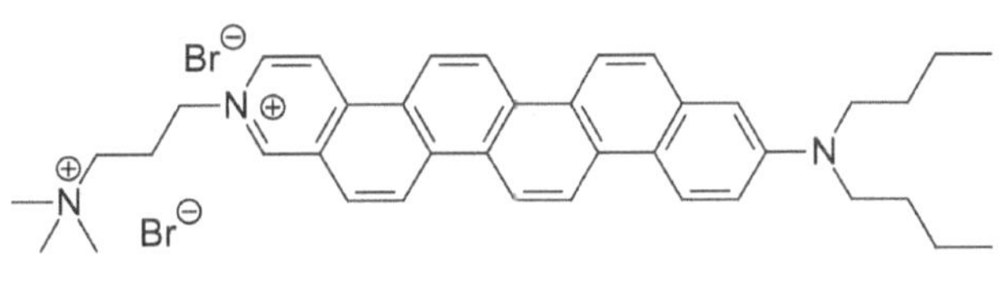

34

Hemicyanine dyes have also been proposed as sensitizers in photo-polymerization to expand the sensitivity of photopolymer process in the visible spectral range [37, 38].

Due to the large Stokes Shift and the high sensitivity of the absorption spectra of the dye's environment there are many papers dealing with hemicyanine dyes and claiming various technical applications. A chapter of this encyclopedia is space limited and so necessarily incomplete. It can give only an introduction into the fundamentals, the history, the most important developments and applications and further developments are described in Ref [39].

References

1. Hamer FM. The cyanine dyes and related compounds. In: Weissberger A, editor. The chemistry of heterocyclic compounds. vol. 18. New York: Interscience, 1964.
2. Griffiths J. Colour and constitution of organic molecules. London: Academic Press, 1976.
3. Sturmer DM. Syntheses and properties of cyanine and related dyes. In: Weissberger A, Taylor EC, editors. The chemistry of heterocyclic compounds. vol. 30. New York: Wiley, 1977:441–587.
4. Mishra A, Behera RK, Behera PK, Mishra BK, Behera GB. Cyanine dyes during the 1990s: a review. Chem Rev. 2000;100:1973.
5. Mustroph H. Cyanine dyes. Phys Sci Rev. 2020;5. DOI:10.1515/psr-2020-0145.
6. White FL, Keyes GH. US 2 263 749. 23 Apr 1937.
7. Picus N, Spoerri PE. Bridge substituted hemicyanines. J Am Chem Soc. 1948;70:3073.
8. Berneth H. Methine dyes and pigments. In: Elvers B, editor. Ullmann's encyclopedia of industrial chemistry. Weinheim: Wiley-VCH, 2012.
9. Hübener G, Lambacher A, Fromherz P. Anellated hemicyanine dyes with large symmetrical solvatochromism of absorption and fluorescence. J Phys Chem B. 2003;107:7896.
10. König W. Über den Begriff der „Polymethinfarbstoffe" und eine davon ableitbare allgemeine Farbstoff-Formel als Grundlage einer neuen Systematik der Farbenchemie. J Prakt Chem. [2] 1926; 112: 1.
11. Sturmer DM, Diehl DR. Polymethine dyes. In: Kirk-Othmer encyclopedia of chemical technology, 3rd ed. vol. 18, Wiley, New York, 1982:848–74.
12. Tyutyulkov N, Fabian J, Mehlhorn A, Dietz F, Tadjer A. Polymethine dyes – structure and properties. Sofia: St. Kliment Ohridski University Press, 1991.
13. Mustroph H. Polymethine dyes. Phys Sci Rev. 2020;5. DOI:10.1515/psr-2019-0084.
14. Mustroph H. Dyes: quantum chemical calculation of electronic spectra. Phys Sci Rev. 2019;4. DOI:10.1515/psr-2019–0040.
15. Mustroph H, Ernst S, Senns B, Towns AD. Molecular electronic spectroscopy: from often neglected fundamental principles to limitations of state-of-the-art computational methods. Color Technol. 2015;131:9.
16. Mustroph H, Towns A. Fine structure in electronic spectra of cyanine dyes: are sub-bands largely determined by a dominant vibration or a collection of singly-excited vibrations? Chem Phys Chem. 2018;19:1016.
17. Schmidt H, Wizinger R. Über die Vinylenhomologen der Di- und Triphenylmethanfarbstoffe. III. Liebigs Ann Chem. 1959;623:204.
18. Tolbert LM, Zhao X. Beyond the cyanine limit: Peierls distortion and symmetry collapse in a polymethine dye. J Am Chem Soc. 1997;119:3253.

19. Grinvald A, Hildesheim R, Farber IC, Anglister L. Improved fluorescent probes for the measurement of rapid changes in membrane potential. Biophys J. 1982;39:301.
20. Brooker LG, Keyes GH, Sprague RH, VanDyke RH, VanLare E, VanZandt G, et al. Color and constitution. X. Absorption of the merocyanines. J Am Chem Soc. 1951;73:5332.
21. Mustroph H, Reiner K, Senns B, Mistol J, Ernst S, Keil D, et al. The effects of substituents and solvents on the ground-state π-electronic structure and electronic absorption spectra of a series of model merocyanine dyes and their theoretical interpretation. Chem Eur J. 2012;18:8140.
22. Brooker LG, Sklar AL, Cressman HW, Keyes GH, Smith LH, Sprague RH, et al. Color and constitution. VII. Interpretation of absorption of dyes containing heterocyclic nuclei of different basicities. J Am Chem Soc. 1875;67:1945.
23. Mustroph H, Reiner K, Senns B. Bond length alternation in unsymmetrical cyanine dyes and its influence on the vibrational structure of their electronic absorption spectra. Color Technol. 2017;133:469.
24. Mikhailenko FA, Shevchuk LI, Rozhdestvenskaya IT. Mono- and dimethine dyes from 2-dimethylamino-5-formylfurans, -thiophenes, and -selenophenes. Khim Geterotsikl Soedin. 1975:316.
25. König W. Zur Frage der Konstitution der Cyaninfarbstoffe. J Prakt Chem. [2] 1912;86: 166.
26. Mills WH, Pope WJ. 2-p-dimethylaminostyrylpyridine methiodide, a new photographic sensitizer. J Chem Soc Trans. 1922;121:946.
27. König E. DE 395 666. 23 May 1922.
28. James TH, editor. The theory of the photographic process. 4th ed. New York: Macmillan, 1977.
29. Keller K, Kampfer H, Matejec R, Lapp O, Krafft W, Frenken H, et al. Photography. In: Elvers B, editor. Ullmann's encyclopedia of industrial chemistry. Weinheim: Wiley-VCH, 2000.
30. Fujita S. Organic chemistry in photography. Berlin: Springer, 2004.
31. Hunger K, editor. Industrial dyes, chemistry, properties, applications. Weinheim: Wiley-VCH, 2003.
32. Waggoner AS. Dye indicators of membrane potential. Annu Rev Biophys Bioeng. 1979;8:47.
33. Loew LM, Bonneville GW, Surow J. Charge shift optical probes of membrane potential. Theory Bio Chem. 1978;17:4065.
34. Loew LM, Simpson LL. Charge-shift probes of membrane potential: a probable electrochromic mechanism for p-aminostyrylpyridinium probes on a hemispherical lipid bilayer. Biophys J. 1981;34:353.
35. Fluhler E, Burnham VG, Loew LM. Spectra, membrane binding, and potentiometric response of new charge response of new charge shift probes. Bio Chem. 1985;24:5749.
36. Fromherz P, Hübener G, Kuhn B, Hinner MJ. ANNINE-6plus, a voltage-sensitive dye with good solubility, strong membrane binding and high sensitivity. Eur Biophys J. 2008;37:509.
37. Kabatc J, Jurek K. Synthesis, properties, and application of new benzothiazole-based sensitisers in polymer chemistry. Color Technol. 2015;131:183.
38. Lalevée J, Fouassier J-P, editors. Polymer chemistry series. Photopolymerisation initiating systems, vol. 29. London: RSC, 2018.
39. Deligeorgiev T, Gadjev N, Kaloyanova S, Lesev N, Vasilev A, Alexiev A. A novel general method for fast and easy preparation of cationic, neutral dimethinehemicyanine and dimethine dyes by uncatalysed Knoevenagel condensation. Color Technol. 2012;128:417.

Heinz Mustroph

36 Hemioxonol dyes

Abstract: Following the classification of hemicyanine dyes Leslie G. S. Brooker suggested the term hemioxonol dye be applied to the case of an oxonol colorant in which one of its terminal unsaturated heterocycles has been replaced by an open chain $N(R^1)R^2$ group or saturated heterocyclic ring. For historical reasons, dyes formed from the reaction of 4-*N,N*-dialkylaminobenzaldehydes or 4-*N,N*-dialkylaminocinnamaldehydes with a heterocycle containing an active methylene group adjacent to a carbonyl group are often called benzylidene dyes and cinnamylidene dyes, respectively. In terms of systematic nomenclature, their proper classification is as hemioxonol dyes. They are used as filter dyes and antihalation dyes in silver halide photography. Their current main usage is in dye diffusion thermal transfer printing (D2T2).

Keywords: benzylidene dyes, cinnamylidene dyes, dye diffusion thermal transfer printing, hemicyanine dyes, hemioxonol dyes, streptocyanine dye

In *oxonol dyes* the two charge-carrying terminal oxygen atoms are constituents of unsaturated heterocycles [1, 2]. The simplest possible type of cyanine dye family is the *streptocyanine dye* class, where two charge-carrying terminal nitrogen atoms do not form part of an unsaturated heterocyclic ring, but they are part of an open chain $N(R^1)R^2$ group or of saturated heterocyclic rings [3–5]. Following the classification of *hemicyanine dyes* Leslie G. S. Brooker suggested the term *hemioxonol dyes*, when in oxonol dyes one of its terminal unsaturated heterocycle with a charge-carrying terminal oxygen atom is replaced by an open chain amino group $N(R^1)R^2$ or a saturated heterocyclic amine [6].

The hemioxonol dyes can be described by the general formula **1** and their phenylogous derivatives of formula **2**, respectively,

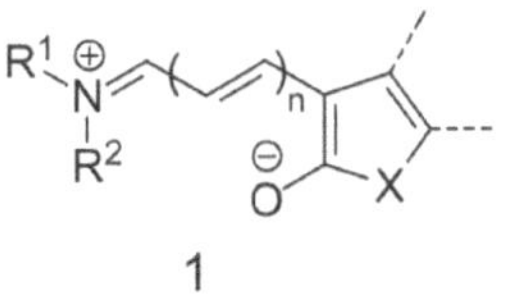

1

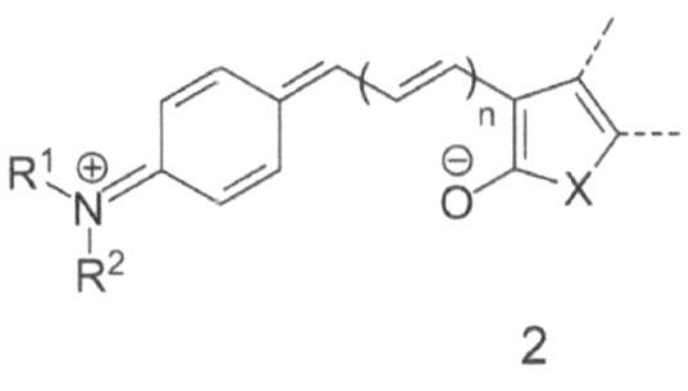

2

This article has previously been published in the journal Physical Sciences Reviews. Please cite as H. Mustroph Hemioxonol Dyes *Physical Sciences Reviews* [Online] 2021, 6. DOI: 10.1515/psr-2020-0175

https://doi.org/10.1515/9783110587104-036

with e. g. $N(R^1)R^2 = NH_2$, NHC_2H_5, NHPh, $N(CH_3)_2$, $N(C_2H_5)$, or saturated heterocyclic rings like e. g.

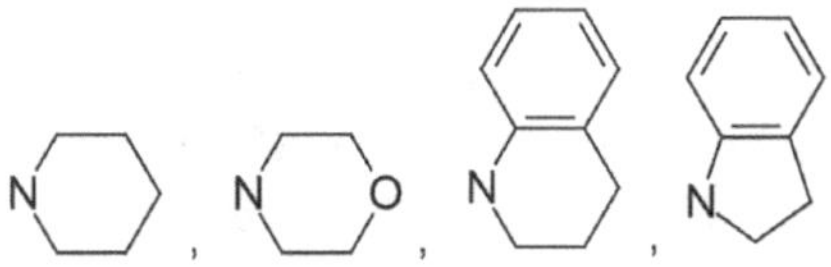

and the second component is an unsaturated heterocyclic ring (where X represents one or two heteroatoms like X = N, O, S) with a charge-carrying terminal oxygen atom.

Some of the first described hemioxonol dyes **3** and **4** were synthesized from 3-ethylrhodanine and either, respectively, diphenylformamidine or malonodianil hydrochloride [6].

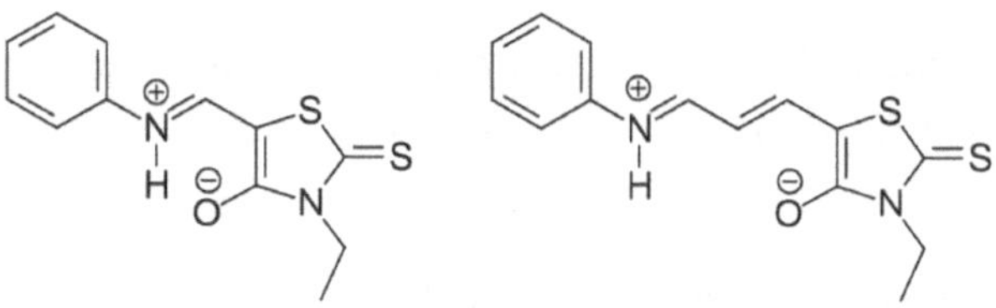

3 **4**

The anilido compounds like, e. g. in **3** and **4** can be easily converted into the acetanilido compounds as exemplified by dye **5** [6].

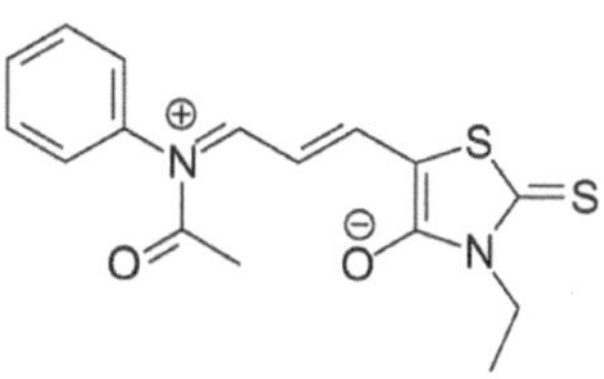

5

Grafton H. Keyes reacted a primary or secondary aliphatic amine with the acetanilido compounds and extended the class of the hemioxonol dyes by the non-aromatic amine derivatives. So, **7** is obtained from **5**, but **6** is synthesized from *N*-acetylated **3** [7]

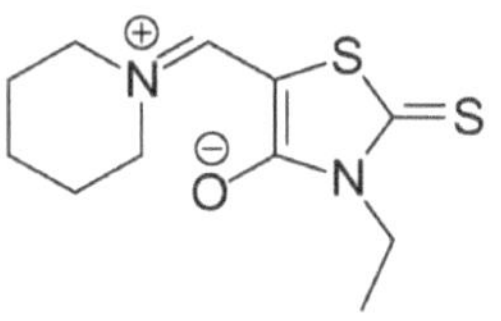

6

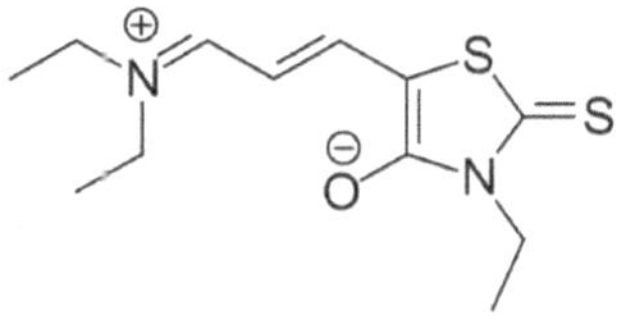

7

In addition, isophorone was used as chain builder to access new hemioxonol dyes with improved properties as exemplified by dyes **8** and **9** [8, 9].

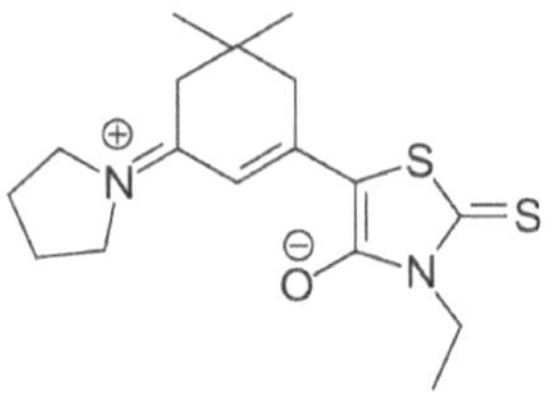

8

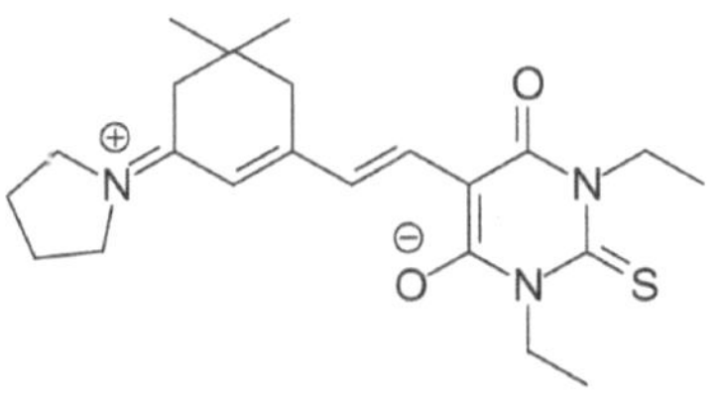

9

König classified the *diarylmethine dyes* as *phenylogous streptocyanine dyes* [5]. Also, with hemioxonol dyes the phenyl group in the 4-*N*,*N*-dialkylaminophenyl group represents a phenylogous extension of the polymethine chain. Nevertheless, the dyes formed from the reaction of 4-*N*,*N*-dialkylaminobenzaldehydes or 4-*N*,*N*-dialkylaminocinnamaldehydes with a heterocycle containing an active methylene group adjacent to a carbonyl group are often called benzylidene dyes, as exemplified by dye **10**, and cinnamylidene dyes like e. g. **11** [10]. In terms of systematic nomenclature, their proper classification is as hemioxonol dyes.

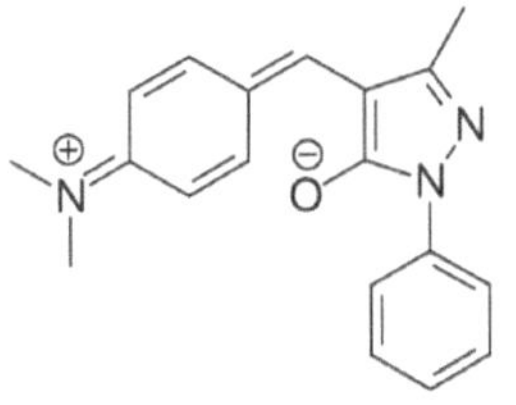

10

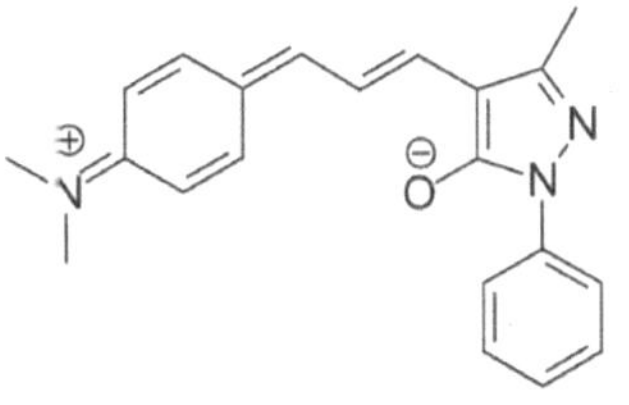

11

In the patent literature, hemioxonol dyes are claimed as spectral sensitizers in silver halide photographic materials. However, there was no real technical application in silver halide photography for them as spectral sensitizers. They are used as *filter dyes* added to the photographic layer or *antihalation dyes* located in various layers of photographic films [11, 12].

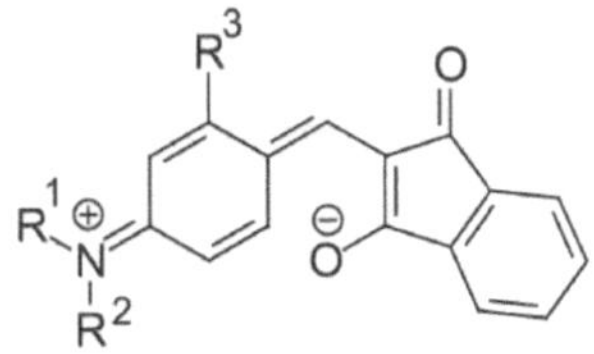

12

Gether Irick and James S. Straley described yellow to red dyes derived from 1,3-indandione **12** and claimed these dyes have excellent affinity for cellulose ester and polyester fibers and exhibit fair to very good lightfastness [13]. Due to the many alternative dyes for this spectral range, they have not found any real technical application in textile coloration [14].

To obtain prints from pictures that have been generated electronically from a digital camera thermal dye diffusion or sublimation transfer systems have been developed [15]. Actually *dye diffusion thermal transfer printing* (D2T2) has found major uses in novelty markets [15]. Basically, the components are dye-donor coating and dye-receiver coating. Thermal dye diffusion transfer works by transmitting heat through the dye-donor coating from the backside to the dye-donor layer. When the dyes in the dye-donor layer are heated sufficiently, they diffuse, transferring to the adjacent dye-receiving layer of the dye-receiver coating.

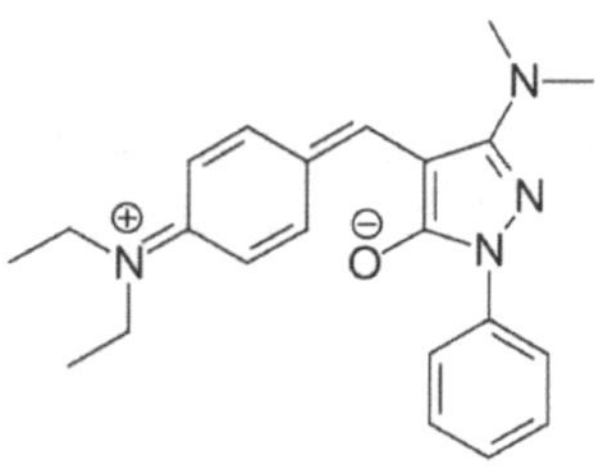

13

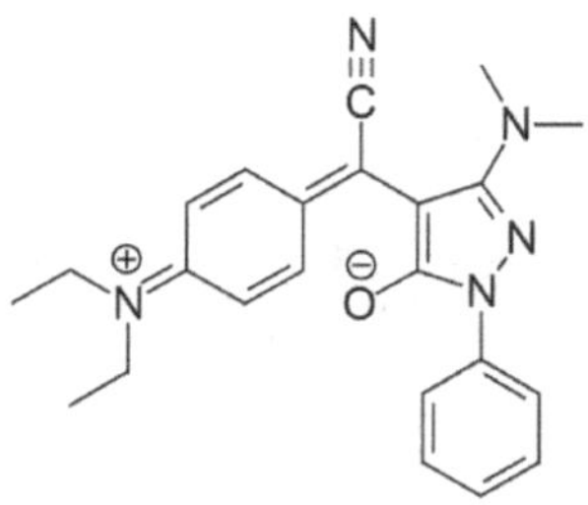

14

The imaging dyes needs to be bright and stable to light and heat. The absorption spectral bands of certain special hemioxonol dyes are narrow and so weakly absorb, if at all in unwanted regions of the spectrum. However, most hemioxonol dyes are unstable to light and heat to a greater or lesser degree. Improvements of lightfastness were obtained by 3-amino-substitution of pyrazol-5-ones like e. g. in the yellow dye **13** and the purple dye **14** [16].

References

1. Brooker LG, Kreyes GH. US 2 241 238. 29 Apr 1939.
2. Mustroph H. Oxonol dyes. Phys Sci Rev. Submitted. DOI:10.1515/psr-2020-0189.
3. Tyutyulkov N, Fabian J, Mehlhorn A, Dietz F, Tadjer A. Polymethine dyes – structure and properties. Sofia: St Kliment Ohridski University Press; 1991.
4. Mustroph H. Polymethine dyes. Phys Sci Rev. 2020;5. DOI:10.1515/psr-2019-0084.
5. Mustroph H. Streptocyanine dyes. Phys Sci Rev. Submitted. DOI:10.1515/psr-2020-0198.
6. Brooker LG. US 2 165 339. 16 Sep 1936.
7. Keyes GH. US 2 186 608. 4 Jun 1937.
8. Brooker LG, Fumia A, Heseltine DW. US 3 976 661. 18 Jan 1974.
9. Van Lare EJ. US 2 956 881 12 Mar 1957.
10. Brooker LG, Keyes GH, Sprague RH, VanDyke RH, VanLare E, VanZandt G Jr, et al. Color and constitution. X. Absorption of the merocyanines. J Am Chem Soc. 1951;73:5332.
11. James TH, editor. The theory of the photographic process. 4th ed. New York: Macmillan, 1977.
12. Keller K, Kampfer H, Matejec R, Lapp O, Krafft W, Frenken H, et al. Photography. In: Elvers B, editor. Ullmann´s encyclopedia of industrial chemistry. Weinheim: Wiley-VCH, 2000.
13. Irick G, Straley JM. Yellow to red methine dyes derived from 1,3-indandione. Text Chem Color. 1969;1:178.
14. Berneth H. Methine dyes and pigments. In: Elvers B, editor. Ullmann´s encyclopedia of industrial chemistry. Weinheim: Wiley-VCH, 2012.
15. Bradbury R. Dyes for dye diffusion thermal transfer (D2T2) printing. In: Freeman HS, Peters AT, editors. Colorants for non-textile applications. Amsterdam: Elsevier; 2000.
16. Diehl DR, Eiff SL, Landry-Coltrain CJ, Russo GM. US 7 781 373. 25 Jan 2007.

Magdalene Gärtner

37 Historical pigments, dyes and binders

Abstract: Artistic creation has accompanied the development of mankind since the beginning. Color as the appearance of things perceptible with the eye is fundamental for the creative process. For the longest time in the past, the use of color was characterized by often very complex manufacturing processes that required the greatest inventiveness and complex knowledge of comprehensive interrelationships from the user. Existing knowledge was passed on and used again and again, but the basic color materials were always further developed and sensitively adapted to the differentiated areas of application and needs. Today, paint coats and, in some cases, extensive collections of recipes provide us with evidence and ideas about these processes in previous times. In this context, it is important to understand that paint recipes were treated like valuables and secrets in the past. The present overview offers insights into the world of historical pigments, dyes and binders. However, a complete overview is not possible at this point due to the diversity of sources, the individual work processes and the different application requirements.

Keywords: historical pigments, historical dyes, historical binders, earth colors, mineral pigments, imprimatur, colored grounds, paintings, gold plating, metal plating, panel painting, natural organic binder, plant rubber, drying oils, resin, wax

37.1 Fundamentals

Color as a means of expression is one of the oldest cultural achievements of mankind. The preparation of colors, the collection and production of pigments, dyes and binders are among the earliest verifiable creative processes. It is a basic prerequisite for people to come to terms with what they have experienced and to capture it in a picture. This cultural asset can be traced as early as the Upper Paleolithic and appears in the cave paintings, for example, of Lascaux (France) or Altamira (Spain). The following remarks refer mainly to the occidental cultural area.

When dealing with historical pigments, dyes and formulations, the reconstruction from our present perspective is often ambiguous. The former manufacturing processes reflect the view of the practicing artist or craftsman in past time. A scientific approach as we know it today is an achievement of modern times.

This article has previously been published in the journal Physical Sciences Reviews. Please cite as M.Gärtner Historical Pigments and Binders *Physical Sciences Reviews* [Online] 2021, 6. DOI: 10.1515/psr-2020-0176

https://doi.org/10.1515/9783110587104-037

Different sources are used for the reconstruction of historical painting materials today. In addition to original coatings of paint on various picture carriers, which can be traced far back in different cultural circles, we have today particularly technical writings, recipe notes or painters' manuals that reflect the everyday situation of the former craftsmen. Early directly transmitted written sources date back to the Middle Ages. The Lucca Manuscript from the ninth century or the Mappae Clavicula from the twelfth century is among the oldest monastic manuscripts that have, however, more the character of unsystematic collections [1]. Since ancient times, there have been indications how painters worked and what technical tools they had. The interest in these sources arose with the beginning of classical studies in the eighteenth and nineteenth centuries [2].

Surviving works of art are also complex testimonies for the materials used in the respective times and regions or in the corresponding cultural spheres. Modern nondestructive scientific examination methods such as neutron diffraction, X-ray diffraction or infrared spectroscopy help to identify historical painting materials. The multilayered structure of a work of art can also be traced using these methods.

It is possible to investigate how the artists selected, procured and processed the materials, but it is almost impossible to translate the recipes into the language of today's chemistry. The chemically complex raw materials that were used are of low purity in the modern chemical sense and are not always comparable due to the natural locations where they were found.

In addition to the knowledge about the processing of these raw materials by the former craftsmen, extensive experience of botany is required, for example, for the plants that were used for art purposes. Different properties of a plant raw material are attributable to the time of harvesting or to the place of growth. In the same way, the knowledge of zoology for the animal raw materials, mineralogy for the mineral pigments and also metallurgy is required.

The old master recipes often require extensive interpretation, since the terminologies and the manufacturing processes described do not correspond to today's materials and production processes. Natural raw materials of mineral, animal or vegetable origin provide the basis for the painting materials produced in more or less complex processes.

Particular attention will be paid here to the extraction and processing of the historical pigments and dyes. The processing of the binders, oils, tempering agents, varnishes and lacquers as well as the leaf metals is closely related to this area, and it is not always possible to make a clear distinction between the individual areas when examining the historical painting materials. When considering pre-industrial paint production, different standards must be applied than in industrial paint production since the nineteenth century.

Other sources of raw materials, low resources, different proportions, manual preparation methods and an individual wealth of experience on the part of the artists, fresh processing and a mostly shorter shelf life of the paints resulted in a completely different approach to the paint production in the past. It is self-evident that, based on these numerous imponderables, exact reproducibility is often not possible. Nevertheless, it is a great field of research, of differentiations and nuances, not least also of accidental discoveries. This completely different approach and various changed conditions explain why the industrial color production is comparable to a revolution in the fields of pigments, dyes and binders.

37.1.1 Raw materials

Raw materials were initially mostly collected and processed by the painters. It is obvious that the daily environment and the geographical proximity were preferred collection areas because of their easier accessibility. Nevertheless, the exchange of raw materials or of finished products was a common practice, especially since this became institutionalized at trade fairs beginning with the Middle Ages. Of course, the sourcing range of painters could thus be expanded and differentiated. The wandering of the painters also brought an enrichment of raw materials and manufacturing habits, which, today, can only be proven in individual cases.

Since the middle of the thirteenth century, for example, the Frankfurt Fair was an important place to go, as unusual and rare materials were offered and raw materials from far away were traded there. This exchange was not always only about raw materials, there were also partially processed or finished materials to purchase, which facilitated the workshop process. It has been proven that the annual trip to the Frankfurt Fair was one of the recurring events for the painters. In most cities, gem dealers or apothecaries also sold raw materials for paint production. The painter Lucas Cranach, for example, had his own pharmacy and a large painting workshop with numerous employees.

The preparation of the painting materials usually took place in own workshops and was carried out according to proven recipes and with a wealth of experience, usually by journeymen. This included crushing the raw material, cleaning processes and rubbing the pigments or dyes with the desired binder on a rubbing stone. It was not always possible to produce the colors in large quantities, because the binders did not have a very long shelf life. Even if preservatives such as clove oil were used, the decomposition process, frequently associated with odor development, resulted in a change of color, although this did not affect the painting properties [3]. Sometimes a "ripening" was also a desired and quality-improving process step.

37.1.1.1 Earth colors

Colored earths such as ocher, Terra di Siena, umbra, Veronese green earth or Bohemian green earth are obtained by cleaning and grinding intensely colored minerals. By strong firing, the color changes to a reddish-brown shade (burnt ocher, burnt Terra di Siena, burnt umber). Earth colors are particularly lightfast and, due to their good availability, they are among the most inexpensive pigments. In addition, earth colors belong to the oldest painting materials of mankind.

37.1.1.2 Minerals

Mineral pigments are obtained by pulverizing of minerals found in nature. They are insoluble. In most cases, complex purification processes, such as the separation of the dewy rock, contribute to the high quality of the precious pigments. Minerals such as lapis lazuli, malachite, cinnabar or realgar are among these raw materials. Like earth colors, they have very good lightfastness.

The coloring properties of mineral pigments depend on their chemical composition, crystal structure and particle size. Historical pigments up to the early nineteenth century usually have particle sizes of 8–20 µm. The particle size of these pigments is nonuniform in most cases.

37.1.1.3 Vegetable raw materials

Vegetable raw materials were preferably used for dyeing and could be processed into paint colors only via detours. For this purpose, vegetable raw materials from roots (madder root), fruits (buckthorn berry), flower or blossom stigmas (acacia blossoms or saffron), color resins (dragon blood tree), barks and color woods (plum tree) and charred seeds (plant black) were used.

The raw materials were often obtained by boiling or dissolving in water, by dissolving in alcohol, or as an acid or alkaline extract. Because of their sensitivity to atmospheric oxygen, light, aging processes and washing processes, they had to be stabilized. This was done by varnishing [3]. This process allowed the originally aqueous dye solution to be used in the pigment form, for example, in oil painting. For this purpose, the aqueous dye solution was mixed mostly with acidic metal salts and an alkaline precipitating agent [4]. Chalk, white lead and clay (kaolin) were used as substrates for the precipitates [5]. Nevertheless, pigments from vegetable raw materials are not as stable as mineral pigments.

37.1.1.4 Animal raw materials

There is also evidence of direct use of animal raw materials for dyeing. The secretion of the purple snail, for example, or the decoction of the cochineal louse were typically used as aqueous solutions that had to be lacquered in addition to be applied in painting. Cockchafer brown, for example, was obtained from the organs of this insect. In the case of Indian yellow, which was used until the beginning of the 1920s, the luminous dye was obtained from the urine of Indian cows. Pigments from animal raw materials are like vegetable raw materials less stable than mineral pigments.

37.1.2 Preparation processes

Painters produced their paints in their own workshops according to individual or traditional recipes. In many cases, unusual and self-explored recipes were well-kept workshop secrets. In the Middle Ages, the tried-and-tested recipes were mainly handed down orally in the mostly monastic workshops or summarized very succinctly. In the early modern period, there was still the handing down from master to apprentice, but experiences and habits were recorded and written down in a more differentiated way.

It was usually the journeymen who were entrusted with the laborious cleaning or rubbing processes. After the pigments were produced, the long-lasting manufacturing process of the paints followed with the rubbing of the pigments, the dyes and the binders. The quality of the color was particularly evident in the long-lasting bonding of the colorant and the binder between rubbing stone and runner.

37.1.3 Recipe collections and painter's manuals

One of the earliest compilations of regulations in which the production of colors is reported is a manuscript from the transition from the eighth to the ninth centuries, the "Compositiones ad tingenda musiva." It is a combination of art-technical and, from the perspective of the time, scientific-technical texts. Medieval transcripts of this kind can be found in the Mappae Clavicula [6]. It is a complex of treatises that came together in this collection. Many older recipes can be found in the Mappa Clavicula, for example, ancient texts of Alexandrian chemistry and descriptions from Pliny, Vitruvius, Dioskudides and Theophrastus.

Many manuscripts of the Middle Ages are not completely preserved. Fragments can be found in various European libraries. A manuscript of the twelfth century in Madrid, in which a late Carolingian original was copied, belongs to these. The wording of the manuscript is reminiscent of both the Lucca manuscript and the Corning manuscript. The text goes probably back to England because the painting instructions contain traditions of English scriptoria of the tenth to twelfth centuries.

There are also similarities with the treatise of Heraclius "De coloribus et artibus Romanorum" in which instructions for painting are combined with instructions for the production of colors and technical regulations.

The Theophilus treatise "De diversis artibus," a manuscript from the early twelfth century, contains painting instructions together with a series of paint production recipes. The painting instructions of the book can be related to the book painting of the eleventh century on the island of Reichenau [7].

In the case of the Theophilus treatise, it is striking that it is in all three extant books the work of a single author, who had own knowledge of the art he writes about. Several transcriptions of this manuscript still exist.

After the twelfth century, a change can be seen in the painting manuscripts. New processes or changes in the color production formulas are developed and written down. Also, the texts gradually move from Latin to the national languages. Nevertheless, the traditions of the early medieval manuscripts did not completely disappear. The "Liber de coloribus illuminatorum sive pictorum" [8], written in the fourteenth century, repeats recipes from the Mappae Clavicula and the Heraclius treatise.

Texts from the following centuries used the existing scripts and were enriched by new recipes. For example, new types of processing and additives are now mentioned. From Italy, the collection of regulations for book illumination "De arte illuminandi" from the fourteenth century, also still written in Latin, has been preserved. The "Libro dell'arte" by Cennino Cennini is written entirely in Italian. The procedures of wall and panel painting of Giotto are recorded here. Rules of book painting show similarities with the Naples manuscript.

The Strasbourg manuscript contains the most comprehensive collection of instructions for the production of colors and the execution of book painting [9]. Nothing is written about oil painting in this source. The first part of the three sections of the Strasbourg manuscript deals with colors for book painting, binders and gold grounds. Heinrich von Lübegge is named as the teacher. The second part also deals with book illumination, but with instructions for gold and silver lettering. Here the instructions go back to Master Andreas of Colmar. Both painter names cannot be historically determined until today. The third part instructs about "tempering" and "flourishing." It talks about the use of glue and oil for painting, as well as about the combination of colors in mixtures. Brightening and shading are also mentioned in this context. Everything is presented in a very sophisticated way [7]. In connection with the Strasbourg Manuscript, there are numerous collections of sources, in which some of the recipes from the Strasbourg manuscript recur individually or in groups.

In the "Mittelrheinisches Musterbuch" (Middle Rhenish pattern book) from the middle of the fifteenth century, it is unique that there is talk of painting tendrils as well as of checkerboard and diamond patterned grounds in various stages of execution. There is also talk of outlining preliminary drawings over which to paint in differentiated layers. In this way, the practiced painting layer structure becomes clear. With this manuscript, one can also assume that it is not a singular writing, but that it belongs in a context of several manuscripts from Mainz.

The importance of the latter source lies above all in the fact that it contains comprehensive instructions for use. In this regard, it differs from the much more concise painter's instructions of earlier times. The secular context of this source allows the assumption that in a workshop, painters were active for whom such instructions were necessary. The situation was still different in the early Middle Ages. The artists were trained in the monastery and learned their art through long exercises in an oral tradition of recipes. For this purpose, the painter's instructions that had been handed down were merely written down in a concise form, in key words, sufficient to capture the essentials.

Jean Le Begue, licentiate of rights and "Greffier" of the Paris Mint completed in 1431 a collection that contains not only individual regulations for the production of colors, but also transcripts of entire treatises. He also comments on the authors of the older manuscripts. Regarding the treatise of Theophilus [10], he writes that the author is an "admirabilis et doctissimus magister de omni scientia picturae artis." Furthermore, he mentions Heraclius [11] as "sapentissimus vir." He also adds to his collection contemporary recipes that he took over, for example, from Johannes Alcherius, who was associated with the librarian of the famous duke Jean de Berry. Le Begue created two alphabetical registers of keywords as well as a compilation of color experiments. With these written sources, which made the knowledge of that time lexicon-like comprehensible, a new consciousness of the traditions becomes clear.

37.1.4 Source situation

Definitions on painting techniques, which are still valid today, date back to the Quattrocento, in particular to Cennino Cennini. This is because most of the main inventions in the painting techniques took place at the end of the fifteenth century. Nevertheless, artistic activities underwent a change compared to the largely closed situation of the late medieval workshop type, with Italy having a time lead of about half a century. Individual experimentation can be found, for example, in artists such as Leonardo da Vinci, Sebastiano del Piombo or Rembrandt van Rijn. The professional status of artists underwent an increasing freedom in the practice of their profession, legitimized since the sixteenth century by the creation and attendance of the academies. In the transalpine area, guild rules and controls prevailed at this point until the seventeenth century. These developments affected also the materials used. With fluent transitions, a separation of the artist from the craftsman took place until the eighteenth century. It should be noted that already in the fifteenth century a development to an increasing specialization in the production of painting utensils begun [[12–16]].

Numerous insights can be gained from the historical written sources. They provide information not only about materials, material procurement, recipes and work processes, but also they provide information about the production of painting media, painting grounds, painting materials and, last but not least, tools.

Another essential basis for research into historical pigments, dyes, painting materials and artistic habits can be derived from the products of the painters' workshops. In recent decades, the examination of paintings has provided valuable additional information.

37.1.4.1 Paintings, painted sculptures, surfaces

Over the centuries, a work process can be traced in the production of painted surfaces, which is built up in numerous layers. This is particularly evident in paintings. It is only through the differentiated interaction of the individual, usually thinly applied layers of primer and paint that the final impression is created. This impression is also characterized by the depth of light and results from the optical interaction of the individual layers.

37.1.4.1.1 Wooden panels

In the Middle Ages, panel painting was done exclusively on wooden panels. The raw material was available everywhere and could be easily processed into stable panels. Studies have shown that the carpenters who supplied the painters with the panels selected proven wood species from the local forests. It is noticeable that the quality of the wood used had to be selected very carefully. This is also confirmed by related guild regulations [17].

37.1.4.1.2 Fabric on wooden panels

An intermediate layer of fabric, fibers (hemp) or parchment and leather (untanned horse, donkey, or cowhide) was often applied to the smoothed surface to conceal the joints of the individual boards or to cover knotholes. The laminations also buffered the movements of the wood.

The over-gluing with hemp fibers is found both on the picture side and on the back of the board. In most cases, the image sides were only pasted over in places that made it necessary. Thus, in the rarest of cases, the glue was applied over the entire surface. Nevertheless, the care taken in over-pasting can be related to the quality of the workmanship [7].

37.1.4.1.3 Textiles

Early (high medieval) paintings on textiles can be found in the application for antependia. These are painted cloths made of linen, which were hung on a wooden frame. It was not until the late Middle Ages that the textiles began to be stretched on wooden frames. The majority of the textiles used as picture carriers are fine flax fabrics in plain weave. Silk can also be found in isolated cases.

In contrast to wooden panels, it was no longer the weight but the width of the loom that set the limits of the format for textile picture supports. From the sixteenth century onwards, it became common practice, first in Venice, to sew lengths of fabric together and thus to have almost no restrictions on the size of the format [7].

37.1.4.1.4 Primers

The primers bound with animal glue usually contain white or gray pigments as colorants and fillers. The materials used are chalk and gypsum. They were built up in numerous layers, and the appropriate grinding or scraping processes were used to achieve an extremely smooth and polishable surface, which could also be covered with leaf metals. The uniform absorbency is an essential prerequisite for the painting process. On wooden panels, the primers are applied more than on canvases. On the latter, they were primarily intended to level the fabric structure and produce a smooth texture. High-quality paintings had thicker layers of primer, especially in the area of painting on wooden panels. The smooth primers had above all the function for the sometimes very thin or even semi-transparent painting layers to increase the luminosity by their reflective effect. In practice, however, the light primers are found only until the Renaissance and are replaced in the Baroque by colored primers. Chalk and white pigments were then replaced primarily by earth colors such as ocher, burnt earths or even umber tones. These were also sometimes applied in several layers. Of course, the painting layers built on top of these had to be applied in a different way due to the dark initial layer. The mostly strongly colored Baroque primers were increasingly mixed with oil-based binders to obtain a higher elasticity, since most of the paintings were produced on canvases [7].

37.1.4.1.5 Imprimatur and colored grounds

Sanding the painting background was followed by impregnation with an insulating character, which was intended to limit or even eliminate the absorbency. In some paintings, it could be proven that underdrawings were applied before this insulating layer of unpigmented painting oil, an oil-resin mixture, glue or an emulsion was applied. Cennino Cennini [18] explains that if the plaster ground is scraped smooth, the first work the painter should do is to draw the picture. Drawing charcoal, metal pencil (silver pencil), or ink diluted with water or paint slurry made with red pigments were used for underdrawings.

Sometimes the isolating layers were also tinted a little. This allowed the underdrawings to be somewhat recessed. Lead white imitations had the advantage of reflecting the incident deep light more strongly than it is the case with chalk or plaster.

37.1.4.1.6 Gold plating, metal plating

For gilded parts of paintings, the primer was impregnated with hot glue water to make it less absorbent. Poliment, a particularly finely whitewashed clay layer, with small amounts of beaten hen's egg, was applied in the following step. Greasiness and suction capacity of the poliment were intended to hold the extremely thin metal leaflets in place and provide a smooth elastic base during polishing. The poliment had to be wetted with egg white strongly diluted in water in order to attach the gold leaf on the surface of the painting. The fine gold platelets were then applied to the moistened surface with special brushes. Due to the elasticity and softness of the raw material, the gold platelets could be spread almost as thin as desired. To save material, intermediate gold was also used, in which very thin gold leaf was applied to silver leaf that had been struck in a preliminary step. This is a kind of cold welding, in which the two metals are inseparably joined. However, the consequence of this is that a transformation of the silver below into brown argentous sulfide cannot be prevented. Silver was also processed in form of platelets, but could not be struck as thinly because of its lower elasticity. Varnish coatings or glue coatings were used to delay the transformation into brown caused by the formation of silver sulfide [19].

The finished metal coatings were sometimes decorated with fine ornaments. Varnish-like substances preserved the luster and at the same time gave the metal plating a differentiated appearance. Binders such as gum Arabic or fish bile can be found in this context as well as vinegar or urine solution of ammonia with an addition of egg white, camphor, mastic and others.

37.1.4.1.7 Panel painting

After the gilding was completed, the preparation of the paints began. The following rubbing of the paints was an essential step. An intimate mixing of pigments and binders resulted in a paint that flowed well from the brush. At the same time, rubbing was intended to optimize the color intensity. Some colorants, such as vermilion, become more brilliant through fine rubbing; azurite or smalt should not be rubbed too hard, otherwise the color intensity would weaken. Some pigments and dyes were rubbed dry first, but water was mostly used as a lubricant for the first operation. Small quantities that were processed immediately could be coated in shells mixed with binder. For oily binders, the watery pastes had to be dried first.

Multilayered paint applications have been common over the centuries. Depending on the specific order, the structure of the painting layers had to be chosen in a way to achieve the desired quality of the work. The cost of executing a work and the materials to be used were precisely determined beforehand between the customer and the painter. If, for example, the extremely precious lapis lazuli was used, clients sometimes even had to make advance payments.

While the painting layer in the medieval panel painting was built up on the light primer and the effect of various colors was intensified or nuanced by appropriate underpainting, the painting concept changed in the Baroque. On the color-intensive primers, the painting layer was applied in such a way that the color tones of the primer had an effect on the painting, e.g., as a shadow system, and were included. The painting layer had fewer layers and could be executed more quickly, which was important because relatively large painting formats became common.

The increasingly used oil binders in the paints had a decisive influence on the painting systems. The pigments, on the other hand, varied little over the centuries of the Middle Ages and even in modern times there is only in the nineteenth century a decisive change due to industrial paint production. Some pigments, such as lapis lazuli, azurite and lead tin yellow, were preferably used in aqueous systems, even if the rest of the painting was predominantly oil-bound.

37.1.4.1.8 Varnishes

According to their composition, varnishes were divided into oil varnishes (drying oils), oil-resin varnishes or oil lacquers (drying oils with melted resins), or egg white varnishes (chicken egg white). They were of great importance as protection against atmospheric influences and also because of their optical function. The varnish coating produces a film with a smooth surface, which reflects light or absorbs part of the light rays (depth light). The smooth surface of binder-rich oil or oil-resin paints changes little during varnishing. Paints with aqueous binder systems form porous layers in which the pigments are not as strongly embedded as in oily binders. Oil or oil-resin varnishes penetrate the paint layer and change significantly the appearance of a painting. This enables the artist to add luster and deep light to the work, depending on the intention [7].

37.1.4.2 Terminology

The terminology of historical painting materials differs significantly from modern chemical terms. The historical paint manufacture was associated with the investigation of material properties and the use of the acquired knowledge for technical painting applications. A specific terminology was developed, which was not used in the same manner over the following centuries. A reconstruction of the recipes is in many cases not exactly possible today.

Some terms refer only to the optical appearance of a color and have nothing to do with the specific components contained in the pigments or dyes. For example, the term “glaze” in the treatise of Theophilus is a designation for lapis lazuli, while various late medieval texts interpret it as azurite. Later, the term is used for different blue colors, sometimes even for mixed colors, e.g., for indigo mixed with lead white or for blue flowers. A systematization of terms in the modern sense was not known and had to be developed in the centuries that followed.

37.1.5 Scientific methods of investigation

With the support of scientific methods such as X-ray fine structure analysis, spectral analysis, chromatography and microscopy, it is possible to identify pigments, dyes and binders contained in the paints without destroying them. These methods can also be used to determine when and in what context a painting was created at the earliest or latest. Specific habits of individual artists or schools of painters can be detected as well. Scientific methods of investigation are also useful for distinguishing copies, imitations or forgeries from originals. Later revisions and retouching on original paintings can be proven as well. Binders are usually difficult to analyze in paintings. They are therefore only rarely able to provide clues for determining the age and the origin of a paint.

Since the beginning of the nineteenth century, the scientific examination of paintings has been a subfield of art science.

37.1.5.1 X-ray examination

One of the most important methods of examining the composition of paintings is done using X-rays. These energetic rays penetrate the painting completely. In the X-ray investigation, all components and layers are indicated in a summation image. In addition to the specific types of wood used for the painting carriers, primers, underdrawings, pigments, dyes and pentimenti can be detected. Paintings are examined with super soft and soft radiation. Depending on the image carrier, X-ray voltages between 10 and 30 kV are used [20].

37.1.5.2 Examination with infrared radiation

The examination of an artwork with infrared radiation allows up to a certain degree the visualization of the layers used. For example, underdrawings and conceptual indications of the painter on the painting carrier, such as color indications, can be made visible. Overpainted areas can also be detected with this examination method. Thickness and opacity of the investigated layers are decisive for the result of the examination. The contrast of the underdrawings to the background is also relevant for the result. Reddish and brown underdrawings are not visible in the infrared. The wavelength used for this examination method is in the range from 760 nm to 2 μm [20].

37.1.5.3 Examination with UV radiation

Changes in the uppermost layers of paintings appear with ultraviolet (UV) radiation. Statements can be made about the state of preservation of the painting. Injuries to the coating varnishes, retouching or overpainting can appear in the UV fluorescence image, since the short-wave invisible UV light stimulates varnishes, binders, pigments or dyes to luminesce. Changes on the surface differ from the original surface of the paintings. The near-UV range from 320 to 400 nm is used for the examination of paintings. With the help of luminescence examinations, only changes on the surface can be made visible [20].

37.1.6 Classification of historical pigments and dyes

Historical pigments can be classified according to their basic materials or their origin:
- Colorants derived from the inorganic nature (natural inorganic pigments such as colored earths, ocher, lapis lazuli, cinnabar).
- Colorants derived from the organic nature of plants and animals (natural organic dyes and pigments such as purple, indigo, sepia, Indian yellow)

- Artificial inorganic colorants produced using chemical syntheses (synthetic inorganic pigments such as lead white, cobalt blue, red lead, ultramarine, titanium dioxide)
- Artificial organic colorants produced using chemical syntheses (synthetic dyes and organic pigments, such as naphthopyran dyes, polymethine dyes, diketopyrrolopyrrole pigments, phthalocyanine pigments)
- When used as a paint, dyes are usually deposited on a solid substrate and thus converted into pigments. In historical sources, they are often referred to as color lakes.

37.2 Historical pigments and dyes

The following listed compilation is intended to be a representative selection of the pigments and dyes used in historical painting.

37.2.1 White colorants

37.2.1.1 Lead white, Kremser white, Cremnitz white, silver white, Venetian white

Lead white (basic lead carbonate, 2 $PbCO_3{\cdot}Pb(OH)_2$, lat.: cerosa, cerussa) is one of the oldest artificially produced pigments. Extensive literature exists on this important white pigment [3, 7, 14, 15, 19, 21–29]. In ancient Greece, lead white was already produced starting from metallic lead. Written sources from the fourth century BC onward describe the manufacture using lead and vinegar. Lead strips rolled into spirals were placed in stoneware pots, and vinegar was added. These vessels were loosely covered and embedded in horse manure. The putrefaction produced heat and carbon dioxide, which converted the metallic lead into lead white.

Until the nineteenth century, lead white was the only white pigment used in easel painting. Almost every painting made before 1835 contains lead white. This can be demonstrated with X-ray examinations because of the strong absorption of X-rays by the lead.

Lead white was mainly applied as a white pigment in oil painting. It was also used in watercolor, pastel and distemper techniques, but here blackening frequently occurred due to conversion into lead sulfide in the presence of hydrogen sulfide. When used in mural painting, lead white proves to be unstable [24].

37.2.1.2 Zinc white, Chinese white, eternally white, pompholyx, snow white, spodium, zinc bloom

In ancient times, zinc white (zinc oxide, ZnO, lat.: lana philosophica, nihilum album, pompholyx) was already known as spodium and pompholyx [3]. The artificial mineral pigment was prepared by oxidation of metallic zinc, by "blowing" zinc ores in the vents of metallurgical furnaces, or by precipitation from zinc solutions.

There is also evidence of its use in medieval medicine. However, its high price stood in the way of a wider use as a painting color. Shortly before 1800, zinc white is occasionally found on paintings. In the 1930s and 1940s, the pigment became increasingly widespread.

It was then used as a pigment for paints. In artists' oil paints, it was often mixed with lead white to increase opacity. It was also used in water-based techniques such as glue, watercolor and gouache paints. Further information on zinc white in this context can be found in the literature, especially in [7, 22–24, 26–29].

37.2.2 Yellow colorants

37.2.2.1 Gold ocher, yellow earth, yellow ocher, Annaberg earth, Chinese yellow, ocher from Derbyshire, French ocher, Italian ocher, Siena earth

Ocher (mainly iron(III) oxide hydrate, FeOOH, and iron(III) oxide, Fe_2O_3, lat.: ochra nativa, luteum montanum) is one of the oldest colorants, which was already used in cave paintings and is still used today [3, 7, 18, 23, 24, 28]. It is absolutely lightfast. However, the coloring capacity varies greatly, depending on the content of iron oxide hydrate. The natural earth pigment was prepared from mined rock by slurrying, grinding and air sifting. The pigment is compatible with all pigments. By firing, red varieties of brick colored and pranced ocher were created.

37.2.2.2 Indian yellow

Indian yellow (magnesium–calcium salt of the euxantinic acid, ital.: gallo indiano) is a dye that is obtained from the urine of cows fed with mango leaves. Mérimée [22] writes that the dye comes from a shrub called Memecylon tinctorium. In India, this plant was also used to dye yellow. The cows fell ill and excreted urine of a strong yellow color as a result of pathological metabolic processes. During heating of the urine, the substance known as Indian yellow was formed. To be used as a pigment, it did not need to be varnish-treated, i.e., deposited on an insoluble substrate. Nevertheless, its use as a glaze color is documented. Especially in the nineteenth century, Indian yellow was valued as a relatively lightfast watercolor [3]. Today, its production is prohibited by law. Additional information on Indian yellow in the context of these remarks can be found in the literature, especially in [7, 23, 24, 30–32].

37.2.2.3 Saffron

The dye saffron (lat.: croceum, croceus) was extracted with water from the dried stigmas of various varieties of crocus growing in Southern Europe and in the Orient [10, 11]. Since ancient times it was used as a spice and as a dye.

Pliny reports about yellow saffron whitewash on stucco in the Minerva temple at Elis [24]. In the "Farbenbelustigung" [33], there is to read: "Saffran färbet sehr hoch gelb, wird in Seidenfarben mit Citronen-Safft vermischet, er zieht auch auf Baumwolle und Seide auf und ist in Wasser und Alkohol löslich." (Saffron dyes very high yellow, is mixed in silk colors with citron saffron, it attracts also on cotton and silk and is soluble in water and alcohol.) White spirit of wine for extraction is found in paintings in Padua, with egg white or gum as a binder [14]. By boiling in rainwater followed by mixing with gum water, it serves as a painter's color [21].

Typically, it was not varnished but processed either dry or in concentrated solution directly with the binder. In written sources, it appears preferentially as a colorant for book painting or as a yellow luster color on tin foils [10, 32]. Further information on saffron in this context can be found in the literature, especially in [3, 7, 15, 19, 26, 28, 31, 34–41].

37.2.2.4 Auripigment, China yellow, kings yellow, operment, orpiment, Persian yellow, noisy yellow, sulfur arsenic, Spanish yellow

Auripigment (arsenic(III) sulfide, As_2S_3, lat.: auripigmentum, arsenicon, auricon) is one of the oldest pigments. It was already found in the Egyptian Amarna. In the ground state, it appears under the microscope as fine gold glittering flakes. Its hardness and toxicity made rubbing difficult. The "Liber de coloribus illuminatorum sive pictorum" [8] recommends a pepper mill for this purpose. Respiratory protection is also recommended on several occasions.

A paint containing auripigment is suitable as a covering coat and also for imitation of gold. Despite its toxicity, it was used as a colorant for textiles. Alum and vinegar were used for this purpose. By burning the pigment, an orange-red color with somewhat coarser crystals was obtained. In the nineteenth century, it was replaced by nontoxic yellow pigments. Additional information on auripigment in the context of these remarks is available in the specialized literature [3, 7, 14, 23, 24, 27, 28, 36, 38, 42–48].

37.2.2.5 Realgar, rushing red, ruby sulfur, operment, Rossz yellow

From ancient times until the Renaissance, realgar was used in large quantities [3]. Different varieties, one yellowish and one reddish, were used on wall paintings in Pompei. Vitruvius and Pliny already reported on these two varieties [49]. The chemical composition of realgar corresponds to that of arsenic sulfide with the formula As_4S_4.

Schreger [50] speaks of a highly reddish-yellow painter's color made of arsenic sublimated with the fourth or fifth part of sulfur; the more often this is done, the more perfect the product turns out. Further information on realgar in the context of these remarks can be found in [7, 14, 18, 23, 26, 36, 46, 51].

37.2.2.6 Lead tin yellow, lead stannate, Canary yellow, giallino

Lead tin yellow (lead stannate, Type I: Pb_2SnO_4, Type II: $PbSn_2SiO_7$, lat.: ochra plumbaria) was fired from lead oxide and tin oxide at 600–800°C [3, 24]. Its main use in panel painting and mural painting can be traced from 1300 until around 1750, when the color disappeared from the artist's palette. Later, the pigment was often replaced by Naples yellow. Color-like yellow pigments were often called massicot.

The color is very lightfast, has very good hiding power and equally very good dyeing power. Both types (Type I and Type II) are alkali-resistant, only in combination with hydrogen sulfide there is blackening due to the formation of lead sulfide. Further information on lead tin yellow in this context is available in the literature [7, 23, 45, 52–54].

37.2.2.7 Massicot, litharge, kings yellow

Lemery [55] states that massicot (lead(II) oxide, PbO) is manufactured by firing at moderate temperatures. De Mayerne [15] writes that massicot in mixtures with other colors spoils easily. He mentions lighter and darker varieties.

In addition to its use as a yellow pigment, massicot appears in earlier technical writings on painting as a pigment with a siccative effect in oily binders. It was also applied as a pigment in primers for weather-resistant oil gilding. It was also used as a flux in glass and porcelain painting [24]. Additional information on massicot in this context can be found in [3, 7, 14, 36, 53].

37.2.2.8 Naples yellow, antimony yellow, giallino

Already on Babylonian bricks about 2500 BC, the pigment antimony yellow ($Pb(SbO_3)_2$ and $Pb_3(SbO_4)_2$, lat.: luteolum neapolitanum) was found [24]. The true Neapolitan yellow, giallino, is a type of yellow stone found in the Naples area.

Cennino Cennini [18] writes that the stone comes from the volcanic peaks of the mountains. However, auripigment could also be meant here, since Naples yellow has not been detected in the area of Mount Vesuvius [3].

Naples yellow has a good hiding power and a very good lightfastness. In addition, the pigment is largely resistant to acids and alkalis. It is compatible with all pigments and binders. For further information on Naples yellow in this context, refer to the specific technical literature [7, 14, 21, 23, 26, 28, 48, 56–58].

37.2.2.9 Poured yellow, arzica, giallo santo

Poured yellow (buckthorn berry extract, $C_{16}H_{12}O_7$, methyl ether of a pentahydroxyflavone, rhamnazin and quercetin, lat.: gilb plumen, giallo santo) represents a group of yellow dyes, such as those obtained from dye broom, immature poplar flowers or yellow wood. Cröker [56] mentions roots (curcuma), berry yellow, yellowwood, dyer's woof, spring birch leaves or "gilb plumen." Schulze [59] writes that of the yellow paint colors, poured yellow is one of the most common. It is obtained by boiling young birch foliage with alum and water followed by pulverizing the strained dry matter and adding slurried chalk (lacquering). Schmidt [26] describes alternatively the preparation of poured yellow by coloring chalk or argillaceous white earth with any yellow plant colorant. Additional information on poured yellow in this context can be found in various related publications [3, 15, 21, 23, 32, 34, 37, 40, 44, 60–64].

37.2.3 Red colorants

37.2.3.1 Red earth pigment, red bolus, red ocher, rubrica, Spanish red, terra di Pozzuoli, terra di Treviso, terra rossa, sinople

The red clays were and are usually named after their origin. Since early Egyptian painting until today, they are very common pigments in all forms of painting. The necessary working processes are mainly grinding, washing and slurrying. Red earth is absolutely lightfast, and compatible with all pigments and binders. It has a very high coloring capacity and is very opaque. Red earth pigments (lat.: rubeum, rubrica) consist mainly of iron(III) oxide with changing amounts of clays and quartz.

A mixed color of red earth and a white pigment was called Cinabrese in Italian and could be applied as one of the brightest sinople pigments useful for skin colors in wall paintings [18]. Additional information on red earth pigments in this context can be found in [3, 7, 19, 23, 28, 38].

37.2.3.2 Madder varnish, madder carmine, madder purple

Madder varnish is a plant dye (alizarin, purpurin, xanthin, ruberythric acid, rubiazine, lat.: rubia) that has been known since ancient. Dioscorides refers to it as "eurythrodanon," Pliny mentions this dye as "rubia."

It is obtained from the roots of the madder plant. The color substances were extracted with sulfuric acid, the extract heated with alum solution and the madder varnish precipitated with soda. Various color shades were obtained by different additives (additions of tin salts provided a purple-red color, iron salts provided blackish-purple shades, and chromium salts provided purple-brown shades). The dyes were varnished on chalk, clay or plaster. Madder varnish has been very important in painting since ancient times until today. Bright glaze colors were produced in different binders, which were applied transparent or semi-transparent over cinnabar, for example.

Madder varnish is not very resistant to alkali or lime. The light resistance of natural root madder varnish is also rather poor. The synthetic alizarin madder varnish has much better lightfastness. Further information on madder varnish in this context is available in the literature [2, 7, 11, 65, 22, 24, 26, 28, 31, 36, 38, 39, 50, 54, 59, 64–68].

37.2.3.3 Cinnabar, mercury blende, minium, mercury sulfide red, vermilion

Cinnabar (red mercury sulfide, HgS, lat.: cinnabaris, minium) is one of the mostly used red pigments and occurs naturally as a mineral. It was already known to the ancient Egyptians and Hebrews. Cinnabar can also be found in Roman wall paintings and has been used over the centuries. It has been artificially produced from mercury and sulfur for centuries, as evidenced by recipes for cinnabar production. In the nineteenth and twentieth centuries, it was displaced by new and more durable pigments such as cadmium red in paints. Industrial production of the synthetic pigment from mercury and sulfur began at the end of the 1770s. Mercury sulfide has good hiding and coloring power. Its lightfastness is moderate. It darkens under strong illumination. Additional information on cinnabar in this context can be found in [2, 3, 7, 9, 14, 19, 22, 24, 27, 28, 34, 36, 44, 46, 47, 54, 56, 59, 61, 65, 69–71].

37.2.3.4 Dragon blood, dragon blood resin, dracorhodin

The red alcohol-soluble color resin of the dragon tree (color resin of complex composition, e.g., dracorhodin, lat.: calamus, resina draconis) was mainly used as a resin or luster color on gold leaf or silver leaf. The red resin, which was already known in ancient, was used in book painting as a watercolor. For the coloring of instrument varnishes, it is processed partly up to the present time. One collects the leaked and solidified resin drops ("tear dragon blood"), which are formed into small balls and traded as "grain dragon blood." A larger yield is obtained by heating the fruits. The melted-out resin was formed into fist-sized balls, sticks or cakes.

As a film former, the colored resin can be used without the addition of an oil. The most precious dragon blood resins are particularly color intensive. The light fastness of the red resin is, however, rather low [24]. Further information on dragon blood in this context can be found in the literature, especially in [3, 7, 26, 28, 31, 32, 34, 37, 41, 54, 63, 72–77].

37.2.3.5 Carmine, carmoisine, cochineal, Florentine lacquer

The dried female cochineal scale insects of the family Dactylopiidae contain about 14% of the animal dye carmine (calcium and aluminum salts of carminic acid, $C_{22}H_{20}O_{13}$, lat.: carminium). Since the sixteenth century, the scale insect was introduced into Europe from Mexico and other Central American areas. Caucasian and Persian species of the scale insect contain the same dye and were also used for dyeing and coloring. The dried scale insects are boiled in water mixed with some acid. It is therefore also called acidic extract with lemon juice or garlic juice [14]. From the carminic acid solution, lime and alum are used to precipitate the carmine. As a result of its low lightfastness, carmine is now largely replaced by synthetic dyes in stains. The dye is not alkali-resistant, so it is not found in lime-based wall paints, but in oil-based and resin-based paint systems or as watercolor. The coloring capacity is very high. For further information on carmine in this context, refer to the specific technical literature [3, 21, 22, 31, 41, 50, 59, 67, 70, 77, 78].

37.2.3.6 Purple

Mediterranean snails supplied the highly prized and extremely valuable purple (animal dye, dibromindigo, lat.: purpura) even in ancient times. For this purpose, the secretion was obtained from the hypobranchial gland of the snail. The gland secretion is initially water-clear, gradually turning yellow, then brown and finally purple. The dye is absorbed from a substrate to produce the final colorant. Even for ancient manuscripts, the parchment was refined with the dye over a large area. The most precious manuscripts of biblical texts or documents were often written with gold inks on these purple parchments. The yield is so low that it takes the secretions of about 10,000 snails to obtain 1 g of purple. Additional information on purple in this context can be found in [3, 7, 15, 23, 24, 75].

37.2.3.7 Red lead, lead red, saturn red

Red lead oxide (lead plumbate, lead(II, IV) oxide, Pb_3O_4, lat.: minium, minium secundarium) is described by Pliny [19] as red color obtained by heating of lead white. Occasionally there is confusion with cinnabar. In the Trier and Brussels painting schools], it is described that red lead is formed by heating over the fire in an earthen vessel [14]. This produces lead yolk as an intermediate stage.

Red lead was often used because it could be used in all artistic techniques. It was also common as watercolor in book painting with egg white [10] or gum water [14]. It is now banned as an anti-rust paint in oily binders. Further information on read lead in this context can be found in the literature, particularly in [3, 7, 9, 11, 18, 21–24, 28, 37–39, 54, 56, 73, 74, 79, 80].

37.2.4 Blue colorants

37.2.4.1 Egyptian blue, Alexandrian blue

Egyptian blue (calcium copper silicate, $CaCuSi_4O_{10}$, lat.: caeruleum) is a vitreous composition made of sand, lime and copper minerals with portions of soda ash added as a melting aid. The pigment can be traced back to as early as 2600 BC and became the most important blue pigment of ancient times. Knowledge of its production was probably lost during the time of the great migration.

Egyptian blue is largely resistant to acids and alkalis, and is therefore also suitable for lime-based wall paintings. Like smalt, it should only be coarsely struck, as it becomes paler with increasing fineness. The pigment appears almost black when bound in oil or varnish with resin or wax [7]. For further information on Egyptian blue in this context, refer to the specific technical literature [3, 31, 48, 58, 66].

37.2.4.2 Natural ultramarine, azure blue, azurum ultramarinum, lapis lazuli, lazur, oriental blue

Lapis lazuli (complex sulfide-containing aluminum silicate, $Na_6Al_6Si_6O_{24}(NaS_n)$, lat.: lapis lazuli, lazurium) was already used as a gemstone in early Egypt. 3500 BC, the pigment was found among the Sumerians on cult objects. Natural ultramarine, extracted from high-quality lapis lazuli, has always been a very precious and highly prized pigment. The best varieties were as expensive as gold, so its use was limited to particularly noble representations or motifs. Madonna cloaks, for example, were often painted with lapis lazuli. Numerous recipes for processing have survived and describe an always similar procedure. Cennini [18] mentions pulverization, annealing and kneading of the so-called pastille. Lapis lazuli is particularly lightfast, stable to alkalis but sensitive to acids. It is compatible with all other pigments with the exception of lead and copper-containing pigments [53].

In 1806, the composition of natural ultramarine was determined. A synthetic pigment corresponding to the natural one in color, chemical composition and properties was produced in a factory shortly after 1830 [24]. The inexpensive production led to a wide distribution of the synthetic pigment even as wall paint. Additional information on natural ultramarine in this context can be found in [3, 7, 23, 28, 36, 54, 62, 70, 78].

37.2.4.3 Smalt, blue glass, Bohemian blue, cobalt glass, scatter blue, kings blue, zaffre, zaffer

Smalt (lat.: smalto) is a glass mass colored with cobalt oxide and made from silica sand, potash and cobalt oxide (SiO_2: 66–72%; K_2O: 10–21%; As_2O_3: 0–8%; CoO: 2–18%). It was crushed by pouring the melt into water. Further crushing of the glass fragments quickly causes the color to fade. Therefore, smalt is left relatively coarse, washed and elutriated. Because of the coarse grain, the colorant has poor hiding power. It was not only applied by brush, but also by scattering techniques. The so-called scatter blue was applied to adhesive substrates [56]. As an enamel powder, it was used in a finer trituration. Similar to Egyptian blue, smalt usually darkens considerably in oil. It is very lightfast and resistant to binders. Further information on smalt in this context can be found in the literature, particularly in [3, 7, 14, 22–24, 28, 48, 54, 59, 78, 81, 82].

37.2.4.4 Indigo, indigo carmine

Indigo is a plant dye containing the glucoside indican (lat.: indico, glastum, vitrum, lulax, lapis indicus pictorum, atramentumindicum pauli). The indigo plant was mainly cultivated in East and West India, China, Mexico, South America and Egypt. For extraction, the fresh leaves, soaked in water, were left to ferment in vats. In a rotting process by means of an alkali, such as old urine or woad ash, the colorless glycoside indican (indigo white) $C_{16}H_{12}N_2O_2$ was formed. This was poured into tubs and oxidized to the dye by vigorous beating. The clear liquid was drained off and the slurry boiled.

The slower the pieces were dried, the finer the structure became. 100 kg of dry plants yielded 1.5–2 kg of indigo. Indigo was imported in large quantities by sea from India and China only since the sixteenth century. In paintings, it can be traced in isolated cases as early as the fourteenth century.

Stöckel [21] notes that indigo untreated does not produce a beautiful color, but dissolved with oleum vitrioli (sulfuric acid), dried again and mixed with white lead it produces a beautiful coating in oil. Because of the relatively high price, it was occasionally recovered even from dyed wool using sulfuric acid [32]. For further information on indigo in this context, refer to the specific technical literature [3, 7, 14, 15, 23, 24, 26, 28, 31, 36, 39, 41, 42, 48, 56, 59, 61, 64, 73, 75, 83–86].

37.2.4.5 Azurite, azure blue

The natural mineral pigment azurite (basic copper carbonate, 2 $CuCO_3 \cdot Cu(OH)_2$, lat.: lapis armenus, azurrum almaneum sive teothonicum) with a bright blue color was a valued blue pigment since ancient times until the seventeenth century. Poorer varieties contained malachite. The mineral came from Silesia, Hungary, Tyrol, Poland or Armenia. It was widely used in mural painting until the nineteenth century and is most stable as a water-based paint. Schreger [50] is critical of its use with oil as a binder. He writes that azurite already becomes green under the brush [32]. In the nineteenth century, the synthetic pigment azurite, which was produced by precipitation from copper salts, came onto the market. It is much more unstable than natural azurite.

The mineral was processed into the pigment by grinding and sludging. Separation of the mineral from remnants of gangue and dirt was done by washing with lye and honey [87], glue, egg or viscous gum in aqueous solution, with impairing particles floating on top and poured off. Liber writes [5]: "Nimm Lazurstein, und zerreibe den mit feinem und nicht allzu starkem Leim. Je sorgfältiger du reibst, desto besser wird er. Lass ihn über Nacht oder noch ein bißchen länger im Leim stehen. Dann gieße die lauwarme Lauge über das geriebene Lazur und rühre darin eine zeitlang. Gieße dann die obere Materie ab, so hast du Lazurasche." (Take lazur stone, and grind that with fine and not too strong glue. The more carefully you rub, the better it will be. Leave it in the glue overnight or a little longer. Then pour the lukewarm lye over the grated lazur and stir it for a while. Then pour off the upper matter, and you have lazur ash.) The process was repeated and with each new infusion and skimming of the affecting particles, the quality was improved [3].

Azurite was particularly color-intensive in coarse rubbing, but could not be painted in this form with a brush. The pigment was distributed in some cases as a powder on sticky surfaces. Since the Middle Ages, people tried to imitate azurite because of its high value. However, only copper complexes were formed. Additional information on azurite in this context can be found in [7, 14, 18, 21, 23, 24, 28, 38, 54, 82].

37.2.4.6 Prussian blue, Paris blue, Antwerp blue, Berlin blue, Chinese blue, Delft blue, iron blue, iron cyan blue, Diesbach blue

In 1706, Prussian blue (iron(III) hexacyanoferrate(II), $Fe^{III}_4[Fe^{II}(CN)_6]_3$) was discovered by Diesbach in Berlin. Schulze describes the production process of Berlin blue analogously as follows [59]: to prepare it, three parts of blood evaporated to dryness and one part of potash are annealed until a blue flame bursts out of the furnace or melting vessel; then the annealed mass, after it has cooled, is leached out with boiling water, whereby the so-called blood lye, i.e., potash dissolved in water, is formed. A solution of two parts of alum and one part of iron vitriol with 10 parts of water is poured into this solution until there is no more effervescence. The precipitate thus obtained is repeatedly washed with boiling water and then treated with twice as much hydrochloric acid as iron vitriol was used, whereby the free iron oxide, which produced the green color, is dissolved and the precipitate now appears beautifully blue. The precipitate is then washed out again several times and then dried in mild heat. If the alum is completely omitted in this preparation, the so-called Paris blue is obtained.

Prussian blue is very stable and compatible with almost all pigments. At the same time, it has extraordinary coloring power in all binders and mixtures. Due to its lack of alkali resistance, it does not appear in lime-bound wall paintings [24]. For further information on Prussian blue in this context, refer to the specific technical literature [3, 23, 88].

37.2.4.7 Cobalt blue, cobalt ultramarine, Leithner blue, Thénard blue

Cobalt inks can probably be traced back to Schneeberg in Saxony from the first third of the sixteenth century. There is no evidence that these were already being used as artists' pigments at that time. In 1775, a recipe by Leithner in Vienna became known, which was then used for the industrial manufacture from 1795 on [7, 24]. The pigment was produced by heating cobalt compounds with aluminum hydroxide.

Cobalt blue (cobalt aluminum oxide, $CoAl_2O_4$) has excellent lightfastness and weather resistance. In addition, it is resistant to acids, alkalis and heat stress. Therefore, it can also be used as a porcelain paint. It is compatible with all pigments and all binders. Further information on cobalt blue in this context can be found in the literature, particularly in [3, 22, 23, 45].

37.2.4.8 Woad

Woad (lat.: lulax, guado, flores crici ver isatidis, vitrum, gadus herba) is a plant dye. Over 50 species of woad have been used in Europe since ancient times for dyeing wool and linen. Various painter's manuscripts since the Middle Ages mention woad and its processing. Schreger [50] defines woad as a foam that forms on the dye broth. The dried woad, which was fermented moistened, is called woad coal [39]. The Brussels painting school states that the extract is obtained using old urine [14]. The dye is formed when the dyed cloth is exposed to the air, analogous to indigo. In France, Italy, Lombardy and Tuscany, woad was cultivated in large quantities until 1560, when the import of indigo from India and America reduced production in Europe [3]. Additional information on woad in this context can be found in [7, 31, 36, 38, 44, 59, 64, 89–91].

37.2.5 Green colorants

37.2.5.1 Green earth, Belgian green earth, Bohemian green earth, Cyprian green earth, Kadan green earth, Tyrolean green earth, Veronese green earth, celadon green, stone green

Green natural earth pigments (lat.: prasinus, creta cirina, verde terra) are weathering products consisting of iron oxide hydrates, magnesium and alkali silicates, essentially of the mica-like minerals glauconite and celadonite. Good color-intensive varieties come from the Verona area (Veronese green earth), from Bohemia (Bohemian green earth) or Cyprus (Cyprian green earth). The colorants were already widespread in ancient wall paintings and were used in all periods of the European panel painting.

Green earths are light and weather resistant in most of the application systems and compatible with all pigments. The use of oil as a binder results, however, in a low hiding power [24]. In early Italian panel painting, green earths served as underpainting for the incarnates. For further information on green earth in this context, refer to the specific technical literature [3, 27, 34, 70, 86, 92].

37.2.5.2 Verdigris, aerugo, copper acetate, copper green, Spanish green

Verdigris (lat.: aerugo, verdigris, viride hispanicum) is the best-known artificial green pigment and has been produced since ancient times [19, 69]. It consists of copper acetates. Neutral or basic copper acetate, respectively, neutral verdigris (green) can be described with the formula $Cu(CH_3COO)_2 \cdot H_2O$, basic verdigris (blue), on the other hand, with the formulas $2\ Cu(CH_3COO)_2 \cdot Cu(OH)_2 \cdot 5\ H_2O$, $Cu(CH_3COO)_2 \cdot Cu(OH)_2 \cdot 5\ H_2O$, and $Cu(CH_3COO)_2 \cdot Cu(OH)_2$. The numerous historical manufacturing processes vary, but the interaction of copper sheets or chips with vinegar, wine pomace, cream of tartar, caustic lime, Roman vitriol and other substances is always similar [6, 11, 14]. In many cases, the use of a dung heap or oak vessel could be found. Additives such as honey, salt, urine and ammonium chloride (sal ammoniacum) were used for the preparation [93]. As incompatible copper compounds were formed, mixing with other pigments was avoided. Formed verdigris layers were isolated using intermediate varnishes.

Basic verdigris was mainly produced in wine-growing regions, e.g., in Southern France, by alternately piling up wine marc and copper plates and allowing them to ferment. The resulting acetic acid formed verdigris on the surface of the copper, which was scraped off. In aqueous binders, it can turn dark when mixed with sulfur-containing pigments.

The Brussels Ms describes verdigris as a very good color in fatty oil [14]. It is recommended to fire it immediately so that it does not fade [15, 32]. Its lightfastness is good, but it is very sensitive to alkali and therefore unsuitable for lime-bound murals. In aqueous binders, it is incompatible with sulfide pigments. This does not apply to oil-based binders and tempera [24]. Further information on verdigris in this context can be found in the literature, particularly in [3, 7, 9, 18, 23, 24, 28, 36, 38, 39, 45, 47, 54, 56, 62–64, 70, 82, 87, 94, 95].

37.2.5.3 Malachite, copper green, malachite green

The pigment malachite (basic copper carbonate, $CuCO_3 \cdot Cu(OH)_2$, lat.: crysocolla) is produced by crushing the corresponding mineral [19]. Since ancient times until modern times, the pigment was appreciated. Just as this mineral is often found with azurite in its natural occurrences, a pigment mixture of malachite and azurite is found under the name Verde azzuro.

Malachite decomposes already in diluted mineral acids under the influence of CO_2. It is lightfast and resistant to dilute alkalis. Malachite was used almost exclusively in aqueous binders. It is not resistant to hydrogen sulfide and sulfide pigments [24]. Additional information on malachite in the context of these remarks can be found in the literature, especially in [3, 7, 14, 23, 49, 82, 96].

37.2.5.4 Schweinfurt green, Basel green, Brunswick green, Brixen green, Eisleben green, Eisenach green, Kassel green, Munich green, Neuwied green, Paris green, Vienna green, Würzburg green, Zwickau green, emerald green, mitis green, smaragd green

In 1778, Rust discovered the luminous bright Schweinfurt green (copper arsenite acetate, $Cu(CH_3COO)_2 \cdot Cu(AsO_2)_2$). It was produced commercially on a larger scale as early as 1805. It was given different names according to where it was produced. Because of its high toxicity, it has not been in use since 1920. Nevertheless, the intense light green (poison green) coloration cannot be replaced by any other pigment.

Schweinfurt green is unstable to acids, bases and hydrogen sulfide. It is well processable with oil and also with glue and was used as wallpaper and wall color. The pigment is incompatible with sulfide pigments, especially in aqueous binders. For further information on Schweinfurt green in this context, refer to the specific technical literature [3, 7, 22–24, 28, 45, 58, 78].

37.2.6 Brown colorants

37.2.6.1 Umber, Caledonian brown

The natural earth pigment umber (iron-containing aluminosilicates with manganese oxides, lat.: umbra) has been known since ancient times. It is a clay silicate with a greenish tint. Because of the manganese content, umber has a siccative effect. Before the sixteenth century, brown pigments played only a minor role. The development of painting on the basis of clays in the baroque period led to a growing importance of brown tones. Burnt varieties of earth pigments take on reddish hues. Umbra can be used in all techniques and is compatible with all pigments and in almost all binders [24]. Additional information on umber in this context can be found in [3, 7, 23].

37.2.6.2 Mummy

Mummy (lat.: mumia arabum, mumia sepulcrum, mumia factitial) was made from Egyptian mummies. There are rubbery, bituminous or resinous mummies depending on the embalming technique. The treatment of the mummies has an influence on the properties of the mummy made from them. Mummy served as an intense black-brown glaze material with transparent properties. Armenini notes that the pigment has no body and can therefore be excellently rubbed with chip green [15]. Schmidt makes the following comment about mummy: “Man benutzt hier Stücke, welche schwarz und glänzend sind” (pieces are used here, which are black and shiny) [26]. A highly resinous or even bituminous material is certainly meant here. Meat, bone and fabric were processed also, as these should give the color more intensity [3]. Further information on mummy in this context can be found in the literature, particularly in [14, 23, 28, 36, 97].

37.2.6.3 Van-Dyck brown, Cassel brown, Cologne earth, charcoal brown, sap brown, Spanish brown

Van-Dyck brown is an earthy amorphous lignite consisting of about 90% organic humus and humic acids. The remaining parts consist of iron oxides, clays, sand, etc. The largest deposits are near Kassel and Cologne. The pigment was already known in the sixteenth century, but was probably not widely used until the seventeenth century. The pigment was very popular for clay painting with the masters Peter Paul Rubens and Anthonis van Dyck. Especially with oily binders, it provided an appreciated glaze color. However, very fine-particle grades tend to “bleed”. The pigment has low lightfastness, but is compatible with all pigments. It is not acid and alkali resistant. For further information on van Dyck brown, refer to the corresponding literature [3, 14, 23, 24, 28].

37.2.6.4 Sepia, cuttlefish

Sepia (lat.: sepia) is produced from various squid species from whose ink sacs, after dissolution in ammonia and precipitation by addition of hydrochloric acid, a finely dispersed ink was produced [50]. It also served as watercolor.

Ossa sepiae, the cuttle-bone, was used as an agent for primers and as an additive in the cooking of oils [3]. Additional information on sepia in this context is available in [23, 31, 49].

37.2.6.5 Asphalt, Antwerp brown, bitumen, earth pitch, mineral pitch

In ancient Egypt, asphalt (natural hydrocarbon mixture with various organic and inorganic impurities, lat.: asphaltum, bitumen babylonicum, cera montana, cedria terrestris) was used as an adhesive, sealant and preservative. The natural evaporation residue of petroleum oils was found in the soil or in lakes. Used in thin glaze with oily binders, asphalt is very suitable for shading because it has very glazing properties. It has been highly prized since Rembrandt van Rijn and came into greater use in Dutch Baroque panel painting. It proved to be unstable. As an underpainting, the colorant migrates into the overlying layers and darkens them, a process that cannot be controlled.

Asphalt was already used by Leonardo da Vinci to blend green and seems to have caused cracking in the surface layer of the painting "Gineva de Benci." It is also mentioned in the paintings of Titian and Jacopo Tintoretto [14]. De Mayerne [32] remarks that it is difficult to grind asphalt. In the "Lacquir-Kunst" (lacquer art) [72], asphalt turns up for a resin varnish made of two parts asphalt and four parts shellac in spirit. Further information on asphalt in this context can be found in the literature, particularly in [3, 23, 24, 28, 36, 37, 50, 58, 62, 97, 98].

37.2.7 Black colorants

37.2.7.1 Ivory black, leg black, bone charcoal, bone black, sammet black

The pigment ivory black (lat.: elephantinum) has been known and used as "elephantinum" since ancient times. It is a slightly blue-tinged black obtained by charring ivory waste or bone rests. In some cases, the ivory pieces were also burned with linseed oil [44]. The term expanded to include bone black and charcoal black.

Approximately, 10–20% of the pigment is carbon as the coloring component; the rest consists of inorganic impurities (mainly calcium phosphate) from the mineral skeleton of the starting material. Ivory black is compatible with all pigments and resistant in most of the binders. It is lightfast and resistant to acids and alkalis [24]. For further information on ivory black, refer to the corresponding literature [3, 14, 15, 23, 32, 70, 99].

37.2.7.2 Lamp black, lamp soot, acetylene soot, soot black

Lamp black (almost pure carbon, usually with traces of tar or inorganic impurities, lat.: atramentum, nigrum lampadis, fumus) has been used for ink production in China since around 2000 BC. In Europe, larger plants for soot production were built in the eighteenth century. The starting materials are gas, oil, wood, resins or pine. Burning the starting materials produced finely divided carbon black. The pigment has a high binder requirement, but is compatible with all binders. It is completely lightfast and resistant to chemicals. Because of the fine particle size and the very strong coloring capacity, lamp black was mainly used for printing inks, watercolors and ink [24]. Additional information on lamp black in this context is available in [3, 14, 23].

37.2.7.3 Iron oxide black, oxide black, universal black, magnetite

The natural magnetite (iron(II,III) oxide, magnetite Fe_3O_4, $FeO{\cdot}Fe_2O_3$, lat.: ferrum oxidatum nigrum) was probably not used as a pigment. The production of synthetic iron oxide black began after 1920. The pigment has meanwhile displaced almost all black pigments in painting technology and in the area of arts. It is absolutely light-fast, weather-resistant and alkali-resistant. In addition, it is compatible with all pigments and is compatible with most of the binders [24]. Further information on iron oxide black in this context can be found in the literature, particularly in [3, 7, 23].

37.2.7.4 Vine black, blue black, plant black, charcoal black

Vine black (lat.: vinea nigrum) is a black pigment that has been in use since ancient. It was created in the early days by charring withered grapevines [19]. The pigment consists of 95% carbon; the residual components are calcium carbonate, potash and iron oxide, which were not completely removed during the washing process.

Vine black was often used in various wall techniques, especially in the stucco-lustro technique, where a mixture of wax, mastic and oil was applied to the wet lime plaster. It is also used as drawing charcoal and as grinding aid. Vine black is compatible with all pigments, has a very good lightfastness and a high chemical stability. Additional information on vine black in the context of these remarks can be found in the literature, especially in [3, 15, 18, 23, 26, 79, 99, 100].

37.2.8 Metal powders

37.2.8.1 Shell gold

The name shell gold derives from the gold powder rubbed with gum Arabic as a binder, which was sold coated on the inside of shells. The fine gold powder was also occasionally used in the Middle Ages with hen's egg white as a binder, especially for writing. After drying, the writing could be polished with a hard material, such as an agate or a dog's tooth. To make gold powder, gold leaf was usually ground in a mortar with an addition of honey. The honey was washed out after grinding. Gold powder with a mercury content was obtained by the preparation of a mercury amalgam followed by heating. Shell gold in this context is described in certain publications, particularly in [3, 5–7, 10, 14, 15, 18, 34, 44, 56, 79, 100].

37.2.8.2 Shell silver

Shell silver was used less than shell gold. In most cases, the silver powder used in artworks turned partially or completely into silver sulfide in the course of a relatively short time under the influence of sulfur compounds. Such undesirable discolorations could not be prevented by any coating. Further information on shell silver can be found in [3, 14, 56].

37.2.9 Natural fish silver

Natural fish silver also called silver luster, mother of pearl essence or pearl essence belongs to the class of effect pigments. By rubbing the silvery scales of whitefishes with water, shiny platelets settle on the bottom of the vessel used. Ammonia is added for leaching. Approximately 3500–4000 fishes are necessary for the production of 100 g of the pigment. Fish silver is the oldest known pearlescent pigment. In 1655, Jacquin, a rosary manufacturer, probably produced this pigment for the first time, Réaumur improved the process. Industrial production of natural fish silver began in North America in 1910. The pigment was initially used mainly for imitating pearls and for introducing light reflections in paintings until it found application also in the cosmetics industry [101].

37.2.10 Artists' pigments in the nineteenth century

Numerous pigments were produced synthetically as a result of the advance of industrialization and the resulting industrial color manufacture. Initially, the aim was to imitate the precious natural and elaborately produced historical pigments. In the nineteenth and twentieth centuries, the color palette of painters expanded considerably. This took place in two stages from 1850 and 1920. The newly created pigments, thanks to industrial production, found widespread use in a short time. Some older pigments fell into oblivion (lead tin yellow) and others were replaced by synthetic substitutes (natural ultramarine, azurite, auripigment, malachite). The composition and the particle size distribution of the pigments could be controlled more and more by chemical and mechanical processing steps. The particle sizes of the pigments became more uniform and tended to be finer as a result of the new production methods.

Starting in England, there was a transition to almost exclusive industrial production of colorants. While the colors prepared in the painters' workshops were stored in parchment or oilcloth bags, so-called paint blisters, in 1840 the Robertson company introduced the tin tubes that became standard since that time. Other paint companies soon followed Robertson's example. Tube paints became the prerequisite for the painters to move the studios into the open air and plein air painting was born. New color shades were developed to complement and enhance the existing ones.

37.2.10.1 Artificial ultramarine blue, peacock blue, ultra blue, universal blue

In 1822, Gmelin succeeded in producing artificial (synthetic) ultramarine (complex sulfide-containing aluminum silicate, $Na_6Al_6Si_6O_{24}(NaS_n)$). This was followed in 1829 by the factory production of artificial ultramarine. The starting materials were soda, kaolin, quartz sand and sulfur. The result was convincing in its high purity, which extremely enhanced the luminosity. The low production costs now also allowed it to be used for wall coatings, which would have been unaffordable with natural ultramarine. Further information on artificial ultramarine blue is available in [3, 7, 23, 24, 28, 36, 54, 62, 70, 78, 82].

37.2.10.2 Titanium white, rutile white, titanium dioxide

In 1908, titanium dioxide (TiO_2) was discovered in Norway and in the USA. Just 1 year later, industrial production took place in both countries under the name Kronos Titanium White. In Germany, the production of the white pigment started in the 1920s under the name Degea Titanium White. The titanium white grades produced until 1938 contained TiO_2 in the anatase form, which was found to be incompatible with oil binders and organic dyes. Subsequently, the production of the rutile form succeeded, which proved to be extensively successful as a pigment. Rutile TiO_2 has the highest opacity of all white pigments. It is lightfast and compatible with all binders [24]. Further information on titanium white in this context is available for the interested reader in [3, 23, 28, 29, 48].

37.2.10.3 Cadmium yellow, cadmium orange), sulfur cadmium, cadmium sulfide

In 1818, Stromeyer discovered the pigment cadmium yellow (CdS) by precipitation in aqueous solution starting from cadmium salts and hydrogen sulfide or metal sulfides. The pigment was produced from 1825 onwards [24]. It has very good hiding power and coloring intensity. The lightfastness of cadmium sulfide pigments is also good. Cadmium yellow is resistant to weak acids or alkalis. It can be used in all binders. Additional information on cadmium yellow in this context is found in [3, 7, 22, 23].

37.2.10.4 Chromium oxide green, Arnaudon green, Casalis green, Plessy green, Schnitzer green

In 1809, Vauquellin produced chromium oxide green (Cr_2O_3) by the reduction of potassium dichromate with sulfur and charcoal. The pigment was initially used in ceramic glazes and 20 years later also in stains. It is completely light and heat resistant. The chemical resistance of chromium oxide green makes it compatible with all other pigments and binders. The beginning industrial production led to a much cheaper pigment with the advantage of a narrower particle size distribution. This characteristic is still a distinguishing feature today, visible in the microscopic images of many paint applications.

In the painting of the late nineteenth century, an effect of the ready availability of painting materials for the users can be seen. The impressionists, and later the expressionists, preferred to use pure unmixed colors, so also chromium oxide green, straight from the tube. Color mixing only took place on the canvas through the juxtaposition of appropriate dabs of color. Further information on chromium oxide green in this context is available in [3, 7, 23, 45].

37.2.11 Historical painting materials and painting techniques of the twentieth century until today

The color palette, which was always expanded over the centuries, changed in the twentieth century to the extent that the manually produced colors were almost no longer used. From then on, the artistic creative process no longer starts, or only in exceptional cases, with the production of the painting material, but begins at the point where the artist already has the selection of painting materials available in a prepared form.

Painters of the twentieth century used mostly a specific palette or a technical specificity of their colorants, not knowing the production processes for the pigments, dyes, and binders. The expressionists used a few clear colorful colors, black and sometimes white. The cubists, on the other hand, focused on a selection of tonal earth colors. In both cases, the colors were applied in one layer and the painting process was completed in a short time, comparable with the painting techniques of former centuries. The futurists proclaimed in their manifestos around 1910 the absolute freedom in the choice of materials to simultaneously capture and convey modern life. Suddenly, new materials and technologies were used, which had never been imaginable before in a painting context: glass, electricity, sounds, smells, sheet metal, trivial objects, and many more.

Nevertheless, in the twentieth and twenty-first centuries there are artistic developments in which the creative processes are combined with new procedures in the production of the painting materials. In the past, the procurement of the raw material for the artist's paints and the manufacturing processes required a great deal of know-how, care, effort, and creativity from the artist. The production of painting materials was an essential part of artistic creation. Today, much has been taken away from contemporary artists in this regard, and they can direct all their creativity and energy to the realization of their artistic ideas. Thus, the artistic process has changed in a fundamental way.

37.3 Historical binders

37.3.1 Natural organic binders

Natural organic binders include waxes (beeswax, carnauba wax, montan wax), drying oils (linseed oil, poppy seed oil, walnut oil), resins of plant origin (amber, copal, dammar, mastic, sandarac, rosin color resins, balsams), resins of animal origin (shellac), vegetable glues (starch, dextrin, gums) and animal glues (casein, hen's egg, gluten glues).

37.3.1.1 Proteins

37.3.1.1.1 Glue, gluten glue, protein glue, bone glue, hide glue, leather glue, warm glue

To produce the various hot glues (lat.: gluten corii, colla), the raw materials are degreased, watered and treated with lime milk. After washing out, the glue decoction is boiled and then allowed to solidify into a jelly, which is then dried.

There are examples of the use of glue already in the tomb of the prefect Rekhmire of Thebes, 1470 BC. In other graves, glue tablets were found [102]. It is therefore a transformation product of collagens, which are found in animal skin, bones and cartilage. A distinction is made according to the starting materials (bones and hides of cows, sheep, goats, horses, deer, rabbits, hares and whales, deer antlers, leather, tendons, scrapings from tanning, bone glue, hide glue, leather glue, parchment glue, fish glue, hare glue and sturgeon glue). The different types of glue vary in color, drying time, adhesiveness and elasticity. The highest quality has sturgeon glue. It is also the most expensive among the glues [7].

Similar glue materials appear in all painting recipes from ancient to the nineteenth century. Le Begue [14] writes that a glue made of fine leather has a higher elasticity than egg white. Pliny mentions glue as a binder for paints, book painting and cloth painting. Glue was also used almost exclusively as a binder in chalk and plaster primers. Since glue solutions dry under stress, a too high binder content causes the paint to flake off. The hygroscopic property of glue causes it to swell at higher humidity and shrink in dry conditions. Glue is offered since many years in the form of tablets, pellets, and beads. For further information on glue in this context, refer to the corresponding literature [3, 23, 24, 27, 28, 49, 59, 89, 90, 95, 98, 103–107].

37.3.1.1.2 Casein, cold glue, casein glue

Casein (lat.: caseinum) precipitated from milk is insoluble in water and, in order to be used as a binder, it must be broken down with the aid of lime, borax or ammonium carbonate. It flocculates due to lactic acid formed by bacteria. The coagulated product is filtered off, washed, dried and pulverized.

Digested caseins are insoluble in water when exposed to air. Casein-containing films dry up with tension. The cheese glue made from alkali-treated white cheese (curd) was used by the old masters as early as the eighth century. A preparation mostly consisting of lime casein with slaked lime can be proven in many old paintings [10]. The use for secco paintings is widespread. Digested casein has emulsifying properties and can be used with resins, drying oils and wax soaps as casein tempers. Additional information on casein in this context can be found in [3, 7, 23, 28, 66, 103, 105, 107–109].

37.3.1.1.3 Egg white, chicken egg white

Both the egg yolk and the egg white are used in painting. While the yolk is fat, only the egg white is used in lean applications. It is often an essential component of a tempera system. As a binder, egg white has been known since ancient times for wall and panel painting. Written sources also mention the use of egg white as a binder or adhesive for shell gold or shell silver. It can also be traced as a binder in colored alumina for bright metal coatings or for bolus below gildings. Finally, egg white is also mentioned as a binder for medieval book illumination. The transparent and brittle films are initially water-soluble and become water-insoluble over time [7].

Egg white (lat.: clarum, ovi albugine, albumen) as a binding agent for book painting was prepared by pressing and by beating [63]. The material pressed and filtered through cloths or a bath sponge is said to be somewhat more brittle than the binder obtained by whisking. Freshly beaten egg whites have a pH value of 10. To neutralize them, acidic substances were often added to prevent the alteration of sensitive dyes. Moreover, stale material had a more favorable viscosity during painting [5]. Further information on egg white in this context is available in [3, 11, 14, 79, 87].

37.3.1.1.4 Egg yolk

Egg yolk (lat.: vitellum ovi) differs from the egg white primarily in its content of oil and lecithin, which acts as an emulsifier. This provides an ideal condition to produce tempera. Cennino Cennini [18] describes the use of egg yolk as a binder in early Italian painting. He recommends different shades of the paler yolk of city hens or that of country hens, which is more reddish, for painting incarnates. Because of its inherent color, egg yolk is mentioned in various painter's manuscripts as a preferred binder for yellow pigments [11, 14]. In the Strasbourg manuscript, it is also recommended for glazing, especially as it loses its inherent color when exposed to light [9]. For further information on egg yolk in this context, refer to [3, 18, 63].

37.3.1.2 Plant rubber

Plant gums are resinous, water-soluble or water-swellable plant exudates (sticky, viscous or stringy) that harden in air and form a resin [28]. As binders and as adhesives, they are already proven in ancient. Thus, their use is already found in paintings on Egyptian mummy coffins.

37.3.1.2.1 Gum Arabic, Arabic gum, gum Saracenicum, gum Senegal, gum embavi, gum acacia

Gum Arabic (lat.: gummi arabicum) is obtained from the sap of different acacia species (mimosaceae). Schreger [50] describes its use for inks, for black colors on leather and hair, for mixing watercolors, for gloss varnishes and others. He mentions also the appearance of gum Arabic as brittle, quite light, glassy and highly soluble in cold and hot water. However, it is insoluble in ethyl alcohol and oils.

Gum Arabic has been found in Egyptian mummy paintings since the first century. It was regarded as a material, which makes colors brilliant [14]. It has also been mentioned as a fabric finish, thickening agent for printing inks or a kind of varnish. Films containing gum Arabic dry up very brittle, so that plasticizers such as honey or glycerol are added [3]. If the concentration is too high, the films flake off completely. Gum Arabic is particularly valued as a binder for watercolor paints. It is also used in inks and for marbling paints. Additional information on gum Arabic can be found in [23, 24, 28, 36, 44, 84].

37.3.1.2.2 Cherry gum, gum cerasorum, cerasa, cerasin, cherry resin

Cherry gum (lat.: cerasus) is similar to gum Arabic but less pure and has a lower adhesive strength. The excrement of various stone fruit trees is sometimes summarized under the term cherry gum. The inferior properties of cherry gum led to its frequent mixing with gum Arabic. Cherry gum was also used with an addition of honey for the design of book illustrations [5].

The binder always remains swellable, which affects the appropriate application of the gum [74]. In medieval manuscripts, the advantage of quick drying compared with oil paints is emphasized. Cherry gum in this context is also described in [3, 10, 23, 24, 30, 79].

37.3.1.2.3 Tragacanth, gum tragacanth, tragacantha, bassoringummi

Tragacanth (lat.: gummi tragacanthae) is derived from various Near Eastern and Mediterranean species of Astragalus, as well as from species of Syrian and Persian origin. The excretion of various papilionaceous plants solidifies into white-yellowish pieces and is usually traded in powdered form [3]. In the "Kreuterbuch" [110], tragacanth is mentioned as the gum of a thorny plant. The best tragacanth is described as white, clean and clear. A poorer quality is noted as reddish or earth-colored. Bonani [111] mentions in this context that tragacanth comes from Spain and Apulia. It is obtained there from the sap of injured roots. Scherger [50] states that this plant rubber consists of hard, dry, fragile, and almost transparent pieces, which swell strongly in water and present a beautiful, shiny, transparent jelly.

Tragacanth was used as a binder especially for book painting, also in mixtures with gum Arabic and cherry gum as well as with oils and proteins. Pastel crayons are usually bound with tragacanth [7]. A recipe for a moldable mass of tragacanth with white lead and honey can be found at Merrifield [14]. It is also used as a primer for drawing paper as a base for oil painting and as a binder for watercolors or for pastel crayons. Further information about tragacanth can be found [23, 24, 36, 48].

37.3.1.3 Drying oils

Drying oils form high-molecular solid products by polymerization. This process can take decades, during which the oil can harden considerably. Pigment particles are embedded in the oil and are thus protected from external influences. The addition of siccatives such as lead, cobalt, or manganese, for example, accelerates the drying of the oil paint. Soot or asphalt delays the drying process.

Yellowing of the oil, which is more pronounced in the dark than in the light, changes the color scheme in a painting over a long period of time, sometimes considerably, depending on the place of storage or the incidence of light.

Pigments mixed with painting oil were rubbed on in lengthy processes in former times. Partly processed mixtures were used as prepared colors that could be stored for longer periods. For this purpose, the mixtures were covered with water in pots and filled in shells or in animal bladders. Oilcloth bags were also suitable for storing the oil colors. After 1840, tin tubes appeared on the market in which the oil colors were sold [3].

Only the so-called drying oils, such as linseed oil, walnut oil, poppy seed oil, are suitable as binders. In the medical writings of Aetius from the fifth and sixth centuries AD, oil is mentioned in connection with artistic techniques of gilding. As a binder for paints, oils appear only in written sources of the twelfth century [7].

37.3.1.3.1 Linseed oil, flax oil

Linseed oil (lat.: oleum lini) was obtained from the linseeds by cold or hot pressing, but better varieties were obtained by cold pressing. The seeds contain 30 to 40% oil. Accompanying substances were removed by post-treatment.

There is evidence that linseed oil paints have been used since about 3000 BC, especially in the north, for painting ships and utensils. Particularly north of the Alps, the use of linseed oil as a painting oil is the preferred method. In the drying processes, the binder polymerizes in the air over a period of years to decades to form high molecular weight structures. In the Baroque period, when oil painting was most widespread, there is evidence in written sources that the oil was freed from mucilage and purified by the addition of sorrel, in the sun [21, 56], by boiling with water [14] and also by mucilage-binding addition of bread crusts and onions [61] or sawdust. Despite the purification processes, linseed oil tends to yellow.

Boiling produces a thick oil that draws threads [100]. This fast-drying binder is called linseed oil varnish. Blends with resins shorten the drying time and produce a less rich oil-resin color. Emulsions with protein glues are called tempera. The oil paint allowed specific applications of different thickness and opacity up to semi-transparent glazes [3]. Thus, a differentiated structure of the painting layer was possible. The transparent oil glazes were also suitable for luster painting on shiny metal bases, because the long drying times ensured the fine dispersion of the colors. Chemical analyses today allow the conclusion that tempera and oil-bound painting layers lie on top of each other. In order to avoid cracking, it had to be taken into account that, following an old painter's rule, “fat over lean” was used. Some of the lower painting layers were even applied with purely water-based binders. Tempera layers were applied on top of these, and finally purely oil-based painting layers were used [15]. The binder often differed from one paint to another.

Theophilus [10] and Heraclius [11] already describe formulations for oil paints, but these are to be understood more as paint colors than as paints for panel painting, which was still largely done in tempera. In order to make oil paints usable for artistic applications, many intermediate steps were still necessary with regard to the structure of the painting layer [3]. Information on linseed oil in this context is also found in [23, 28, 37, 38, 44, 54, 105, 112].

37.3.1.3.2 Walnut oil, nut oil, oleum de nucibus

The nut oil (lat.: oleum iuglans) is obtained from the Persian walnut (English walnut, Juglans regia). For this purpose, dried walnuts are pressed. The oil is whiter than linseed oil, but does not dry as easily [74]. Nevertheless, it dries faster than poppy seed oil. Written sources and binder analyses indicate that walnut oil was used primarily in Italy, it is mentioned, for example, by Leonardo da Vinci. Besides the addition of various siccatives, Prange [61] recommends reduction of the oil by burning it off: “Ohne wirkliches Anzünden erhält das Öl die Leichtigleit zu trocknen niemals.” (without igniting, the oil never acquires the ease with which it dries). Additional information on walnut oil in this context is available in [3, 31, 78, 113].

37.3.1.3.3 Poppy seed oil, oleum papaveris, papaverolia, mags seed oil

Poppy seed oil (lat.: oleum papaveris) was probably used on a larger scale from the seventeenth century onwards [7]. The seeds of the poppy plant from the Mediterranean area and from Asia were mainly used for extraction. The seeds contain 40–50% oil. It is described by Watin [74] as the whitest of all oils. Therefore, it is particularly suitable for lead white. Schreger [50] states that it is more popular in oil painting than any other oil. He notes that the old 1-year-old light oil is better than the very fresh one, which appears more yellowish. However, it dries relatively poorly. The painter Wilhelm Leibl took advantage of this for his specific painting style [3]. For further information on poppy seed oil in this context, refer to [23, 28, 37, 105].

37.3.1.3.4 Hemp oil, hanif sat oll, oleum cannabium

Hemp oil (lat.: oleum cannabium) is obtained from hemp seed. In painting application, it dries slowly and tends to form wrinkles [105]. Cooked, it can be used instead of linseed oil or nut oil for the preparation of varnish. In the "Farbenbelustigung" [33], a prescription is given in which hemp oil is boiled with pumice stone, burned sheep bone (ossa comnusta), and sandarac. Information on hemp oil in this context is also found in [3, 14, 30, 32, 34, 79, 100].

37.3.1.4 Resins, resina

Resins, excretory products of plants, are not soluble in water, unlike plant gums. They dissolve in organic solvents, such as alcohol, turpentine oil or acetone. Solutions of resins are found mainly in coatings and glaze paints. When added to oily or even aqueous binders, they increase gloss and transparency of the paint. Resins dissolved in volatile solvents dry physically by evaporation of the solvent. The remaining resin film oxidizes and polymerizes. The aging process is accompanied by yellowing and embrittlement [7].

37.3.1.4.1 Turpentine balsam, Venetian turpentine (larch turpentine), Strasbourg turpentine (silver fir balsam), Bordeaux turpentine (pine balsam), Cyprian turpentine

Turpentine balsam (lat.: terebinthinae) was harvested from living trees. Approximately 100 pine species were used for this purpose. These include white pines, larches, spruces and silver firs. In the past, turpentine was often understood to be the resinous sap of pistacia terebinthus. The "Kreuterbuch" describes the resin of the turpentine tree Terebinthus in the way that it provides the best quality of turpentine balsam: white, clear, light and fragrant [114].

Turpentine balsam is used, often with alcohol and mastic, mainly for the preparation of varnishes [24]. Turpentine oil is one of the most important solvents in painting. Venetian turpentine is still used today as a natural resin solution, while Strasbourg turpentine is no longer extracted. Venetian turpentine is used as an emulsifier in various painting techniques, e.g., in egg tempera or casein tempera. When added to oil paints, it produces an enamel-like effect [24]. Additional information on turpentine balsam in this context is available in [3, 14, 19, 28, 36, 73, 92, 115].

37.3.1.4.2 Mastic, mastic gum, mastix

The resin mastic (lat.: resina mastix) is obtained from the mastic shrub Pistacia lentiscus (cashew family) as Levantine mastic or mastic electa [3]. It is traded in form of tears. The best resin qualities come from the island of Chios and the Levant. The extraction is done by lightly scratching the bark. The resin drops solidify in 2–3 weeks. Mastic is highly soluble in alcohol, turpentine oil and hot oil. It is widely used in painting for the preparation of alcohol or turpentine oil varnishes, but it tends strongly to yellowing. In an early mention of Boltz [34], mastic appears boiled in linseed oil. Watin [74] states that mastic is more transparent but much more expensive than sandarac. For further information on mastic in this context, refer to [23, 28, 30, 31, 36, 37, 62, 76, 105, 114, 115].

37.3.1.4.3 Dammar, cat's eye dammar, dammar resin, dammar gum

In the eighteenth century, dammar is not yet mentioned in written sources [61]. Nevertheless, it could have been in trade under other names. It is a light transparent resin of the dammar tree, which is a dipterocarpaceae native in Southeast Asia. Siddon [98] mentions that it is not dissimilar to sandarac. The brittle material is comparable in hardness with rosin. It is soluble in turpentine and is still widely used as a varnish for paintings. Dammar has only little tendency to yellowing, but on dark backgrounds it tends to turn blue. Siddon [98] mentions also that an oil-resin varnish called "Dammarharz" mixed with two parts of poppy oil and dissolved at 40°C of heat leads to a better retouching varnish than mastic. Dammar varnishes dry elastic and with only little yellowing. They are therefore very well suited for painting applications. Additional information on dammar can be found in [3, 14, 23, 24, 28, 37, 41, 76, 82, 105, 116].

37.3.1.4.4 Sandarac, sandarach, sandarac resin

The North African resin sandarac (lat.: resina sandaraca) of Tetraclinis articulata resembles dammar, but painting films with this resin dry with a slightly reddish discoloration. It was considered to be juniper resin until the nineteenth century [116]. In the "Kreuterbuch" [110] it is compared with mastic. It says further that from the cut bark of a kind of juniper a juice emerges, which becomes hard like rubber. Dissolved with oil, spirit of wine, spikenard and turpentine, it is used as a varnish, but this is very hard and brittle. Additions of gum elemi or camphor are said to make it more elastic. In the Bologna Ms., it is mixed with linseed oil [14]. Alcohol dissolutions are also often mentioned Sandarac in this context is also described in [3, 15, 18, 23, 24, 36, 37, 41, 65, 76, 115, 117].

37.3.1.4.5 Gum lac, shellac, stick lac, tree varnish, rubber varnish, button varnish, resin varnish, flat varnish

Gum lac (lat.: resina laccae) is the secretion of the East Asian hollyhock lice (lacquer scale lice) Kerria lacca (Kerr), Coccus lacca. After fertilization, the females excrete the resin and die. The twigs with the resin secretions are cut off and processed by melting and washing. The dried secretion is waxy, brittle and transparent. In alcohol, a reddish-brown solution is obtained, which is used to close open-pored surfaces. In combination with other resins, the reddish-brown shade can be lightened. Cröker [56] states that sandarac, rosin, amber and mastic are used in varying proportions. Siddon [98] mentions the bleaching of shellac with chlorinated water, whereby the quality suffers heavily. Varnishes containing gum lac dry hard. They are described as waterproof and very resistant. It is mentioned that the varnish surface blooms and darkens considerably in a humid atmosphere. Further information on gum lac in this context can be found in [3, 14, 21, 23, 24, 28, 31, 36, 37, 48, 59, 73, 76, 111].

37.3.1.4.6 Amber, agtstein, ambergris

The fossil resin amber (lat.: succinum), which was mainly used for the preparation of oil-resin varnishes, mostly comes from Baltic regions. Around 300 types of amber have been discovered so far. They were formed by natural polymerization and oxidation processes. The exact origin and age of ambers can often not precisely be determined.

The most common Baltic amber is of the best quality and has a melting point of 290°C. Since the Middle Ages, amber varnishes have been produced by boiling amber in drying fatty oils [14]. The varnishes are very hard and are used, among other things, as a coating for boats, carriages and to protect iron [56]. Numerous attempts to process amber in different solvents have shown that this is a difficult task. In alcohol, it leads to unsatisfactory results. At Mayerne [32], attempts are reported to boil crushed amber in water and to dissolve it by heating in spiked oil and spirit of wine. Stöckel [21] recommends to melt amber followed by the addition of turpentine oil in small amounts. Amber varnishes over oil paint applications cause the brush marks to melt, which was popular in the old painting technique [118]. For further information on amber in this context, refer to [3, 23, 24, 28, 36, 98, 105].

37.3.1.5 Waxes

Numerous different waxes occur in nature. However due to their usually very low occurrence, they can hardly be used for painting. Beeswax is the most important wax for the art technique.

37.3.1.5.1 Beeswax cera alba, yellow wax

Beeswax has a melting point of 61–68°C. It contains proportions of resin [19]. The so-called virgin wax is the best unincubated quality [92]. Bleached wax is known in several variations. Bleaching of beeswax cera alba can not only be performed in salt water [95] but also in the sun [100]. In ancient times, the wax was used as a binder for encaustic. In the binder, beeswax and its sodium soaps can be traced. By boiling with alkalis, the wax is partially saponified [24]. The formed wax soaps have an emulsifying effect and allow painting with resins, balsams, drying oils and glues. Pliny [19] describes wax–resin mixtures for shipbuilding, as the mixtures become harder in salt water. Wax was still used as a painting agent in form of wax soaps in the Middle Ages. Cröker [56] describes a mixture with tallow, which, when mixed with white lead, had the appearance of alabaster. As an additive in varnishes, the wax serves for matting and for the attainment of a greater smoothness. The same applies to the addition of wax in oil paints. The wax was also used as a gilding agent on fabrics. Additional information on beeswax cera alba can be found in [3, 7, 23, 98, 105, 119].

37.3.1.5.2 Carnauba wax, palm wax, vegetable wax

Carnauba wax (lat.: cera carnaubae) comes from the Brazilian palm Copernicia purnifera and is significantly harder than beeswax. The melting point is 83–86°C. The additive in beeswax serves for hardening of the softer raw material. Thus, the painting use of carnauba wax takes place typically together with beeswax, with which it can be homogeneously melted together. The increase in melting temperature is associated with higher resistance and gloss when polishing coatings [24]. For further information on van carnauba wax, refer to the corresponding literature [3, 7, 23, 24, 28, 56, 98, 105, 119].

37.3.1.5.3 Montan wax, mountain tallow, mountain wax, ceresin, earth wax, mineral wax, neftgil, ozokerite

Montan wax (lat.: cera montanus) comes from mineral deposits of plant sediments. In the past, it was extracted by mining, but today it is an extraction product in coal refining. The sulfur content of the humic acids in montan wax is the reason why this wax is not compatible with all pigments [24].

37.3.2 Synthetic organic binders

More recently, numerous synthetic resins have been developed to replace the previously used natural resins. Synthetic resins do not imitate natural resins. In numerous synthetic resins, chemical and physical properties occur, which are not found in natural resins. Aqueous dispersions that have some similarity to tempera binders are often used as binders for artists' paints. Acrylic resin dispersions are by far the most common synthetic binders [7].

37.3.3 Inorganic binders

37.3.3.1 Lime, slaked lime, lime slurry, hydrated lime, milk of lime, lime water, white lime

Both unburnt and burnt lime find art technical application. Slaked lime (lat.: calx viva, calvaria usta) or leather lime was used as lime whitewash and lime plaster. Burnt lime was used to improve painting oils. The lime water settled in lime pits was used to prepare paints. Lime was also obtained from burnt oyster shells, river mussels or egg-shells [3].

Lime was already widely used as a mortar and binder in ancient Egypt, ancient Greece and Rome. Lime pigments were bound with lime whitewash alone or with additions of casein or skim milk. Caustic lime as an additive in casein solutions formed a fast-hardening filler material for holes and cracks.

37.3.3.2 Water glass, mineral paint, silicate paint

Water glass is a syrupy clear solution of sodium silicate, potassium silicate or lithium silicate in water. Depending on whether sodium, potassium or lithium silicates are predominantly contained, the water glass is called sodium water glass, potassium water glass or lithium water glass. The production takes place via melting of silica sand with soda, potash or lithium carbonate. These starting components are melted, and the glassy melt is dissolved in water after cooling. The cooled glass is pulverized and dissolved in hot water to form the water glass. Since the nineteenth century, it can be traced as a binder for weather-resistant wall paintings. Around 1878, Adolf Wilhelm Keim developed silicate paints based on water glass, which are used until today [7]. Silicate paints are not only free of solvents and other substances, but also they are harmful and have exceptional good application properties.

References

1. Liber illuministarius. bearbeitet und veröffentlicht. In: Berger E, editor. Beiträge zur Entwicklungs-Geschichte der Maltechnik, III. Folge. München, 1912:191–201. Reprint Liechtenstein 1982.
2. Berger E. Die Maltechnik des Altertums, München, 1904. Reprint Vaduz 1992.
3. Brachert T. Lexikon historischer Maltechniken. München, 2001.
4. Kittel H. Pigmente. Stuttgart, 1960.
5. Liber illuministarum pro fundamentis auri et coloribus ac consimilibus collectus ex diversis. Kunsttechnisches Handbuch, Kloster Tegernsee, Münchner Staatsbibliothek, Cgm 821, Transkription und Übersetzung A. Bartel und M. Lautenschlager (unveröff.).

6. Clacicula M. Smith CS, Hawthorne JG, Hrsg. A little key to the world of medieval techniques. Philadelphia: Transactions of the American Philosophical Society, 1974.
7. Kühn H, Roosen-Runge H, Straub RE, Koller M. Reclams Handbuch der künstlerischen Techniken. Bd. 1. Stuttgart: Farbmittel, Buchmalerei, Tafel- und Leinwandmalerei, 1984.
8. Liber de coloribus illuminatorum sive pictorum, from Sloane Ms No. 1754. Hrsg. D. Thompson jr., Cambridge (Mass.). In: Speculum I. 1926.
9. V., Borradaile R, Hrsg. Das Straßburger Manuskript, Handbuch für Maler des Mittelalters. München. (I, 1966). Auch in: E. Berger, Beiträge zur Entwicklungs-Geschichte der Maltechnik, III. Folge, München 1912, S. 155–190, Reprint Liechtenstein 1982.
10. Theophilus Presbyter und das mittelalterliche Kunsthandwerk. Gesamtausgabe der Schrift De artibus in zwei Bänden, Malerei und Glas. Brepohl E, Hrsg. Bd. I, Weimar/Wien, 1999. Theophilus Presbyter und die mittelalterliche Goldschmiedekunst, Buch III, Hrsg. E. Brepohl, Wien/Köln/Graz 1987, Bd. II (Beide Bände wurden hier in der Zählweise von Brepohl zitiert.).
11. Heraclius, Von den Farben und Künsten der Römer. Eitelberger R, Edelberg V, Hrsg. Quellenschriften für Kunstgeschichte und Kunsttechnik des Mittelalters und der Renaissance. Bd. 4. Wien, 1873. 166–257. Reprint Osnabrück 1970. Auch in: Merrifield 1967.
12. Schlosser JV. Quellenbuch zur Kunstgeschichte des abendländischen Mittelalters. Wien, 1896.
13. Eastlake CL. Methods and materials of painting of the great schools and masters. Bd. 2. London, 1847. Reprint New York 1960.
14. Merrifield M. Original treatises on the arts of painting, London 1849. Bd. 2. New York: Neuausgabe, 1967.
15. Berger E. Quellen für Maltechnik während der Renaissance und deren Folgezeit, XVI.—XVII. Jahrhundert, München, 1901. Reprint Vaduz 1984.
16. van de Graaf JA. Het de Mayerne Manuscript als bron voor der schildertechnik van de barok. Mijdrecht, 1958.
17. Marette J. Connaissance des primitifs par l'étude du bois. Paris, 1961.
18. Cennini C. Il Libro dell Arte, A Cura di E Brunello. Vicenca, 1982.
19. Plinius Secundus C. Naturalis historia. 6 Bde., Leipzig 1870–1898. Nachdruck Stuttgart, 1967–70.
20. Nicolaus K. DuMont's Handbuch der Gemäldekunde. Köln, 1986.
21. Stöckels HF. Hofschreiners zu Schleiz, Praktisches Handbuch für Künstler, Lackirliebhaber und Oehlfarben-Anstreicher. 2nd Aufl, Nürnberg, 1799. Reprint Rottenburg 1981.
22. Mérimée JF, Vergnaud AD. Handbuch für Kunst- und Zimmermaler, Vergolder, Restaurateurs alter Oelgemälde, Farben- und Tapetenfabrikanten (bearb. von H. C. Hebra). Ulm, 1839.
23. Wehlte K. Werkstoffe und Techniken der Malerei, Mit einem Anhang über Farbenlehre. Ravensburg, 1967, ungekürzte Ausgabe.
24. Schramm HP, Hering B. Historische Malmaterialien und ihre Identifizierung. Graz, 1988.
25. Doerner M. Malmaterial und seine Verwendung im Bilde. 19th Aufl. Leipzig, 2001.
26. Schmidt CH. Vollständiges Farbenlaboratorium. Weimar, 1857.
27. Maltechnik-Restauro, Internationale Zeitschrift für Farb- und Maltechniken. Restaurierung und Museumsfragen. München. (bis 1987).
28. Schießl U. Die deutschsprachige Literatur zu Werkstoffen und Techniken der Malerei von 1350 bis ca. Worms, 1989. 1950.
29. Hering MH. Weiße Farbmittel, Fürth. 2000
30. Doerner M. Malmaterial und seine Verwendung im Bilde. 3rd Aufl. München, 1933.
31. Schweppe H. Handbuch der Naturfarbstoffe, Vorkommen, Verwendung, Nachweis. Landsberg a. Lech, 1992.

32. De Mayerne T, Turquet P. sculptoria et quae subalternarum artium, 1620. In: Berger E, editor. Quellen für Maltechnik während der Renaissance und deren Folgezeit. München. 1901: 107–333. Reprint Vaduz/ Liechtenstein 1993.
33. Anonymus. Die mit allerhand schönen und curiösen und geheimen Wissenschaften angefüllte Farbebelustigung, eingeteilt in zwei Teile. Nürnberg, 1741. nach Boltz von Ruffach 1549.
34. Boltz von Ruffach V. Illuminierbuch. Wie man allerlei Farben bereiten, mischen und auftragen soll, Bern 1549. Hier zit. in der Neuausgabe München 1913. Reprint Walluf. 1976
35. Schurtz GN. Neu-eingerichtete Material-Kammer: Das ist Gründliche Beschreibung aller fürnehmsten Materialien und Specereyen. Nürnberg, 1673.
36. Pomet P. Der aufrichtige Materialist und Specerey-Händler ausgefertiget von Peter Pomet. Leipzig, 1717.
37. Winckler E. Die Lack- und Firniß-Fabrikation in ihrem ganzen Umfange Nebst einer Anleitung zur Lackirkunst. 2nd Aufl. Leipzig, 1860.
38. Ploss E. Studien zu den deutschen Maler- und Färberbüchern des Mittelalters, Ein Beitrag zur deutschen Altertumskunde und Wortforschung. München (masch.), 1952, Diss. phil.
39. Ploss E. Ein Buch von alten Farben, Technologie der Textilfarben im Mittelalter mit einem Ausblick auf die festen Farben. Heidelberg/ Berlin, 1962.
40. Stoll A, Pflanzenlacke G, Schüttgelb, Safran. Eine Übersicht über Herstellung und Verwendung anhand historischer Quellen. In: MT 1981, H. 2. 73–110
41. Fontana E, Hellwig F, Martius K. Historische Lacke und Beizen. Nürnberg: Germanisches Nationalmuseum, 1992.
42. Compositiones ad tingenda musiva, herausgegeben. Uppsala, 1932. übersetzt und philologisch erklärt von H. Hedfors.
43. Vielheuer C. Gründliche Beschreibung Fremder Materialien und Specereyen Ursprung, Wachstum, Herkommen und deroselben Natur und Eigenschaften. Leipzig, 1676.
44. Anonymus. Wieder neu aufgerichtete und vergrösserte in zwey Theilen angewiesene Curieuse Kunst- und Werck-Schul (Bd. 1). Der Neu-aufgerichteten und Vergrösserten in sechs Bücher und Theilen verfasten curieusen Kunst- und Werck-Schul sehr verlangter nunmehr erfolgter Anderer Theil J. K. Chymiae ac aliarum Artium Cultore, Nürnberg 1707, (Ausgaben zur Kunst- und Werck-Schul: 1696, 1705, 1707, 1732, 1759, 1763, 1782, 1784).
45. Kühn H. Pigmente in den Gemälden der Schack-Galerie. München, 1969.
46. Goltz D. Studien zur Geschichte der Mineralnamen in Pharmazie, Chemie und Medizin von den Anfängen bis Parcelsus. In: Sudhoffs Archiv Zeitschrift für Wissenschaftsgeschichte. Beihefte, Heft 14. Wiesbaden, 1972.
47. Höhle G. Zur ober- und mittelrheinischen Buchmalerei um die Mitte des 15. Jhs — ausgehend von 2 Maltraktaten, Berlin, 1984, Diss.
48. Restauro, Zeitschrift für Kunsttechniken. Restaurierung und Museumsfragen. München. (ab 1988, vor 1988: MT).
49. Wehlte K. Werkstoffe und Techniken der Malerei. 5th Aufl, Ravensburg, 1985. (posthum überarbeitete und gekürzte Ausgabe).
50. Schreger CH. Tabellarische Uebersicht der rohen und künstlich zubereiteten Farben und Farbenmaterialien zur Prüfung ihrer Güte und Aechtheit. Nürnberg/Sulzbach, 1805.
51. Kunckel J. Ars vitraria experimentalis oder vollkommene Glasmacher-Kunst. Frankfurt und Leipzig, 1689. Reprint Hildesheim und New York 1972.
52. Colmarer Kunstbuch. Bern, Burgerbibl. Ms. hist. helv. XII. 45, Transkription D. Oltrogge (unveröff.).
53. Lindgren U, Hrsg. Europäische Technik im Mittelalter. Berlin: Tradition und Innovation, 1996.

54. Dürer A. Die Gemälde der Alten Pinakothek, Bayerische Staatsgemäldesammlungen. München, 1998. Mit technischen Beiträgen von B. Heimberg, A. Burmester, C. Krekel, J. Koller, I. Fiedler und U. Baumer.
55. Lemery N. Vollständiges Materialien-Lexicon, Darinnen Alle und jede Simplicia, vorgestellet sind, welche aus denen so genannten drey Reichen, der Thiere, der Kräuter und der Mineralien, hauptsächlich zu Dienste der Medicin und Apothecker-Kunst genommen und gebrauchet werden. Leipzig, 1721.
56. Cröker JM. Der wohl anführende Mahler. Jena, 1736. Reprint München 1982 mit Einführung und Glossar von U. Schießl.
57. Harley R. Artists brushes — historical evidence from the sixteenth to the nineteenth century. London/Boston, 1972:61–6.
58. Neumüller OA, Hrsg. Römpps Chemielexikon. 8th Aufl, Stuttgart, 1988.
59. Schulze A. Technologische Chemie und Materialienkunde. Oder gründlicher Unterricht für Künstler. Quedlinburg und Leipzig: Fabrikanten und Handwerker, 1826.
60. Colerus J. Calendarium oeconomicum und perpetuum, Vor die Haußwirt, Ackerleut, Apotecker und andere gemeine Handwercksleut, Wittenberg, 1591. Reprint Leipzig 1988.
61. Prange CF. Der Weltweisheit und freyen Künste Magister Farbenlexicon, worinnen die möglichsten Farben der Natur durch die wirkliche Ausmahlung enthalten sind. Halle, 1782.
62. Bouvier MP. Handbuch der Oelmalerei für Künstler and Kunstfreunde. 4th Aufl. Braunschweig, 1861.
63. Anonymus Bernensis. Über die Bindemittel und das Coloriren von Initialen. In: Hagen H, Hrsg. mit einer Notiz über die Quellenliteratur der Eitemperatechnik, Quellenschriften für Kunstgeschichte und Kunsttechnik des Mittelalters und der Renaissance, Bd. VII, Wien, 1874.
64. Woudhuysen-Keller R. Das Farbbüchlein, Codex 431 aus dem Kloster Engelberg. Diss Zürich, 2000. (unveröff.).
65. von Lippmann EO. Entstehung und Ausbreitung der Alchemie, Berlin, 1919. Hildesheim/New York 1978.
66. Technische Mitteilungen für Malerei. München.
67. Stößel I. Rote Farblacke in der Malerei, Herstellung und Verwendung im deutschsprachigen Raum zwischen 1400 und 1850, Diplomarbeit, Staatliche Akademie der Bildenden Künste. Stuttgart, 1985.
68. Arent P. Die Farbstofflieferanten Alkannawurzel, Krappwurzel, Sandelholz und Rotholz, Semesterarbeit. FH Hildesheim, 1995. (unveröff.).
69. Vitruvius Pollio M. De Architectura Libri X. Fensterbusch C, Hrsg, 3rd Aufl. Darmstadt, 1981.
70. Eyßvogel FG. Neu-Eröffnetes Magazin bestehend In einer Versammlung allerhand raren Künsten und besonderen Wissenschaften. Bamberg: Zweyter Theil, 1756.
71. Soukup RW, Mayer H. Alchemistisches Gold, Paracelsistische Pharmaka, Laboratoriumstechnik im 16. Jahrhundert. Wien, 1997.
72. Anonymus. Neu entdeckte Lacquir-Kunst oder Gründliche Anweisung, Wie man nicht nur unterschiedliche bißher geheim gehaltene kostbare Lacquen sondern auch den biß ietzo unbekannten Gummi Copal leichtlich auflösen könne. Leipzig; 1708.
73. Krünitz DJ. Oeconomische Encyclopaedie, oder allgemeines System der Staats-. Bd. 242. Haus-, und Landwirtschaft, 1787–1858.
74. Watin JF. Kunst des Staffiermalers, Vergolders, Lackierers und Farbenfabricanten. Ilmenau, 1824.
75. Roosen-Runge H. Farbgebung und Technik frühmittelalterlicher Buchmalerei, Studien zu den Traktaten, “Mappae clavicula” und, “Heraclius. Bd. 2. München, 1967.
76. Kühn H. Erhaltung und Pflege von Kunstwerken und Antiquitäten. München, 1974.

77. Weber J. Die Farbstofflieferanten Cochenille, Kermes, Lac-Dye, Drachenblut, Orlean und ihre Verwendung zum Färben von Hölzern, Semesterarbeit, FH Hildesheim, 1995. (unveröff.).
78. Prechtl JJ. Technologische Encyklopaedie oder alphabetisches Handbuch der Technologie, der technischen Chemie und des Maschinenwesens. Bd. 20. Stuttgart, 1830–1855.
79. Berger E. Quellen und Techniken der Fresko-, Oel- und Tempera-Malerei des Mittelalters. 2nd Aufl. München, 1912. Reprint Walluff 1973.
80. Georgica Curiosa Aucta I. Das ist umständlicher Bericht und klarer Unterricht von dem Adelichen Land- und Feld-Leben, Nürnberg 1687 II Georgica Curiosa. Oder: Des auf alle in Teutschland übliche Land- und Haus-Wirthschafften gerichteten, hin und wieder mit vielen untermengten raren Erfindungen und Experimenten versehenen, auch einer merklichen Anzahl Kupfer gezierten Adelichen Land- und Feld-Lebens Anderer Theil. Nürnberg, 1687.
81. Kunst-Büchlein oder gründlicher Gebrauch von Etz Arbeit, in und außerhalb des feuers wie auch jede Farben zubereiten, erhalten, bessern und wiederbringen. Frankfurt und Leipzig, 1687.
82. Studies in Conservation. The journal of the international institute for conservation of historic and artistic works. London.
83. Ruland M, Alchemiae L. Sive Dictionarium Alchemisticum, Frankfurt, 1612. Reprint Hildesheim 1964.
84. Beckmann J. Vorbereitung zur Waarenkunde oder zur Kenntniß der vornehmsten ausländischen Waaren, Göttingen 1793. Weitere Editionen. 1794;1796:1800.
85. Anonymus. Enthüllte Geheimnisse der Farbenbereitung für Oel-, Wasser-, Tusch-, Pastell-, Emaille-, Seiden- und Wachsmalerei. Quedlinburg und Leipzig, 1824.
86. Oltrogge D. Das Blau in der mittelalterlichen Buchmalerei, Quellenschriften als Basis naturwissenschaftlicher Farbuntersuchungen, Farbmittel. In: Blau: Farbe der Ferne. Heidelberg, 1990:104–30.
87. Manuskript 610/1957 der Stadtbibliothek Trier. Transkription, Übersetzung und Kurzkommentar von D. Oltrogge (unveröff.).
88. Welch. Particle characteristics of Prussian blue in an historical paint. J Am Inst Conserv. 1988;27
89. Gottfried August Hoffmann IV. die Chymie zum Gebrauch des Haus-Land und Stadtwirthes, des Künstlers, Manufacturiers, Fabricantens und Handwerkers. Leipzig, 1757.
90. Lic. Gottfried August Hoffmanns Chymischer Manufacturier und Fabricant. Gotha. 1758
91. Händel A. Der Erfurter Waidhandel, Magisterarbeit. Universität Erlangen-Nürnberg, 1994. (Phil. I).
92. Zedler JH. Großes vollständiges Universallexicon Aller Wissenschafften und Kuenste, Bd. 64. Halle/Leipzig, 1732–51. Reprint Graz.
93. Kertzenmacher MP. Alchimia, daß ist alle Farben, Wasser, Olea, Salia, und Alumina, damit man alle Corpora, Spiritus und Calces Praeparirt, Sublimiert und Fixirt, zubereyten. Franckfurt am Mayn, 1613. (1539, 1563].
94. Ploss E. Ein Malerbüchlein aus dem Bamberger Karmeliterkloster. In: 100. Bericht des Historischen Vereins für die Pflege der Geschichte des ehemaligen Fürstbistums Bamberg. Bamberg, 1964:331–46.
95. Birelli JB. Alchimia Nova. Frankfurt, 1654.
96. Gettens RJ, Stout GL. Painting materials, a short encyclopaedia. New York, 1966.
97. Bothe CI. Der grösste Kehricht aller Farben? Über Asphalt und seine Verwendung in der Malerei. Mainz; 1999.
98. Siddons GA. praktischer und erfahrener englischer Ratgeber ein aus langjähriger Erfahrung geschöpftes Rezeptbuch Neuer Schauplatz der Künste und Handwerke. Bd. 82. Weimar, 1842.

99. Carl Friedrich August Hochheimers. Neueste Chemische Farben-Lehre; oder: Sammlung deutlicher, und durch angestellte Versuche bewährt gefundener Vorschriften, Bd. 2 . Leipzig, 1809.
100. Kunstbuch des Wolfgang Sedelius I, Cgm 4117. München: Staatsbibliothek.
101. Pfaff G. Spezielle Effektpigmente. 2nd Aufl. Hannover: Vincentz Network, 2007.
102. Reich G. Kollagen. Dresden, 1966.
103. Duhamel du Monceau. Die Kunst verschiedene Arten von Leim zu machen. In: Schauplatz der Kuenste und Handwerke, Bd. 11. Leipzig und Königsberg, 1772.
104. Schäfer L. Gründlicher Unterricht im Polieren und Beizen sowie im Lackieren, Vergolden, Versilbern und Bronzieren aller Arten von Meubles und metallenen Gegenständen. Quedlinburg und Leipzig, 1844.
105. Doerner M. Malmaterial und seine Verwendung im Bilde. 8th Aufl. Stuttgart, 1944.
106. Greber JM. Tierische Leime. Heidelberg, 1950.
107. Willers H. Herstellung von tierischem Leim und seine Verwendung im Bereich der Tafel- und Faßmalerei nach Angaben deutschsprachiger Quellenliteratur des 16. bis 19. Jahrhunderts, Diplomarbeit, Kunstakademie Stuttgart, 1980. (unveröff.).
108. Scherer R. Das Kasein. Wien, Leipzig, 1905.
109. Sutermeister E. Das Kasein, Chemie und technische Verwendung. Berlin, 1932.
110. Lonicerus A. Kreuterbuch, Kunstliche Conterfeytunge der Bäume, Stauden, Hecken, Kräuter, Getreyd, Gewürtze. Ulm, 1679. Reprint Grünwald 1962.
111. Reinbold JJ. Neuer Tractat von Firniß-Laquir und Mahlerkünsten, Nach dem Original des berühmten Pater Bonani in Rom. Berlin und Leipzig, 1735.
112. Keller R. Leinöl als Malmittel. MT 1973, H. 2. 74–105.
113. Leuchs JK. Allgemeines Waaren-Lexikon oder vollständige Warenkunde. 3rd Aufl. Nürnberg, 1843.
114. Kreutterbuch deß hochgelehrten und weitberühmbten Herrn D. Petri Andreae Matthioli. Jetzt widerumb mit vielen schönen newen Figuren, auch nützlichen Artzneyen, und andern guten Stücken, zum vierdten mal auß sonderm Fleiß gemehret, und verfertiget, durch Ioachimum Camerarium. Franckfurt am Mayn, 1626. Reprint Grünwald o. J.
115. Walch K, Koller J. Lacke des Barock und Rokoko. In: Arbeitshefte des Bayerischen Landesamtes für Denkmalpflege. Bd. 81, 1997.
116. Schneider G. Färben mit Naturfarben. Ravensburg, 1979.
117. Reinbold JJ. Neuer Tractat von Firniß- Laquier- und Mahlerkünsten, Nach dem Original des berühmten Pater Banani in Rom. Mit vielen neuen Arcanis, unterschiedlichen Beschreibungen des Gummi Copals und des Bernsteins. 2nd Aufl. Berlin und Leipzig, 1730.
118. da Vinci L. Trattato della Pittura. Condotto sul Cod. Vaticano Urbinate 127. Vicenza, 2000.
119. Büll R. Vom Wachs. Frankfurt, 1968.

Further literature

I Source papers

G. Agricola, Zwölf Bücher vom Berg- und Hüttenwesen, Neuausgabe der De Re Metallica Libri XII, Basel 1556, VDI Düsseldorf 1971, 5. Aufl.

Amberger Malerbuch, Staatl. Provinzialbibliothek Amberg, Ms. 77. Transkription von D. Oltrogge (unveröff.)

L. E. Andés, Die Technischen Vollendungsarbeiten der Holz-Industrie. Das Schleifen, Beizen, Poliren, Lackiren, Anstreichen und Vergolden des Holzes nebst der Darstellung der hiezu verwendbaren Materialien in ihren Hauptgrundzügen, Wien/Pest/Leipzig 1895, Reprint Köln 1985

Anonymus, Der curiöse Schreiber. von allerley künstlichen und erdenklichen Arthen zu schreiben samt dem curiösen Mahler, Dresden 1679

Anonymus, Vollständige Haus- und Landbibliothek Regensburg 1699

Anonymus Bernensis, Über die Bindemittel und das Coloriren von Initialen, Hrsg. H. Hagen, mit einer Notiz über die Quellenliteratur der Eitemperatechnik, Quellenschriften für Kunstgeschichte und Kunsttechnik des Mittelalters und der Renaissance, Bd. VII, Wien 1874

R. Straub (Hrsg.), Der Traktat de Clarea in der Burgerbibliothek Bern, in: Jahresbericht des Schweizerischen Instituts für Kunstwissenschaft Zürich 1984, S. 89–114

Didier de Arclais de Montamy, Traité des Couleurs pour la Peinture en Email et sur la Porcelaine, Paris 1765, Reprint Hildesheim/New York 1981

Das "Büchlein über Pflanzen von Baumen" des Tegernseer Abtes Konrad Ayrinschmalz vom Jahr 1479, in: Beiträge zur altbayerischen Kirchengeschichte, Bd. 38, 1989, S. 65–177

T. Bardwell, The Practise of Painting and Perspective made easy, London 1756

D. J. Bausch, Schediasma Posthumum di Coeruleo & Chrysocolla, Jena 1668

J. Beckmann, Vorbereitung zur Waarenkunde oder zur Kenntniß der vornehmsten ausländischen Waaren, Göttingen 1793. Weitere Editionen 1794, 1796, 1800

A. Rauch und B. Jokusch, Anweisungen für die Behandlung von Holzoberflächen in einer Quelle in Bergamo um 1525, in: Zeitschrift für Kunsttechnik und Konservierung, Jg. 10, 1996, H. 1, S. 57–61

Liber illuministarius, bearbeitet und veröffentlicht in: E. Berger, Beiträge zur Entwicklungs-Geschichte der Maltechnik, III. Folge, München 1912, S. 191–201, Reprint Liechtenstein 1982

J. B. Birelli, Alchimia Nova, Frankfurt 1654

V. Biringuccio, De la Pirotechnica, Venezia 1540, Reprint Mailand 1977

H. Blümner, Technologie und Terminologie der Gewerbe und Künste bei den Griechen und Römern, I—IV, Leipzig 1875

Gottfried Christian Bohns Waarenlager, oder Woerterbuch der Produkten und Waarenkunde, 2 Bde., Hamburg 1805

V. Boltz von Ruffach, Illuminierbuch. Wie man allerlei Farben bereiten, mischen und auftragen soll, Bern 1549. Hier zit. in der Neuausgabe München 1913, Reprint Walluf 1976

J. J. Reinbold, Neuer Tractat von Firniß-Laquir und Mahlerkünsten, Nach dem Original des berühmten Pater Bonani in Rom, Berlin und Leipzig 1735

F. de Bondaroy, Die Kunst das vergoldete und versilberte Leder zu verfertigen, Hrsg. J. H. G. Justi, Berlin, Stettin, Leipzig 1763

M. P. L. Bouvier, Handbuch der Oelmalerei für Künstler and Kunstfreunde, Braunschweig 1861, 4. Aufl.

W. Brozowsky, Waarenkunde in systematischer Darstellung der wichtigsten Natur- und Kunstprodukte des Handels, Wien 1869, 2. Aufl.

I. Bruder, Handbuch der Orgelbaukunst und Spieluhren, Simsonswald 1829

H. Brunschwig, Liber de arte distillandi de compositis, Das Buch der waren Kunst zu destillieren, Straßburg 1507

Catalogus Oder Register, aller Apoteckischen Simplicien und Compositen, so in den beyden Messen, zu Franckfurt am Mayn, durch die Materialisten, Kauffleut, Würtzelträger und Kräutler, auch durch die Apothecker daselbst verkaufft werden, Frankfurt 1582

Cennino Cennini, Il Libro dell Arte, A Cura di E Brunello, Vicenca 1982

Bayerische Staatsbibliothek München, Cod. Germ. 824, fol. 66–72. Transkription D. Oltrogge (unveröff.)

Colmarer Kunstbuch, Bern, Burgerbibl. Ms. hist. helv. XII. 45, Transkription D. Oltrogge (unveröff.)
Bayerische Staatsbibliothek München, Cod. Clm. 20174, aus den Jahren 1464–1473, Transkription und kurzer Kommentar von D. Oltrogge, FH Köln (unveröff.)
J. Colerus, Calendarium oeconomicum und perpetuum, Vor die Haußwirt, Ackerleut, Apotecker und andere gemeine Handwercksleut, Wittenberg 1591, Reprint Leipzig 1988
J. Colerus, Oeconomia Ruralis et Domestica. Das ist ein sehr nützliches allgemeines Hausbuch, Mainz 1645. Eine weitere Ausgabe Colerus 1651
J. Cracau, Warenkunde für den praktischen Drogisten, 2 Bde., Zittau 1895
J. M. Cröker, Der wohl anführende Mahler, Jena 1736, Reprint München 1982 mit Einführung und Glossar von U. Schießl
E. Darmstaedter, Die Alchemie des Geber, Berlin 1922, Reprint Hildesheim 1978
E. Darmstaedter, Berg-, Probier- und Kunstbüchlein, mit einer Bibliographie, München 1926
P. Dioscurides, De materia medica. Hrsg. M. Wellmann, 3 Bde., Berlin 1906–14, Reprint Berlin 1958
Pedanios Dioscurides, Arzneimittellehre, Hrsg. J. Berendes, Reprint Wiesbaden 1970
G. V. C. Dreme, Der Firniß- und Kittmacher; eine Anleitung zur vorteilhaftesten Bereitung der besten Lack- und Oelfirnisse, Brünn 1821
Duhamel du Monceau, Die Kunst verschiedene Arten von Leim zu machen, in: Schauplatz der Kuenste und Handwerke, Bd.11, Leipzig und Königsberg 1772
J. Dumas, Handbuch der angewandten Chemie, Nürnberg 1832
J. C. Erxleben, Anfangsgründe der Chemie, 1775
F. G. Eyßvogel, Neu-Eröffnetes Magazin bestehend In einer Versammlung allerhand raren Künsten und besonderen Wissenschaften, Zweyter Theil, Bamberg 1756
Anonymus, Die mit allerhand schönen und curiösen und geheimen Wissenschaften angefüllte Farbebelustigung, eingeteilt in zwei Teile. Nürnberg 1741 [nach Boltz von Ruffach 1549)
Anonymus, Enthüllte Geheimnisse der Farbenbereitung für Oel-, Wasser-, Tusch-, Pastell-, Emaille-, Seiden- und Wachsmalerei, Quedlinburg und Leipzig 1824
Anonymus, Neu-ausgefertigtes und mit vielen raren Kunst-Stücken geziertes Farben Büchlein, 1748
G. Francus v. Frankenau, Flora Francica, Leipzig 1753, 5. Aufl.
H. Gautier, L'Art de Laver ou Nouvelle Maniere de Peindre sur le Papien Lyon 1687, Reprint The Printed Sources of Western Art, Portland, Oregon 1972
I Georgica Curiosa Aucta, Das ist umständlicher Bericht und klarer Unterricht von dem Adelichen Land- und Feld-Leben, Nürnberg 1687 II Georgica Curiosa. Oder: Des auf alle in Teutschland übliche Land- und Haus-Wirthschafften gerichteten, hin und wieder mit vielen untermengten raren Erfindungen und Experimenten versehenen, auch einer merklichen Anzahl Kupfer gezierten Adelichen Land- und Feld-Lebens Anderer Theil, Nürnberg 1687
Herrn von Hohbergs Georgica Aucta. Oder: Adelichen Land- und Feld-Lebens, Dritter Theil, Nürnberg 1715
C. Glaser, Neu-eröffnete Chymische Artzney- und Werck-Schul, nach der Ausgabe von Nürnberg 1677 neu herausgegeben von H. Poeckern, Leipzig 1988
J. A. van de Graaf, Het de Mayerne Manuscript als bron voor der schildertechnik van de barok, Mijdrecht 1958
J. K. Guetle, Gründlicher Unterricht zur Verfertigung guter Firnisse nebst Kunst zu Lackieren und Vergolden, Nürnberg 1806, 2. Aufl.
J. K. Guetle, Auf Erfahrung gegründete Vorschrift für Fabrikanten, Künstler und Handwerker, die mit Verfertigung und dem Gebrauch von Farben und Firnissen auf Gegenständen aller Art zu tun haben, Nürnberg 1807 (1. Aufl. 1792)
P. Hackert, Ueber den Gebrauch des Firnis in der Mahlerey. Sendschreiben des berühmten Philipp Hackert an den Ritter Hamilton, Dresden 1800

J. S. Halle, Werkstätte der heutigen Künste oder die neue Kunsthistorie, 4 Bde., Brandenburg und Leipzig, 1761-1765
Anonymus, Neues Handbuch für Künstler, Bamberg/Leipzig 1765
Praktisches Handbuch für Künstler. Aus dem Englischen nach Imison, The School of Art, Dresden 1792/1793
Anonymus, Praktisches Handbuch für Kunst und Fabrikwesen, Grätz 1803
Das mittelalterliche Hausbuch, nach dem Originale im Besitze des Fürsten von Waldburg-Wolfegg-Waldsee, Hrsg. H. Th. Bossert und W. F. Storck, Leipzig 1912
Anonymus, Hellots Färbekunst oder Unterricht Wolle und wollene Zeuge zu färben, Altenburg 1765
Heraclius, Von den Farben und Künsten der Römer, Hrsg. R. Eitelberger v. Edelberg, in: Quellenschriften für Kunstgeschichte und Kunsttechnik des Mittelalters und der Renaissance, Bd. 4, Wien 1873, Reprint Osnabrück 1970. Auch in: Merrifield 1967, S. 166–257
Dionysius von Phurna, Hermenaia, Malerbuch des Malermönchs Dionysius vom Berge Athos, München 1960
Carl Friedrich August Hochheimers, Neueste Chemische Farben-Lehre; oder: Sammlung deutlicher, und durch angestellte Versuche bewährt gefundener Vorschriften, Leipzig 1809, 2 Bde.
Gottfried August Hoffmann, I. V. L., die Chymie zum Gebrauch des Haus-Land und Stadtwirthes, des Künstlers, Manufacturiers, Fabricantens und Handwerkers, Leipzig 1757
Lic. Gottfried August Hoffmanns Chymischer Manufacturier und Fabricant, Gotha 1758
Anonymus, Richtige und vollständige Beschreibung der Holzfärberey oder Beitzen von verschiedenen Sorten, Leipzig 1790
J. Hübner, Curieuses Natur-Kunst-Gewerck und Handlungslexicon, Leipzig 1712
J. C. Jacobsen, Technologisches Wörterbuch, Berlin 1781-1793
J. Q. Jahn, Abhandlung über das Bleichen und die Reinigung der Oele, Dresden 1803
K. Karmasch, Grundriß der mechanischen Technologie, 2 Bde., Hannover 1841
Anonymus, Drey schoner kunstreicher büchlein, Leipzig 1532
Anonymus, Kunstbüchlin, Gründtlichen rechten gebrauches, aller Kunstbaren Werckleut, Frankfurt 1574
M. P. Kertzenmacher, Alchimia, daß ist alle Farben, Wasser, Olea, Salia, und Alumina, damit man alle Corpora, Spiritus und Calces Praeparirt, Sublimiert und Fixirt, zubereyten, Franckfurt am Mayn 1613 (1539, 1563]
F. A. Kirsch, Neu-verfertigtes und in zwey Theil eingerichtetes Kunst- Hauß- Arzney- und Wunder-Buch. Oder Der zum viertenmal wieder aufgelegte Curiose Künstler, Bd.1, Nürnberg 1719
J. G. F. Klein, Ausführliche Beschreibung der Metall-Lothe und Löthungen, Berlin 1760, Reprint Düsseldorf 1987
F. Knirim, Die Harzmalerei der Alten, Leipzig 1839
A. Lonicerus, Kreuterbuch, Kunstliche Conterfeytunge der Bäume, Stauden, Hecken, Kräuter, Getreyd, Gewürtze, Ulm 1679, Reprint Grünwald 1962
Kreutterbuch deß hochgelehrten und weitberühmbten Herrn D. Petri Andreae Matthioli, Jetzt widerumb mit vielen schönen newen Figuren, auch nützlichen Artzneyen, und andern guten Stücken, zum vierdten mal auß sonderm Fleiß gemehret, und verfertiget, durch Ioachimum Camerarium, Franckfurt am Mayn 1626, Reprint Grünwald o. J.
D. J. G. Krünitz, Oeconomische Encyclopaedie, oder allgemeines System der Staats-, Haus-, und Landwirtschaft, 242 Bde., 1787-1858
J. Kunckel, Ars vitraria experimentalis oder vollkommene Glasmacher-Kunst, Frankfurt und Leipzig 1689, Reprint Hildesheim und New York 1972
J. Kunckel von Löwensterns Collegium Physicochymicum experimentale oder Laboratorium Chymicum, Hamburg/Leipzig 1716, Reprint Hildesheim/New York 1975

Anonymus, Kunstbüchlein gerechten gründtlichen Gebrauchs aller kunstbaren Werckleut, Augsburg 1535 und Frankfurt 1624

Kunst-Büchlein oder gründlicher Gebrauch von Etz Arbeit, in und außerhalb des feuers wie auch jede Farben zubereiten, erhalten, bessern und wiederbringen, Frankfurt und Leipzig 1687

Anonymus, Sehr geheim gehaltene und nunmehro frey entdeckte experimentierte Kunst-Stücke, Zittau und Leipzig 1763

Anonymus, Die wohl eingerichtete Kunst und Werkschule oder Handbuch für Künstler Zwei Theile, Nürnberg 1784

Anonymus, Wieder neu aufgerichtete und vergrösserte in zwey Theilen angewiesene Curieuse Kunst- und Werck-Schul (Bd. 1). Der Neu-aufgerichteten und Vergrösserten in sechs Bücher und Theilen verfasten curieusen Kunst- und Werck-Schul sehr verlangter nunmehr erfolgter Anderer Theil J. K. Chymiae ac aliarum Artium Cultore, Nürnberg 1707, (Ausgaben zur Kunst- und Werck-Schul: 1696, 1705, 1707, 1732, 1759, 1763, 1782, 1784)

Anonymus, Gmndmäßige und sehr deutliche Anwendung zu der schönen Laccir- und Schildkrotten-Arbeit und zu allerley erdencklichen Horn- und fürtrefflichen Holz-Arbeiten, Nürnberg 1703

Anonymus, Neu entdeckte Lacquir-Kunst oder Gründliche Anweisung, Wie man nicht nur unterschiedliche bißher geheim gehaltene kostbare Lacquen sondern auch den biß ietzo unbekannten Gummi Copal leichtlich auflösen könne, Leipzig 1708

Anonymus, Neu-entdeckte Lacquir-Kunst, oder gründliche Anweisung, wie man nicht nur unterschiedliche bißher geheim gehaltene kostbare Lacquen verfertigen könne, Leipzig 1709

N. Lemery, Vollständiges Materialien-Lexicon, Darinnen Alle und jede Simplicia, vorgestellet sind, welche aus denen so genannten drey Reichen, der Thiere, der Kräuter und der Mineralien, hauptsächlich zu Dienste der Medicin und Apothecker-Kunst genommen und gebrauchet werden, Leipzig 1721

J. C. Leuchs Farben- und Färbekunde, 2 Bde., Nürnberg 1828

J. K. Leuchs, Allgemeines Waaren-Lexikon oder vollständige Warenkunde, Nürnberg 1843, 3. Aufl.

Die Alchemie des Andreas Libavius, ein Lehrbuch der Chemie aus dem Jahre 1597, Reprint Weinheim 1964

Liber illuministarum pro fundamentis auri et coloribus ac consimilibus collectus ex diversis, Kunsttechnisches Handbuch, Kloster Tegernsee, Münchner Staatsbibliothek, Cgm 821, Transkription und Übersetzung A. Bartel und M. Lautenschlager (unveröff.)

Liber de coloribus illuminatorum sive pictorum, from Sloane Ms. No. 1754, Hrsg. D. Thompson jr., Cambridge (Mass.), in: Speculum I, 1926

G. P. Lomazzo, Trattato della Pittura, Scultura et Architettura, Milano 1585

Fr. Lucanus, Anleitung zur Restauration alter Oelgemälde, Leipzig 1828, Reprint Stuttgart 1996

Compositiones ad tingenda musiva, herausgegeben, übersetzt und philologisch erklärt von H. Hedfors, Uppsala 1932

Mappae Clacicula, A little Key to the World of medieval Techniques, Hrsg. C. St. Smith und J. G. Hawthorne, in: Transactions of the American Philosophical Society, Philadelphia 1974

J. J. Marx, Teutsche Material-Kammer, Nürnberg 1687

De Mayerne, Theodor Turquet, Pictoria, sculptoria et quae subalternarum artium, 1620, in: E. Berger, Quellen für Maltechnik während der Renaissance und deren Folgezeit, München 1901, S. 107–333, Reprint Vaduz/ Liechtenstein 1993

J. F. L. Mérimée und A. D. Vergnaud, Handbuch für Kunst- und Zimmermaler, Vergolder, Restaurateurs alter Oelgemälde, Farben- und Tapetenfabrikanten (bearb. von H. C. Hebra), Ulm 1839

M. Merrifield, Original Treatises on the Arts of Painting, London 1849, Neuausgabe New York 1967, 2 Bde.

J. G. Meusel (Hrsg.), Museum für Künstler und für Kunstliebhaber, Zweites Stück, 1788

H. Miehr, Praktisches Handbuch der Lackir-, Vergoldungs-, Bronzir, Beiz-, Färbe- und Polirkunst, Weimar 1852, Reprint Hannover 1995, mit Einführung und Glossar von H. Michaelsen
J. C. Müller, Praktische Anweisung zum Lakkieren und die dazu gehörigen Lakke und Firnisse zu verfertigen, Leipzig 1801
D. Nohl, Etliche Anmerkungen über etliche Punckten wegen des Orgelwercks welche Joh. Theodore Nolle, ein in diesen Künsten erfahrener Meister, beygefüget habe wollen, o. O. 1731
Oeconomische Encyklopädie 1787
Kunstbuch, Deß wolerfarnen Herrn Alexij Pedemontani von mancherley nützlichen und bewerten Secreten oder Künsten, Basel 1671 (1616)
A. J. Pernety, Dictionaire portativ de peinture, sculpture, et gravure, Paris 1757
A. J. Pernety, Handlexicon der bildenden Künste, worinnen alles was beym Zeichnen, Malen, Bildhauen, Kupferstechen u. s. w. üblich ist, erkläret wird, Berlin 1764
C. Perrot, Praktisches Handbuch der Farbenbereitung. Für Farbenbereiter und Maler von Claude Perrot, Quedlinburg und Leipzig 1833, 2. Aufl.
J. B. Pictorio, Die mit vielen raren und curiösen Geheimnissen angefüllte Illuminier-Kunst, Nürnberg 1729
Le Pileur d'Alpigny, Traité des couleurs matérielles, Paris 1779
C. Plinius Secundus, Naturalis historia. 6 Bde., Leipzig 1870–1898, Nachdruck Stuttgart 1967–1970
P. Pomet, Der aufrichtige Materialist und Specerey-Händler ausgefertiget von Peter Pomet, Leipzig 1717
C. F. Prange, Der Weltweisheit und freyen Künste Magister Farbenlexicon, worinnen die möglichsten Farben der Natue durch die wirkliche Ausmahlung enthalten sind, Halle 1782
J. J. Prechtl, Technologische Encyklopaedie oder alphabetisches Handbuch der Technologie, der technischen Chemie und des Maschinenwesens, 20 Bde., Stuttgart 1830–1855
Theophilus Presbyter, Schedula diversarium artium, Hrsg. R. Eitelberger von Edelberg, Quellenschriften zur Kunstgeschichte Bd. 7, Wien 1874, Reprint Osnabrück 1970
Reformatio Und erneuwerte Ordnung der Apotecken, und wie es mit den Ordinarijs Physicis oder Stadtärtzten, ihres Verdiensts halben, in deß Heiligen Reichs Freystadt Wormbs, hinfürter gehalten werden soll, Frankfurt 1582
J. J. Reinbold, Neuer Tractat von Firniß- Laquier- und Mahlerkünsten, Nach dem Original des berühmten Pater Banani in Rom. Mit vielen neuen Arcanis, unterschiedlichen Beschreibungen des Gummi Copals und des Bernsteins, Berlin und Leipzig 1730, 2. Aufl.
H. Roosen-Runge, Die Buchmalereirezepte des Theophilus, Quellengeschichtliche Untersuchungen zur "Schedula Diversarum Artium" des Theophilus, in: Münchner Jahrbuch der Bildenden Kunst, 3. Folge, Bde. III/IV, 1952/1953, S. 159–171
G. E. Rosenthal, Litteratur der Technologie, Berlin und Stettin 1795, Neudruck 1972
L'art du meunisier, Encyclopédie ou dictionnaire des sciences, des arts et des metiers, Genève/ Neufchâtel, Bd. 21, 3. Aufl. 1779
J. Roux, Die Farben. Ein Versuch über Technik alter und neuer Malerei, Heidelberg 1824
M. Ruland, Lexicon Alchemiae, sive Dictionarium Alchemisticum, Frankfurt 1612, Reprint Hildesheim 1964
J. von Sandrart, Teutsche Academie der Bau- Bild- und Malerey-Künste, Nürnberg 1675
L. Schäfer, Gründlicher Unterricht im Polieren und Beizen sowie im Lackieren, Vergolden, Versilbern und Bronzieren aller Arten von Meubles und metallenen Gegenständen, Quedlinburg und Leipzig 1844
G. Schäfer, Das Handbuch der Malerei vom Berge Athos, Trier 1885. Neubearbeitung: Slawisches Institut München 1960
Allgemeines Journal fuer die Handlung oder Gemeinnuetzige Auffsaetze, Versuche und Nachrichten, Schwerin, Wismar, 2 Bde., Buetzow 1787

O. Schluttig und G. Neumann, Die Eisengallustinten, Grundlagen zu ihrer Beurteilung, Dresden 1890
C. H. Schmidt, Vollständiges Farbenlaboratorium, Weimar 1857
H. Schrader, Praktisches Lehrbuch der gesamten Wollen- und Schönfärberei, Berlin 1832
C. H. T. Schreger, Tabellarische Uebersicht der rohen und künstlich zubereiteten Farben und Farbenmaterialien zur Prüfung ihrer Güte und Aechtheit, Nürnberg/Sulzbach 1805
A. Schulze, Technologische Chemie und Materialienkunde. Oder gründlicher Unterricht für Künstler. Fabrikanten und Handwerker, Quedlinburg und Leipzig 1826
G. N. Schurtz, Neu-eingerichtete Material-Kammer: Das ist Gründliche Beschreibung aller fürnehmsten Materialien und Specereyen, Nürnberg 1673
Johannes scultetus, Das Wundartzneyische Zeughaus, Frankfurt 1666, Reprint Blaubeuren 1974
Kunstbuch des Wolfgang Sedelius I, Cgm 4117, Staatsbibliothek München
G. A. Siddons praktischer und erfahrener englischer Ratgeber ein aus langjähriger Erfahrung geschöpftes Rezeptbuch Neuer Schauplatz der Künste und Handwerke, Bd. 82, Weimar 1842
D. H. Soxhlet, Die Kunst des Färbens und Beizens von Marmor, künstlichen Steinen, von Knochen, Horn und Elfenbein und das Färben und Imitieren von allen Holzsorten, Wien/Pest/Leipzig 1899, Reprint Köln 1985
E. Spränger, Färbebuch, Erlenbach/Zürich o. J.
J. Stalker und G. Parker, A Treatise of Japaning and Varnishing, 1688, Reprint London 1960/1971
H. F. A. Stöckels, Hofschreiners zu Schleiz, Praktisches Handbuch für Künstler, Lackirliebhaber und Oehlfarben-Anstreicher, Nürnberg 1799, 2. Aufl., Reprint Rottenburg 1981
V. und R. Borradaile (Hrsg.), Das Straßburger Manuskript, Handbuch für Maler des Mittelalters, München (I, 1966). Auch in: E. Berger, Beiträge zur Entwicklungs-Geschichte der Maltechnik, III. Folge, München 1912, S. 155–190, Reprint Liechtenstein 1982
J. M. Teuber, Vollständiger Unterricht der gemeinen und höhern Drehkunst Nebst einem Anhang von der Lacquier-Kunst, Regensburg/Wien 1756
Theophilus Presbyter und das mittelalterliche Kunsthandwerk. Gesamtausgabe der Schrift De artibus in zwei Bänden, Malerei und Glas, Hrsg. E. Brepohl, Weimar/Wien 1999, Bd. I, Theophilus Presbyter und die mittelalterliche Goldschmiedekunst, Buch III, Hrsg. E. Brepohl, Wien/Köln/Graz 1987, Bd. II (Beide Bände wurden hier in der Zählweise von Brepohl zitiert.)
G. C. F. Thon, Thons vollständige Anleitung zur Lackierkunst. Ein notwendiges und nützliches Handbuch, Ilmenau 1822, 2. Aufl.
T. Thon, Der Gebäudemaler und Decorateur. Neuer Schauplatz der Künste und Handwerke, Bd. 18, Ilmenau 1826
C. F. G. Thon, Vollständige Anweisung zur Lackierkunst, 2. Aufl. Weimar 1835, 6. Aufl. 1855
C. F. G. Thon, Die Holzbeizkunst, Weimar 1840 (1. Aufl.: Die Holzbeizkunst oder Holzfärberei in ihrem ganzen Umfang, Leipzig 1822)
Manuskript 610/1957 der Stadtbibliothek Trier, Transkription, Übersetzung und Kurzkommentar von D. Oltrogge (unveröff.)
M. B. Valentini, Schaubühne oder Natur- und Mineralienkammer. Auch Ost-Indisches Sendschreiben und Rapportien. Museum Museorum oder vollständige Schau-Bühne aller Materialien und Specereyen nebst deren natürlichen Material- Kunst- und Naturalienkammern, Ost- und Westindischen Reisebeschreibungen, Curiösen Zeit- und Tag-Registern, Natur und Artzney-Kündigung wie auch selbst eigenen Erfahrung, Frankfurt, 1. Aufl. 1704, 2. Aufl. 1714
Vasari on Technique, Hrsg. G. B. Brown, New York 1960
C. Vielheuer, Gründliche Beschreibung Fremder Materialien und Specereyen Ursprung, Wachstum, Herkommen und deroselben Natur und Eigenschaften, Leipzig 1676
M. Vitruvius Pollio, De Architectura Libri X, Hrsg. C. Fensterbusch, Darmstadt 1981, 3. Aufl.
J. P. Voit, Fassliche Beschreibung der Künste und Handwerke, Nürnberg 1804

J. F. Watin, Der Staffirmaler oder die Kunst anzustreichen, zu vergolden und zu lackieren, wie solche bey Gebäuden, Meublen, Galanteriewaaren, Kutschen usw. auf die beste, leichteste und einfachste Art anzuwenden ist, Leipzig 1774, Neuauflage 1779
J. F. Watin, Kunst des Staffiermalers, Vergolders, Lackierers und Farbenfabricanten, Ilmenau 1824
E. Winckler, Die Lack- und Firniß-Fabrikation in ihrem ganzen Umfange Nebst einer Anleitung zur Lackirkunst, Leipzig 1860, 2. Aufl.
Anonymus, Der zu vielen Wissenschafften dienstlich-anweisende und von neuem wieder aufgelegte Curiose Künstler Kunst-Hauß-Artzney- und Wunder-Buch, Nürnberg 1710
Anonymus, Neues Kunst- und Wunderbuch oder verborgene Geheimnisse welche ein sterbender Vater seinen Kindern übergeben, Nürnberg 1820, 5. Aufl.
Anonymus, Der in vielen Wissenschaften reich versehene Curiöse Künstler oder wohl eingerichtetes Haus- Arzney- Kunst- und Wunderbuch, Nürnberg o. J.
J. M. Würsch, Gründliche auf eigene praktische Anwendung gestützte Abhandlung über Porträtmalerei in Oelfarben, Rapperswyl 1834
J. H. Zedler, Großes vollständiges Universallexicon Aller Wissenschafften und Kuenste, 64 Bde., Halle/Leipzig 1732–1751. Reprint Graz

II General literature

ABC Geschichte der Chemie, Leipzig 1989
L. E. Andés, Praktisches Handbuch für die gesamte Lack- und Farbenindustrie, Wien/Leipzig 1904
K. Anheuser, Im Feuer vergoldet, Geschichte und Technik der Feuervergoldung und Amalgamversilberung. AdR-Schriftenreihe zur Restaurierung und Grabungstechnik, Bd. 4, Stuttgart 1999
Arbeitsblätter für Restauratoren, Hrsg. Arbeitsgemeinschaft der Restauratoren, Mainz
D. Arends, E. Hickel und W. Schneider, Das Warenlager einer mittelalterlichen Apotheke, Braunschweig 1960
P. Arent, Die Farbstofflieferanten Alkannawurzel, Krappwurzel, Sandelholz und Rotholz, Semesterarbeit, FH Hildesheim 1995 (unveröff.)
M. Bachmann und T. Hofmeier, Geheimnisse der Alchemie, Basel 1999
Aus der Arbeit der Badischen Anilin- und Sodafabrik, Werkszeitung, seit 1951
H. Baumann, Die griechische Pflanzenwelt, München 1982
J. Beckmann, Beyträge zur Geschichte der Erfindungen, 3 Bde., Leipzig Further literature 1782-92
E. Berger, Quellen für Maltechnik während der Renaissance und deren Folgezeit, XVI.—XVII. Jahrhundert, München 1901, Reprint Vaduz 1984
E. Berger, Die Maltechnik des Altertums, München 1904, Reprint Vaduz 1992
E. Berger, Fresko- und Sgraffito-Technik nach älteren und neueren Quellen, München 1909, Reprint Wiesbaden 1979
E. Berger, Quellen und Techniken der Fresko-, Oel- und Tempera-Malerei des Mittelalters, München 1912, 2. Aufl., Reprint Walluff 1973
M. Berthelot, Die Chemie im Altertum und im Mittelalter. Aus dem Französischen übertragen, Leipzig/Wien 1909, Reprint Hildesheim 1970
C. Blänsdorf, Blaue Farblacke in Staffeleimalerei und Skulpturenfassung, Untersuchungen zu Herstellungs- und Verwendungsmöglichkeiten, mit Schwerpunkt auf den Anthocyanfarbstoffen, Diplomarbeit, Schule für Gestaltung, Bern 1997
C. I. Bothe, Der grösste Kehricht aller Farben? Über Asphalt und seine Verwendung in der Malerei, Mainz 1999

T. Brachert, Patina, München 1985
T. Brachert (Hrsg.), Beiträge zur Konstruktion und Restaurierung alter Möbel, München 1986
G. Brink (Hrsg.), Farbstoffe und Färbetechniken, Hannover 1988
Bulletin de l'Institut Royale du Patrimoine Artistique, Brüssel
M. Butor, Die Alchemie und ihre Sprache. Essays zur Kunst und Literatur, Frankfurt a. M./Paris 1984
C. Cession, Vergoldung und Metallauflagen im 18. Jahrhundert, in: Restauratorenblätter Bd. 11, 1990
V. Cordier, Die chemische Zeichensprache einst und jetzt, Graz 1928
Unsichtbare Meisterzeichnungen auf dem Malgrund, Cranach und seine Zeitgenossen, Hrsg. I. Sandner, Regensburg 1998
J. A. van de Graaf, Het de Mayerne Manuscript als bron voor de schildertechnik von de barok, Nijdrecht 1958
M. Doerner, Malmaterial und seine Verwendung im Bilde, München 1933, 3. Aufl.
M. Doerner, Malmaterial und seine Verwendung im Bilde, Stuttgart 1944, 8. Aufl.
Albrecht Dürer, Die Gemälde der Alten Pinakothek, Bayerische Staatsgemäldesammlungen, München 1998. Mit technischen Beiträgen von B. Heimberg, A. Burmester, C. Krekel, J. Koller, I. Fiedler und U. Baumer
C. L. Eastlake, Methods and Materials of Painting of the great Schools and Masters, 2 Bde., London 1847, Reprint New York 1960
F. Ellwanger-Eckel, Herstellung und Verwendung künstlicher grüner und blauer Kupferpigmente, Staatliche Akademie der Bildenden Künste, Stuttgart 1979 (Diplomarbeit)
Allgemeine Enzyklopädie der Wissenschaften und Künste, (Hrsg. J. Ersch und J. Gruber) 168 Bde., Leipzig 1818–1889, Reprint ADEVA
Firnis, Material-Ästhetik-Geschichte, Internationales Kolloquium, Braunschweig 15. bis 17. Juni 1998, AdR-Schriftenreihe zur Restaurierung und Grabungstechnik (1999)
H. Fischer, Mittelalterliche Pflanzenkunde, München 1929
E. Fontana, F. Hellwig und K. Martius, Historische Lacke und Beizen, Germanisches Nationalmuseum Nürnberg, 1992
R. Fuchs und D. Oltrogge, Colour material and painting technique in the Book of Kells, in: Proceedings, Oxford 1994
H. Genaust, Ethymologisches Wörterbuch der botanischen Pflanzennamen, Basel/Boston/Stuttgart 1983, 2. Aufl.
G. W. Gessmann, Geheimsymbole der Alchemie, Arzneikunde und Astrologie des Mittelalters, 1959, 2. Aufl.
R. J. Gettens and G. L. Stout, Painting Materials, A Short Encyclopaedia, New York 1966
D. Goltz, Studien zur Geschichte der Mineralnamen in Pharmazie, Chemie und Medizin von den Anfängen bis Parcelsus, in: Sudhoffs Archiv Zeitschrift für Wissenschaftsgeschichte, Beihefte, Heft 14, Wiesbaden 1972
H. Kühn, H. Roosen-Runge, R. E. Straub und M. Koller, Reclams Handbuch der künstlerischen Techniken, Bd. 1, Farbmittel, Buchmalerei, Tafel- und Leinwandmalerei, Stuttgart 1984
A. Händel, Der Erfurter Waidhandel, Magisterarbeit, Universität Erlangen-Nürnberg 1994 (Phil. I)
R. D. Harley, Artists Pigments c. 1600-1835, London 1970
R. Harley, Artists Brushes — Historical Evidence from the Sixteenth to the Nineteenth Century, London/Boston 1972, S. 61–66
B. Hartwieg und M. von der Goltz, Bildproduktion. Über die Herstellung niederländischer Gemälde im 17. Jahrhundert, Hannover 2000
M. H. B. Hering, Weiße Farbmittel, Fürth 2000

G. Heydenreich, Herstellung, Grundierung und Rahmung in den Werken Lucas Cranachs d. Ä., in: Katalog der Ausstellung "Unsichtbare Meisterzeichnungen auf dem Malgrund, Cranach und seine Zeitgenossen", Wartburg 1998

E. Hickel, Chemikalien im Arzneischatz deutscher Apotheken des 16. Jahrhunderts, unter besonderer Berücksichtigung der Metalle, Diss. Braunschweig 1963

E. Hickel, Salze in den Apotheken des 16. Jahrhunderts, Braunschweig 1965

J. Hofenk-de Graaff, Natural Dyestuffs - Origin, Chemical Constitution, Identification, ICOM Plenary Meeting, Amsterdam, Sept. 1967

R. Hofmann, Färbepflanzen und Färbedrogen, in: Restauratorenblätter, Bd. 13, 1992, S. 39–63

G. Höhle, Zur ober- und mittelrheinischen Buchmalerei um die Mitte des 15. Jhs — ausgehend von 2 Maltraktaten, Diss. Berlin 1984

H. Huth, Künstler und Werkstatt der Spätgotik, Augsburg 1923, Neudruck Darmstadt 1981

Kodex 189 der Erfurter Wissenschaftlichen Allgemeinbibliothek, De coloribus, 13.-14. Jh. Hrsg. J. Kadenback, Mittelalterliche Farbrezepte für Erfurter Maler, in: Beiträge zur Erhaltung von Kunstwerken I, Berlin 1982

R. Keller, Leinöl als Malmittel, in: MT 1973, H. 2, S. 74–105

H. Kittel, Pigmente, Stuttgart 1960

U. Körber-Grohne, Naturpflanzen in Deutschland, Kulturgeschichte und Biologie, Stuttgart 1987

O. P. Krämer, Rezepte für die Metallfärbung und Metallüberzüge ohne Stromquelle, Saulgau 1970, 7. Aufl.

O. Krätz, 7000 Jahre Chemie, München 1990

B. und G. Krumbiegel, Bernstein — Fossile Harze aus aller Welt, Fossilien, Sonderband 7, Weinstadt 1994

H. Kühn, Farbe, Farbmittel, in: Reallexikon zur deutschen Kunstgeschichte, Bd. 4, Stuttgart 1958

H. Kühn, Pigmente in den Gemälden der Schack-Galerie, München 1969

H. Kühn, Erhaltung und Pflege von Kunstwerken und Antiquitäten, München 1974

U. Lindgren (Hrsg.), Europäische Technik im Mittelalter, Tradition und Innovation, Berlin 1996

E. O. von Lippmann, Abhandlungen und Vorträge zur Geschichte der Naturwissenschaften, 2 Bde., Leipzig 1906

E. O. von Lippmann, Entstehung und Ausbreitung der Alchemie, Berlin 1919, Hildesheim/New York 1978

E. O. von Lippmann, Beiträge zur Geschichte der Naturwissenschaften und der Technik, Berlin 1923

E. O. Lippmann, Beitrage zur Geschichte der Naturwissenschaften und Technik, Bd. 2, Wenheim 1953, Reprint Liechtenstein 1971

D. von Manteuffel-Szoege, Systematisches Glossar zum Straßburger Manuskript, Magisterarbeit, Universität München 1978/1979

L. Masschelein-Kleiner und L. Maes, Ancient techniques in eastern mediterranean regions, ICOM Committee for Conservation, 5th Triannual meeting, Zagreb 1978

H. Michaelsen, Transparente Überzüge auf Holzoberflächen in den Quellenschriften zwischen 1750 und 1850, in: ZfK 1998, H. 1, S. 332–351

A. Milner, Handbuch der Färbetechniken, Bern/Stuttgart/Wien 1994

Maltechnik-Restauro, Internationale Zeitschrift für Farb- und Maltechniken, Restaurierung und Museumsfragen, München (bis 1987)

L. Nencki, Die Kunst des Färbens mit natürlichen Stoffen, Geschichte - Methoden - Rezepte, Bern/Stuttgart 1934

K. Nicolaus, DuMont's Handbuch der Gemäldekunde, Köln 1986

D. Oltrogge, Das Blau in der mittelalterlichen Buchmalerei, Quellenschriften als Basis naturwissenschaftlicher Farbuntersuchungen, Farbmittel, in: Blau: Farbe der Ferne, Heidelberg 1990, S. 104–130

I. Origo, Im Namen Gottes und des Geschäfts, Lebensbild eines toskanischen Kaufmanns der Frührenaissance, Berlin 1997
H. Ottenjahn, Farbige volkstümliche Möbel, Museumsdorf Cloppenburg 1979, S. 25
R. L. Parker, Mineralienkunde, Thun/München 1963
L. Pauling, Chemie, Weinheim 1958
G. Pfaff, Spezielle Effektpigmente. 2. Auflage. Vincentz Network, Hannover 2007
E. Ploss, Studien zu den deutschen Maler- und Färberbüchern des Mittelalters, Ein Beitrag zur deutschen Altertumskunde und Wortforschung, Diss. phil., München (masch.) 1952
E. Ploss, Zur Geschichte des Wortes, "Salmiak", in: Die BASF, Aus der Arbeit der Badischen Anilin- und Sodafabrik, H. 1, Febr. 1956, 6. Jg., S. 4–7
E. Ploss, Die Fachsprache der deutschen Maler im Mittelalter, in: Zeitschrift für deutsche Philologie, 79. Bd., 1960, S. 70–83, S. 315–324
E. Ploss, Ein Buch von alten Farben, Technologie der Textilfarben im Mittelalter mit einem Ausblick auf die festen Farben, Heidelberg/ Berlin 1962
E. Ploss, Ein Malerbüchlein aus dem Bamberger Karmeliterkloster, in: 100. Bericht des Historischen Vereins für die Pflege der Geschichte des ehemaligen Fürstbistums Bamberg, Bamberg 1964, S. 331–346
E. Ploss, H. Roosen-Runge, H. Schipperges und H. Buntz, Alchimia. Ideologie und Technologie, München 1970
E. Ploss, Das Amberger Malerbüchlein, Zur Verwandtschaft der spätmhd. Farbrezepte, in: Festschrift für Hermann Heimpel, zum 70. Geburtstag am 19. September 1971, Veröffentlichungen des Max-Planck-Institutes für Geschichte 36/III, Göttingen 1972, S. 693–703
C. Priesner, Alchemie, München 1998
E. Reinkowski-Häfner, Tempera, Zeitschrift für Kunsttechnik und Konservierung, Jg. 8, 1994, H. 2, S. 297–317
Restauro, Zeitschrift für Kunsttechniken, Restaurierung und Museumsfragen, München (ab 1988, vor 1988: MT)
A. Riechers, Die Geige und ihr Bau, Leipzig 1896
G. Rink, Farbstoffe und Färbetechniken, Hannover 1988
H. Roosen-Runge, Farbe. Farbmittel der abendländischen mittelalterlichen Buchmalerei, in: Reallexikon zur deutschen Kunstgeschichte, Bd. 4, Stuttgart 1958, Sp. 1463–1492
H. Roosen-Runge, Farbgebung und Technik frühmittelalterlicher Buchmalerei, Studien zu den Traktaten, "Mappae clavicula" und, "Heraclius", 2 Bde., München 1967
L. Roth, K. Kormann und H. Schweppe, Färbepflanzen und Pflanzenfarben, Landsberg/Lech 1992
J. Ruska, Das Buch der Alaune und Salze, Berlin 1935
I. Sandner, B. Bünsche, G. Meier, H. Schramm und J. Voss, Konservierung von Gemälden und Holzskulpturen, München 1989
G. Schäfer, Blauholz, Rothölzer, Gelbholz, Ciba AG, Basel 1937
U. Schießl, Rokokofassung und Materialillusion, Mittenwald 1979
U. Schießl, Techniken der Faßmalerei in Barock und Rokoko, Worms 1989
U. Schießl, Die deutschsprachige Literatur zu Werkstoffen und Techniken der Malerei von 1350 bis ca. 1950, Worms 1989
G. Schneider, Färben mit Naturfarben, Ravensburg 1979
M. Schniering, Die Farbstofflieferanten Blauholz, Waid und indischer Indigostrauch, Semesterarbeit, FH Hildesheim 1995 (unveröff.)
A. Scholtka, Theophilus Presbyter. Die maltechnischen Anweisungen und ihre Gegenüberstellung mit naturwissenschaftlichen Untersuchungsbefunden, in: Zeitschrift für Kunsttechnik und Konservierung, H. 1, 1992, 6. Jg., S. 1–53
R. Scholz, Aus der Geschichte des Farbstoffhandels im Mittelalter, Diss. München 1929

H. P. Schramm, B. Hering, Historische Malmaterialien und ihre Identifizierung, Graz 1988
H. Schweppe, Handbuch der Naturfarbstoffe, kommen, Verwendung, Nachweis, Landsberg a. Lech 1992
S. Seidel, Die Curieuse Kunst- und Werck-Schul, Diplomarbeit, Staatliche Akademie der Bildenden Künste, Stuttgart 1992
Studies in Conservation. The Journal of the International Institute for Conservation of Historic and Artistic Works, London
V. Simons, Die Farbstofflieferanten Gelbholz, Curcuma, Gelbbeeren, Berberitze, Wau, Quercitron und ihre Verwendung, Semesterarbeit, FH Hildesheim 1995
R. W. Soukup und H. Mayer, Alchemistisches Gold, Paracelsistische Pharmaka, Laboratoriumstechnik im 16. Jahrhundert, Wien 1997
K. Sprengel (Hrsg.), Theophrasts Naturgeschichte der Gewächse, Darmstadt 1971
M. Stappel, Firnis und Holzbeize um 1700, Eine restauratorische Gegenüberstellung von Holzoberflächen und alten Rezepten, Diplomarbeit, FH Köln 1996, München 1997
E. Stock, Das Buch der Farben, Göttingen 1956
A. Stoll, Gelbe Pflanzenlacke, Schüttgelb und Safran, Eine Übersicht über Herstellung und Verwendung anhand historischer Quellen, in: MT 1981, H. 2, S. 73–110
I. Stößel, Rote Farblacke in der Malerei, Herstellung und Verwendung im deutschsprachigen Raum zwischen 1400 und 1850, Diplomarbeit, Staatliche Akademie der Bildenden Künste, Stuttgart 1985
R. Straub und T. Brachert, Konservierung und Denkmalpflege, Teil I (Tafelbild) und Teil II (Gefaßte Holzskulptur und Schnitzaltar), Zürich 1965
R. Straub, Der Traktat "De Clarea", Zürich 1965
L. Suhling, Erdöl und Erdölprodukte in der Geschichte, Deutsches Museum, Abhandlungen, 43. Jg., 1975, H. 2/3
E. Sutermeister, Das Kasein, Chemie und technische Verwendung, Berlin 1932
D. V. Thompson, The Materials and Techniques of Medieval Painting, New York 1956, London 1936, 1. Aufl. Technische Mitteilungen für Malerei, München
V. Trost, Gold- und Silbertinten, Technologische Untersuchungen zur abendländischen Chrysographie und Argyrographie von der Spätantike bis zum hohen Mittelalter, in: Beiträge zum Buch- und Bibliothekswesen 28, Wiesbaden 1991
S. Waetzold, Systematisches Verzeichnis der Farbnuancen: Quellengeschichtliche Untersuchung zur "Schedula diversarum artium", Münchner Jahrbuch der Bildenden Kunst, 3. Folge, Bd. III/IV, 1952/1953, S. 150
C. Wagner, Arbeitsweisen und Anschauungen in der Gemälderestaurierung um 1800, München 1988
K. Walch und J. Koller, Lacke des Barock und Rokoko, in: Arbeitshefte des Bayerischen Landesamtes für Denkmalpflege, Bd. 81, 1997
J. Weber, Die Farbstofflieferanten Cochenille, Kermes, Lac-Dye, Drachenblut, Orlean und ihre Verwendung zum Färben von Hölzern, Semesterarbeit, FH Hildesheim 1995 (unveröff.)
K. Wehlte, Werkstoffe und Techniken der Malerei, Mit einem Anhang über Farbenlehre, Ravensburg 1967 (ungekürzte Ausgabe)
K. Wehlte, Werkstoffe und Techniken der Malerei, Ravensburg 1985, 5. Aufl. (posthum überarbeitete und gekürzte Ausgabe)
E. C. Wellmer, Goldfarbene Lacke in den Quellenschriften zur Maltechnik, in: Arbeitsblätter für Restauratoren 1996, H. 2, S. 1–13
H. Willers, Herstellung von tierischem Leim und seine Verwendung im Bereich der Tafel- und Faßmalerei nach Angaben deutschsprachiger Quellenliteratur des 16. bis 19. Jahrhunderts, Diplomarbeit, Kunstakademie Stuttgart 1980 (unveröff.)
J. Wolters, Zur Geschichte der Löttechnik, Degussa 1975

R. Woudhuysen-Keller, Das Farbbüchlein, Codex 431 aus dem Kloster Engelberg, Diss. Zürich 2000 (unveröff.)
C. H. Wunderlich, Zur Diskussion gestellt: Kupfercalciumacetat, in: Restauro 1998, H. 1, S. 22–24
G. Zerr und R. Rübenkamp, Handbuch der Farbenfabrikation, Berlin 1922
Zeitschrift für Kunsttechnik und Konservierung, mit Mitteilungen des deutschen Restauratorenverbandes e. V., Worms

III Lexic

Encyclopédie Roret, Nouveau manuel complet, ohne Ort und Jahr, Reprint Inter-Livres, Paris o. J. Fabricant de Vernis de tout espèce, par A. Romain Nouveau Manuel complet de la Dorure sur bois a l'eau et à la Mixtion, par M. J Saulo Nouveau Manuel complet du Fabricant de cadres, Passe-Partout, Chàssis, Encadrements, etc., par J. Saulo et de Saint-Victor
M. Haeffner, Dictionary of Alchemy, London 1991
G. Hunnius, Pharmazeutisches Wörterbuch, Berlin 1950
Lexikon zur Arzneimittelgeschichte, 7 Bde., Frankfurt a. M. 1968–1975
W. Schneider, Lexikon alchemistisch-pharmazeutischer Symbole, Verlag Chemie 1962
L. Mackensen, Ursprung der Wörter, Etymologisches Wörterbuch, Wiesbaden 1985
H. Marzell, Wörterbuch der deutschen Pflanzennamen, Leipzig 1937-1972
Meyers Lexikon, Leipzig 1924-1933
R. Palla, Das Lexikon der untergegangenen Berufe, Frankfurt a. M. 1998
Römpps Chemielexikon, Hrsg. O. A. Neumüller, Stuttgart 1988, 8. Aufl.
Schmeil-Fitschen, Flora von Deutschland und seinen angrenzenden Gebieten, Heidelberg 1982, 87. Aufl.
W. Schneider, Wörterbuch der Pharmazie, Bd. IV, Geschichte der Pharmazie, Stuttgart 1985
G. Seufert, Farbnamenlexikon von A-Z, Göttingen 1955
Ullmanns Enzyclopädie der technischen Chemie, 1976, 4. Aufl. Großes vollständiges Universal Lexicon, Leipzig/Halle 1747

Heinz Mustroph and Andrew Towns

38 Indophenol and related dyes

Abstract: This article covers three closely related types of colorant: the indophenol class of dyes plus those of their indaniline and indamine counterparts. Their roots stretch all the way back to the mid-nineteenth-century period during which synthetic dyes were successfully commercialized for the first time. The early history and subsequent development of these classes are briefly surveyed along with their chemistry. Two of their most important applications involve the synthesis of the dyes as part of the coloration process: the development of images from photographic film and the dyeing of hair. This introduction to indo dyes closes by describing their role in these and other coloration processes.

Keywords: Berthelot reaction, Bindschedler's Green, coupler, developer, indamine dyes, dye diffusion thermal transfer printing, D2T2, indaniline dyes, indoaniline dyes, indophenol dyes, Indophenol Blue, oxidative coupling, silver halide photography

38.1 Fundamentals

The phenomenon in which oxidation of *p*-aminophenol **1** and *p*-phenylenediamine **4** in the presence of phenols **2** or anilines **6** yields *indophenol* **3**, *indaniline* (sometimes written as indoaniline) **5** and *indamine dyes* **7** is well known [1, 2].

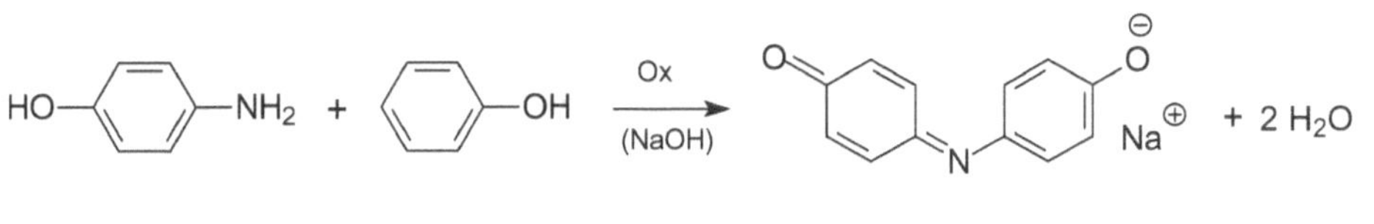

1 **2** **3**

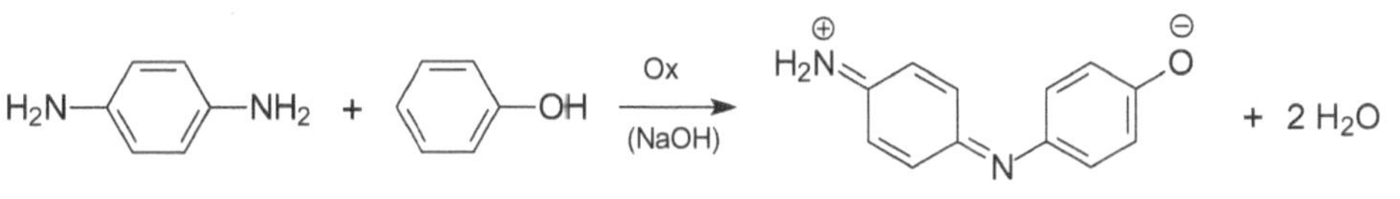

4 **2** **5**

This article has previously been published in the journal Physical Sciences Reviews. Please cite as H. Mustroph, A. Towns, Indophenol and Related Dyes *Physical Sciences Reviews* [Online] 2021, 6. DOI: 10.1515/psr-2021-0017

https://doi.org/10.1515/9783110587104-038

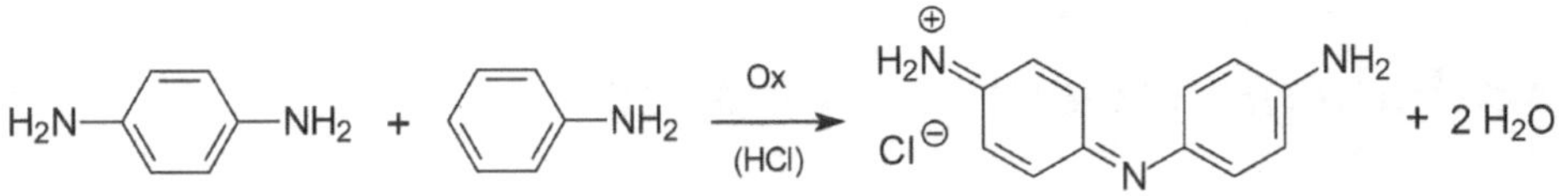

4 6 7

In the above fundamental reactions, Ox represents the oxidizing agent. It stands for all possible oxidizing agents such as hypochlorite, dichromate, peroxide, peroxydisulfate, ferricyanide, nitroferricyanide or even aerial oxygen. Given its heavy dependence on this type of chemistry, in silver halide photography *p*-aminophenol, *p*-phenylenediamine and their derivatives are referred to as *developers*. In other fields of technology they are called *primary intermediates* or *oxidation bases*, for example in hair coloration or fur dyeing. Phenol, aniline and their derivatives are called *couplers* or *secondary intermediates*. The process of oxidation of a primary intermediate and its reaction with a secondary intermediate is known as *oxidative coupling*.

In the case of *p*-aminophenol, *p*-benzoquinone imine is formed upon oxidation, while *p*-benzoquinone diimine is formed from *p*-phenylenediamine. The oxidized species undergo an electrophilic substitution reaction with couplers to generate colored products. So, the indophenols can be considered as *N*-phenylsubstituted *p*-benzoquinone imines and the indanilines and indamines as *N*-phenylsubstituted *p*-benzoquinone diimines.

Considering the *Indophenol* sodium salt **3** [3], the inner salt *Indaniline* **5** and *Phenylene Blue* **7** [4], respectively, the structural similarity to oxonol, merocyanine and cyanine dyes is immediately noticeable. The *N,N*-bis methylated derivative of Indaniline **5** is *Phenol Blue* **8** [5], whereas the *N,N,N',N'*-tetrakis methylated derivative of the simplest indamine dye, Phenylene Blue **7**, is *Bindschedler's Green* **9** [6].

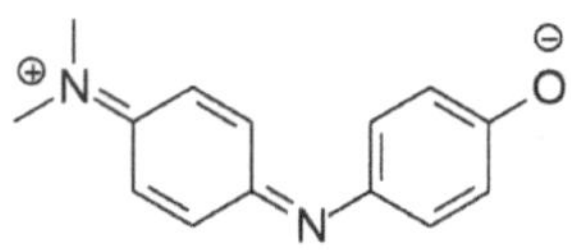

8

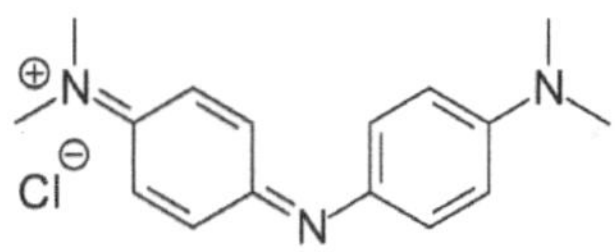

9

Like all other polymethine dyes, the salts of indophenols **3**, indamines **7** and the inner salts of indanilines **5** are characterised by an odd number $2n + 3$ of π-centres and an even number $2n + 4$ of π-electrons (where n is the number of vinylene groups –CH=CH–), in which the central methine group –CH= is replaced by the strong electron acceptor –N= [7–9].

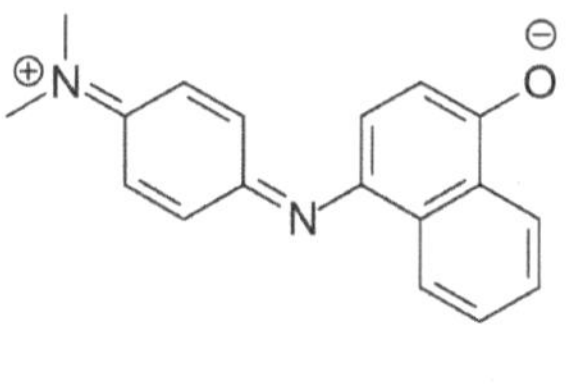

10

λ_{max} = 583 nm
ε_{max} = 11,000 M^{-1} cm^{-1} ($CHCl_3$) [10]

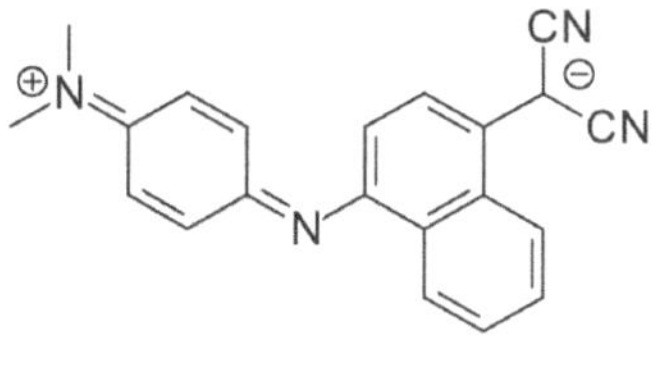

11

λ_{max} = 722 nm
ε_{max} = 25,500 M^{-1} cm^{-1} ($CHCl_3$) [11]

Of all these three dye classes, the indanilines have by far the greatest commercial application. Of particular industrial importance to non-hair coloration applications are the derivatives of 1-naphthol. Dye **10** is still on the market today as *Indophenol Blue*. Analogously to the streptomerocyanine dyes [12], replacing 1-naphthol by 1-naphthylmalononitrile resulting in **11** leads to a substantial bathochromic shift.

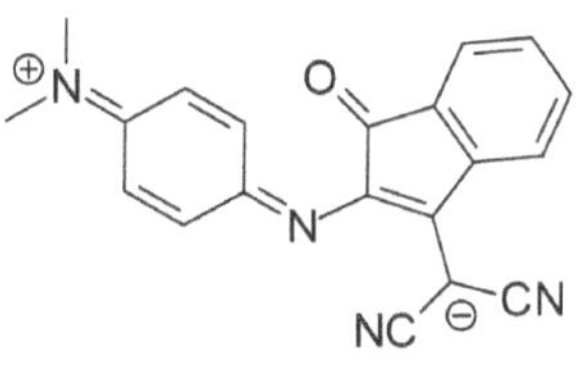

12

λ_{max} = 609 nm
ε_{max} = 48,200 M^{-1} cm^{-1} (CH_2Cl_2) [13]

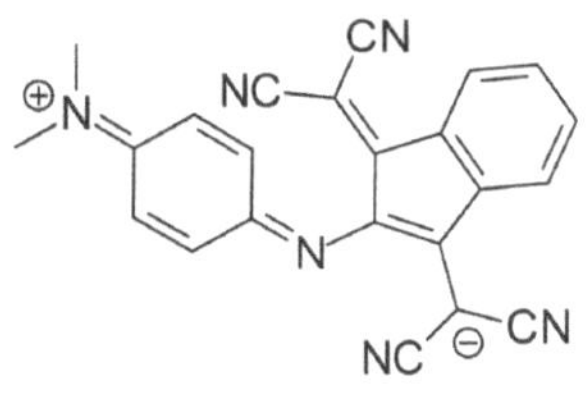

13

λ_{max} = 755 nm
ε_{max} = 25,500 M^{-1} cm^{-1} (CH_2Cl_2) [13]

Further red-shifting in absorption occurs in indaniline dyes derived from 3-dicyano-vinylindan-1-one **12** and 1,3-bisdicyanovinylindane **13**, offering not only another route to blue dyes, but also to NIR absorbers.

38.2 History of the indophenol and related dyes

It requires some deeper research to shed light on the historical background of the indophenol dye class. We should remember that the genesis of synthetic organic dye chemistry lies in the year 1856. During this year, three dyes (cyanine, mauveine, fuchsine) were prepared or their synthesis was published. The indophenol **3** was first synthesized by Marcel Berthelot in 1859 [3]. He used an alkaline solution of phenol and sodium hypochlorite, today called Berthelot's reagent. Ammonia reacts with Berthelot's reagent to form a deep blue product. Today this reaction is well known as the Berthelot test to determine whether a sample contains ammonia. Of course, Berthelot could not give a reaction equation and the chemical structure of the dye. One must bear in mind that Friedrich A. Kekulé suggested the cyclic hexagonal structure for benzene only in 1865. Although Berthelot did not know it, the blue dye **3** formed in his test is the parent structure of the indophenol class.

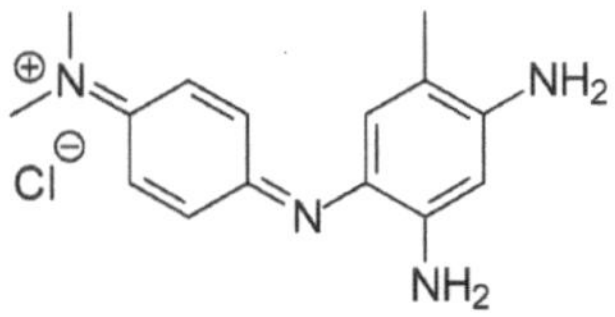

14

The first indamine dyes were described by Otto N. Witt, derived from meta-tolylenediamine (toluene-2,4-diamine) and became popular as *Toluylene Blue* **14** [14]. The basic structure of all indamine dye belongs to *Phenylene Blue* **7**, first described by Rudolf Nietzki [4]. Its *N*-tetrakis methylated derivative was reported by Robert Bindschedler [6].

The synthesis of an "indophenol" (in the guise of **10**) was first described by Horace Köchlin and Witt [15] and the dye class was claimed in a series of patents in various countries. They first used the term "indophenol" in a patent [16] and soon after in papers [5, 17]. It is interesting to read Witt's first report regarding the discovery of the "indophenols" and giving the generic name:

> *The discovery of indophenol was an accidental one. Our original intention was to form the dyestuff resulting from the action of nitrosodimethylaniline hydrochloride upon α naphthol on the fibre by printing a mixture of the ingredients and steaming. These experiments were all made in the presence of acetic acid. Once, however, an alkaline mixture was used, and on that occasion Mr. Horace Köchlin observed that a blue color was obtained instead of a violet one, both in a liquid mixture and after steaming on the tissue. This unexpected occurrence was closely studied by Mr. Köchlin and myself, and finally resulted in the production of a new colouring matter. The reaction which gives rise to this blue is applicable to almost all phenols, and in most cases results in the formation of intensely blue substances, which, for this reason, have received the generic name of indophenols. The indophenols are the phenolic analogues of tolyulene blue and its relatives; they were the missing link in the chain of evidence which here, as in all other colour reactions, indicates the existence of two parallel series of dyes of acid and basic character.* [5]

Also interesting, at the end of Witt's paper there is a note where Heinrich Caro from BASF claims that he already knew of the indophenol-type of dye. However, Caro's claim of precedence was given no credence by Witt:

In reply to the foregoing remarks, Dr. Otto N. Witt desires to make the following statement, in as much as because he was unable to be present at the meeting and therefore could not take part in the discussion, he feels that silence on this part may be construed as an admission of Dr. Caro's claims.

Dr. Witt is of opinion that Dr. Caro cannot have prepared indophenol in a pure state, or else he would have seen that it is an interesting and useful colouring matter. Nor could he have recognised its nature or composition, for then he would have seen that it was quite worthy of scientific investigation. Dr. Witt moreover, is of opinion that 'claims to priority, such as the one brought forward by Dr. Caro, are neither admitted by the patent laws of any civilised country nor by the rules of good scientific society.' [5].

On the one hand Witt's statement is no surprise given the need by Witt to defend his and Köchlin's patents. But it is also correct in terms of scientific content up to the accidental discovery. At this time it was known, without an additional oxidation step, that *p*-nitroso-*N*,*N*-dimethylaniline hydrochloride **15** reacts with 1-naphthol **16** in acetic acid to afford a violet dye. Today the structure of this dye is known to be **17**.

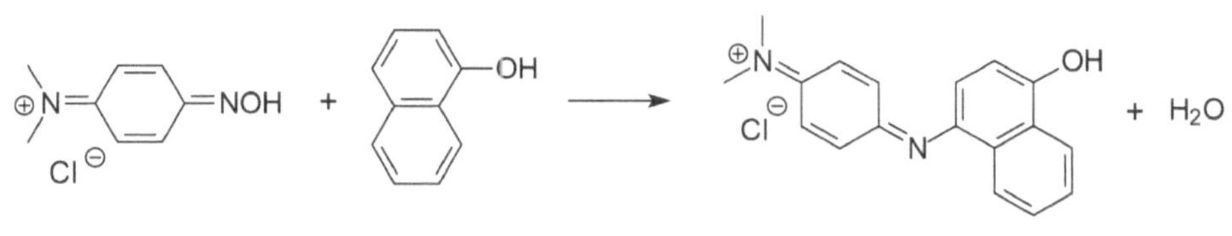

15 **16** **17**

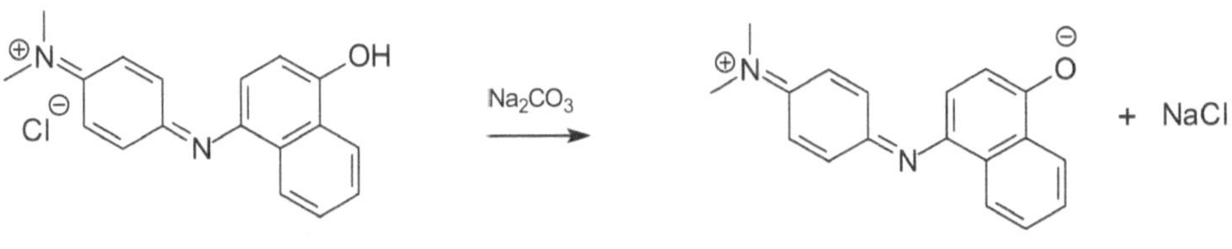

17 **10**

In the alkaline solution the deep blue inner salt **10** forms, whereas in the acetic acid solution the violet salt **17** arises.

In dye chemistry all three classes are commonly called "indophenols" [1] according to Witt [5, 17], because it has historically been the case that he has considered all three subtypes to be thought of collectively as indophenols.

However, already in 1883 Richard Möhlau noticed:

Da die indophenolartigen Farbstoffe einen ausgeprägten Phenolcharakter besitzen, welcher den sogenannten Indophenolen völlig abgeht, so dürfte es wohl richtiger sein, erstere als Indophenole und letztere, weil sie von einem Derivat des Anilins abstammen, als Indoaniline zu bezeichnen. [18]

(Since the indophenol-like dyes have a pronounced phenolic character, which is completely absent from the so-called indophenols, it would probably be more correct to designate the former as indophenols and the latter as indoanilines, because they are derived from a derivative of aniline.)

Witt had read this paper and mentioned, that he will refrain from discussing Möhlau's communication on this subject [17]. So, in the following both terms were used in a confusing manner.

A few years later, Nietzki coined the term indamine dyes [19]. Therewith, in the following years and step by step a more strictly correct approach prevailed to think of indamines and indanilines as separate from indophenols given the differences in their chemistry, color and technical properties. An exception lives on in **10**, where the name Indophenol Blue remains popular owing to Witt's papers [5, 17]. It is for this reason that this chapter is entitled "Indophenol and related dyes".

38.3 General synthetic routes to indophenol and related dyes

Berthelot's original reagent is an alkaline solution of phenol and sodium hypochlorite.

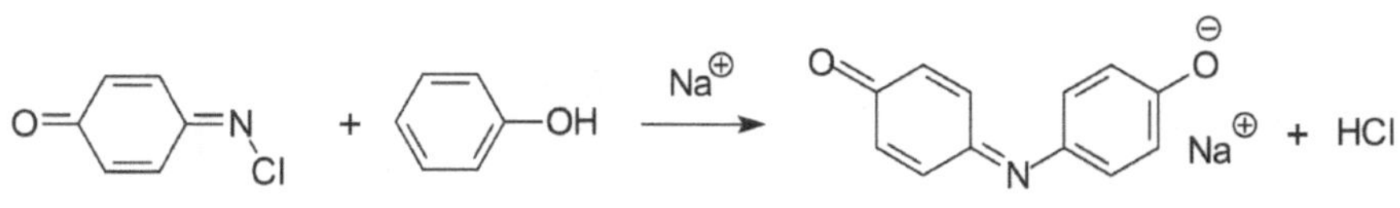

18 **2** **3**

Ammonia reacts with hypochlorite in basic conditions to form monochloramine at basic conditions, which reacts with a phenol to produce *N*-chloro-*p*-benzoquinone imine **18**. This also reacts with phenol **2** to form the blue indophenol anion **3**. Nowadays nitroferricyanide is added to increase the reaction rate.

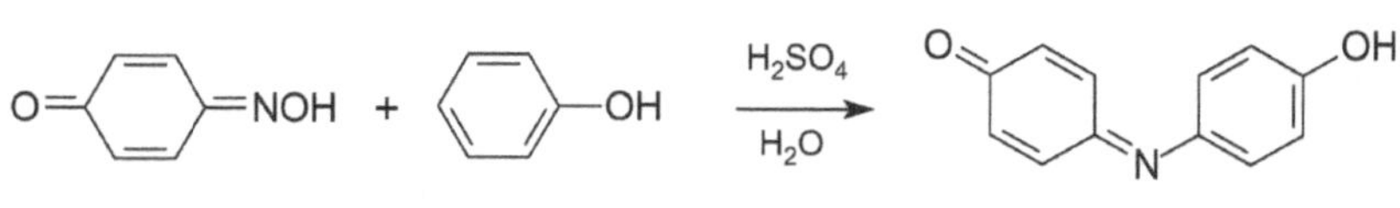

19 **2** **20**

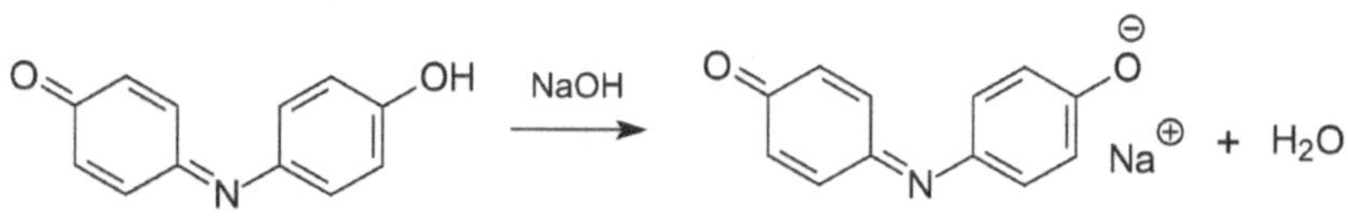

20 **3**

The most popular synthetic method is to oxidize p-aminophenol **1** to *p*-benzoquinone imine **19** which reacts with phenol **2** in sulfuric acid. After dilution with water, the red indophenol **20** product then converts under basic conditions to the blue indophenol anion **3.**

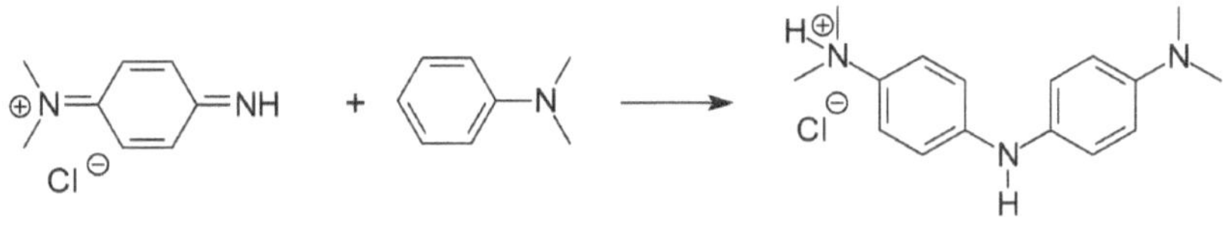

21 22 23

23 9

It is characteristic of the *p*-benzoquinone diimines **21** that they add a molecule of aniline or *N,N*-dialkylaniline **22** to afford a *p*-phenylenediamine **23**. This leuco dye can be easily oxidized to the indamine, which in this example is Bindschedler's Green **9**.

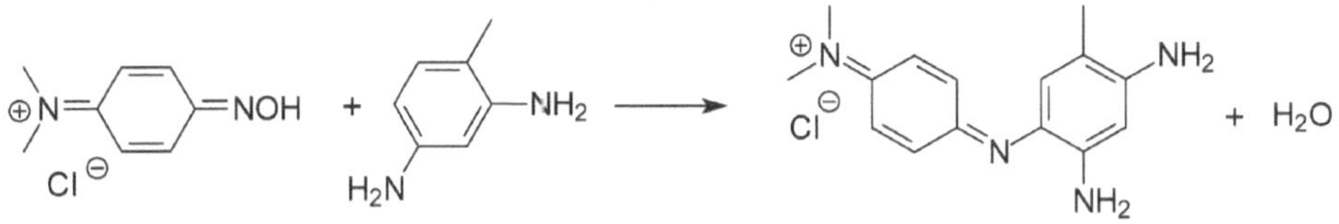

15 24 14

Without additional oxidation step one can use *p*-nitroso-*N,N*-dialkylaniline hydrochlorides **15** in hot water for condensation with e.g. *N,N*-dialkylanilines or meta-tolylenediamines **24** to indamine dyes like e.g. Toluylene Blue **14** [14].

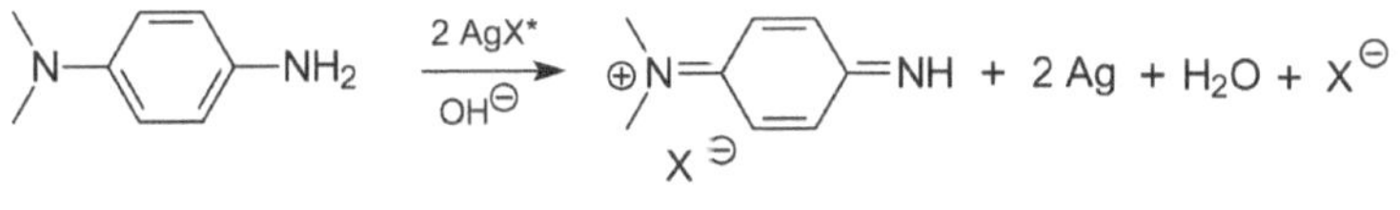

25 26

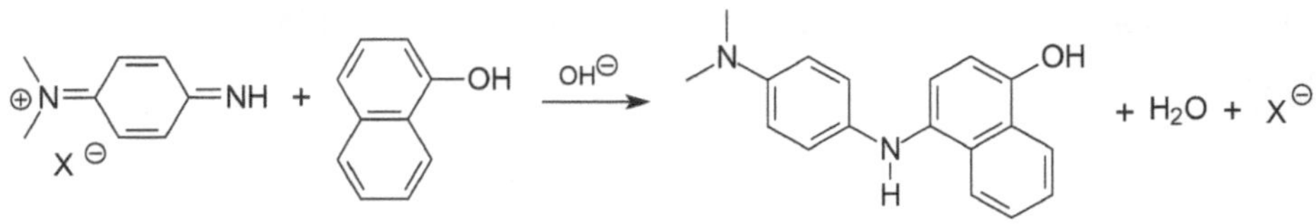

26 **16** **27**

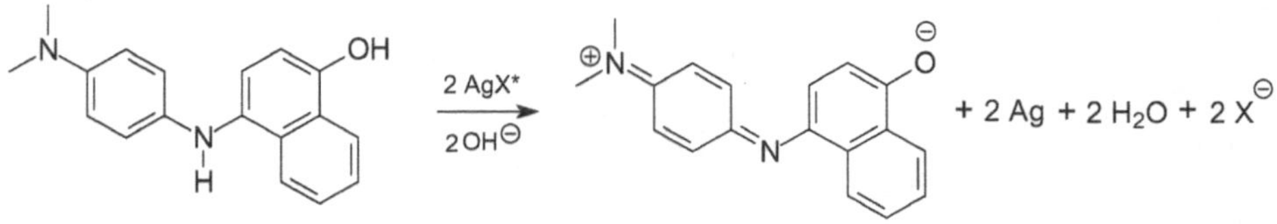

27 **10**

In the company Neue Photographische Gesellschaft (NPG) Rudolf Fischer and his coworker Hans Siegrist discovered that exposed silver halide AgX* also can act as an oxidizing agent for *N,N*-dialkyl-*p*-phenylenediamines. The exposure of silver halide crystals forms a few silver aggregates on their surface, called a "latent image". In an alkaline solution, containing a developer, the silver atoms of the latent image catalyze the reduction of the silver cations, whereby the developer is oxidized. Fischer and Siegrist discovered, that from *N,N*-dialkyl-*p*-phenylenediamines *N,N*-dialkyl-*p*-benzoquinone diimines are formed, which add a molecule of the coupler to a leuco dye, which is oxidized by two further silver cations to dyes. Depending on the coupler employed, one gets yellow, purple and cyan image dyes [20, 21].

With this invention they laid the chemical basis for subtractive silver halide color photography.

38.4 Commercial uses of indophenol and related dyes

The colors of indophenol dyes depend in part on the pH of their environment. In an acidic solution the red phenol form **20** predominates whereas in alkaline solution the blue phenolate **3** exists. The light stability of the indophenols is low. In the presence of acids and alkalis they hydrolyze. These properties limit technical applications. One application of indophenol dyes is in the analytical chemistry using the Berthelot [3] or similar reactions [22].

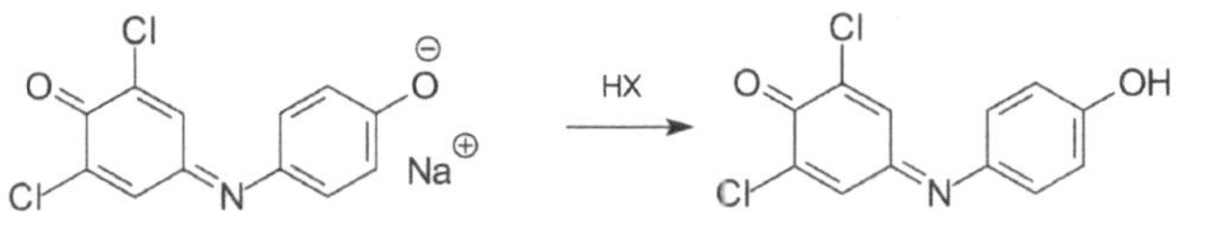

28 **29**

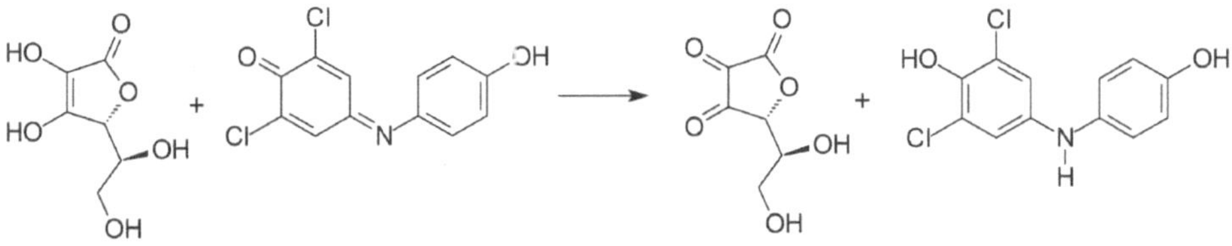

30 **29** **31** **32**

Another application is the quantitative determination of ascorbic acid **30** (Vitamin C) with *Tillman's Reagent* (2,6-dichloroindophenol sodium salt) **28**. This indophenol dye is used as a redox indicator dye. The measurement is carried out in the acidic range (pH 3.5–4) in order to keep the disturbance from oxidized secondary products of ascorbic acid low. At this pH the deep blue enolate **28** is protonated to the red phenol form **29** with λ_{max} = 518 nm. During this process ascorbic acid **30** is oxidized to **31** and the red dye **29** is reduced to the colorless leuco dye **32**. This allows the determination of the content of ascorbic acid in e.g. fruits and juices.

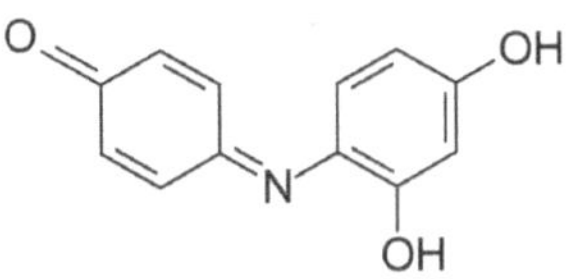

33

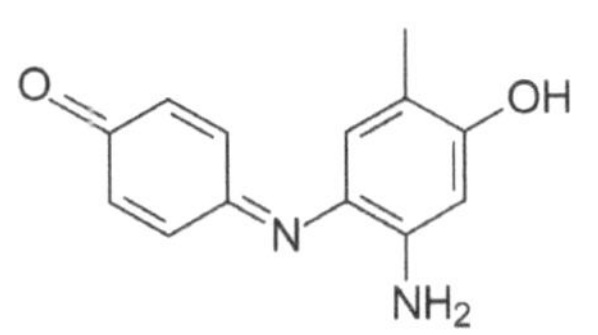

34

Oxidative dyeing is the most efficient process for the permanent coloration of hair. Of commercial use for over a century, its longevity is a consequence of the process's ability to partly remove the natural coloration of hair whilst simultaneously depositing colored species in a rapid and economic manner under physiologically tolerable conditions. In this technology, complex mixtures of indophenol, indaniline and/or indamine dyes are formed in situ within the hair fiber [23–26]. Primary intermediates are always oxidized in the presence of couplers to prevent formation of self-coupled products (i. e. reaction of oxidized primary intermediate with its unoxidized form to generate toxicologically undesirable products). As an important primary intermediate *p*-aminophenol is oxidized by hydrogen peroxide and coupled under alkaline conditions with substituted phenols to generate indophenols. For example, reaction with resorcinol initially forms the indophenol **33**, which can undergo further coupling to produce more drably colored products – the mixture furnishes light brownish coloration. When the coupler 5-amino-2-methylphenol is employed instead, its methyl group blocks further coupling, so that only the orange-red **34** indophenol dye results, giving brighter, more chromatic coloration. Some markets prefer use of 4-amino-3-methylphenol instead of *p*-aminophenol historically for toxicological reasons. Both are valued by industry as means of creating warm shades by producing indophenols as orange or reddish components of oxidative coloration effects. Of huge importance in hair coloration is the much-used combination of *p*-phenylenediamines and resorcinol, because in addition to generating a magenta indoaniline species, further coupling occurs to create oligomeric products that produce a valuable greenish-brown base coloration. Commercial oxidative hair dye products for home use are usually formulated with more than one primary intermediate and several couplers, so complex mixtures of indo dyes as well as larger colorant species (some oligomeric) are formed within the hair shaft. The choice of bases and couplers is strongly dictated by regulatory and safety considerations in many jurisdictions [26].

Bindschedler's Green **9** itself has no practical applications, because it hydrolyzes very easily. However, technically it is used as intermediate for the synthesis of 1,4-diazine, 1,4-oxazine and 1,4-thiazine dyes, where the two phenyl rings of **9** are bridged at the 2,2´-positions by –NR–, –O– and –S–, respectively.

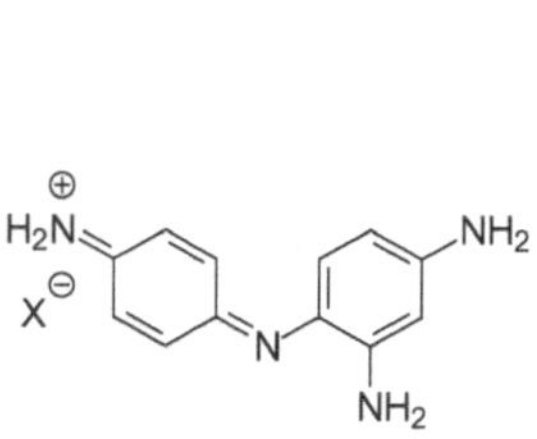

35

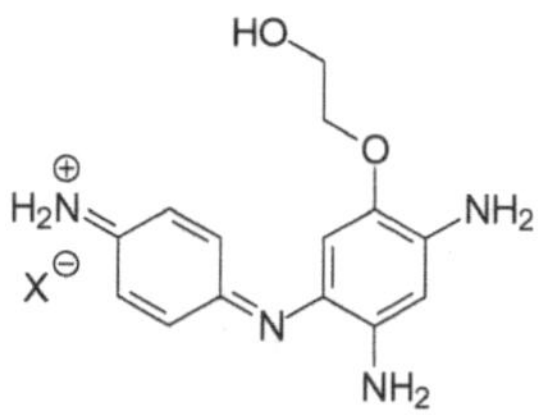

36

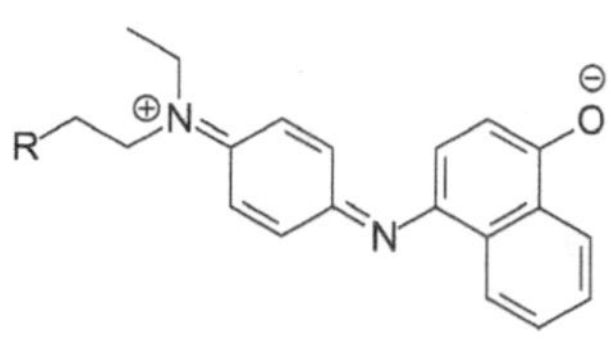

37

Some indamine dyes are very important blue and green products of oxidative coupling in hair coloration, useful in constructing neutral hues by balancing warm reddish species or setting up cool ash tones. However, the simplest version of this chemistry, not used in hair coloration, entails coupling of *p*-benzoquinone diimine with aniline to furnish the relatively unstable indamine dye **7**. Coupling with *m*-phenylenediamine gives the more stable blue-violet dye **35**, which may experience subsequent coupling to give bluish-black species. However, these dyes exhibit the tendency to undergo intramolecular cyclization to 2,7-diaminophenazines. This reaction can be inhibited by introducing an electron donating alkoxy substituent such as in **36**. Also of value in creating blue coloration is the well-known combination of a *p*-phenylenediamine and 1-naphthol that forms indophenol **37** (R = OH) which is an analogue of **10**. Indanilines are another important part of oxidative hair coloration. Coupling of *p*-phenylenediamines with *m*-aminophenol creates magenta indanilines which, along with the oligomeric products formed from further reaction, furnishes useful purplish-brown coloration: it serves as a valuable counter to the greenish-brown shades produced from resorcinol when targeting natural-looking browns. Use of formulations based on a *p*-phenylene-diamine, *m*-aminophenol and resorcinol is thus common.

The *N,N*-diethyl derivative of **10** (Indophenol Blue), i. e. **37** (R = H), is called Fettblau Z (C. I. Solvent Blue 22) and used today as cheap, oil-soluble blue dye. It is also the example 3 in Fischer's patent [27].

Based on their invention Fischer and Siegrist suggested that color couplers for producing yellow, purple and cyan dyes could be incorporated into the appropriate layers of a subtractive three-color multi-layer film system [28]. However, since all three layers were in direct contact with each other, their color couplers tended to diffuse between single layers during processing. So, only monochromatic photographic materials were possible and the NPG brought so-called monochrome *Chromalpapier* onto the market in 1913.

Accurate color reproduction requires that each image dye is formed exclusively within its proper layer in a multilayer system. After reading about Fischer's work with color couplers the professional musicians Leopold Mannes and Leopold Godowsky developed with support from Kodak a practical subtractive three-color multi-layer film system. This film went on sale under the brand name Kodachrome in 1935.

Like Fischer, they had the same difficulty in preventing the color couplers diffusing between the photographic layers. They solved this by putting the color couplers in the developer instead of in the single photographic layers. This process involved development of all three layers with the same coupler, followed by selective bleaching of the two layers, new development of the remaining two layers with the next coupler, selective bleaching of the third layer followed by the third color development. This process was extremely complex, required about 30 different stages that could only be carried out in laboratory conditions. For this reason, photographers had to send their exposure films back to Kodak for development [29, 30].

In the more general system the couplers are put into their proper layer during manufacturing of the color photographic film or paper. Incorporation of the couplers during the manufacture of the film or paper simplifies subsequent processing. Technically two types of couplers were used. Both types contain oleophilic groups which hinder their diffusion through gelatin.

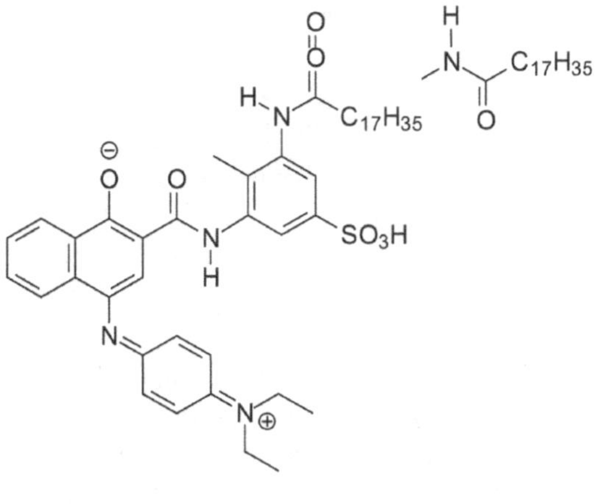

39

In the Agfa system the non-diffusing couplers, that produces dyes such as **38**, were substituted with water-solubilizing groups (COOH, SO_3H) to facilitate their direct introduction into the aqueous gelatin system where dyes such as **39** are formed [29–31].

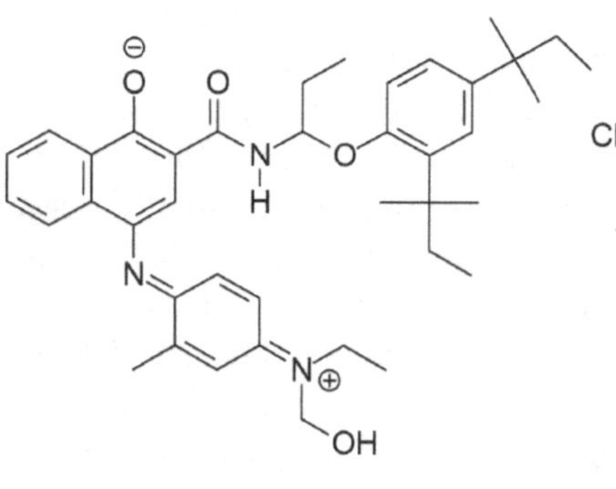

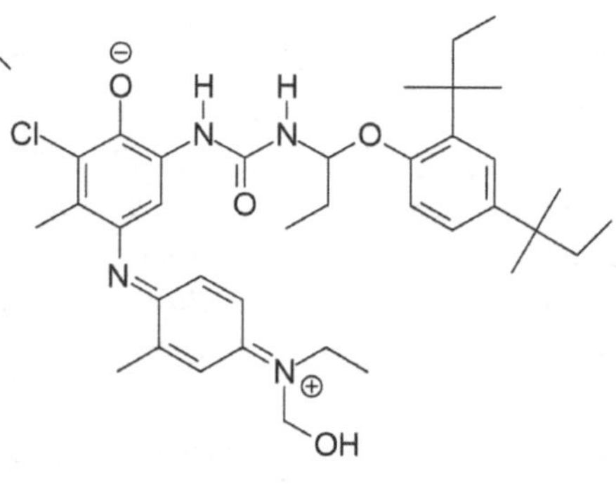

40 **41**

In the Kodak system solvent soluble couplers were dissolved in a high-boiling solvent and dispersed in the aqueous gelatin system. The couplers derived from 1-naphthol were used to develop cyan image dyes such as **40** in color photographic negative films, whereas the 2-acylamidophenols are used to develop cyan image dyes such as **41** in color photographic papers [29, 30].

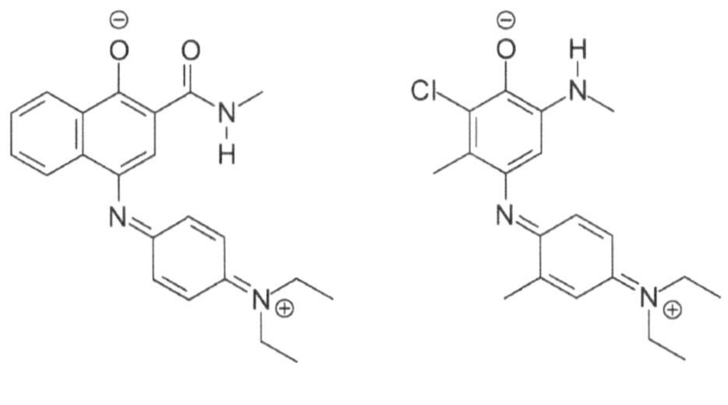

42 **43**

Thermal dye transfer systems are among the best way to obtain high quality color hard copies from digital stored images. Here, the image dye is transferred through a thermal induced dye transfer from a dye donor to a dye receiver layer. With the development of low-cost laser diodes the elegant laser induced dye transfer (LIDT), often called dye diffusion thermal transfer (D2T2) printing was developed [32, 33]. It is not surprising the leading photographic companies used their knowhow for the development of image dyes as illustrated with the two indaniline cyan dyes **42** and **43**.

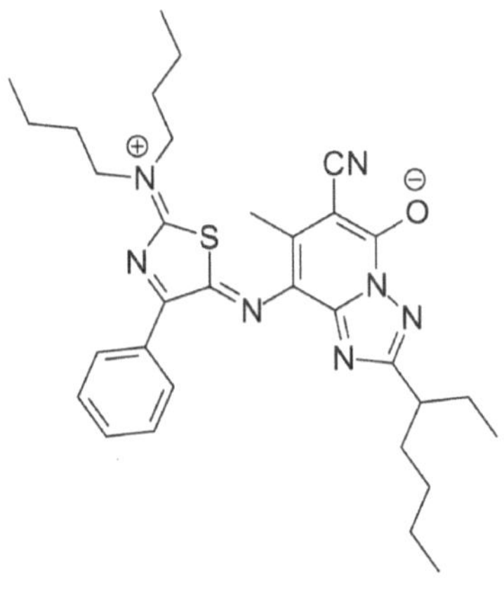

44

In the 1990s BASF developed and commercialized the novel blue disperse dye Dianix Cyanine B **44** [34]. It has the same absorption maximum as Foron Brilliant Blue S-R **45** [35, 36], but a substantially smaller absorption band width, conferring **44** with a more brilliant blue shade. Surprisingly, **44** is also claimed for the application as thermal transfer dye [37].

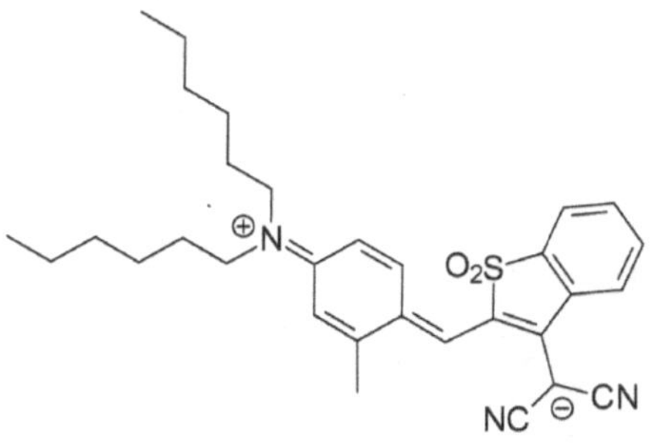

45

References

1. Hunger K, editor. Industrial dyes, chemistry, properties, applications. Weinheim: Wiley-VCH, 2003.
2. Towns A. Colorants: general survey. Phys Sci Rev. 2019;4. DOI:10.1515/psr-2019-0008.
3. Berthelot M. Violet d´aniline. Rép Chim Appl. 1859;1:284.
4. Nietzki R. Ueber die Farbstoffe der Safranin Reihe. Ber Dtsch Chem. 1883;16:464.
5. Witt ON. On the application of indophenol blue to calico printing and dyeing. J Soc Chem Ind. 1882;1:255.
6. Bindschedler R. Ueber gemeinschaftliche Oxydation aromatischer Diamine und Monoamine, ein Beitrag zur Kenntnis der Safranine. Ber Dtsch Chem Ges. 1883;16:864.
7. Griffiths J. Colour and constitution of organic molecules. London: Academic Press, 1976.
8. Tyutyulkov N, Fabian J, Mehlhorn A, Dietz F, Tadjer A. Polymethine dyes – structure and properties. Sofia: St. Kliment Ohridski University Press, 1991.
9. Mustroph H. Polymethine dyes. Phys Sci Rev. 2020;5. DOI:10.1515/psr-2019-0084.
10. Kim SH, Matsuoka M, Yodoshi T, Suga K, Kitao T. Synthesis and absorption spectra of some indonaphthol infra-red absorbing dyes. J Soc Dyers Colour. 1989;105:212.
11. Kubo Y, Mori F, Komatsu K, Yoshida K. Novel synthesis of naphthoquinone methide near-infrared dyes. J Chem Soc Perkin Trans I. 1988;2439.
12. Mustroph H. Streptomerocyanine dyes. Phys Sci Rev. 2021; in press. DOI:10.1515/psr-2020-0211.
13. Bello KA, Cheng L, Griffiths J. Near-infrared absorbing methine dyes based on dicyanonvinyl derivatives of indane-1,3-dione. J Chem Soc, Perkin Trans II. 1987;815.
14. Witt ON. On a new class of colouring matters. J Chem Soc Trans. 1879;35:356.
15. Koechlin H, Witt ON. DE 15 915. 19 Mar 1881.
16. Koechlin H, Witt ON. US 263 341. 29 Aug 1882.
17. Witt ON. Ueber die Indophenole. Ber Dtsch Chem Ges. 1884;17:76.
18. Möhlau R. Zur Kenntnis indophenolartiger Farbstoffe und der Indophenole. Ber Dtsch Chem. 1883;16:2843.
19. Nietzki R, Otto R. Zur Kenntnis der Indamine und Indophenole. Ber Dtsch Chem. 1888;21:1736.
20. Fischer R, Siegrist H. Über die Bildung von Farbstoffen durch Oxydation mittels belichteten Halogensilbers. Phot Korr. 1914;51:18.

21. Fischer R, Siegrist H. Über die Bildung von Farbstoffen durch Oxydation mittels belichteten Halogensilbers II. Phot Korr. 1914;51:208.
22. Searle PL. The Berthelot or indophenol reaction and its use in the analytical chemistry of nitrogen. Analyst. 1984;109:549.
23. Corbett JF. Hair colouring. Rev Prog Color. 1985;15:52.
24. Corbett JF. An historical review of the use of dye precursors in the formulation of commercial oxidation hair dyes. Dyes Pigm. 1999;41:127.
25. Morel OJ, Christie RM. Current trends in the chemistry of permanent hair dyeing. Chem Rev. 2011;111:2537.
26. Towns A. A review of developments in industrial hair colorant actives for oxidative dyes. Color Technol. 2021;137 in press. DOI:10.1111/cote.12545.
27. Fischer R. DE 253 335. 7 Feb 1912.
28. Neue Photographische Gesellschaft. DE 257 169. 14 Jun 1911.
29. James TH, editor. The theory of the photographic process. 4th ed. New York: Macmillan, 1977.
30. Keller K, Kampfer H, Matejec R, Lapp O, Krafft W, Frenken H. et al. Photography. In: Elvers B, editor. Ullmann's encyclopedia of industrial chemistry. Weinheim: Wiley-VCH, 2000.
31. Gill M, Mustroph H. Zum praktikablen Farbfilm. Chem Unserer Zeit. 2015;49:124.
32. Bradbury R. Thermal printing. In: Gregory P, editor. Chemistry and technology of printing and imaging systems. Glasgow: Blackie Academic, 1996.
33. Bradbury R. Dyes for dye diffusion thermal transfer (D2T2) printing. In: Freeman HS, Peters AT, editors. Colorants for non-textile applications. Amsterdam: Elsevier, 2000.
34. Lange A, Bach V, Senns R, et al. DE 4232557. 29 Sep 1992.
35. Baumann W. CH 636 116. 22 Dec 1978.
36. Baumann W. The world's most brilliant blue. A new dyestuff for polyester fibres. Foron brilliant blue S-R. Int Gazette (Sandoz). 1983;48:4.
37. Senns R, Siegel B, Herrmann M. DE 19 649 802. 2 Dec 1996.

Gerhard Pfaff

39 Inorganic luminescent pigments

Abstract: Inorganic luminescent pigments (luminescent materials, luminophores, phosphors) as synthetically generated crystalline compositions absorb energy followed by emission of light with lower energy, respectively, longer wavelengths. The light emission occurs often in the visible spectral range. External energy is necessary to enable luminescent materials to generate light. Luminescent pigments are divided into fluorescent and phosphorescent pigments. This classification goes back to different energy transitions. Emission based on allowed optical transitions, with decay times in the order of μs or faster is defined as fluorescence. Emission with longer decay times is called phosphorescence. The occurrence of fluorescence or phosphorescence as well as the decay time depend on structure and composition of a specific luminophore. There are four luminescence mechanisms discussed for inorganic luminescent materials: center luminescence, charge-transfer luminescence, donor–acceptor pair luminescence, and long-afterglow phosphorescence. The emission of luminescent light can have its origin in different excitation mechanisms such as optical excitation (UV radiation or even visible light), high-voltage or low-voltage electroluminescence and excitation with high energy particles (X-rays, γ-rays). Inorganic luminescent pigments are used mainly in fluorescent lamps, cathode-ray tubes, projection television tubes, plasma display panels, light-emitting diodes (LEDs) and for X-ray and γ-ray detection. The pigment particles are dispersed for the applications in specific binder systems. They are applied in form of thin layers and by means of luminophore/solvent suspensions, containing adhesive agents, on a substrate.

Keywords: fluorescence, luminescent pigments, luminophores, phosphorescence, phosphors

39.1 Fundamentals and properties

Inorganic luminescent pigments belong to the group of luminescent materials, which are also called phosphors or luminophores. They are synthetically manufactured crystalline compounds that absorb energy followed by emission of light with lower energy, respectively, longer wavelengths. The light emission of luminophores often occurs in the visible spectral range. The nature of luminescence therefore differs from black body radiation. External energy is necessary for luminescent materials to generate

This article has previously been published in the journal Physical Sciences Reviews. Please cite as G. Pfaff, Inorganic Luminescent Pigments *Physical Sciences Reviews* [Online] 2021, 6. DOI: 10.1515/psr-2020-0180

https://doi.org/10.1515/9783110587104-039

light. Luminescence can have its origin in different types of excitation such as photo- or electroluminescence caused by X-rays, cathode-rays, UV radiation or even visible light. Electronically excited states in atoms and molecules are the basis for luminescence processes. The emission process is based on quantum mechanical selection rules [1].

The distinction of luminescence in fluorescence and phosphorescence goes back to different energy transitions. Forbidden optical transitions are generally slower than allowed ones. Emission based on allowed optical transitions with decay times in the order of µs or faster is defined as fluorescence. Emission with longer decay times is called phosphorescence. The time in which the emission intensity decreases to 1/e or 1/10 is defined as the decay time. The occurrence of fluorescence or phosphorescence as well as the decay time depend on structure and composition of a specific luminophore [1].

Crystal lattices (host lattices) suitable for inorganic luminescent materials are based on various colorless compounds, typically on silicates, phosphates, sulfides, oxides, or halides predominantly of alkaline earth metals or zinc. Activators (activator ions such as Mn^{2+}, Ag^{+}, Cu^{+}) are incorporated as emission centers in the crystal structure, partially in combination with sensitizers (sensitizing ions such as Sb^{3+}, Pb^{2+}, Ce^{3+}) in small amounts of 10^{-4} to 10^{-2}%. The activators are located on cation sites (e.g. Ag^{+} or Cu^{+} on Zn^{2+} positions in a wurtzite lattice). The electrical neutrality in the case of the incorporation of Ag^{+} or Cu^{+} in the ZnS lattice is preserved by the additional incorporation of Cl^{-} or Al^{3+} [2].

Color and decay time of the emitted light of a luminescent material depend in particular on the activators (transition metals or rare earth metals in various oxidation states), the lattice geometries and on influences of the crystal field of the host lattice.

Luminophores are used in a large variety of devices, such as fluorescent lamps and tubes, light-emitting diodes (LEDs) lamps, displays, daylight paints, security features, safety signs, and modern medical equipment [3].

Organic luminescent materials have gained considerable interest besides inorganic luminophores. They find an application mainly in organic LEDs (TV screens, computer monitors, smartphones and others) where they are used in form of thin films [4].

39.2 Luminescence mechanisms

Four luminescence mechanisms are used to describe the effects of inorganic luminophores: center luminescence, charge-transfer luminescence, donor–acceptor pair luminescence, and long-afterglow phosphorescence.

Luminescence phenomena start with the absorption of energy by the luminophore. This absorption can take place either by the host lattice or by intentionally incorporated sensitizing ions. The excitation energy is transferred through the crystal lattice by an energy transfer process. The emission of light takes place at the activator ions. The emission color, which corresponds with the emitted wavelengths, can be adjusted usually by choosing the proper activator and sensitizing ions, without changing the host lattice in which the ions are incorporated.

The mechanism of center luminescence is characterized by emission, which is generated at an optical center. This mechanism is distinguished from emission resulting from optical transitions between host lattice band states. An ion or a molecular ion complex can act as a luminescence center. The mechanism of center luminescence leads to so-called characteristic luminescence, which is defined in such a way, that the emission could also occur at an ion in vacuum, i.e., when the optical transition involves electronic states of the ion only. Characteristic luminescence is characterized by relatively sharp emission bands (spectral width typically a few nm), but also of broad bands, which can have widths exceeding 50 nm. Broad emission bands are typically seen when the nature of the chemical bonding in ground- and excited state differs significantly. Broad emission bands are observed for many optical transitions in the partly filled d-energy level of transition metal ions (d → d transitions), for transitions between the 5d-level and the 4f-level of rare-earth ions (d → f transitions), and also for emission on s^2 ions (these ions possess a "lone pair" of s electrons), like Tl^+, Pb^{2+} or Sb^{3+}. Sharp emission bands are typical for optical transitions between electronic states having a chemical bonding character, which is nearly the same for the ground and the excited state. A comparable situation is found for optical transitions between electronic states that hardly participate in the chemical bonding (e.g., f → f transitions on rare-earth ions) [1].

The nature of the bonding (covalent, ionic) and the symmetry of the site at which the emitting ion is incorporated play a very important role for optical processes where electronic states are involved, which participate in the chemical bonding. The ligand field theory is suitable to understand and describe these processes. An example for a broad d → d emission band (in the green part of the spectrum) is the emission of Mn^{2+} in $BaMgAl_{10}O_{17}$:Eu^{2+},Mn^{2+}. The weak blue emission band originates from a d → f optical transition on Eu^{2+}. This luminescent composition can be applied as green phosphorescent material in very high quality fluorescent lamps and also in plasma display panels. An example for a d → d emission, consisting of a few relatively sharp bands, is the emission of Mn^{4+} in $Mg_4GeO_{5.5}F$:Mn^{4+}. The emitting species is again a manganese ion, but its charge (and therefore its electronic configuration) is different. This composition can be used as red phosphor primarily in fluorescent lamps producing deep red colors [1].

Some green Y_2O_2S based phosphors with Tb^{3+} as activator are suitable for application in fluorescent lamps. Y_2O_3:Eu^{3+} is the most important red phosphor with a line emission at about 611 nm. Width and position of the emission bands originating from optical transitions within the f-energy level are almost independent of the chemical environment. The relative intensity of some separate bands, however, depends on the structure of the crystal lattice. Transitions on many rare-earth ions are spin- and parity forbidden. They are therefore rather slow. For a number of rare-earth ions, however, also broad emission bands occur. The reason for this is the d → f emission, which is known from ions like Eu^{2+} or Ce^{3+}. Such transitions are allowed and therefore very fast. Consequently, several important industrial phosphors are based on rare-earth ions [1].

The mechanism of charge-transfer luminescence can be traced back to an optical transition that takes place between different kinds of orbitals or between electronic states of different ions. Width and position of the emission bands also depend in this case on the chemical environment. $CaWO_4$ is an example for charge-transfer luminescence. It shows luminescence originating from the WO_4 group. The transition is explained by charge transfer from the oxygen ions to empty d-levels of tungsten. The chemical bonding situation changes very strongly during the transition process leading to a very broad emission spectrum. There is no need for the incorporation of an activator in the $CaWO_4$ lattice because this material operates by self-activation [1].

The mechanism of donor–acceptor pair luminescence can be discussed in some semi-conducting materials, doped with both donors and acceptors. The band model is suitable for the understanding of this mechanism. Electrons from the valence band are excited and lifted energetically into the conduction band where they are captured by ionized donors. The resulting electron holes in the valence band are caught by ionized acceptors. The emission comprises an electron transfer between neutral donors and neutral acceptors resulting in ionized donors and acceptors. The final state is Coulomb stabilized. The wavelength of the emitted light, respectively, the luminescence color depends on the distance between the donor and the acceptor in a donor–acceptor pair. The shorter this distance is, the higher is the energy of the photon generated. In the crystal lattice of a luminophore, whose effect is based on the mechanism of donor–acceptor pair luminescence, many different donor–acceptor distances are possible. This leads to a relatively broad emission band. Donor–acceptor pair luminescence occurs in the blue and green emitting luminophores $ZnS:Ag^+,Cl^-$, $ZnS:Ag^+,Al^{3+}$, $ZnS:Cu^+,Au^+,Al^{3+}$, and $ZnS:Cu^+,Al^{3+}$. Luminescent materials of this type are used amongst others in color television picture tubes [1].

The mechanism of long-afterglow phosphorescence is based on the phenomenon that optical excitation energy is stored in the crystal lattice by trapping of photo excited charge carriers. The composition $SrAl_2O_4:Eu^{2+},Dy^{3+}$ is an example for this mechanism. Eu^{2+} is oxidized to Eu^{3+} and Dy^{3+} is reduced to Dy^{2+} by optical excitation. Thermal excitation leads to the reoxidation of Dy^{2+} to Dy^{3+}. The resulting free electron is captured by Eu^{3+} and Eu^{2+} is formed again. This reduction step is accompanied by the delayed emission of light. The time delay for the emission is determined by the thermal excitation process for Dy^{2+}. Emission is generated by this material, which is still visible several hours after the excitation. Luminophores based on the mechanism of long-afterglow phosphorescence find application in exit signs for escape routes, watches and safety features. Examples for long-afterglow materials are the compositions $ZnS:Cu^+$ and $SrS:Bi^{3+}$ [1].

39.3 Excitation mechanisms

Three excitation mechanisms are discussed for inorganic luminescent materials: optical excitation of luminescence, electroluminescence, and excitation with high energy particles.

Optical excitation of luminescence occurs when the absorption of UV or visible light leads to the emission of light of a longer wavelength in the visible part of the electromagnetic spectrum. Examples for materials using the mechanism of optical excitation and containing a sensor-activator pair are $BaMgAl_{10}O_{17}:Eu^{2+}$ (blue emitting), $BaMgAl_{10}O_{17}:Eu^{2+},Mn^{2+}$ (green emitting), and $Ca_5(PO_4)_3(F,Cl):Sb^{3+},Mn^{2+}$ (white emitting as a result of orange emission by Mn^{2+} and blue emission by Sb^{3+}). A further technically relevant sensitizer-activator pair is Ce^{3+},Tb^{3+}, used for example in $GdMgB_5O_{10}:Ce^{3+},Tb^{3+}$. All green emitting luminophores in high quality fluorescent lamps contain this combination incorporated in a suitable crystal lattice [1]. Optical excitation is applied not only in fluorescent lamps but also in phosphor converted LEDs in which phosphors are used to change the wavelength of the radiation emitted by the LED. Optical absorption can take place on activator ions or sensitizer ions, the so-called optical centers. Sensitizer ions are preferably used when the optical absorption of the activator ions is too weak to be suitable for practical applications, e.g., because the optical transition is forbidden. Energy transfer from the sensitizer ions to the activator ions is necessary in this case. The optical absorption may also take place by the host lattice itself (band absorption based on host lattice sensitization). Energy transfer from host lattice states to the activator ions is necessary.

Electroluminescence as an excitation mechanism for luminescent materials is divided into high-voltage electroluminescence and low-voltage electroluminescence.

High-voltage electroluminescence is attributed to an electrical breakthrough in a semiconducting material. $ZnS:Mn^{2+}$, $ZnS:Cu^{+}$, and $SrS:Ce^{3+}$ are suitable luminescent materials for high-voltage electroluminescence [1]. Typical voltages used are in the order of 100 V. Generation and acceleration of charge carriers take place in the host lattice. The charge carriers are able to excite activator ions necessary for the generation of luminescence. Electroluminescent devices based on this mechanism have typically a long lifetime. Their efficiency is, however, only in the order of one percent. Consequently, devices working on the basis of high-voltage luminescence are only used in applications where reliability is of high importance and efficiency plays only a minor role, e.g., for emergency and exit signs.

Low-voltage electroluminescence is of high technical importance for the blue LED and also for organic electroluminescent devices. Efficient light-emitting structures using this mechanism are possible, which do not require either high or low pressure. Low-voltage electroluminescent devices may also be a solution to overcome the cascade-based technique, which is used in conventional luminescent devices. In fluorescent lamps, for instance, a discharge is generated initially, and the resulting invisible radiation is converted into visible light. A significant energy loss has to be taken into account here. In cathode-ray tubes, as another example, an electron beam is generated first, which consists of electrons with high kinetic energy. These electrons subsequently impact on the luminescent material, what finally leads to excitations where electrons in the conduction band are coupled to holes in the valence band (excitons). Excitons generated in this way are transferred to activator ions. As a result, the energy efficiency of the luminophores is limited to about 20% and of white light emitting devices to less than 50% [5].

In low-voltage electroluminescent devices, on the other hand, the step leading to emission is the recombination of electrons in the conduction band with holes in the valence band. Only the band gap energy is required in this case to excite the luminescence. The color of the emitted light can be adjusted to a certain extent by selecting the appropriate semi-conductor. In summary, the generation of low-voltage electroluminescence may be very energy efficient [1].

Luminescent materials for LEDs have to meet strict requirements. The Stokes Shift needs to be small and the absorption has to be high. In addition, the materials must operate efficiently also at high temperatures, have to be stable against radiation and should not show a less than linear increase in output power with input power at high excitation densities (saturation) [1].

Excitation with high energy particles as a further mechanism is based on the effect that after absorption of electrons or high-energy photons by the luminescent material electron-hole pairs are generated in the crystal lattice. The electron-hole pairs adjust their speed to the surrounding by collisions (thermalization), whereby band gap excitations can take place. The excitation is transmitted to an activator or sensitizer after thermalization. As a result, light emission can be observed. A large number of electron-hole pairs is generated for each absorbed electron or high-energy phonon. Each electron-hole pair is able to initiate the emission of one photon on an activator ion [1].

39.4 Production of luminescent pigments

The synthesis of luminescent materials, respectively, pigments proceeds usually by mixing and firing processes. Homogeneous mixtures of the starting materials are treated at temperatures in the range from 1000 to 1500 °C in oxidizing or reducing atmosphere. The powdery reactants are thoroughly mixed and ground in the dry or wet state followed by calcination. Important for the decision on the calcination conditions are the activator characteristics and the crystal lattice of the luminophore to be formed. Often, a second or even a third calcination step with intermediate cautious comminution is necessary to achieve a homogeneous luminescent material. Too intensive grinding can lead to undesired lattice defects. Additional thermal treatment for curing of such effects is necessary in such cases [6].

Adding of flux agents or melting salts can facilitate the production process and the development of the desired crystallinity of the luminescent pigments formed. Lower reaction temperatures and an optimization of the grain size of the luminophores are possible. The flux agents act by dissolving at least one of the reactants. Two types of fluxes are used for the manufacture of luminescent pigments, a nonvolatile liquid, mostly a molten salt, or a volatile liquid.

Melting salts used for the synthesis of luminescent pigments do not react with the starting materials and maintain their composition during the formation of the pigments. Frequently used melting salts are $Na_2B_4O_7$, $Na_4P_2O_7$, Na_2SiO_3, and Na_2MoO_4.

Flux agents, on the other hand, often react with the starting materials. They are used typically in amounts of less than 10% by weight of the luminophore and undergo decomposition or evaporation during the formation of the luminescent material. Suitable substances acting as flux agents are AlF_3, NaF, and NH_4Cl.

Carbonates, nitrates, or hydroxides are suitable starting materials for the synthesis of luminophores. They decompose during the calcination and can improve the reaction speed. Impurities of the starting materials should be on a very low level and clearly below the concentration of the activators. The problem is that impurities can act as discharge centers. The result can be the reduction of the luminous efficiency of the luminophores. Extremely pure starting materials are formed by precipitation reactions. Co-precipitation can be used for certain compositions, whereby a reaction mixture is obtained, which contains the starting materials in an intimately mixed form.

The manufacture of a luminescent pigment can be explained well using the synthesis of $Y_2O_3{:}Eu^{3+}$. The formation of this important luminophore takes place in most cases via the precipitation of mixed oxalates starting from purified $Y(NO_3)_3$ and $Eu(NO_3)_3$ solutions. The filtered, washed and at 600 °C pretreated precipitates are subsequently calcined for several hours at 1440 to 1500 °C. The required crystal structure is formed in this temperature range [7]. This method can be used whenever suitable precipitates for each of the reaction partners exist.

Spray drying is another process to obtain intimate reaction mixtures. Small water droplets containing the reactants are transported in a gas stream and heated up. The water is evaporated very fast during heating and an optimum reaction mixture for the calcination is obtained.

Some luminescent pigments can be synthesized using the polysulfide flux process. This method is suitable for the compositions $Y_2O_2S{:}Eu^{3+}$ and $Gd_2O_2S{:}Pr^{3+}$ [8]. The oxides of the metals are mixed in this case with excess sulfur and an alkali metal carbonate. The carbonate reacts during heating-up with sulfur to form a liquid polysulfide flux. This flux undergoes a reaction with the metal oxides and oxysulfides are formed. Residues of the flux can be removed easily by washing the reaction product with water.

In cases where the amounts of activator ions are very low, special routes have to be used. This applies for example to specific ZnS based luminescent pigments. The amounts of activator ions in such compositions can be in the order of only 100 ppm. Tiniest amounts of these ions can be precipitated on ZnS particles by preparing a suspension of ZnS in water followed by adding a solution containing a solved activator salt. The precipitation of a composition containing the activator ions on the ZnS particles is done by adding a suitable precipitating agent, e.g. $(NH_4)_2S$.

Some pigment compositions need a reducing atmosphere to incorporate activator ions in a nonpreferred low oxidation state, e.g., Eu^{2+}, or to prevent oxidation of the host lattice, e.g., during the synthesis of ZnS based pigments. Reducing gases used in such cases are hydrogen in form of hydrogen/nitrogen mixtures or carbon monoxide [1].

39.5 Pigment properties and uses

Compositions and application fields of selected inorganic luminophores are summarized in Table 39.1 together with information on activator ions, emission wavelengths, and colors. Luminescent pigments are applied in form of thin layers and by means of luminophore/solvent suspensions, containing adhesive agents, on a substrate [2, 6].

Table 39.1: Composition and application fields of selected luminophores [6].

Activator	Composition of the luminophore	Emission [nm]	Color	Application field
Mn^{2+}	Zinc orthosilicate Zn_2SiO_4	525	Green	Fluorescent lamps, projection television tubes, plasma display panels
Mn^{2+}/Sb^{3+}	Calcium fluoro chloro apatite $Ca_5(PO_4)_3(Cl,F)$	480 and 580	White (blue and yellow)	Fluorescent lamps
Mn^{4+}	Magnesium fluoro germanate $Mg_2GeO_4 \cdot ½\ MgO \cdot 0{,}5\ MgF_2$	710	Red	Mercury high-pressure lamps
Mn^{4+}	Magnesium arsenate $Mg_6As_2O_{11}$	700	Red	Fluorescent lamps
Sb^{3+}	(Sr,Ca) chloro bromo phosphate $(Sr,Ca)_5(PO_4)_3(Cl,F)$	480–500	Blue–green	Fluorescent lamps
Sn^{2+}	(Sr,Mg) orthophosphate $(Sr,Mg)_3(PO_4)_2$	630	Pink	Fluorescent lamps, mercury high-pressure lamps
Pb^{2+}/Ce^{3+}	Strontium thiogallate $SrGa_2S_4$	614	Orange–red	Flying-spot scanners
Ce^{3+}	Calcium aluminum silicate $Ca_2Al_2SiO_7$ Yttrium aluminate $Y_3Al_5O_{12}$	405 550	Blue Yellow	Flying-spot scanners, fluorescent lamps

Table 39.1 (continued)

Activator	Composition of the luminophore	Emission [nm]	Color	Application field
Eu^{2+}	Barium fluoride bromide Ba(F,Br)	440	Blue	X-ray detection
Eu^{3+}	Yttrium oxide Y_2O_3	625	Red	Fluorescent lamps, cathode-ray tubes, projection television tubes, plasma display panels
Eu^{3+}	Yttrium vanadate YVO_4	630	Red	Plasma display panels
Tb^{3+}	Yttrium oxysulfide Y_2O_2S	525	Green	Cathode-ray tubes, projection television tubes
Ag^{+}/Cl^{-}	zinc sulfide ZnS	440	Blue	Cathode-ray tubes, projection television tubes
Cu^{+}/Al^{3+}	Zinc sulfide ZnS	525	Green	Cathode-ray tubes, projection television tubes
Mn^{2+}	Zinc sulfide ZnS	585	Yellow	Radar tubes
Zn^{2+} (excess)	zinc oxide ZnO	505	Green	Flying-spot scanners
without	Ca,Mg tungstate $CaWO_4$ $MgWO_4$	415 480	Blue–Violet Blue–green	Fluorescent lamps

Typical particle sizes of luminophores are in the range from 0.005 to 0.5 µm. The thickness of a pigment containing layer on the inside of a television screen is in the order of 10–30 µm. Luminophores can be applied in all common pigment-relevant systems but also in very special formulations used only for this type of materials.

Blue and red emitting cathode ray tube phosphors (CRT phosphors) are often coated with a daylight absorbing pigment in order to enhance the contrast in daylight viewing conditions. Blue emitting phosphors are therefore coated with $CoAl_2O_4$ and red emitting phosphors with α-Fe_2O_3. Coating of green emitting phosphors with a green absorbing pigment is not necessary because of the high sensitivity of the human eye for green light. Liquid crystal displays (LCDs) used in television, computer monitors, notebooks, digital cameras, and mobile telephones also need luminescent materials. All these devices require a backlight, which works either with thin or flat fluorescent lamps or with LEDs [1].

Fluorescent lamps work with mixtures of luminescent materials with different emission wavelengths to achieve a special color (tricolor phosphor mixtures). Even white light generating lamps with higher brightness than that of metal halide lamps are possible based on this mixing principle. The resulting emission color is adapted to the sensitivity of the human eye. The problem of fluorescent lamps is the need of mercury for their function. This environmental disadvantage can be reduced by lowering the mercury content in fluorescent lamps. Mercury-free fluorescent lamps, based on a Xe/Ne-discharge may be an alternative here, but the efficiency of the lighting device is lower in this case. The use of vacuum UV radiation, however, could increase the efficiency of such mercury-free devices to the level of currently applied fluorescent lamps.

Some specific luminescent materials are suitable for X-ray and γ-ray detection. This phenomenon is utilized for computed tomography (CT) and position emission tomography (PET). X-rays in the case of CT or γ-rays in the case of PET are converted into visible light by these luminophores. Photo multipliers or photodiodes are used for the detection of the so generated visible light.

Luminescent materials used for X-ray detection can also be used in X-ray intensifying screens. In this case, the X-rays are converted into visible photons, which subsequently irradiate a photographic film. This film is mostly sandwiched between two luminophore sheets. Typical thicknesses of such luminescent layers are in the order of a few hundred micrometers [1].

LEDs are of highest interest for luminescent materials. LED lamps generating white light with a low color temperature are of special interest. One way to produce such lamps is the combination of red, green, and blue LEDs in one LED device housing (RGB principle). A disadvantage of this approach is a color shift at higher temperatures and an aging of the LED. A more advantageous way is the use of the so-called phosphor principle. In this case, a white LED can be produced by the combination of a blue emitting LED with one or more phosphors. An example for the use of only one phosphor is the combination of the blue LED with the yellow emitting $(Y,Gd)_3Al_5O_{12}:Ce^{3+}$. Alternatively, the blue emitting LED can be combined with a green (e.g., $CeMgAl_{11}O_{19}:Tb^{3+}$) and a red (e.g., $Y_2O_3:Eu^{3+}$) luminophore. Some nitridosilicate phases such as $LaSi_3N_5:Eu^{2+}$ (emission in the blue-green region) and $Ba_2Si_5N_8:Eu^{2+}$ (emission in the yellow-orange region) are interesting for the application as conversion phosphors in LEDs, too [5, 9].

Luminescent materials operate in most of their applications at physical limits. This refers to the absorption of the exciting radiation and the quantum efficiency (number of visible photons generated divided by the number of photons absorbed) with which luminescence is generated. The quantum efficiency in plasma display panels, fluorescent lamps, and LEDs is at almost 100%. The energy efficiency of the phosphors in cathode-ray tubes is in the order of 20% [5]. In contrast, the average energy efficiency of luminescent devices is rather low (plasma display panels 1–2%, fluorescent lamps 15–25%, phosphor converted LED lamps 20%, cathode-ray tubes 1–2%) [1].

Cascade phosphors and quantum dots are two further groups of materials with possible potential for broader applications in solid-state lighting (SSL) and other uses.

Cascade phosphors are luminescent materials characterized by more than one absorbed or emitted photon. A well-selected combination of energy levels, normally f-levels of lanthanide ions, is necessary to establish a cascade transition. $CaWO_4:Yb^{3+},Er^{3+}$ is an example for cascade phosphors, where the absorption of two photons with smaller energy (preferably in the infrared range of the spectrum) is succeeded by the emission of one photon with higher energy (in the visible range of the spectrum) [10]. Such behavior is called wavelength up-conversion. It is known for the summation of up to five photons. The efficiency of such cascade processes is with <% very low. Applications are therefore limited to areas where energy efficiency is not important.

The so-called wavelength down-conversion approach is the opposite effect to the up-conversion for SSL. It uses a down-conversion material to generate visible light when excited by near UV radiation or blue emission from InGaN LEDs [11]. A suitable material for down-conversion processes is $YF_3:Pr^{3+}$ [12, 13]. Two photons with lower energy are emitted per photon absorbed at a wavelength below 220 nm. Down-conversion phosphors are necessary to enable energy efficient mercury-free lamps using Xe/Ne discharge.

Quantum dots act on the basis of size quantization effects, which occur in semiconductors (e.g. InGaAs, CdSe, GaInP, or InP) with radii typically less than 10 nm [14–16]. The spectral position of the emitted light can be shifted over a large range including the complete visible spectral range. The adjustment of the emission is possible just by varying the particle size. Tunable narrow band emission can be realized based on this size variation. Quantum dots with diameters of 2 to 10 nm are small enough to enable transparent nonscattering luminescent layers. Based on this, electroluminescent displays and flexible light sources are possible to produce. Quantum dots may also be used as luminescent markers for optical imaging in medical devices [17–19].

Anti-counterfeiting of valuable documents, currency and branded products is a further application field for luminescent materials. Security ink formulations based on lanthanide doped luminescent nanomaterials or quantum dots are examples for the use of luminescent materials in anti-counterfeiting applications [20, 21].

Based on the criterion of composition, some luminescent pigments can be regarded as harmless, others as potentially hazardous. From the occupational health and safety perspective, especially the operation with pigment dust requires attention. The inhalation of the dust is most critical and must therefore be avoided by all available means.

Special care is necessary when the pigments contain larger amounts of zinc, cadmium, barium, or strontium. However, there is only a little risk during application due to the low solubility of most luminescent materials. Pigments with particle sizes in the nanometer range have to be handled very carefully. Measures must be taken here to minimize the exposure to these materials for employees affected.

References

1. Ronda C. In industrial inorganic pigments. Buxbaum G, Pfaff G, editors. 3rd ed. Weinheim: Wiley-VCH Verlag, 2005:269.
2. Pfaff G. In Winnacker-Küchler: Chemische Technik, Vol. 7 Industrieprodukte. Dittmeyer R, Keim W, Kreysa G, Oberholz A, editors. 5th ed. Weinheim: Wiley-VCH, 2004:373.
3. Feldmann C, Jüstel T, Ronda CR, Schmidt P. Inorganic luminescent materials: 100 years of research and application. J Adv Funct Mat. 2003;13:511.
4. Kalyani NT, Dhoble SJ. Organic light emitting diodes: energy saving lighting technology – a review. Renew Sust Energ Rev. 2012;16:2696.
5. Robbins DJ. On predicting the maximum efficiency of phosphor systems excited by ionizing radiation. J Electrochem Soc. 1980;127:2694.
6. Pfaff G. Inorganic pigments. Berlin/Boston: Walter de Gruyter GmbH, 2017:274.
7. Kotera Y, Higashi T, Sugai M, Ueno A. The effect of La substitution in Eu^{3+}-activated Y_2O_3 phosphors on their luminescent property. J Lumin. 1984;31:709.
8. Patent US. 3,418,246 (M. R. Royce). 1968.
9. Uheda K, Takizawa H, Endo T, Yamane H, Shimada M, Wang C-M, Mitomo M. Synthesis and luminescent property of Eu^{3+}-doped $LaSi_3N_5$ phosphor. J Lumin. 2000;87–89:967.
10. Auzel F. Advances in nonradiative processes in solids. NATO ASI Series 1991;249:135.
11. McKittrick J, Shea-Rohwer LE. Review: down conversion materials for solid-state lighting. J Am Ceram Soc. 2014;97:1327.
12. Sommerdijk JL, Bril A, de Jager AW. Two photon luminescence with ultraviolet excitation of trivalent praseodymium. J Lumin. 1974;8:341.
13. Piper WW, de Luca JA, Ham FD. Cascade fluorescent decay in Pr^{3+}-doped fluorides: achievement of a quantum yield greater than unity for emission of visible light. J Lumin. 1974;8:344.
14. Colvin VL, Schlamp MC, Alivisatos AP. Light-emitting diodes made from cadmium selenide nanocrystals and a semiconducting polymer. Nature. 1994;370:354.
15. Empedocles SA, Neuhauser R, Shimizu K, Bawendi MG. Photoluminescence from single semiconductor nanostructures. Adv Mater. 1999;11:1243.
16. Talapin DV, Rogach AL, Kornowski A, Haase M, Weller H. High luminescent monodisperse CdSe and CdSe/ZnS nanocrystals synthesized in a hexadecylamine-trioctylphosphine oxide-trioctylphosphine mixture. Nano Lett. 2001;4:207.
17. Bera D, Qian L, Tseng T-K, Holloway PH. Quantum dots and their multimodal applications: a review. Materials. 2010;3:2260.
18. Shipway AN, Katz E, Willner I. Nanoparticle arrays on surfaces for electronic, optical, and sensor applications. Chem Phys Chem. 2000;1:19.
19. Dujardin E, Mann S. Bio-inspired materials chemistry. Adv Mater. 2002;14:775.
20. Andres J, Hersch RD, Moser JE, Chauvin AS. A new anti-counterfeiting feature relying on invisible luminescent full color images printed with lanthanide-based inks. Adv Funct Mater. 2014;24:5029.
21. Kumar P, Singh S, Gupta BK. Future prospects of luminescent nanomaterial based security inks: from synthesis to anti-counterfeiting applications. Nanoscale. 2016;8:14297.

Gerhard Pfaff

40 Iron blue pigments

Abstract: Iron blue pigments are representatives of the inorganic blue pigments. They are characterized by a sufficient hue, high tinting strength, and suitable rheological properties. Iron blue pigments are subdivided into "soluble iron blue" and "insoluble iron blue". Both pigments are synthesized using precipitation techniques. Iron blue pigments are mainly used in printing applications. Special uses of the pigments are agriculture and medicine.

Keywords: iron blue pigments, "insoluble iron blue", "soluble iron blue"

40.1 Fundamentals and properties

Iron blue, also known as Paris blue, Prussian blue, Berlin blue, Milori blue, or Turnbull's blue is the name for blue microcrystalline Fe(II)Fe(III) cyano complexes. Specific qualities of these complexes are used as inorganic blue pigments [1–4].

Iron blue with the nomenclature C.I. Pigment Blue 27:77510 (so-called "insoluble iron blue") was synthesized for the first time in 1704 by the chemist Diesbach in Berlin using a precipitation reaction. The obtained precipitate can be regarded as the oldest synthetic coordination compound. Milori was the first to produce this coordination compound as a pigment on an industrial scale in the early nineteenth century [5]. Another iron blue pigment is the so-called "soluble iron blue" with the nomenclature C.I. Pigment Blue 27:77520.

X-ray and infrared spectroscopic investigations have shown that "soluble" iron blue pigments can be described by the formula $M^I[Fe^{II}Fe^{III}(CN)_6] \cdot x\ H_2O$ [6]. M^I stands for sodium, potassium or ammonium ions. Potassium and ammonium ions are used mostly in industrial manufacturing processes for iron blue pigments because they produce excellent hues [3].

The name "soluble iron blue" for $M^I[Fe^{II}Fe^{III}(CN)_6] \cdot x\ H_2O$ goes back to colloidal particles, which are colloidally solved after the synthesis. Excessive Fe^{2+} and Fe^{3+} ions lead to the precipitation of "insoluble iron blue". The composition of this precipitate is reflected in the formula $Fe^{III}_4[Fe^{II}(CN)_6]_3 \cdot x\ H_2O$ or $Fe^{III}[Fe^{II}Fe^{III}(CN)_6]_3 \cdot x\ H_2O$ (x = 14–16).

This article has previously been published in the journal Physical Sciences Reviews. Please cite as G. Pfaff, Iron Blue Pigments *Physical Sciences Reviews* [Online] 2021, 6. DOI: 10.1515/psr-2020-0178

https://doi.org/10.1515/9783110587104-040

Figure 40.1 shows the $Fe^{II}Fe^{III}(CN)_6$ lattice as a part of the crystal structure of M^I $[Fe^{II}Fe^{III}(CN)_6] \cdot x\ H_2O$ [7]. Fe^{2+} ions (Fe^{II}) form a face-centered cubic lattice, which is interlocked with another face-centered cubic lattice of Fe^{3+} ions (Fe^{III}). A three-dimensional cubic lattice is formed with the corners and face centers occupied by Fe^{2+} ions. The CN groups are located at the edges of the cubes, each between a Fe^{2+} and the neighboring Fe^{3+} ion. The carbon atom of the CN group is bonded to the Fe^{2+} ion whereas the nitrogen atom is linked to the Fe^{3+} ion. The M^+ ions (M^I) and the water molecules have their positions inside the cubes formed by the iron ions. The intensive color of iron blue is explained by a charge-transfer of electrons between Fe^{2+} and Fe^{3+} ions.

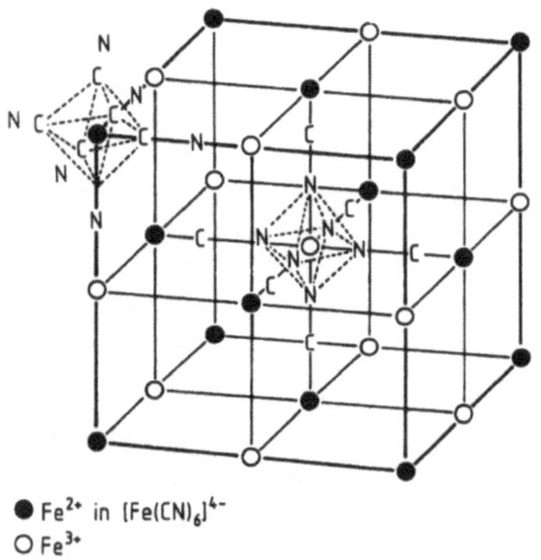

Figure 40.1: Section of the crystal lattice of iron blue [7].

The crystal lattice of $Fe^{III}{}_4[Fe^{II}(CN)_6]_3 \cdot x\ H_2O$ consists also of a three-dimensional cubic Fe^{3+}–NC – Fe^{2+} framework. One fourth of the Fe^{2+} sites (and cyanide positions as well) are vacant. Water molecules at empty nitrogen positions complete the coordination sphere of the Fe^{3+} ions in the crystal structure [8].

Coordinative water as expressed in the formula $M^I[Fe^{II}Fe^{III}(CN)_6] \cdot x\ H_2O$ is absolutely necessary for the stabilization of the crystal lattice. Iron blue loses its pigment properties when this water is removed.

Further investigations on the structure of iron blue pigments used analytical methods like neutron diffraction and X-ray photoelectron spectroscopy [9–11].

40.2 Production of iron blue pigments

Iron blue pigments ("insoluble iron blue") are synthesized by a precipitation process in aqueous solution starting from potassium or sodium hexacyanoferrate(II) and iron(II) sulfate. A white iron(II) hexacyanoferrate(II) precipitate ($M^I_2Fe^{II}[Fe^{II}(CN)_6]$, Berlin white) is formed.

$$K_4\left[Fe^{II}(CN)_6\right] + FeSO_4 \rightarrow K_2Fe^{II}\left[Fe^{II}(CN)_6\right] + K_2SO_4 \tag{40.1}$$

The use of iron(II) chloride instead of iron(II) sulfate is possible. The precipitate is separated from the soluble reaction products by decantation. The precipitate is aged in a next step by heating up the aqueous suspension followed by oxidation with alkali chlorate, hydrogen peroxide, or alkali dichromates in the presence of hydrochloric acid. A blue pigment of the composition $Fe^{III}_4[Fe^{II}(CN)_6]_3$ is formed under these conditions.

When sodium hexacyanoferrate(II) solution is used for the process, the pigment properties are adjusted by adding a potassium or ammonium salt during the precipitation of Berlin white or prior to the oxidation reaction.

Industrial performing of the precipitation is done batch-wise in stirred tanks by simultaneous or sequential addition of aqueous solutions of alkali hexacyanoferrate(II) and the iron(II) compound to a diluted acid. Temperature and concentration of the starting solutions have a significant influence on size and shape of the precipitated particles. The ageing time for the white precipitate varies in duration and temperature depending on the required properties of the finished pigment. The suspension of the blue pigment obtained during the oxidation step is pumped onto filter presses. The resulting filter cake is washed with water and dried using tunnel or belt dryers. Spray- or spin-flash dryers are likewise possible for the drying step. The dried solid is ground to achieve the desired product quality. The pigment can be supplied as a powder, but also in form of rods or pellets, which are obtained by extruding the washed filter cake with a granulator. Iron blue pigments treated in this way are dust-free after drying [3].

Iron blue pigments can be improved with respect to their dispersion behavior by adding organic compounds to the pigment suspension before filtering to prevent the particles from agglomeration during the drying step [3].

"Soluble iron blue" pigments are synthesized using the reaction of potassium hexacyanoferrate(II) with an iron(III) salt or by the reaction of potassium hexacyanoferrate(III) with an iron(II) salt in aqueous solution.

$$K_4\left[Fe^{II}(CN)_6\right] + Fe^{3+} \rightarrow K\left[Fe^{II}Fe^{III}(CN)_6\right] + 3\ K^+ \tag{40.2}$$

$$K_3\left[Fe^{III}(CN)_6\right] + Fe^{2+} \rightarrow K\left[Fe^{II}Fe^{III}(CN)_6\right] + 2\ K^+ \tag{40.3}$$

In both cases, colloidal iron blue is formed when the mole ratio 1:1 is used. The formation of the insoluble $Fe^{III}_4[Fe^{II}(CN)_6]_3$ out of $K[Fe^{II}Fe^{III}(CN)_6]$ by addition of excessive Fe^{2+} and Fe^{3+} ions is basically possible, but not favorable for an industrial process.

40.3 Pigment properties and uses

Iron blue pigments are characterized by a sufficient hue, high tinting strength, and suitable rheological properties. The pure pigments show a black color impression. The chroma of the pigments is recognizable for the human eye only by adding of light scattering colorants.

Iron blue pigments consist of very fine particles in the size range from 0.01 to 0.1 μm (primary particles). Due to their small sizes, the pigments are difficult to disperse and tend to agglomeration, in some cases also to aggregation.

Light- and weather fastness of the iron blue pigments are excellent. These properties can be affected negatively when iron blue is mixed with white pigments [12]. Iron blue pigments are stable against diluted mineral acids and oxidizing agents. They are unstable against concentrated acids and hydroxides. The pigments are thermally stable for short periods up to temperatures of 180 °C. They are combustible with an ignition temperature in air above 140 °C [7].

Main applications for iron blue pigments are the printing ink industry and agriculture. Micronized iron blue pigments are used for coloring of fungicides. Iron blue pigments are mostly applied in pure form, e. g. in printing inks. Well dispersed iron blue pigments give printing inks a black tone. The pigments are very interesting for printing applications, especially for rotogravure printing. Their deep hue, good hiding power and advantageous cost/performance-situation are decisive for the use in this specific printing method. Iron blue pigments are often mixed with phthalocyanine pigments in the case of multicolor printing applications [3]. Other application fields of the pigments are single- and multiple-use carbon papers and blue copying papers. The pigments are used here for toning the carbon black [7].

Micronized iron blue pigments are used to color organic fungicides that are applied for treating vines, olives or citrus fruits. The high color intensity of the pigments, even in small amounts, make the fungicides visible in their application. The fungicides are mixed in this case with micronized iron blue pigments [13].

"Water-soluble" iron blue pigments find application directly in the aqueous phase for the manufacturing of blue paper. It is also possible to use a suitable iron blue pigment, which is ground together with a water-soluble binder, applied to the paper, dried, and glazed.

Iron blue is used in medicine for the decontamination of persons or animals having ingested radioactive material. The isotope ^{137}Cs, which would otherwise be freely absorbed via the human or animal digestive tract, is exchanged with the iron(II) of the iron blue. ^{137}Cs is excreted after this exchange through the intestines [14–16].

Iron blue pigments do not exhibit acute toxicity (LD50 value rat oral: >5000 mg/kg). They are not irritating to skin or mucous membranes. Long-term tests have shown that there are no harmful effects of iron blue pigments on humans and animals. The adsorption of iron blue pigments in living beings is very low. The decomposition of iron blue to toxic cyanide in aqueous media is found to be very low [3, 4].

Medical studies on the application of $Fe^{III}{}_4[Fe^{II}(CN)_6]_3$ in doses of up to 20 g/d for the decontamination of persons exposed to radioactive cesium have shown no toxicological effects [17].

Investigations on fishes have found that the effect of iron blue pigments is harmless. Slight toxic effects have been detected on bacteria when iron blue is suspended in the water [3].

References

1. Endriss H. Aktuelle anorganische Buntpigmente. Zorll U, editor. Hannover: Vincentz Verlag, 1997;114.
2. Winkeler H. In Kittel - Lehrbuch der Lacke und Beschichtungen. Spille J, editor. vol. 5. 2nd ed. Stuttgart/ Leipzig: S. Hirzel Verlag, 2003:122.
3. Böhland T. In industrial inorganic pigments. Buxbaum G, Pfaff G, editors. 3rd ed. Weinheim: Wiley-VCH Verlag, 2005:145.
4. Pfaff G. Inorganic pigments. Berlin/Boston: Walter de Gruyter GmbH, 2017:144.
5. Clauss C, Gratzfeld E. In Pigmente. Kittel H, editor. 3rd ed. Stuttgart: Wissenschaftl. Verlags GmbH, 1960.
6. Dix MF, Rae AD. The structure of Prussian blue and analogues. J Oil Colour Chem Assoc. 1978;61:69.
7. Wertheim GK, Rosencwaig A. Characterization of inequivalent iron sites in Prussian blue by photoelectron spectroscopy. J Chem Phys. 1971;54:3235. Buser HJ, Ludi A, Petter W, Schwarzenbach D. Single-crystal study of Prussian blue: $Fe_4[Fe(CN)_6]_2$, $14H_2O$. J Chem Soc Chem Commun. 1972;23:1299.
8. Buser HJ, Schwarzenbach D, Petter W, Ludi A. The crystal structure of Prussian blue: $Fe_4[Fe(CN)_6]_3$, xH_2O. Inorg Chem. 1977;16:2704.
9. Ivanov VD. Four decades of electrochemical investigation of Prussian blue. Ionics 2020;26:531.
10. Emrich RJ, Traynor L, Gambogi W, Buhks E. Surface analysis of electrochromic displays of iron hexacyanoferrate films by X-ray photoelectron spectroscopy. J Vac Sci Techn A. 1987;5:1307.
11. Herren F, Fischer P, Ludi A, Haelg W. Neutron diffraction study of Prussian blue, $Fe_4[Fe(CN)_6]_3$ xH_2O. Location of water molecules and long-range magnetic order. Inorg Chem. 1980;19:956. Ludi A. Berliner Blau. Chem unserer Zeit. 1988;22:123.
12. Müller-Fokken L. Licht- und Wetterbeständigkeit von Eisenblaupigmenten in Lacken. Farbe + Lack. 1978;84:489.
13. Vossen-Blau zur Färbung von Fungiziden. Schriftenreihe Pigmente, Nr. 50. Frankfurt/M: Degussa AG, 1985.
14. Madshus K, Strömme A. Increased excretion of ^{137}Cs in humans by Prussian blue. Z Naturforsch B. 1968;23:391.
15. Nigrovic V. Retention of radiocaesium by the rat as influenced by Prussian blue and other compounds. Phys Med Biol. 1965;10:81.
16. Dresow B, Nielsen P, Heinrich HC. Efficacy of different hexacyanoferrates(II) in inhibiting the intestinal absorption of radiocaesium in rats. Z Naturforsch C. 1990;45:676.
17. Verzijl JM, Joore HC, van Dijk A, Wierckx FC, Jean Savelkoul TF, Glerum JH. In vitro cyanide release of four Prussian blue salts used for the treatment of cesium contaminated persons. Clinical Toxicol. 1993;31:553.

Gerhard Pfaff

41 Iron oxide pigments

Abstract: Natural and synthetic iron oxide pigments are by far the most important colored pigments. Their high importance is based on the variety of stable colors ranging from yellow via orange, red and brown to black. Iron oxide yellow (α-FeOOH), iron oxide red (α-Fe_2O_3) and iron oxide black (Fe_3O_4) are the most important representatives of the iron oxide pigments. Synthetic iron oxide pigments are produced industrially on a large scale by solid-state processes, precipitation processes and by the Laux process. Main advantages of synthetic iron oxide pigments compared with natural types are their pure hue, the consistent, reproducible quality and their tinting strength. Iron oxide pigments are mainly used in construction materials, paints, coatings, and plastics, but also in cosmetics, pharmaceuticals and special applications such as ceramics, magnetic coatings and toners.

Keywords: iron oxide pigments, iron oxide black, iron oxide red, iron oxide yellow, goethite, hematite, magnetite, solid-state process, precipitation process, Laux process

41.1 Fundamentals and properties

The high importance of iron oxide pigments is based on their wide variety of colors ranging from yellow via orange, red and brown to black, the chemical stability, the nontoxicity, and the good performance/price ratio. The relatively low production costs for iron oxide pigments and accordingly low prices are further reasons for the market success of these colorants. Iron oxide pigments are divided in natural and synthetic iron oxide types. All iron oxides used for pigment purposes consist of well-defined compounds with known crystal structures [1–6]:

α-FeOOH (goethite): diaspore structure, pigment iron oxide yellow C.I. Pigment PY 42, pigment particles with preferably acicular shape, color change with increasing particle size from green-yellow to brown-yellow.

γ-FeOOH (lepidocrocite): boehmite structure, use only as magnetic pigment, color change with increasing particle size from yellow to orange.

α-Fe_2O_3 (hematite): corundum structure, pigment iron oxide red C.I. Pigment PR 101, pigment particles with spherical, acicular and cubical shape, color change with increasing particle size from light red to dark violet.

This article has previously been published in the journal Physical Sciences Reviews. Please cite as G. Pfaff, Iron Oxide Pigments *Physical Sciences Reviews* [Online] 2021, 6. DOI: 10.1515/psr-2020-0179

https://doi.org/10.1515/9783110587104-041

γ-Fe_2O_3 (maghemite): spinel superlattice, ferromagnetic, brown color.

Fe_3O_4 (magnetite): inverse spinel structure, ferromagnetic, pigment iron oxide black C.I. Pigment PB 111, black color.

The production of synthetic iron oxides started at the end of the nineteenth century. Increasing demands and quality requirements led to the replacement of the previously used natural iron oxide pigments by synthetic types. Inventive effort has been directed towards developing of processes for the technical synthesis of iron oxide pigments since this time. After more than 100 years, only less than 20% of all iron oxide pigments are natural ones.

41.2 Natural iron oxide pigments

The use of natural iron oxides and iron oxide hydroxides as colorants goes back to prehistoric times. Examples for the application of iron oxides colors can still be found in the form of cave paintings (Altamira cave, Grotte de Lascaux, Grotte Chauvet). The ancient cultures in Egypt, Greece and in the Roman Empire as well as the developed countries in later times used natural iron oxides as pigments for coloring purposes [7–9].

The iron oxide content of natural iron oxides varies strongly and is dependent on the mineral deposit. Hematite from natural sources is the basis for the important class of natural red pigments. Natural goethite, on the other hand, is often suitable for the production of yellow pigments, whereas umber and sienna can be used for brown pigments. It is self-evident that only deposits with high iron oxide content are used as raw materials for natural iron oxide pigments. Magnetite from natural sources has found only little use as raw material for black pigments because the tinting strength of the resulting black powders is poor.

Hematite is found in larger quantities and with sufficient quality in numerous deposits all over the world, e.g., in the surroundings of Malaga in Spain (Spanish red) and near the Persian Gulf (Persian red). Many of the deposits are of only local relevance. Natural hematite used for the production of red pigments has Fe_2O_3 contents ranging from 45 to 95%. Spanish reds, for example, have a Fe_2O_3 content of more than 90%. A high hematite content is the reason for the red color of many ores and rocks. A special variety of natural iron oxide is micaceous iron oxide (MIO), which occurs in the form of a mineral cleavable to thin platelets. Discovery sites for larger quantities of MIO are located in Austria. The cleaved mineral is mainly used in corrosion protection.

Goethite is the color giving component of yellow ocher. Ocher occurs in the form of light yellow to light brown weathering products of iron containing ores. The iron oxide content of ocher ores ranges from 20 to 80%. Important deposits for the production of natural iron oxide hydroxide are found in South Africa and France. The Fe_2O_3 content corresponds to the iron oxide hydroxide content of the ocher. It is ca. 20% in the French deposits and ca. 55% in the deposits of South African [5].

Umber is a weathering product of iron and manganese containing clays. Important deposits for umber are found in Cyprus. The iron oxide amount of umbers varies from 45 to 70%, the manganese dioxide amount ranges from 5 to 20%. Umbers have deep brown to greenish brown colors in the raw state. They become dark brown with a red undertone after calcination an are then referred to as burnt umbers.

The term Terra di Siena (sienna) includes a group of brown to orange colored weathering products of iron containing ores with an iron oxide content of 45 to 80% and a manganese content of <1%. Calcined Terra di Siena is characterized by a red-brown color.

Magnetite is the most widespread iron ore. It is a black mineral with an iron oxide content of 80 to 95%. Black magnetite pigments can be produced out of magnetite ore only by grinding.

The production of natural iron oxide pigments depends mainly on the composition of the raw materials. These are typically ground, washed, converted into a slurry, dried and then ground in ball mills. Instead of ball mills, disintegrators or impact mills can be used for the grinding step. Sienna and umber ores are calcined after grinding to remove the water. The hue of the obtained iron oxide pigments depends on the calcination temperature, the time of the thermal treatment and the composition of the raw materials.

Today, natural iron oxide pigments are used only in a limited number of applications. These include inexpensive marine coatings, coatings with a glue, oil, or lime base, coloring of cement, artificial stone and wallpaper as well as the production of crayons, drawing pastels, and chalks.

41.3 Synthetic iron oxide pigments

Synthetic iron oxide pigments have compositions corresponding to that of the naturally found minerals hematite, goethite, lepidocrocite, and magnetite in their pure forms. They exhibit typically higher and more constant iron oxide contents than the iron oxide ores. Synthetic iron oxide pigments contain converted Fe_2O_3 amounts of 95% and more. Other components of iron oxide pigments are Al_2O_3 and SiO_2 (together up to 5%) and water-soluble salts, mostly sulfates and chlorides (up to 0.5%). Particle sizes of synthetic iron oxide pigments are typically in the range from 0.1 to 1.0 µm. Iron oxide yellow pigments crystallize mostly in an acicular shape with a length of 0.3 to 0.8 µm and a diameter of 0.05 to 0.2 µm. Iron oxide black pigments are often characterized by cubical particles. The diameters of these pigments are in the range from 0.1 to 0.6 µm.

The main advantages of synthetic iron oxide pigments compared with natural types are their pure hue, the consistent, reproducible quality and their tinting strength. The industrial production of iron oxide pigments has the objective to synthesize almost pure iron oxides and oxide hydroxides with red, yellow, orange, and black color shades. Brown iron oxide colorants consist typically of mixtures of red and/or yellow and/or black pigments. They are not available by a direct manufacturing process. However, it is possible to synthesize homogeneous brown oxide phases for pigment purposes. This is done in comparatively small quantities compared to the pigment mixtures, particularly for $(Fe,Mn)_2O_3$ and γ-Fe_2O_3. It has to be mentioned here that γ-Fe_2O_3 is used as magnetic pigment, not for color applications.

41.4 Production of iron oxide pigments

Three main processes are used in present times for the production of high-quality iron oxide pigments. All these processes allow the exact control of the most important parameters for the final pigments, which are the particle size distribution, the particle shape and the composition:
- Solid-state processes for red, black, brown pigments
- Precipitation processes for yellow, red, orange, black pigments
- Laux process for yellow, red, black pigments

The iron containing raw materials used for the production of iron oxide pigments are mainly byproducts from other industrial processes, especially iron(II) sulfate and iron(III) chloride from the steel industry and iron scrap from the metal working industry.

41.4.1 Solid-state processes

Thermal decomposition of iron compounds is the basis for the production of high quantities of iron oxide red pigments. Main part of the process is the oxidative calcination of decomposable iron compounds, particularly of iron(II) sulfate and α-FeOOH, at high temperatures. Red pigments are also available by the oxidation of Fe_3O_4 powders. The agglomerated and aggregated pigments obtained after the high-temperature reaction are typically ground to the desired particle size in pendular mills, pin mills, or jet mills.

The iron(II) sulfate used for the solid-state process has the composition $FeSO_4 \cdot 7\,H_2O$ in almost all cases. The heptahydrate is dehydrated at the beginning to the monohydrate at elevated temperatures. The monohydrate is roasted hereafter in an oxidizing atmosphere at temperatures above 650 °C. Red-colored α-Fe_2O_3 (copperas red) is formed under these conditions. As an intermediate product, iron(III) sulfate is formed. The diagram of the solid-state process is shown Figure 41.1.

$$6\,FeSO_4 \cdot H_2O + 11/2\,O_2 \rightarrow \alpha - Fe_2O_3 + 2\,Fe_2(SO_4)_3 + 6\,H_2O \qquad (41.1)$$

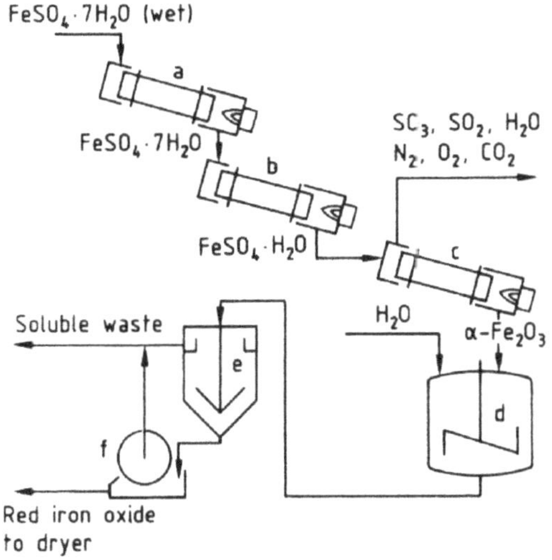

Figure 41.1: Production of copperas red: (a) dryer; (b) rotary kiln (dewatering); (c) rotary kiln; (d) tank; (e) thickener; (f) filter [6].

$$2\,Fe_2(SO_4)_3 \rightarrow 2\,\alpha\text{–}Fe_2O_3 + 6\,SO_3 \tag{41.2}$$

The sulfur trioxide formed as a by-product can be processed to sulfuric acid. Fe_2O_3 pigments of lower quality with a relatively poor tinting strength and a blue tinge are obtained by single-stage calcination of $FeSO_4 \cdot 7\,H_2O$ in an oxidizing atmosphere [5].

Industrial-grade iron(III) chloride from pickling plants can basically also be used for the solid-state process. The pigment qualities achieved by using of $FeCl_3$ instead of $FeSO_4$ are lower. They are often pelletized and reused in the manufacture of steel. $FeCl_2$ can also be used for the production of α-Fe_2O_3. In the Ruthner process, a solution of $FeCl_2$ reacts in a combined spray-drying calcination process to α-iron(III) oxide. The resulting powders are suitable for the manufacture of hard- and soft-ferrites.

$$2\,FeCl_2 + 2\,H_2O + 1/2\,O_2 \rightarrow \alpha - Fe_2O_3 + 4\,HCl \tag{41.3}$$

The thermal decomposition of iron(III) oxide hydroxide leads to pure red iron oxide pigments with a high tinting strength:

$$2\,\alpha - FeOOH \rightarrow \alpha - Fe_2O_3 + H_2O \tag{41.4}$$

Fe_3O_4 obtained from the Laux process (see 4.3) or from other processes may be calcined in rotary kilns in an oxidizing atmosphere under countercurrent flow to synthesize α-Fe_2O_3. A large variety of red pigments can be obtained depending on the starting material and the reaction conditions.

$$2\,Fe_3O_4 + 1/2\,O_2 \rightarrow 3\,\alpha - Fe_2O_3 \tag{41.5}$$

Controlled oxidation of Fe_3O_4 at 500 °C can be used for the synthesis of single-phase brown γ-Fe_2O_3 with a neutral hue [10].

Platelet-shaped MIO is obtained by the reaction of iron(III) chloride with iron at 500 to 1000 °C in an oxidizing atmosphere in a tubular reactor [5, 11].

Black Fe_3O_4 pigments with high tinting strength are accessible by thermal treatment of iron salts under reducing conditions [12].

Single-phase brown pigments with the composition $(Fe,Mn)_2O_3$ are obtained by solid-state reactions of iron oxides or iron oxide hydroxides with small quantities of manganese compounds [13]. The calcination of thermally decomposable iron and chromium compounds leads to pigments with the composition $(Fe,Cr)_2O_3$ [14].

41.4.2 Precipitation processes

The three technically most important iron oxide pigments α-FeOOH, α-Fe_2O_3 and Fe_3O_4 are produced by precipitation processes combined with the oxidation of Fe (II) to Fe(III). The precipitation processes start from iron(II) sulfate solution, the oxidizing agent used is air. The reaction conditions are decisive for the achievement of soft pigments with a pure bright hue. In some process executions, nucleation seeds are used. Yellow α-FeOOH is typically produced starting from $FeSO_4 \cdot 7\,H_2O$ as the iron salt source. In other process variants, solutions from iron and steel pickling are used [15]. The free acid usually contained in pickling solutions can be neutralized by a reaction with scrap iron.

The manufacture of iron oxide yellow is executed in a batch process and starts with the addition of alkali, mostly sodium hydroxide, to an iron(II) sulfate solution in open reaction vessels. Air is blown into the suspension of freshly-formed iron hydroxide and oxidizes the iron in the precipitate under formation of α-FeOOH:

$$2\,FeSO_4 + 4\,NaOH + 1/2\,O_2 \rightarrow 2\,\alpha - FeOOH + 2\,Na_2SO_4 + H_2O \qquad (41.6)$$

The pH value in the vessel is kept in the acidic range during the precipitation reaction. A suitable acid is added for this purpose during the whole operation. The reaction temperature is in the range from 10 to 90 °C. The time of up to 100 hours depends on the chosen temperature and on the desired particle size of the final yellow pigments. Highly consistent iron oxide yellow pigments with a pure color are obtained when seeds of α-FeOOH are produced separately before the main precipitation reaction takes place (Figure 41.2, seeding tank a, precipitation tank b) [16].

Black iron oxide pigments with magnetite structure and a good tinting strength can be produced directly by precipitation. The reaction is executed at 90 °C while air is passed into the suspension at a pH value of ≥7 (1-step precipitation). Such Fe_3O_4 pigments are achievable when the reaction is finished at a $FeO : Fe_2O_3$ ratio of 1 : 1:

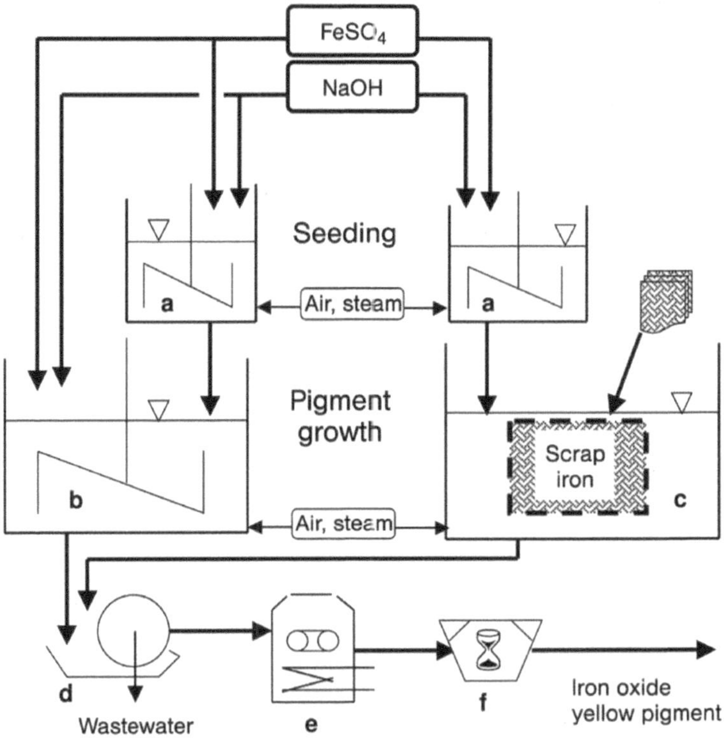

Figure 41.2: Production of yellow iron oxide by the precipitation (left) and Penniman process (right). (a) seeding tank; (b) precipitation pigment reactor; (c) Penniman pigment reactor with scrap basket; (d) filter; (e) dryer; (f) mill [6].

$$3\,FeSO_4 + 6\,NaOH + 1/2\,O_2 \rightarrow Fe_3O_4 + 3\,Na_2SO_4 + 3\,H_2O \tag{41.7}$$

The precipitation process can also be carried out at 150 °C under pressure. Improved Fe_3O_4 pigments are possible to produce under these conditions [17]. Another way to produce iron oxide black pigments consists in the rapid heating to 90 °C of a suspension of iron oxide hydroxide with the necessary quantity of $FeSO_4$ and NaOH. $Fe(OH)_2$ is formed as an intermediate product in this case [18, 19]:

$$2\,\alpha-FeOOH + FeSO_4 + 2\,NaOH \rightarrow Fe_3O_4 + Na_2SO_4 + 2H_2O \tag{41.8}$$

Orange iron oxide pigments with lepidocrocite structure can be synthesized if the precipitation occurs in diluted iron(II) sulfate solutions in the presence of air. Sodium hydroxide or other alkalis are added to adjust neutral pH conditions for the precipitation [20, 21]:

$$2\,FeSO_4 + 4\;NaOH + 1/2\,O_2 \rightarrow 2\;\gamma-FeOOH + 2\,Na_2SO_4 + H_2O \tag{41.9}$$

Soft iron oxide pigments with a pure red color are possible to produce by preparing α-Fe_2O_3 nuclei in an aqueous solution followed by the addition of an iron(II) salt solution and oxidation with air at 80 °C. Hydronium ions formed during the precipitation are neutralized by adding alkali and keeping the pH value constant [22]:

$$2\,FeSO_4 + 4\ NaOH + 1/2\,O_2 \rightarrow \alpha - Fe_2O_3 + 2\,Na_2SO_4 + 2H_2O \qquad (41.10)$$

α-Fe_2O_3 precipitates are also obtained when iron(II) salts react at 60 to 95 °C in aqueous solution with an excess of sodium hydroxide in the presence of air. Small amounts of other cations accelerate the formation of α-Fe_2O_3 under these conditions [23].

The precipitation method that is used mostly for the production of iron oxide yellow pigments is the Penniman process [24, 25]. Iron(II) sulfate, sodium hydroxide solution and scrap iron are the raw materials used in this process. Considerable quantities of impurities in the used iron(II) sulfate must be removed by precipitation reactions in an upstream process step. The iron used must be free of disruptive other metals. The process usually consists of two stages, the formation of α-FeOOH nuclei and the reaction of iron(II) sulfate, metallic iron and oxygen in the presence of these nuclei under the formation of iron oxide yellow.

α-FeOOH nuclei are formed at the beginning of the Penniman process by addition of caustic soda or other alkaline lye at 20–50 °C to an iron(II) sulfate solution (seeding tank a). Air is blown through the suspension with the precipitate (eq. (41.6)). Yellow, orange or red nuclei may be formed depending on the reaction conditions used. The nuclei suspension is pumped into a reaction vessel containing large amounts of scrap iron (reactor with scrap basket c) and water. The subsequent main process step is the completion of the α-FeOOH formation by growing of precipitated iron oxide hydroxide onto the nuclei. The nuclei driven step of the process is responsible for the formation of sulfuric acid:

$$2\,FeSO_4 + 1/2\,O_2 + 3\,H_2O \rightarrow 2\,\alpha - FeOOH + 2\,H_2SO_4 \qquad (41.11)$$

The resulting sulfuric acid reacts with the scrap iron under formation of iron(II) sulfate:

$$2\,Fe + 2\,H_2SO_4 \rightarrow 2\,FeSO_4 + 2\,H_2 \qquad (41.12)$$

Iron(II) sulfate generated according to this reaction is oxidized with air according to eq. (41.11).

The Penniman process can be executed at different intervals, from days to several weeks, depending on the reaction conditions and the desired pigment quality. Finally, metal impurities and too coarse particles are removed from the precipitate using sieves or hydrocyclones. Water-soluble salts are removed during filtration by using a downstream washing procedure (filter d). The following drying is carried out with band or spray dryers (dryer e). The necessary grinding of the dried and partially lumpy product is done using disintegrators or jet mills (mill f). An advantage of the Penniman process compared with other precipitation processes is the small quantity of alkali and iron(II) sulfate required. Hydroxide is used only for the formation of the α-FeOOH nuclei and iron(II) sulfate is only necessary in small amounts for the seed formation and at the beginning of the main process step. The sulfate is then continuously renewed by dissolving of iron by the reaction with the sulfuric acid liberated by hydrolysis (eq. (41.11)).

Iron oxide yellow produced by the Penniman process can be used for the manufacture of iron oxide red pigments according to eq. (41.4). The calcination α-Fe_2O_3 in reducing atmospheres, e.g. under hydrogen or hydrogen containing gases, leads to black Fe_3O_4:

$$3\,\alpha-Fe_2O_3+H_2\rightarrow 2\,Fe_3O_4+H_2O \tag{41.13}$$

41.4.3 Laux process

The Laux process, respectively, the nitrobenzene reduction process or the aniline process consists in the reduction of aromatic nitro compounds with iron under formation of iron oxides and aniline. Iron(II) chloride or aluminum chloride solutions, sulfuric acid, and phosphoric acid act as additives in the process to obtain iron oxides of pigment quality [26]. Various types of pigmentary iron oxides can be produced depending on the reaction parameters. Yellow, red and black pigments are possible to manufacture by using the Laux process. Brown pigments can be produced by mixing of α-FeOOH, α-Fe_2O_3 and Fe_3O_4 in suitable ratios. The choice of the additives used in the process is of high importance for the color of the final pigments. The reduction of the nitro compound, mostly nitrobenzene, in the presence of aluminum chloride leads to high quality yellow pigments [27].

$$2\,Fe+C_6H_5NO_2+2\,H_2O\rightarrow 2\,\alpha-FeOOH+C_6H_5NH_2 \tag{41.14}$$

If iron(II) chloride is used instead of aluminum chloride black pigments with very high tinting strength are obtained [23]:

$$9\,Fe+4\,C_6H_5NO_2+4\,H_2O\rightarrow 3\,Fe_3O_4+4\,C_6H_5NH_2 \tag{41.15}$$

The addition of phosphoric acid results in pigments with good tinting strength [28]. Calcination of products thus obtained leads to light red and dark violet pigments. Figure 41.3 shows the production scheme for the Laux process, named after its inventor.

The reaction rate, which has a high influence on the type and the quality of the resulting pigments depends mainly on the composition and particle size of the iron used, the addition rates of iron and nitrobenzene, and the pH value during the process. The iron is oxidized by the nitrobenzene and iron ions are formed. These ions are responsible for the advancement of the precipitation process accompanied by the aniline formation, when nitrobenzene is used as the nitro compound [5].

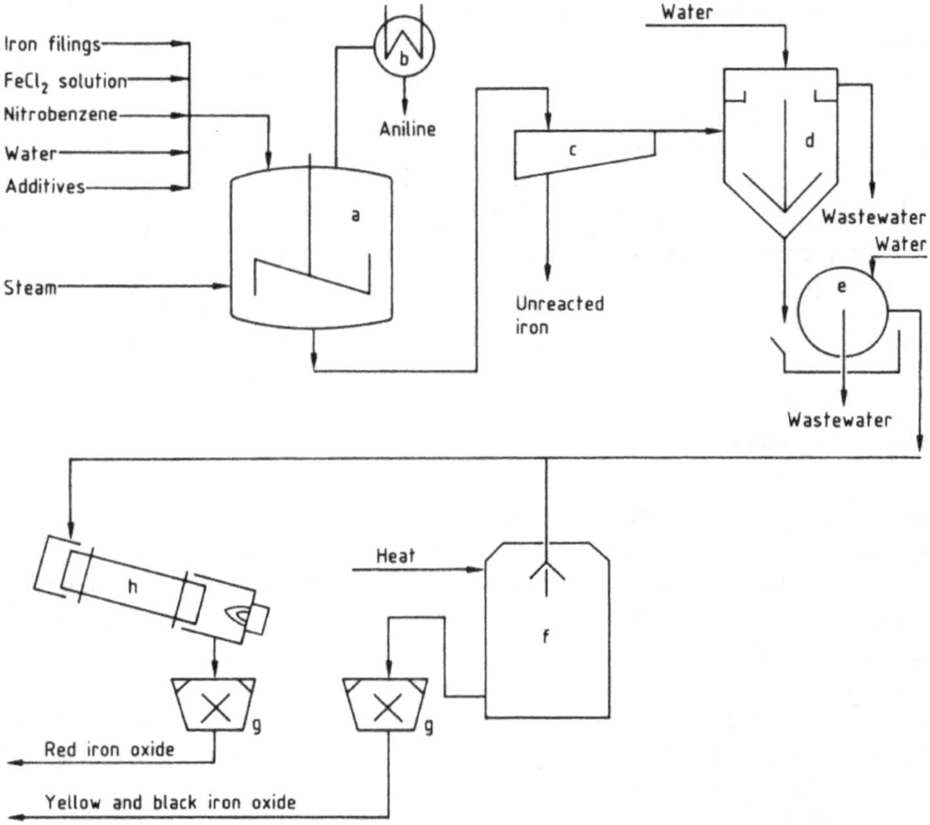

Figure 41.3: Production of iron oxide pigments by the Laux process: (a) reactor; (b) condenser; (c) classifier; (d) thickener; (e) filter; (f) dryer; (g) mill; (h) rotary kiln [6].

The iron used for the Laux process is ground material from iron casting or forging. It should be free of oil and grease. The adjustment of the particle size takes place by grinding the iron in edge runner mills and classification with vibratory sieves. The addition of iron and nitrobenzene is done gradually to a stirred reactor containing an aqueous solution of the other reactants (iron(II) chloride, aluminum chloride, sulfuric acid and/or phosphoric acid). Temperatures of up to 100 °C are used, for the reduction of the nitrobenzene. Aniline is formed under these conditions, which is removed by steam distillation. Unreacted iron is separated using suitable classifiers. The iron oxide, respectively, iron oxide hydroxide slurry is diluted with water in the next step and brought onto a filter. Soluble salts are removed by washing with water before the filter stuff is dried on appropriate dryers to form yellow or black pigments. Calcination of the thus obtained powders in rotary kilns in an oxidizing atmosphere at 700–800 °C leads to red or brown pigments. The tinting strength of the pigment powders can be improved by thermal treatment in a nonoxidizing atmosphere at temperatures in the range from 500 to 700 °C [29]. The pigments are typically ground after the calcination to achieve the desired final particle size.

The special attractiveness of the Laux process consists in the simultaneous manufacture of iron oxide pigments and aniline, respectively, other aromatic amines. The process has the additional advantage that it does not generate byproducts that harm the environment [5].

41.4.4 Other production processes

There are further processes for the production of yellow, red, black or brown iron oxide pigments, which are executed only on a small scale. The thus obtained iron oxide pigments are used mostly for special applications:

- Iron oxide pigments with high analytical purity can be synthesized by the reaction of relatively pure iron(II) sulfate with air in water followed by neutralization with sodium carbonate. Heavy-metal impurities in the sulfate are removed by a preprecipitation step. The resulting pigments are suitable for the application in cosmetics and foods [2].
- Very pure iron oxide red pigments with a primary particle size in the nanometer range can be manufactured by thermal decomposition of iron pentacarbonyl at 600–800 °C in an excess of oxygen. Powders obtained using this route of synthesis belong to the class of transparent iron oxide pigments [2, 3, 5, 6, 30].
- Platelet-shaped iron oxide red pigments are accessible by hydrothermal dehydration and crystallization of iron oxide yellow. α-FeOOH used in this process as starting component is formed previously from iron(II) sulfate in alkaline suspension [2, 5, 6, 31, 32]. Thus synthesized α-Fe_2O_3 platelets find a limited application as effect pigments in paints and coatings.

41.5 Pigment properties and uses

The Fe_2O_3 contents of iron oxide red and black pigments are in the range from 92 to 96 wt %. Some very pure iron oxide powders are characterized by Fe_2O_3 contents of 99.5 to 99.8 wt %. Such pure powders are produced for special applications, e.g., for the manufacture of ferrite materials. The Fe_2O_3 contents of yellow and orange iron oxide pigments are in the range from 85 to 87 wt % corresponding to α-FeOOH contents of 96 to 97 wt %. The amount and nature of adhering water-soluble salts as well as the particle size of the ground powders determine the pigment quality to a decisive extent (effect on hue, tinge and tinting strength). Particle diameters of about 0.1 µm lead to a yellow tinge for red iron oxide, diameters of about 1.0 µm generate a violet tinge of the red pigments [5].

Iron oxide yellow pigments crystallize preferably in the shape of needles. Their optical properties are dependent on the shape of the particles but also on the particle size. Another parameter of importance is the length to width ratio of the particles. The lengths are typically in the range of 0.3 to 0.8 µm and the diameters have values from 0.05 to 0.2 µm. This results in length:diameter ratios of 1.5 to 16. α-FeOOH powders can also consist of spherical particles. The shape always depends on the specific precipitation conditions. The particle diameters of black iron oxide pigments are in the range from 0.1 to 0.6 µm [5].

The thermal stability of iron oxide pigments is different and depends on the composition. α-Fe_2O_3 is stable up to temperatures of about 1200 °C in air. It does not change its red color in this temperature range. Black Fe_3O_4 is converted into brown γ-Fe_2O_3 at about 180 °C in the presence of oxygen. Temperatures of 350 °C and higher lead to the formation of red α-Fe_2O_3. α-FeOOH decomposes above 180 °C to red α-Fe_2O_3. It is possible to increase the thermal stability of iron oxide yellow to nearly 260 °C by incorporation of aluminum in the pigment.

Synthetically produced iron oxide pigments are lightfast, weather and alkali resistant. They are characterized by good tinting strength and, depending on their particle size, excellent hiding power. The refractive indices (yellow 2.37, red 2.87, black 2.42), the specific surface areas (yellow 10–18, red 3–15, black 3–15 m^2/g) and the densities (yellow 4.0–4.2, red 5.0–5.2, black 4.7–4.9 g/cm^3) are further important properties for the optical and the technical behavior of iron oxide pigments.

The reflectance spectra of α-Fe_2O_3 pigments show a selective absorption in the green and the blue spectral range of the visible light. The nonabsorbed, i.e., the reflected part of the visible light (600–720 nm), is responsible for the red color impression. α-FeOOH pigments reflect the light in the range from 520 to 720 nm [6]. The color origin of the iron oxides can be explained plausibly based on the crystal lattice structure [33].

Iron oxide pigments are widely used in many applications. A major part of natural and synthetic iron oxide pigments is produced for coloring of construction materials. Examples for this important application field are concrete roof tiles, paving bricks, fibrous cement, bitumen, mortar, and rendering. Only small amounts of the pigments are necessary in most of these application cases.

Paints and coatings are other large application fields for iron oxide pigments. Mainly synthetically produced pigments are used here, especially in base coats, topcoats, dispersion paints, and wood protection systems. The particle sizes of iron oxide pigments used in paints and coatings are typically in the range of 0.1 to 1.0 µm to fulfill the requirements regarding the hiding power. Table 41.1 shows the relationship of particle size and properties of iron oxide red pigments [6]. Iron oxide pigments are easy to incorporate in paint and coatings systems and exhibit good dispersion behavior. Only the extremely small sized transparent iron oxide reds need a more expensive special procedure for the dispersion in the relevant application media.

Table 41.1: Relationship of particle size and selected properties of α-Fe_2O_3 pigments [6].

Particle size (μm)	0.001–0.01	0.1–1.0	10.0–100.0
Type of pigment	Transparent iron oxide reds	Hiding iron oxide reds	Micaceous iron oxide
Color/appearance brown	Yellowish-red	Yellow-red to violet	Metallic to black
Hiding power	Low	High	Low
Specific surface area (m^2/g)	90–100	3–15	1–3
Dispersibility	Difficult	Easy	Very easy

Iron oxide red pigments are used in a broad variety of plastics. A great advantage is the low tendency of the pigments to migrate and to bleed in the relevant polymer systems. The pigments can be applied in plasticized PVC, stabilized rigid PVC and other plastics because they are thermally stable in the processing area. The low thermal stability of iron oxide yellow pigments is the reason for their limited use in thermoplastics. The stability of the yellow pigment types can be increased up to 260 °C by coating the pigment particles with stable inorganic layers. Pigments stabilized in such a way can also be used in polyolefins and polystyrene. Iron oxide black pigments are stable in most of the polymers. The reductive situation in many polymer melts allows the use of the pigments at processing temperatures up to 280 °C.

The high thermal stability of iron oxide red pigments is the basis for their use in high temperature applications. The pigments can be applied without larger problems for coloring of ceramic bulks as well as for enamels and glazing.

Synthetic iron oxide pigments are also used in cosmetics and personal care products. Examples of cosmetics containing the pigments are makeup and skin care preparations. The pigments have also the approval for the use in pharmaceutical products [34].

Another application field for iron oxide pigments is the paper industry. Larger pigment amounts are used here especially for the fabrication of decor paper for furniture lamination. Some special iron oxide blacks and reds are applied in magnetic coatings (audiotapes, videotapes, magnetic stripes), toners for photocopiers, friction pads (brake lining, clutch lining), and polishing agents for metal and glass surfaces.

Iron oxide pigments are classified as nontoxic substances. The situation is different for some natural iron oxides that are contaminated with heavy-metal compounds. Another problem of these natural products may be a higher proportion of crystalline silicon dioxide. In the case of synthetic iron oxide pigments, the amount of crystalline silica is typically below the detection limit for X-ray diffraction analysis. For cosmetic and pharmaceutical applications, only synthetic iron oxides of very high purity are admitted. National regularity authorities have defined the levels of heavy metals for iron oxide pigments in their use. It is to be noted that the heavy-metal contents of the pigments applied in cosmetic or personal care products must be extremely low. A risk to human health can thus be excluded when using iron oxide pigments in this application segment.

References

1. Schwertmann U, Cornell RM. Iron oxides in the laboratory, preparation, and characterization. Weinheim: VCH Verlagsgesellschaft, 1991.
2. Endriss H. Aktuelle anorganische Buntpigmente. Zorll U, editor. Hannover: Vincentz Verlag, 1997:45.
3. Buxbaum G, Kresse P. In Kittel - Lehrbuch der Lacke und Beschichtungen. vol. 5. Spille J, editor. 2nd ed. Stuttgart/Leipzig: S. Hirzel Verlag, 2003:71.
4. Cornell RM, Schwertmann U. The iron oxides. 2nd ed. Weinheim: Wiley-VCH Verlagsgesellschaft, 2003.
5. Buxbaum G. In industrial inorganic pigments. Buxbaum G, Pfaff G, editor. 3rd ed.Weinheim: Wiley-VCH Verlag, 2005:99.
6. Pfaff G. Inorganic pigments. Berlin/Boston: Walter de Gruyter GmbH; 2017:97.
7. Jolly JL, Collins CT. “Natural iron oxide pigments”, iron oxide pigments, part 2, information circular-bureau of mines 8813. U. S. Government Printing Office. Washington, 1980.
8. Lagroix F, Banerjee SK, Jackson MJ. Geological occurrences and relevance of iron oxides. In Iron oxides: From Nature to applications. Faivre D, editor. Weinheim: Wiley-VCH Verlag GmbH & Co. KGaA, 2016:9.
9. Nicola M, Mastrippolito C, Masic A. Iron oxide-based pigments and their use in history. In Iron oxides: From Nature to applications. Faivre D, editor. Weinheim: Wiley-VCH Verlag GmbH & Co. KGaA, 2016:545.
10. Patent DE. 38 20 499 (Bayer). 1988.
11. Carter EV, Laundon RD. J Oil Colour Chem Ass. 1990;1:7.
12. Patent DE. 26 53 765 (Bayer). 1976.
13. Patent DE. 11 91 063 (Bayer). 1963.
14. Patent DE. 25 17 713 (BASF). 1975.
15. Patent US. 6,689,206 (Bayer). 2004.
16. Patent US. 2,388,659 (Interchem. Corp.). 1943.
17. Patent US. 3,133,267 (C. K. Williams & Co.). 1934.
18. Patent US. 2,631,085 (Reconstruction Finance Corp.). 1947.
19. Patent GB. 668929 (Reymers Holms Gamla Ind.). 1950.
20. Patent DE. 704 295 (Glemser). 1937.
21. Patent DE. 22 12 435 (Pfizer). 1972.
22. Patent DD. 26901 (Hollnagel, Kühn). 1960.
23. Patent US. 2,620,261 (C. K. Williams & Co.). 1947.
24. Patent US. 1,327,061 (West Coast Kalsomine Co.). 1917.
25. Patent US. 1,368,748 (Natioanal Ferrite Co.). 1920.
26. Patent DE. 463 773 (I.G. Farbenind.). 1925.
27. Patent DE. 515 758 (I.G. Farbenind.). 1925.

28. Patent DE. 551 255 (I.G. Farbenind.). 1930.
29. Patent EP. 0 249 843 (Bayer). 1987.
30. Patent DE. 466 463 (I.G. Farbenind.). 1926.
31. Patent EP. 0 014 382 (Bayer). 1980.
32. Pfaff G. In special effect pigments. Pfaff G, editor. 2nd ed. Hannover: Vincentz Network GmbH & Co. KG, 2008:38.
33. Köhler P, Schneider V, Kischkewitz J. Schwarz, Rot, Gelb im Fokus. Farbe + Lack. 2014;120:32.
34. Rowe RC. Synthetic iron oxides: the ideal pharmaceutical colourants. Pharm Int. 1984;9:221.

Heinz Mustroph

42 Merocyanine dyes

Abstract: Merocyanine dyes belong to the class of neutral polymethine dyes, where one terminal component is typically found in cyanine dyes and the second obtained from an active methylene compound. The different electron acceptor/donator abilities of the two terminal components have a marked impact on the electronic structure of a merocyanine dye and its equilibrium structure and electronic spectra. Their first technical application was spectral sensitization in silver halide photography. Today they have numerous of applications in textile dyeing and as membrane potential sensitive fluorescent dyes.

Keywords: bond lengths alternation, Brooker's merocyanine, electronic spectroscopy, Franck–Condon principle, isoenergetic point, merocyanine dyes, neutral polymethine dyes, nonlinear optics, Valence Bond theory, silver halide photography

42.1 Fundamentals

To introduce a concept of importance when considering the influences on the spectra of merocyanine dyes, this chapter will open by examining the case of *N,N*-dimethylformamide (DMF) as a simple model for such colorants. Usually the structural formula that has been used for DMF is **1N**. In the formalism of the *Valence Bond (VB) theory* the electronic structure of molecules can be described by a linear combination of *contributing structures* to the so-called *resonance hybrid*. Linus Pauling postulated that the charge-separated contributing structure **1S** with a nitrogen-carbon double bond instead a single bond contributes significantly to the electronic structure of the ground state [1]. Pauling estimated that the non-charge-separated contributing structure **1N** contributes 60% and **1S** contributes 40% to the electronic ground state. This results in a value of $p_{NC} = 0.4$ for the VB-π bond order of the NC bond. Pauling's prediction was later experimentally confirmed in the early days of dynamic NMR spectroscopy. In the H NMR spectrum of DMF at room temperature two methyl signals are observed. From this it can be concluded that the two methyl groups are differently shielded. With increasing temperature these two signals broaden and on continuing rise in the temperature further the two peaks become a narrow peak, exactly between the two peaks at room temperature. The reason is that the NC bond has a high proportion of double bond character, which causes a

This article has previously been published in the journal Physical Sciences Reviews. Please cite as H. Mustroph, Merocyanine Dyes *Physical Sciences Reviews* [Online] 2021, 6. DOI: 10.1515/psr-2020-0145

https://doi.org/10.1515/9783110587104-042

hindered internal rotation of the amide group at room temperature. With increasing temperature the rotation barrier is overcome and the rate at which the methyl groups exchange their positions is so high that they can no longer be distinguished on the NMR time scale [2]. The free activation enthalpy of the internal rotation of the amide group is of $\Delta G^{\ddagger} = 20.4–22.4$ kcal mol^{-1}.

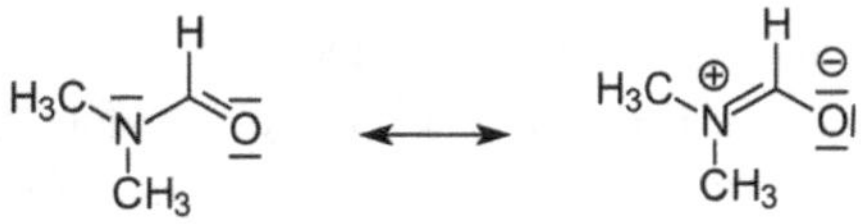

1N **1S**

For the *neutral polymethine dyes* [3], where one terminal component is typically found in *cyanine dyes* and the second obtained from an active *methylene compound,* like e.g. **2**, **3**, Frances M. Hamer suggested the term *merocyanine* (*μεροσ* = part) *dyes* [4].

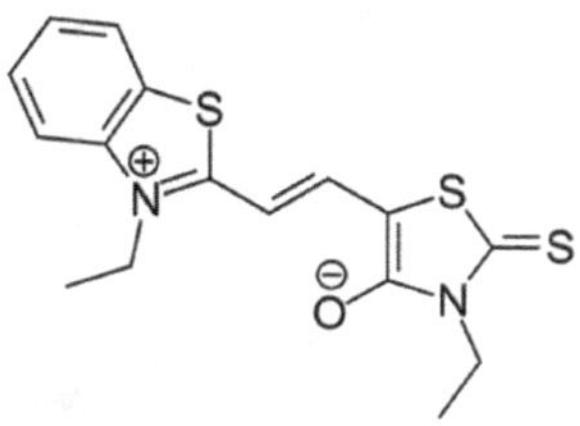

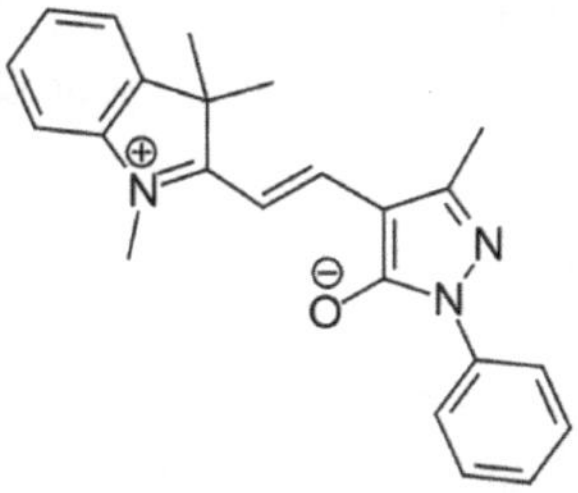

2 **3**

Based on the term *streptocyanine dyes* [3, 5–7] for cyanine dyes with two open-chain R_2N groups (R = H, alkyl, aryl) like e.g. **4**, the open-chain merocyanines **5** are called *streptomerocyanine dyes* [3, 8], which are the simplest model compounds for merocyanines.

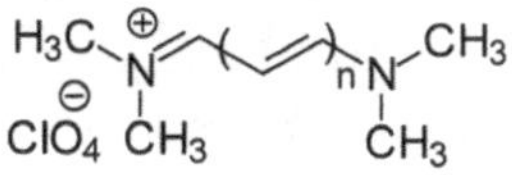

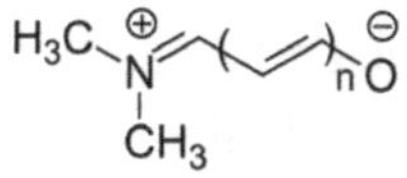

4 **5**

Due to their electronic structure, merocyanine dyes are classified as neutral polymethine dyes [3, 5–9]. As cyanine dyes they are characterised by an odd number $2n + 3$ of π-centres and $2n + 4$ π-electrons (where n is the number of vinylene groups –CH = CH–). Therefore, one would expect them to show the characteristic feature of cyanine dyes like narrow absorption bands in the electronic spectrum, increasing molar absorption coefficient with increasing n and each additional vinyl group in the polymethine chain giving a bathochromic shift of the 0–0 vibronic transition of about 100 nm (the so-called vinylene shift) [5–12].

Table 42.1: Absorption maxima λ_{max} (nm) and molar absorption coefficients ε_{max} (10^{-3} M^{-1} cm^{-1}) of the streptocyanines **4** and the streptomerocyanines **5** in dependence on the number of vinylene groups *n* in dichloromethane [13].

n	**4**		**5**	
	λ_{max}	ε_{max}	λ_{max}	ε_{max}
1	312	64.5	283	37
2	416	119.5	361	51
3	519	207	421	56
4	625	295	462	65
5	734	353	491	68
6	848	(220)	512	72

The streptocyanines **4** confirm these predictions perfectly (Table 42.1), with exception when $n = 6$. The reason for falling ε_{max} in this case is the dye's existence as a complex mixture of geometric isomers.

However, the streptomerocyanines **5** in no way meet both predictions (Table 42.1). There is a much smaller increasing molar absorption coefficient than in the case **4** with increasing *n* and no vinylene shift of λ_{max}.

6N **6S**

The probably most well-known merocyanine is *Brooker´s Merocyanine* **6** with $\lambda_{max} = 605$ nm, $\varepsilon_{max} = 154 \cdot 10^3$ M^{-1} cm^{-1} in pyridine and $\lambda_{max} = 444$ nm, $\varepsilon_{max} = 54 \cdot 10^3$ M^{-1} cm^{-1}) in water [14]. Leslie G. S. Brooker et al. interpreted the electronic spectra of the merocyanines in terms of the VB theory with the assumption that the electronic structure of merocyanines can be described as a *resonance hybrid* between a *non-charge-separated contributing structure* **6N** and *charge-separated contributing structure* **6S** [15, 16]. Each contributing structure has a special electronic energy E^{el}(**N**) and E^{el}(**S**), respectively. In the 2 × 2 matrix representation of the electronic Hamiltonian E^{el}(**N**) – E and E^{el}(**S**) – E represent the diagonal matrix elements and the off-diagonal matrix element β is the electronic interaction matrix element between **N** and **S**. Solving the 2 × 2 matrix gives two electronic Eigenvalues E [17]. In other words, the linear combination of the two contributing structures causes a splitting between them (Figure 42.1).

Then it is assumed that the lowest electronic Eigenvalue corresponds to the electronic energy of the ground state $E^{el}(S_0)$ while the second provides the energy of the first electronic excited state $E^{el}(S_1)$. The energy difference ΔE from the electronic ground state $E^{el}(S_0)$ to the first electronic excited state $E^{el}(S_1)$ is the electronic transition energy [16, 17]:

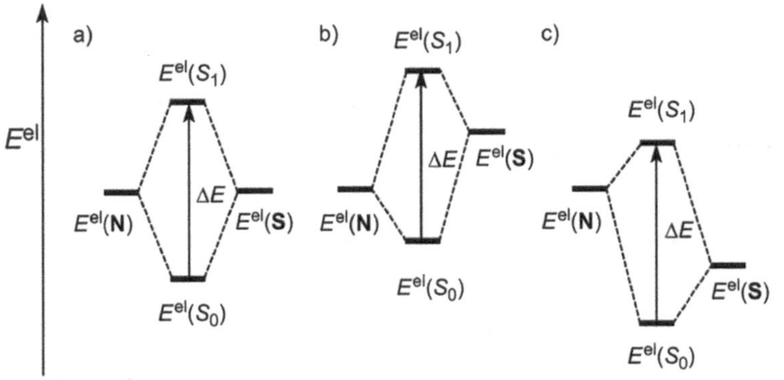

Figure 42.1: VB interaction diagram of the non-charge-separated contributing structure **N** and the charge-separated contributing structure **S** with (a) exact equivalent electronic energies $E^{el}(\mathbf{S}) = E^{el}(\mathbf{N})$ and different electronic energies (b) $E^{el}(\mathbf{S}) > E^{el}(\mathbf{N})$ and c) $E^{el}(\mathbf{S}) < E^{el}(\mathbf{N})$ with the resulting electronic transition energy ΔE from the electronic ground state $E^{el}(S_0)$ to the first electronic excited state $E^{el}(S_1)$.

$$\Delta E = E^{el}(S_1) - E^{el}(S_0) = \{[(E^{el}(\mathbf{S}) - E^{el}(\mathbf{N})]^2 + 4\beta^2\}^{1/2} \tag{42.1}$$

Brooker assumed that β in eq. (42.1) is a constant. In a symmetrical polymethine dye the two principal contributing structures have the same electronic energy, resulting in $\Delta E = 2\beta$. This provides the lowest possible electronic transition energy. Brooker called the electronic states of a molecule, described by linear combination of two contributing structures of exact equivalent electronic energy the *isoenergetic point* (Figure 42.1(a)) [16].

In an unsymmetrical dye the two contributing structures have different electronic energies. With increasing difference between $E^{el}(\mathbf{N})$ and $E^{el}(\mathbf{S})$ ΔE increases, which shall cause a hypsochromic effect (Figure 42.1(b) and Figure 42.1(c)). Furthermore, Brooker assumed, that polar solvents stabilize **S**, whereas **N** remains unaffected and, therefore, $E^{el}(\mathbf{S})$ is lowered and so polar solvents causes a bathochromic effect in the case $E^{el}(\mathbf{S}) > E^{el}(\mathbf{N})$ (Figure 42.1(b)). Vice versa, in the case $E^{el}(\mathbf{S}) < E^{el}(\mathbf{N})$ polar solvents lead to a hypsochromic effect (Figure 42.1(c)).

Already in 1963 John N. Murrell wrote regarding this model: "This is an elegant explanation but it sounds too simple to be true." [18] Systematic experimental studies suggested that this model is indeed too simple [17, 19].

Theodor Förster suggested that the Eigenfunctions of the electronic ground state $\psi(S_0)$ [eq. (42.2)] and first electronic excited state $\psi(S_1)$ [eq. (42.3)] can be approximated as a linear combination of the wavefunctions of the two contributing structures ψ_N and ψ_S [20]. The degree of mixing of **N** and **S** is determined by c^2.

$$\psi(S_0) = (1 - c^2)^{1/2}\psi_N + c\psi_S \tag{42.2}$$

$$\psi(S_1) = c\psi_N - (1 - c^2)^{1/2}\psi_S \tag{42.3}$$

Seth R. Marder et al. built on this model and called the resonance hybrid where the two contributing structures contribute equally ($c^2 = 0.5$) the *cyanine limit*, in which it is assumed that the equilibrium geometry R_e in the ground electronic state S_0 [$R_e(S_0)$] and R_e in the first excited electronic state S_1 [$R_e(S_1)$] are equal [21, 22].

The region $c^2 < 0.5$ was called „before the cyanine limit“ and $c^2 > 0.5$ „beyond the cyanine limit“ [21, 22]. From this model it follows, as c^2 moves away from $c^2 = 0.5$ the equilibrium geometries in S_0 and in S_1 differ increasingly in dependence on the deviation to $c^2 = 0.5$.

As a simplified measure of $R_e(S_0)$ bond length alternation (BLA) in the electronic ground state was introduced [21, 22]. BLA is estimated as the average of the absolute difference between adjacent carbon–carbon equilibrium bond lengths in a polymethine chain. So, the equilibrium geometry in the ground electronic state of polyatomic molecules is transferred to a simplified parameter in a diatomic molecule. Hence the value is BLA = 0 at the theoretical cyanine limit [$c^2 = 0.5$ in eq. (42.2)].

Of course, the solution of the 2×2 matrix representation of the electronic Hamiltonian gives the same results like eq. (42.1). In Brooker´s model β in eq. (42.1) is a constant. Marder et al. set $\beta = -t$ and (different to Brooker´s model) it is said the electronic interaction matrix element between **N** and **S** – t depends on c^2. In the two cases $c^2 = 0$ and $c^2 = 1$ the value of $-t = 0$ and ΔE is given by the difference $E^{el}(\mathbf{S}) - E^{el}(\mathbf{N})$.

The wavefunctions of the two contributing structures ψ_N and ψ_S are purely electronic and so the wavefunctions of the two molecular states $\psi(S_0)$ and $\psi(S_1)$, are purely electronic too. Consequently, as in Brooker's model in this model ΔE is the purely electronic transition energy. However, in molecules there are no pure electronic transitions!

Only in single atoms are there pure electronic transitions. Atoms do not rotate and vibrate and therefore, the spectral bandwidth is narrow in atomic absorption spectroscopy. In contrast to atoms molecules rotate, nuclei vibrate and transitions between electronic states are connected with simultaneous transitions between rotational and vibrational states. Therefore, absorption bands in molecular spectroscopy are broader than those encountered in atomic spectroscopy. The shapes of the former are mainly determined by the spacing and intensity distribution of the vibrational sub-bands.

In both models the pure electronic value ΔE is used and correlated with the easily measured absorption maxima λ_{max} as experimental comparative values. However, the λ_{max} value is nothing more than the intensity maximum in an electronic spectrum. It corresponds to the vibronic (vibrational-electronic) transition with the highest Franck-Condon factor [$S^2_{00,1v}$] as illustrated in Figure 42.2 [12, 23].

The spacing of sub-bands in polymethine dye spectra is primarily determined by a dominant symmetric vibration associated with its polymethine chain and the observed intensities of these sub-bands can be explained in terms of this dominant vibration by the Franck–Condon principle for diatomic molecules [24].

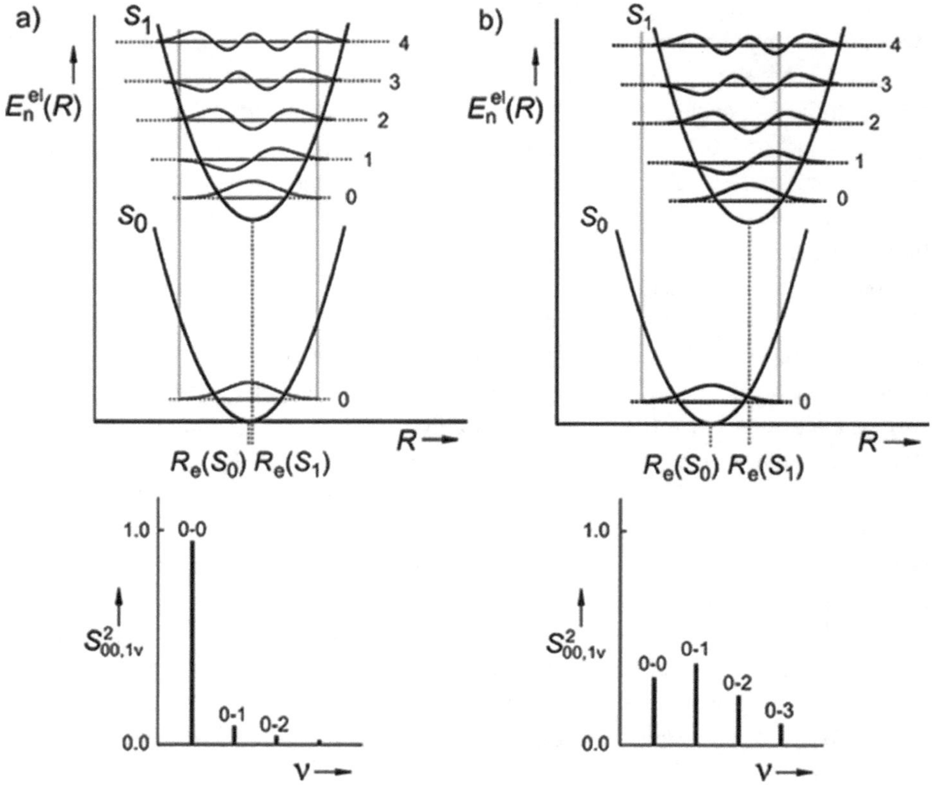

Figure 42.2: Two diagrams providing the effect of different equilibrium geometry in the electronic ground S_0 [$R_e(S_0)$] and first excited state S_1 [$R_e(S_1)$] on the vibronic sub-bands transition intensities of diatomic molecules. The adiabatic potential energy surfaces of S_0 and S_1 are modelled by the quantum harmonic oscillator, its vibrational eigenfunctions assuming equal vibrational frequency in S_0 and S_1 and the relative transition intensities [$S^2_{00,1v}$] in dependence on the differences of $R_e(S_0)$ and $R_e(S_1)$ [12, 23, 24].

At the cyanine limit, where it is assumed that R_e in S_0 and S_1 are equal [$R_e(S_1) = R_e(S_0)$] the only vibronic transition will obviously be the 0–0 vibronic transition [12, 23–25]. That does not correspond to experimental reality. The $S_0 \rightarrow S_1$ electronic transition in polymethine dyes can be ascribed mainly to an electronic transition from the highest occupied molecular orbital (HOMO) to the lowest unoccupied molecular orbital (LUMO). The HOMO has bonding character whereas the LUMO has antibonding character. It therefore follows that $R_e(S_1)$ is basically greater than $R_e(S_0)$. The difference $R_e(S_1) - R_e(S_0)$ depends on the electronic structure in S_1 and S_0.

If the difference between equilibrium geometries is relatively low [$R_e(S_1) \gtrneqqless R_e(S_0)$] as in symmetric cyanine dyes the absorption intensity is largely concentrated in the 0–0 vibronic transition (Figure 42.2(a)) [10, 12, 23]. Already different terminal heterocyclic groups in unsymmetrical cyanine dyes cause an unsymmetrical electronic structure in S_1 and S_0. This leads to increasing differences $R_e(S_1) > R_e(S_0)$ which cause an intensification of higher 0-*v* vibronic transitions (Figure 42.2(b)) [26]. If the change of the equilibrium geometries is much larger in unsymmetrical cyanine dyes [$R_e(S_1) \gtrless R_e(S_0)$] the absorption curve does not exhibit a vibrational fine structure [26], as it is the case in most spectra of merocyanine dyes. In general, if the spectra do not show a vibrational structure, it is not clear which underlying vibronic transitions is represented by the λ_{max} values. Obviously, neither the effects of structural changes nor solvents on the electronic spectra of merocyanine dyes can be accounted for by the simple pure electronic model with two contributing structures.

It is very important not to consider the pure electronic transitions only, but the effects of the changes in equilibrium geometry S_1 and S_0 on the distribution of the spacing and intensity of the vibrational sub-bands. Therefore, the Förster model with the linear combination of the two wavefunctions ψ_N and ψ_S [20] should be used only in combination with the Franck-Condon principle for diatomic molecules [12, 23, 24].

This shall be illustrated with Figure 42.3. The spectra of 4,4´-dimethylpyrido-trimethincyanine show a clear vibrational structure in dichloromethane (DCM) and in MeOH and λ_{max} corresponds to the 0–0 vibronic sub-band. The spectrum of the iso-π-electronic Brooker´s Merocyanine exhibits in DCM a clear vibrational structure and also here λ_{max} is represented by the 0–0 sub-band (Figure 42.3(a)). However, in MeOH the vibrational structure disappears and λ_{max} is substantially hypsochromic shifted (Figure 42.3(b)).

Both contributing structures for 4,4´-dimethylpyrido-trimethincyanine have the same two terminal components with the same electron acceptor/donor abilities and so there is a small solvatochromic effect on the 0–0 transition only. In **6** the two terminal components have different electron acceptor/donator abilities and, therefore, already in DCM there is a change of R_e from S_0 to S_1 [$R_e(S_1) > R_e(S_0)$], so λ_{max} is represented by the 0–0 sub-band, but the intensity of the 0–1 sub-band is substantial higher. Within the Förster model it follows, that the polar solvent MeOH moves c^2 to $c^2 \gg 0.5$ with the result of a big change of R_e from S_0 to S_1 [$R_e(S_1) \gg R_e(S_0)$] the absorption curve does not exhibit a vibrational fine structure and λ_{max} is substantially hypsochromic shifted (Figure 42.3(b)).

Increasing the electron donor ability of the terminal phenolate component by substitution of two t-Bu groups as in 7, leads to a more symmetric electronic structure and spectra with a clear vibrational structure in all aprotic solvents and in most alcohols [17, 27, 28].

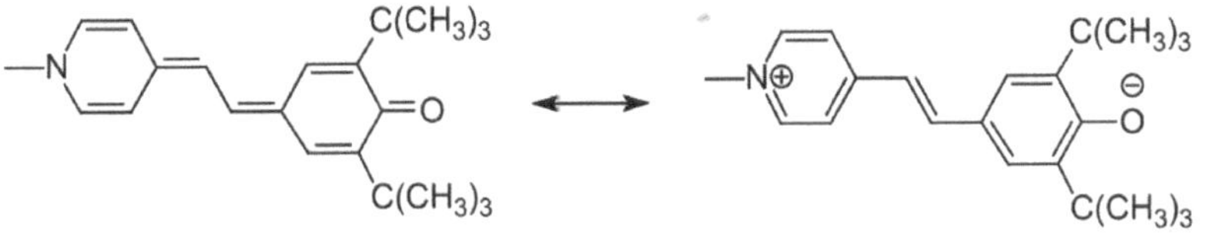

7N **7S**

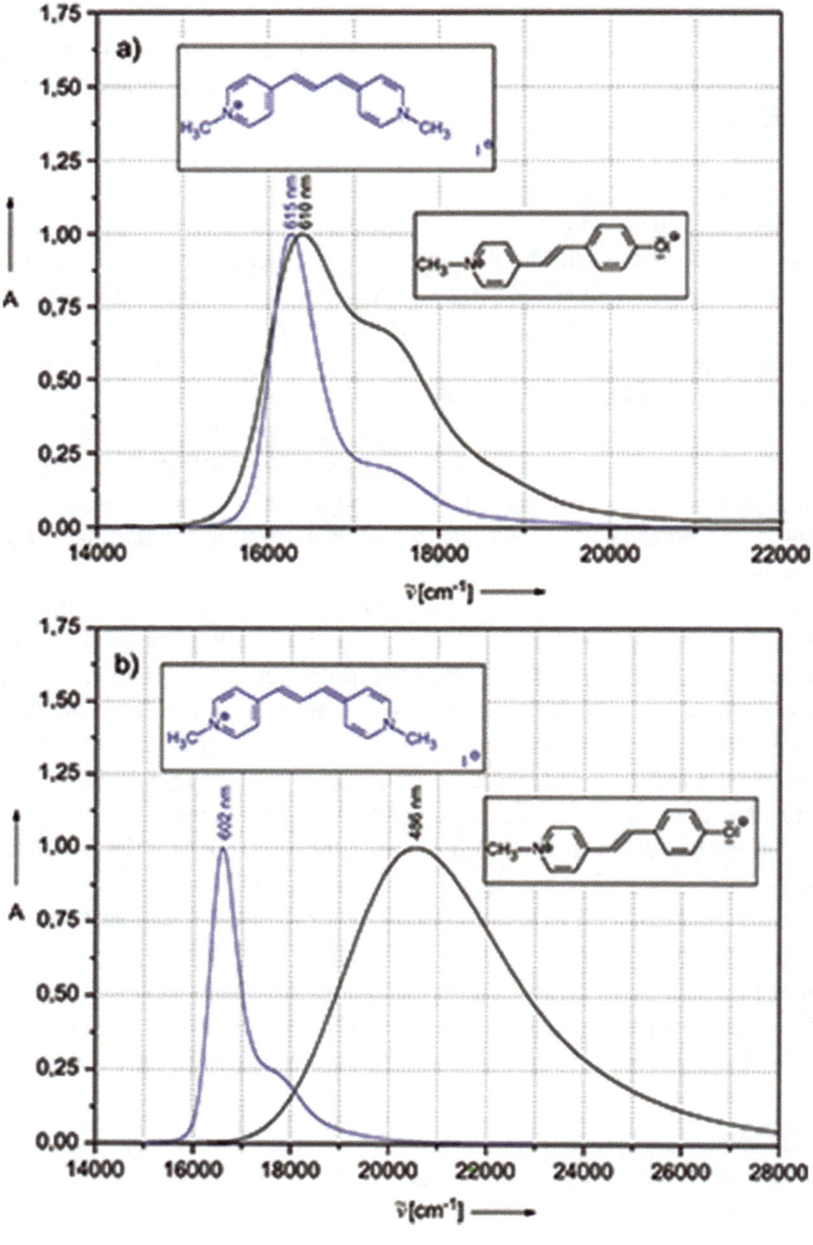

Figure 42.3: Normalized absorption spectra of Brooker's Merocyanine and the iso-π-electronic 4,4´-Dimethylpyrido-trimethincyanine (a) in DCM and (b) in MeOH.

A similar effect can be achieved by replacing oxygen in the streptomerocyanine **5** with malodinitrile as in **8** (Table 42.2).

Table 42.2: Absorption maxima λ_{max} (nm) and molar absorption coefficient ε_{max} (10^{-3} M^{-1} cm^{-1}) of the streptomerocyanines **8** in dependence on the number of vinylene groups *n* in EtOH [29].

n	λ_{max}	ε_{max}
1	372	93
2	473	260
3	579	470
4	670	520

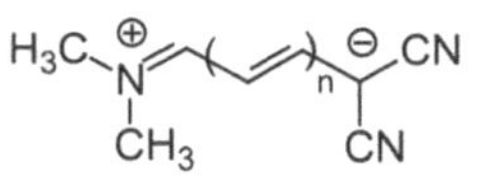

8

The crystal structures of **5** ($n = 3$) BLA = 5.8 pm and of **8** ($n = 3$) BLA = 0.5 pm provide direct evidence for decreased BLA in **8** in comparison with **5** [21].

NMR investigations and related quantum chemical calculations on a number of cyanines, merocyanines and polyenes have revealed that there is an almost linear correlation between the ^3J(H,H) coupling constants for *trans* vicinal protons in the polymethine chain and the carbon–carbon equilibrium bond lengths [28–34]. Accordingly, the degree of bond length alternation in solution can be estimated by the absolute difference ΔJ of the ^3J(H,H) coupling constants between adjacent vinylene groups. For illustration, for **5** ($n = 1$) is the value $\Delta J = 4.8$ Hz in $CDCl_3$ [30], whereas for **8** ($n = 1$) is $\Delta J = 1.0$ Hz in [D6]DMSO [31]. For comparison, for **4** ($n = 2$) is $\Delta J = 0.6$ Hz in $CDCl_3$ and $\Delta J = 0.8$ Hz in [D6]DMSO.

To transfer the structural influence in polyatomic molecules to the Franck-Condon principle for diatomic molecules the BLA (ΔJ) in the electronic ground state can be used. The small BLA and ΔJ in the streptomerocyanines **8** are clear experimental evidence, that the differences between carbon–carbon equilibrium bond lengths in the polymethine chain are small. From eqs. (42.2) and (42.3) it follows that the difference $R_e(S_1) - R_e(S_0)$ is small and it is highly likely λ_{max} values represent the 0–0 vibronic transition energy. So, the streptomerocyanines **8** reflect the above discussed structure influences on λ_{max} and ε_{max} in an excellent manner and all spectra of compounds $n > 0$ exhibit a clear vibrational structure [31].

Vice versa, the great BLA and ΔJ in the streptomerocyanines **5** are clear experimental evidence, that the differences between carbon–carbon equilibrium bond lengths in the polymethine chain are great. With increasing BLA (ΔJ) $R_e(S_0)$ and $R_e(S_1)$ differ increasingly and it is highly likely λ_{max} values represent another 0–v vibronic transition. In addition, all spectra of series **5** exhibit no fine structure [13].

In summary, merocyanines fall within the polymethine dye class, but whoever discusses their absorption spectra should pay attention to fine structure of the absorption bands.

42.2 History of the merocyanine dyes

The first merocyanine dye **9** was synthesized by John D. Kendall at the photographic company Ilford by the reaction of rhodanic acid with 3-ethyl-2-methylthiobenzthiazolium [35].

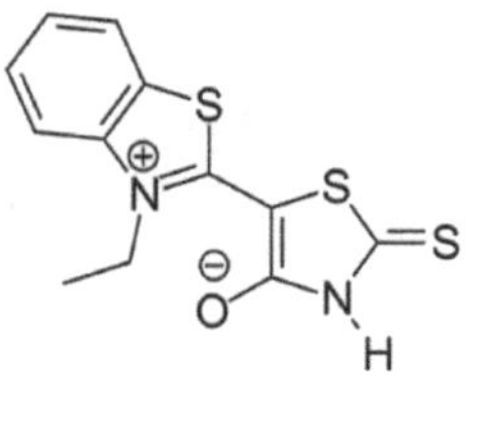

9

As mentioned already Hamer suggested the term merocyanine [4]. Also here, silver halide photography was the driving force to make further developments in the field of merocyanine dyes. A short time later Hamer and Brooker presented a lot of new merocyanines based on components from cyanine dye chemistry and open-chain or cyclic keto-methylene compounds like e.g. acetylacetone, barbituric acids, cyanoacetamides, cyanoacetates, hydantoin, isoxazolones, malodinitrile, malonates, pyrazolones, thiohydantoin, thio-oxazolidinediones and thiazolinones [4–7, 36].

For illustration only, one of the most successful and widely used merocyanines in black and white materials was the green sensitizer **10** (Sto 749).

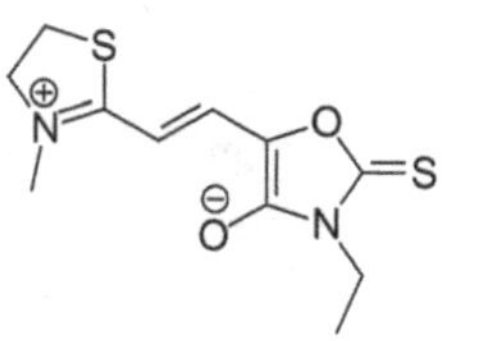

10

Later merocyanines for textile dyeing and other applications were developed. Since completely different technical requirements were demanded of them, e.g. fastness, they were constructed from new components.

42.3 General synthetic routes to merocyanine dyes

The first merocyanine dye **9** was prepared from the reaction between rhodanic acid with 3-ethyl-2-methylthio-benzthiazolium [35]. It is the general way to synthesize zeromethine merocyanines **13** using heterocyclic quaternary salt **11** with a leaving group R and the anions of keto-methylene compounds **12** (Figure 42.4) [4–7, 36].

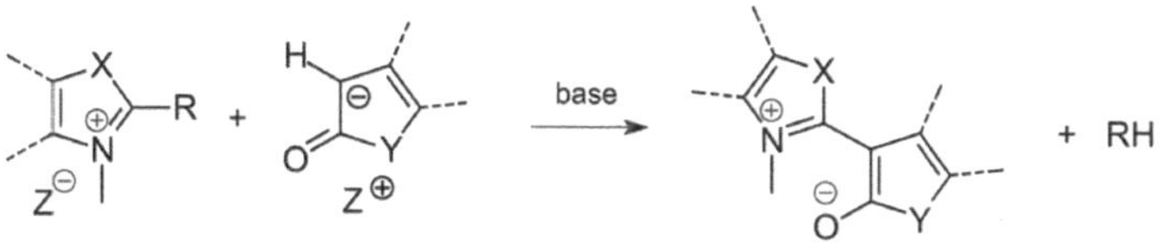

11 **12** **13**

R = Cl, Br, I, SR, $SO_3^{\ominus}$, ...

Figure 42.4: Reaction scheme.

To obtain the vinylogues of the zeromethine merocyanines **18** a 2-methyl heterocyclic quaternary salt **14** reacts with diphenylformamidine hydrochloride **15** (n = 0), malondialdehyde dianil hydrochloride **15b** (n = 1) or glutacondialdehyde dianil hydrochloride **15c** (n = 2) to the intermediate **16** with *β*-acetanilido vinyl groups (sometimes *β*-anilino vinyl groups). groups. Then the intermediate **16** is condensed with anions of keto-methylene compounds **17** to the merocyanines **18** (Figure 42.5) [4–7, 36].

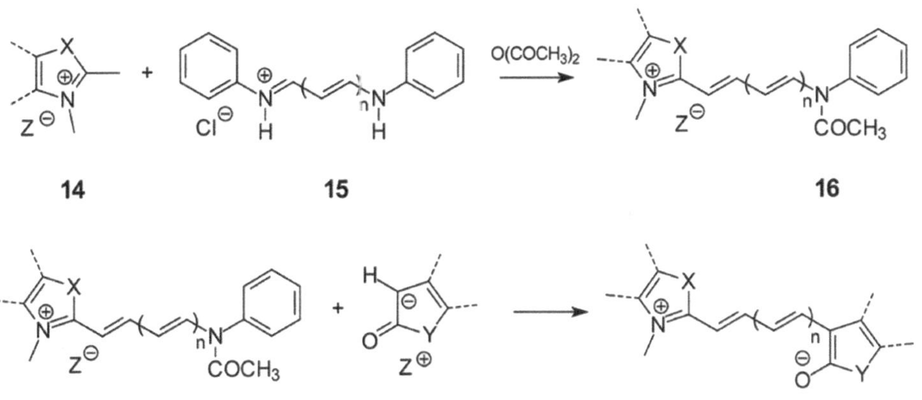

Figure 42.5: Reaction scheme.

Commercially used are only zeromethine **13**, dimethine **18** ($n = 0$) and tetramethine **18** ($n = 1$) merocyanines. Dyes with longer polymethine chain **18** ($n > 1$) are of less technical importance due to their low stability.

Especially for the synthesis of dimethine dyes based on 1-alkyl-2,3,3-trimethylindolenine the corresponding aldehyde (Fischer´s aldehyde) is used. So, e.g. **3** is synthesized by the reaction of 2-(1,3,3-trimethylindolin-2-ylidene)acetaldehyde with 1-phenyl-3-methylpyrazol-5-one.

42.4 Commercial uses of merocyanine dyes

The first and main technical application of merocyanine dyes was in silver halide photography as spectral sensitizers [37–39].

Zeromethine merocyanines like **9**, **19** and **20** are suitable blue sensitizers in the region 450–500 nm. The carboxyl groups in **19** and **20** increase the water solubility, preventing residual coloration after processing.

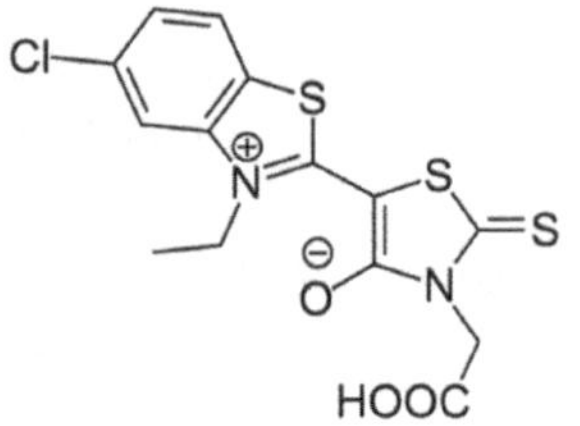

19

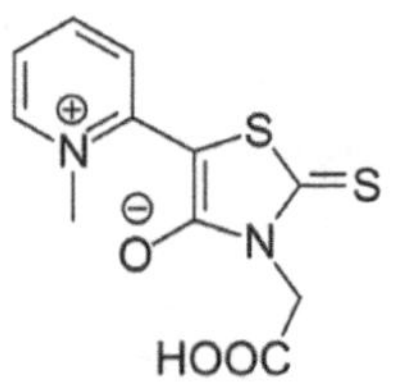

20

Dimethine merocyanine like e.g. **2**, **10** and **21** are used as green sensitizers in the region 500–600 nm.

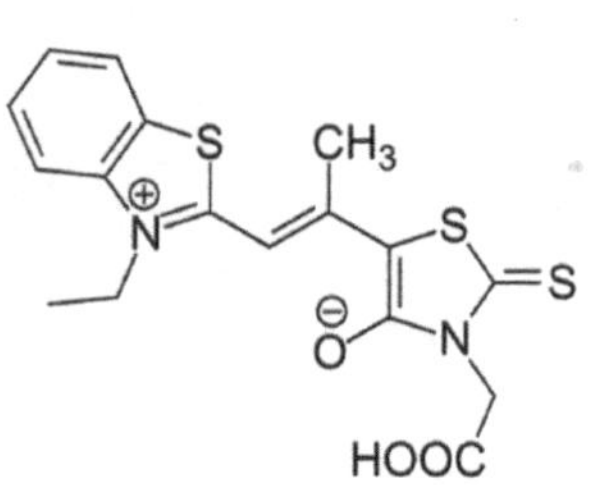

21

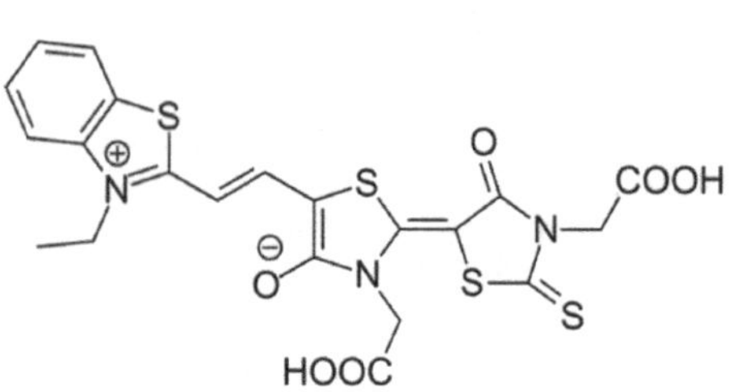

22

As red sensitizers in the region 600–700 nm tetramethine and trinuclear merocyanines **22** are used. In all cases the function of the carboxyl groups is to increase the water solubility, preventing residual coloration after processing.

Due to their low light fastness and thermal stability merocyanines like **2**, **6**, **7**, **9**, **10**, **19–22** are not suitable for textile dyeing. To improve these properties derivatives of N-akyl-1,8-naphtholactam were introduced. By condensation of N-ethyl-1,8-naphtholactam with malodinitrile the yellow dye **23** is obtained. The reaction of the corresponding methylene-ω-aldehyde with ethyl cyanoacetate gives the dimethine merocyanine dye **24**. The extension of the polymethine chain by a vinyl group leads to a red color of **24**.

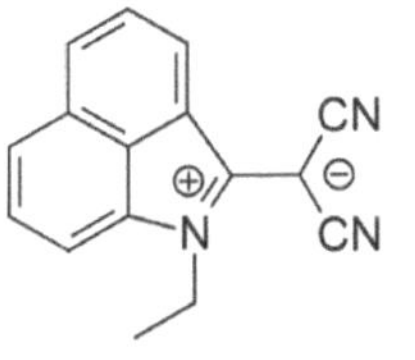

23

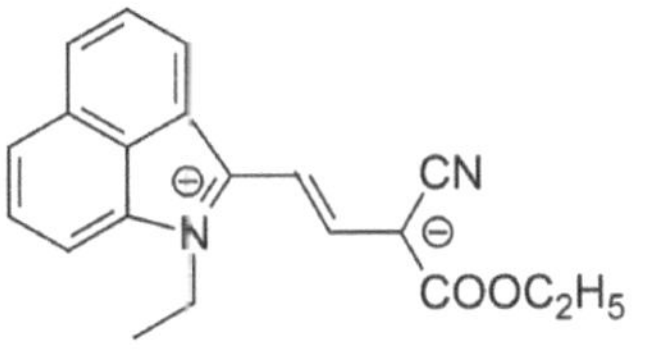

24

The dimethinemerocyanine **25** is obtained by condensation of Fischer´s aldehyde with ethyl cyanoacetate and gives a stable yellow dye. Substitution with a carboxylate group at the 5-position of the indoline heterocycle produces a very lightfast dye.

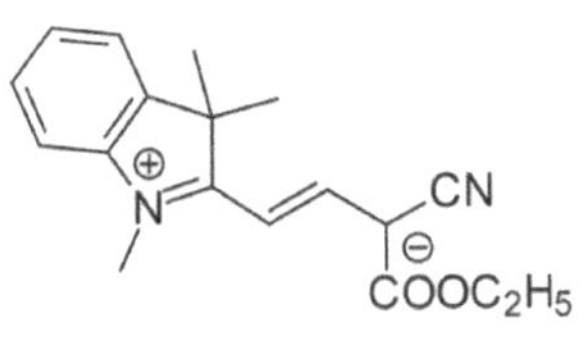

25

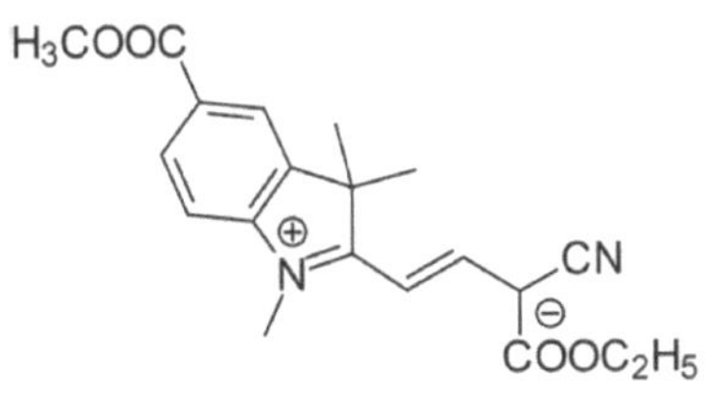

26

Textile dyeing is not the main field of application of merocyanine dyes, however the samples **23–26** illustrate how dyes, developed for use in another technology, provide suggestions for the development of new textile dyes with higher demands on fastness properties.

In the 1990s the field of nonlinear optics (NLO) was inflated with too much hype [21, 22, 40], followed by an abrupt decline. Due to the low light fastness and thermal stability of the functional colorants developed for this field there was no real technical application. Nevertheless, this area of research stimulated many new and interesting dye syntheses [40, 41] like e.g. **27** and **28**.

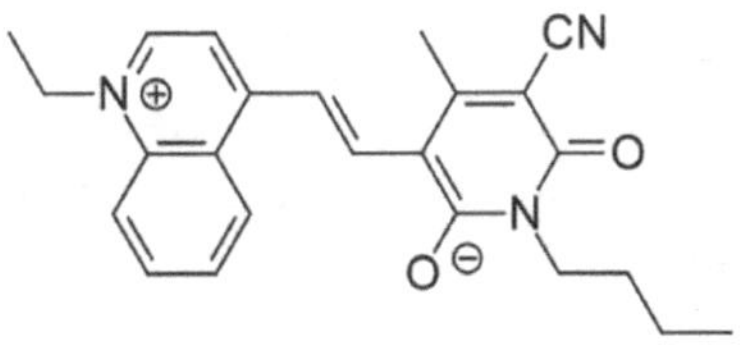

27

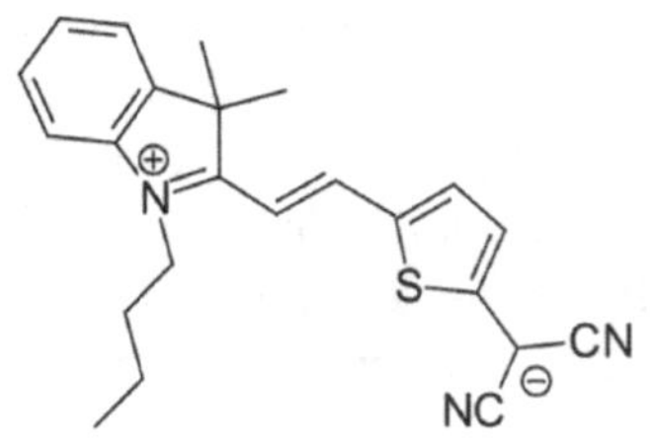

28

Merocyanine 540 **29** was among the first fluorescent dyes to determine membrane potentials in eukaryotic cells and prokaryotic bacteria [42].

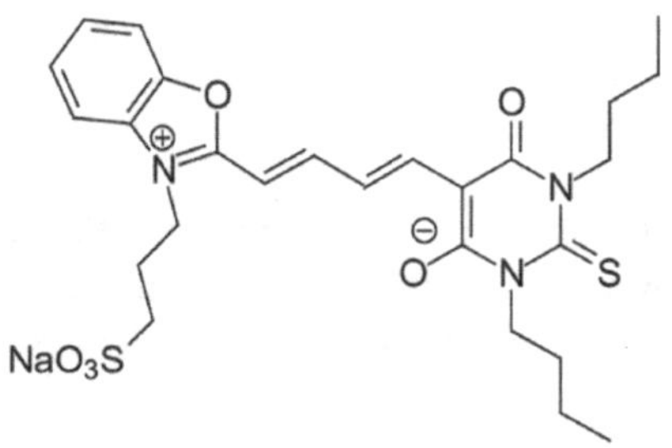

29

However, its use for this application has declined with the advent of superior probes and its extreme phototoxicity. Irradiation of Merocyanine 540 **29** produces both singlet oxygen and other reactive oxygen species, including oxygen radicals. Therefore, it is now more commonly used as a photosensitizer in photodynamic therapy.

References

1. Pauling L. The nature of the chemical bond. 3rd ed. Ithaca: Cornell University Press, 1960:281–2.
2. Friebolin H. Basic one- and two-dimensional NMR spectroscopy. 4th ed. Weinheim: Wiley-VCH, 2005.
3. Mustroph H. Polymethine dyes. Phys Sci Rev. 2020;5. DOI:10.1515/psr-2019-0084.
4. Hamer FM, Winton BS. A Trimethincyanine; a trimethinoxonol and some compounds related to trimethinoxonols; some di- and tetra-methinmerocyanines. J Chem Soc. 1949;1126.
5. Berneth H. Methine dyes and pigments. In: Elvers B, editor. Ullmann's encyclopedia of industrial chemistry. Weinheim: Wiley-VCH, 2012.
6. Berlin L, Riester O. Methoden zur Herstellung von Cyaninen (Polymethinen). In: Müller E, editor. Methoden der organischen Chemie (Houben-Weyl). vol. 5/1d. Stuttgart: Thieme, 1972:231–98.

7. Hamer FM. The cyanine dyes and related compounds. In: Weissberger A, editor. The chemistry of heterocyclic compounds. vol. 18. New York: Interscience, 1964.
8. Tyutyulkov N, Fabian J, Mehlhorn A, Dietz F, Tadjer A. Polymethine dyes – structure and properties. Sofia: St Kliment Ohridski University Press, 1991.
9. Griffiths J. Colour and constitution of organic molecules. London: Academic Press; 1976.
10. Mustroph H. Cyanine dyes. Phys Sci Rev. 2020;5. DOI:10.1515/psr-2020-0145.
11. Hallas G. Effects of extended conjugation in some basic dyes. J Soc Dyers Colour. 1970;86:237.
12. Mustroph H, Ernst S, Senns B, Towns AD. Molecular electronic spectroscopy: from often neglected fundamental principles to limitations of state- of-the-art computational methods. Color Technol. 2015;131:9.
13. Malhotra SS, Whiting MC. Researches on polyenes. Part VII. The preparation and electronic absorption spectra of homologous series of simple cyanines, merocyanines, and oxonols. J Chem Soc. 1960;3812.
14. Brooker LG, Keyes GH, Heseltine DW. Color and constitution. XI. Anhydronium bases of p-hydroxystyryl dyes as solvent polarity indicators. J Am Chem Soc. 1951;73:5350.
15. Brooker LG, Keyes GH, Sprague RH, VanDyke RH, VanLare E, VanZandt G, et al. Color and constitution. X. Absorption of the merocyanines. J Am Chem Soc. 1951;73:5332.
16. Brooker LG. Some recent developments in the chemistry of photographic sensitizing dyes. Experientia Suppl. 1955;2:229.
17. Mustroph H, Reiner K, Senns B, Mistol J, Ernst S, Keil D, et al. The effects of substituents and solvents on the ground-state π-electronic structure and electronic absorption spectra of a series of model merocyanine dyes and their theoretical interpretation. Chem Eur J. 2012;18:8140.
18. Murrell JN. The theory of the electronic spectra of organic molecules. London: Methuen, 1963:86.
19. Mustroph H, Mistol J, Senns B, Keil D, Findeisen M, Hennig L. Relationship between the molecular structure of merocyanine dyes and the vibrational fine structure of their electronic absorption spectra. Angew Chem Int Ed. 2009;48:8773.
20. Förster T. Farbe und Konstitution organischer Verbindungen vom Standpunkt der modernen physikalischen Theorie. Z Elektrochem. 1938;45:548.
21. Marder SR, Perry JW, Tiemann BG, German CB, Gilmour S, Biddle SL, et al. Direct observation of reduced bond length alternation in donor/scceptor polyenes. J Am Chem Soc. 1993;115:2524.
22. Marder SR, Gorman CB, Tiemann BG, Cheng L-T. Stronger acceptors can diminish nonlinear optical response in simple donor-acceptor polyenes. J Am Chem Soc. 1993;115:3006.
23. Mustroph H. Dyes: quantum chemical calculation of electronic spectra. Phys Sci Rev. 2019;4. DOI:10.1515/psr-2019-0040.
24. Mustroph H, Towns A. Fine structure in electronic spectra of cyanine dyes: are sub-bands largely determined by a dominant vibration or a collection of singly-excited vibrations? Chem Phys Chem. 2018;19:1016.
25. Mustroph H, Reiner K, Mistol J, Ernst S, Keil D, Hennig L. Relationship between the molecular structure of cyanine dyes and the vibrational fine structure of their electronic absorption spectra. Chem Phys Chem. 2009;10:835.
26. Mustroph H, Reiner K, Senns B. Bond length alternation in unsymmetrical cyanine dyes and its influence on the vibrational structure of their electronic absorption spectra. Color Technol. 2017;133:469.
27. Catalán J, Mena E, Meutermans W, Elguero J. Solvatochromism of a typical merocyanine: stilbazolium betaine and its 2,6-di-tert-butyl derivative. J Phys Chem. 1992;96:3614.

28. Morley JO, Morley RM, Fitton AL. Spectroscopic studies on Brooker's merocyanine. J Am Chem Soc. 1998;120:11479.
29. Scheibe G, Seiffert W, Hohlneicher G, Jutz C, Springer HJ. Trans-Protonenkopplungskonstanten und C–C-Abstände bei Cyaninen, Merocyaninen und Olefinen. Tetrahedron Lett. 1966;5053.
30. Radeglia R, Dähne S. Lösungsmittelinduzierte Elektronenverschiebungen in einfachen Merocyanin-Farbstoffen. J Mol Struct. 1970;5:399.
31. Scheibe P, Schneider S, Dörr F, Daltrozzo E. Spektroskopische Untersuchungen an Merocyaninen I. UV- und NMR-Spektren von malodinitril-substituierten vinylogen Säureamiden. Ber Bunsenges Phys Chem. 1976;80:630.
32. Wähnert M, Dähne S, Radeglia R. The effect of solvents on the electronic structure of symmetrical polymethinecyanines – ^{1}H- and ^{13}C-NMR spectroscopic investigations. Adv Mol Relax Interact Processes. 1977;11:263.
33. Blanchard-Desce M, Alain V, Bedworth PV, Marder SR, Fort A, Runser C, et al. Large quadratic hyperpolarizabilities with donor-acceptor polyenes exhibiting optimum bond length alternation: correlation between structure and hyperpolarizability. Chem Eur J. 1997;3:1091.
34. Morley JO, Morley RM, Docherty R, Charlton MH. Fundamental studies on Brooker's merocyanine. J Am Chem Soc. 1997;119:10192.
35. Kendall JD. GB 426 718. 3 Oct 1933.
36. Brooker LG, Keyes GH, Sprague RH, VanDyke RH, VanLare E, VanZandt G, et al. Studies in the Cyanine dye series. XI. The merocyanines. J Am Chem Soc. 1951;73:5326.
37. James TH, editor. The theory of the photographic process. 4th ed. New York: Macmillan, 1977.
38. Keller K, Kampfer H, Matejec R, Lapp O, Krafft W, Frenken H, et al. Photography. In: Elvers B, editor. Ullmann's encyclopedia of industrial chemistry. Weinheim: Wiley-VCH, 2000.
39. Fujita S. Organic chemistry in photography. Berlin: Springer, 2004.
40. Würthner F, Wortmann R, Meerholz K. Chromophore design for photorefractive organic materials. Chem Phys Chem. 2002;3:17.
41. Würthner F. DMF in acetic anhydride: a useful reagent for multiple-component syntheses of merocyanine dyes. Synthesis. 1999;2103.
42. Waggoner AS. Dye indicators of membrane potential. Annu Rev Biophys Bioeng. 1979;8:47.

Robert Christie and Adrian Abel

43 Metal complex pigments

Abstract: There are several commercially significant metal complex organic pigments that are based on first row transition metals. The most important of these are the copper phthalocyanine blue and green pigments which find virtually universal use in paints, printing inks, and plastics. These pigments are of such prime importance that they are dealt with separately in three other chapters in this series. This paper describes a group of pigments that are complexes of iron, copper, nickel, and cobalt with polydentate colored ligands of azo, azomethine, oxime, and isoindoline chemical types. The oldest metal complex organic pigment that still finds some use is CI Pigment Green 8, an octahedral oxime iron complex. In the 1970s and 1980s, there was considerable industrial research effort aimed at developing metal complex pigments based on azomethine and isoindoline structures, many of which were found to offer excellent lightfastness, good solvent resistance and thermal stability, although they exhibited rather dull colors. However, several products provide brilliant effects when used in combination with metallic and pearlescent pigments in automotive paints. Many of the pigments introduced have since been withdrawn by the original manufacturers, but a few remain on the market. The synthesis of metal complex pigments generally involves the preparation of the colored ligand, which is then complexed with the transition metal ion

Keywords: metal complex pigments, azo metal complex pigments, azomethine metal complex pigments, oxime metal complex pigments, isoindoline metal complex pigments, transition metal complexes, coordination compounds, chelates, mordants, CI Pigment Yellow 117, CI Pigment Yellow 129, CI Pigment Yellow 150, CI Pigment Yellow 153, CI Pigment Yellow 177, CI Pigment Yellow 179, CI Pigment Orange 59, CI Pigment Orange 65, CI Pigment Orange 68, CI Pigment Red 257, CI Pigment Green 8, CI Pigment Green 10

43.1 Fundamentals

Since a primary requirement of a pigment is insolubility in its application medium, a variety of approaches have been adopted over the years to develop improved organic pigments with reduced solubility. These approaches have produced pigments

This article has previously been published in the journal Physical Sciences Reviews. Please cite as R. Christie, A. Abel, Metal Complex Pigments *Physical Sciences Reviews* [Online] 2021, 6. DOI: 10.1515/psr-2020-0181

https://doi.org/10.1515/9783110587104-043

not only with enhanced fastness to solvents that the pigment may encounter in application but can also lead to improvements in other fastness properties such as to light and heat. The methods used include increasing molecular size and/or incorporating functional groups that provide reduced solubility, such as amide and sulfonamide groups, and halogens. It is also well established that the incorporation of certain metals into dye structures produce insoluble organic pigments by introducing some inorganic character. The use of alkaline earth metal cations (Ca, Mg, Sr, Ba) to insolubilize anionic (acid) dyes and of polyoxoanions of molybdenum and tungsten to insolubilize cationic (basic) dyes are discussed in the chapters entitled Monoazo (monohydrazone) pigments based on derivatives of 2-naphthol, and Cationic (basic) dye complex pigments, respectively. Metal complexes, also known as coordination compounds, are compounds that consist of a central metal bonded to donor molecules or ions known as ligand [1, 2]. An important feature of the chemistry of transition metals (d block elements) is that they readily form metal complexes. A ligand is a molecule that bonds to the metal atom through several donor atoms in the ligand, the complexes thus formed referred to as chelate. Metal complex dyes and pigments are formed by the first transition metal series, in which the coordination number is either four, in which case the geometrical arrangement is either square planar or tetrahedral, or six, in which case the geometry is octahedral. This contribution discusses metal complex organic pigments formed by metals of the first transition series.

43.2 History

An early use of transition metal complexes in the dyeing of textiles involved the use of mordants to "fix" the dye to the fabric and provide enhanced fastness performance. Mordanting agents of a variety of chemical types were generally required when natural dyes were used. The term originates from the Latin *mordere*, meaning "to bite", as it was originally envisaged that a mordant caused the dye to "bite onto the fiber", thus making it fast to washing. As synthetic dyes became dominant, transition metal ions, notably Cr (III), were often used as mordants to form coordination-complexes with certain dyes, referred to as mordant dyes, thus assisting attachment to the fiber. The process can also be used to intensify stains in cell or tissue preparations. Although mordant dyeing is still used, especially by small batch dyers, this method has largely been displaced in industry by other dye application classes [3].

One of the oldest synthetic pigments that can be described as a metal complex is designated as CI Pigment Green 8, usually referred to as Pigment Green B, and sometimes as Iron Green. This pigment is based on CI Mordant Green 4 complexed with iron (II). In 1885, Hoffman originally reported the precipitation of the complex prepared by addition of iron (II) sulfate to an aqueous alcoholic solution of 1-nitroso-2-naphthol [4]. As it was an insoluble product with a dull, uninteresting color, its potential significance was not realised at the time. It was not until 1921 that BASF released the product on to the market as a pigment. Its main application was for printing wallpaper, and it could also be used for the coloration of rubber. Initially, the pigment had a coarse texture, but when methods for improving the texture were developed, it found a market in textile printing in the early 1940s, only to be eclipsed a few years later by the introduction of the much stronger and brighter copper phthalocyanine green (CI Pigment Green 7). The development of metal complex pigments received relatively little further attention in the first half of the twentieth century, until an azo nickel complex, known as Nickel Azo Yellow (CI Pigment Green 10) was developed and patented by Kvalnes and Woodward in 1946 [5]. This pigment was introduced to the market by DuPont in 1947 as Green Gold. It was found to provide excellent lightfastness that was unequalled when it was first introduced, and it offered good value due to its low cost. The unmetallized dye from which the complex was formed had been known since 1905. The only other important addition to the range of azo metal complex pigments was CI Pigment Yellow 150, which is still marketed based on its high tinctorial strength and good fastness properties. In the 1970s through to the 1980s, a significant body of industrial research centered on metal complex pigments, especially those based on azomethine and isoindoline structures. The products developed were mainly complexes of copper, nickel, and cobalt. Although most of those introduced are no longer offered by the original manufacturers, a few remain on the market. One reason for the disappearance of several pigments has been the rationalisation of product ranges by pigment manufacturers, especially as a consequence of the acquisition of Ciba by BASF, both of whom had been active in this area and, to a lesser extent, the acquisition of the Hoechst pigment range by Clariant. An additional factor has been a growing concern over nickel and cobalt compounds, which required investment in additional testing concerning their safety in use. Thus, only pigments that provided a suitable return on the investment were registered under REACH, a necessary requirement for their manufacture or importation into Europe. Also, in most cases, a major disadvantage is that they exhibit rather dull colors.

43.3 Structures and properties

The most important transition metal complexes used as organic pigments are the copper phthalocyanine blue and green pigments [6,7,8]. Copper phthalocyanine (CuPc) (**1**), CI Pigment Blue 15, the structure of which is illustrated in Figure 43.1, is by far the most important blue organic pigment, finding almost universal use as a colorant for paints, printing inks, plastics, and a wide range of other applications. Similarly, its halogenated derivatives, CI Pigments Green 7 and 36, provide the most important green organic pigments. The polycyclic molecular structure is planar and rigid, consisting of four isoindole units connected by nitrogen atoms, thus forming an internal 16-membered ring of alternate carbon and nitrogen atoms. The copper atom is coordinated to the four internal nitrogen atoms of the phthalocyanine ring system in a square planar arrangement. Since these pigments are of such commercial dominance, they are discussed separately in three individual chapters in this series – Phthalocyanine Pigments: General Principles, Phthalocyanine Blue Pigments, and Phthalocyanine Green Pigments. This paper describes metal complex organic pigments of other chemical types which have had some, but generally limited, commercial impact [9–11].

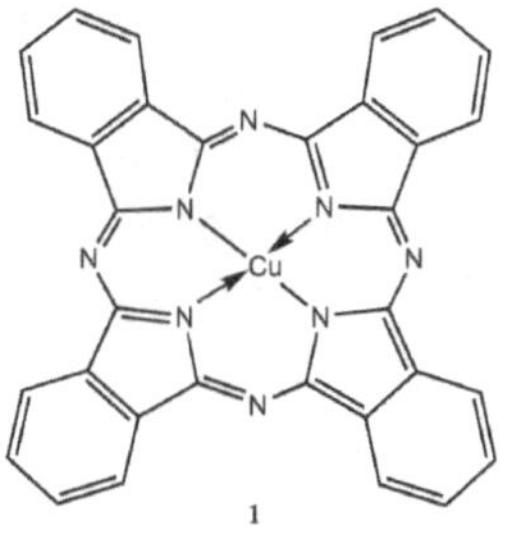

1

Figure 43.1: Structure of copper phthalocyanine (1).

In the chemistry of textile dyes, transition metal complexes frequently exhibit significantly better lightfastness than the unmetallized dye. This feature is demonstrated in the premetallized dye class, which consists of octahedral complexes of chromium (III) or, to a lesser extent, cobalt (III), with the dye as a polydentate ligand. It has been proposed that the photostability of the dye is enhanced by coordination with the transition metal ion because of a reduction in electron density at the chromophoric group, leading to improved resistance to photochemical oxidation [12]. Other factors that may contribute to the lightfastness include steric protection to degrading influences by the metal ion, and the ability of transition metal ions to quench excited states that otherwise might lead to photochemical decomposition. In the second half of the twentieth century, significant effort was devoted to attempts to exploit the potential of metal complex chemistry to provide high-performance organic pigments, particularly in yellow and red shades, aiming to complement copper phthalocyanines, which cannot be extended outside blues and greens. Several interesting products, which offered excellent lightfastness together with reasonably good solvent resistance and thermal stability, were commercialized. The improvement in solvent resistance of the metal complex, compared with the free colored ligand, may be explained by an increase in molecular size, and by molecular aggregation within the crystal lattice structure. Most metal complex pigments are derived from colored polydentate ligands and are thus chelates, a feature that provides enhanced thermodynamic stability. A major disadvantage of metal complex organic pigments is that, in most cases, they exhibit dull colors. This feature has been attributed to a variety of effects that broaden the absorption band compared with the colored ligand. The color of the organic ligand is due to a π–π* electronic transition which can provide bright colors. In the metal complexes, additional visible absorption bands may appear, for example, due to ligand-metal charge transfer or d–d transition electronic transitions, overlapping the absorption bands due to the ligand [13]. Curiously, this feature does not apply to copper phthalocyanines in which the colors of the complexes are indeed brighter than metal-free phthalocyanine. The visible spectrum of metal-free phthalocyanine in solution shows two absorption bands, while the corresponding spectrum of copper phthalocyanine exhibits a single narrow major absorption band, a feature that has been explained by the higher symmetry of the complexed species compared with the metal-free species.

According to the chemical structure of the ligand, metal complex pigments may be categorized as azo, azomethine, oxime, or isoindoline types.

43.3.1 Azo metal complex pigments

In view of the importance of azo metal complex dyes, for example as in the premetallized textile dye class, it may appear rather surprising that there are so few commercially successful azo metal complex pigments. It is essentially the dullness of color that limits their application. The longest-established product is CI Pigment Green 10 (**2**), known as Nickel Azo Yellow, a 2:1 nickel complex in which the azo compound acts as a bidentate ligand (Figure 43.2). Although highly lightfast, this pigment suffers from a dull color and low tinctorial strength, and solvent resistance that is inadequate for certain coatings applications. CI Pigment Yellow 150 (**3**), a 1:1 nickel complex of azobarbituric acid, can provide dull, medium yellow shades, and has been used in paints and printing inks. Although the structure in Figure 43.2 shows only three atoms bonded to the nickel ion, the geometrical arrangement is likely to be square planar, perhaps with a water molecule occupying the remaining position.

43.3.2 Azomethine metal complex pigments

Although azomethine metal complex pigments have been extensively investigated, and have been the subject of many patents, their commercial exploitation appears to be rather limited. The pigments that have given Colour Index designations are either monoazomethine copper complexes, such as CI Pigments Yellow 117 (**4**) and 129 (**5**), or bisazomethine nickel complexes such as CI Pigments Orange 65 (**6**) and 68 (**7**), as illustrated in Figure 43.3. Azomethine metal complex pigments have found use in combination with aluminum pigments in automotive metallic finishes because of their excellent lightfastness properties combined with high transparency.

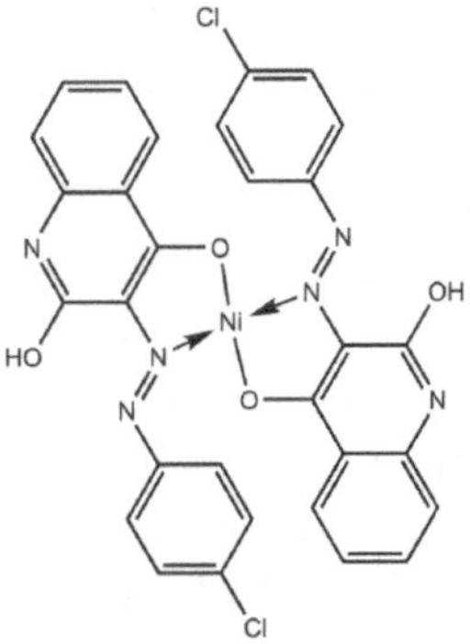

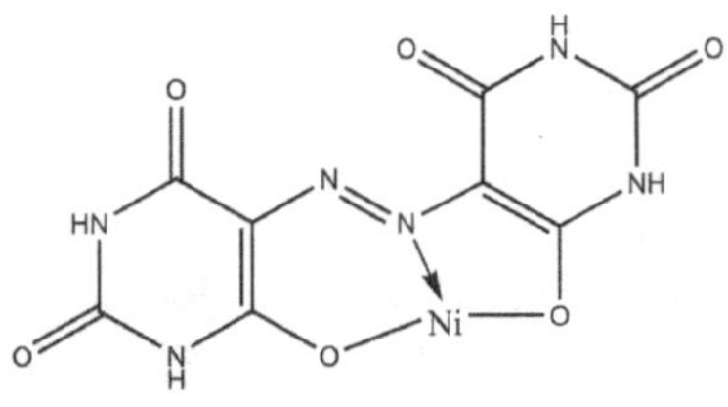

3

2

Figure 43.2: Structures of azo metal complex pigments (**2**) and (**3**).

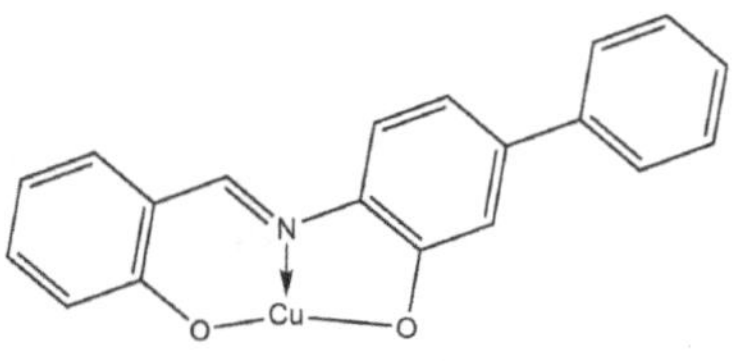

4

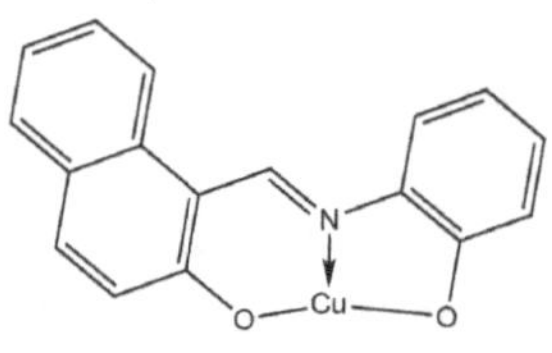

5

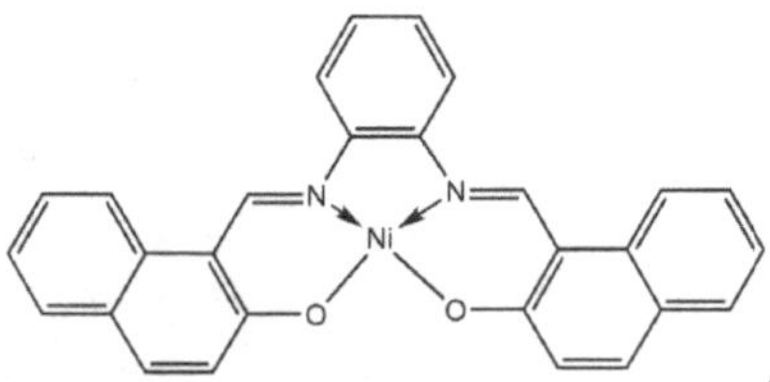

6

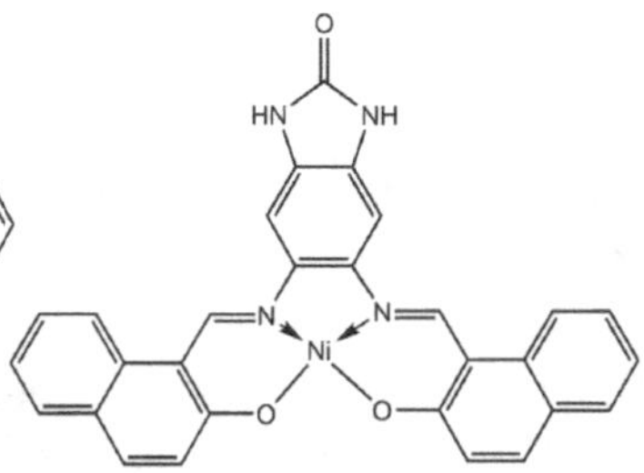

7

Figure 43.3: Structures of azomethine metal complex pigments.

43.3.3 Oxime metal complex pigments

The longest-established oxime metal complex pigment is CI Pigment Green 8, a 3:1 iron (II) complex of 1-nitroso-2-naphthol, usually formulated as the octahedral structure (**8**) (Figure 43.4). This rather dull green pigment has been used in applications requiring high stability to alkali such as in cement, and in rubber where it is resistant to the curing processes, but it is of considerably reduced importance currently. There have been a few square planar 2:1 dioxime nickel complex pigments, exemplified by CI Pigments Yellow 153 (**9a**) and Orange 59 (**9b**), which show good lightfastness and reasonable brightness and intensity of color. The color of these pigments has been

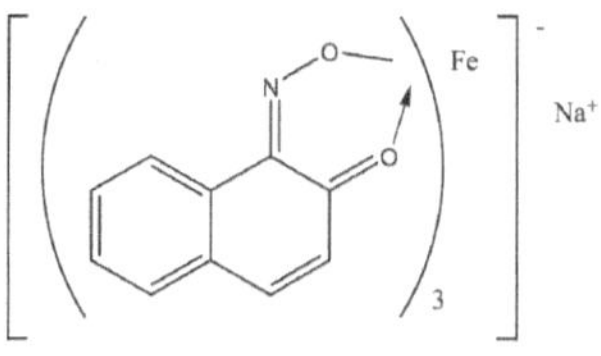

8

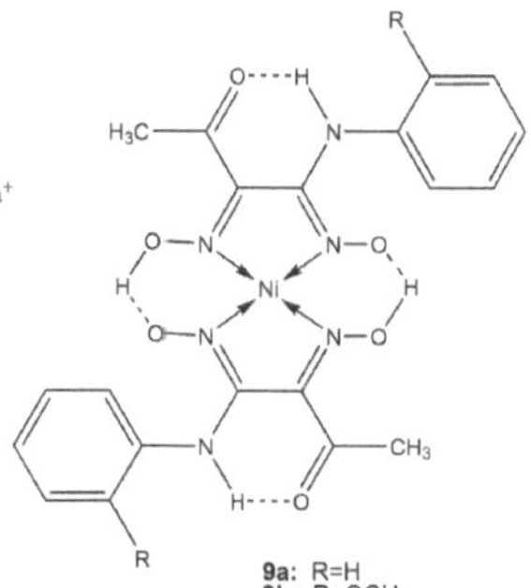

9a: R=H
9b: R=OCH_3

Figure 43.4: Structures of oxime metal complex pigments.

attributed to ligand-metal charge transfer transitions, influenced also by intermolecular metal-metal interactions in the crystal lattice structure [14].

43.3.4 Isoindoline metal complex pigments

Isoindoline pigments are important high-performance carbonyl pigments, which are discussed in another chapter in this series, entitled Isoindoline and isoindolinone pigments. CI Pigments Yellow 177 (**10a**) and 179 (**10b**) have been introduced as dull yellow 2:1 cobalt complex pigments based on isoindolines, in which the cobalt atom adopts tetrahedral geometry (Figure 43.5).

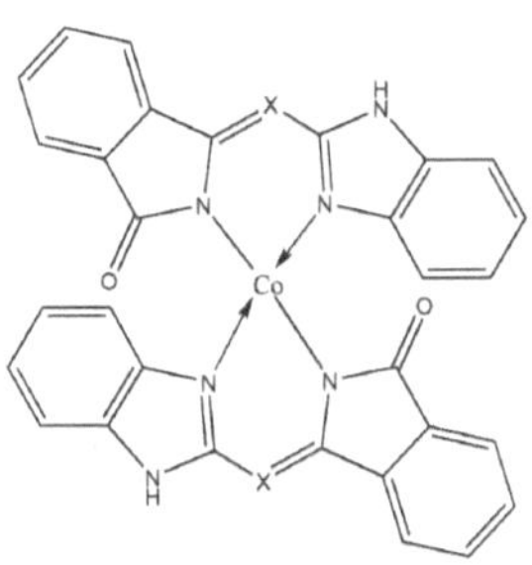

10a: X = C(CN)

10a: X = N

Figure 43.5: Structures of isoindoline metal complex pigments (**10a**) and (**10b**).

43.4 Synthesis and manufacture

The synthesis of metal complex pigments generally involves the preparation of the colored ligand, which is then complexed with a transition metal ion, either Fe^{2+}, Co^{2+}, Ni^{2+}, or Cu^{2+}.

43.4.1 Azo metal complex pigments

CI Pigment Green 10 (**2**) is synthesized by preparing the azo compound using the traditional monoazo pigment synthetic procedure, which involves azo coupling of diazotized 4-chloroaniline with 2,4-dihydroxyquinoline as the coupling component. This is followed by treatment with a nickel (II) salt in appropriate proportions, as illustrated in Figure 43.6. In the synthesis of CI Pigment Yellow 150 (**3**), also illustrated in Figure 43.6, the preparation of the azo ligand is unusual in that it involves initially, rather than a traditional diazotization process, a diazo transfer reaction, involving the reaction of *p*-toluenesulfonyl azide with barbituric acid (**11**) to produce diazobarbituric acid (**12**). This diazo compound is then reacted further by azo coupling with a further molecule of barbituric acid (**11**) to form the azo compound, which is then complexed with a nickel (II) salt.

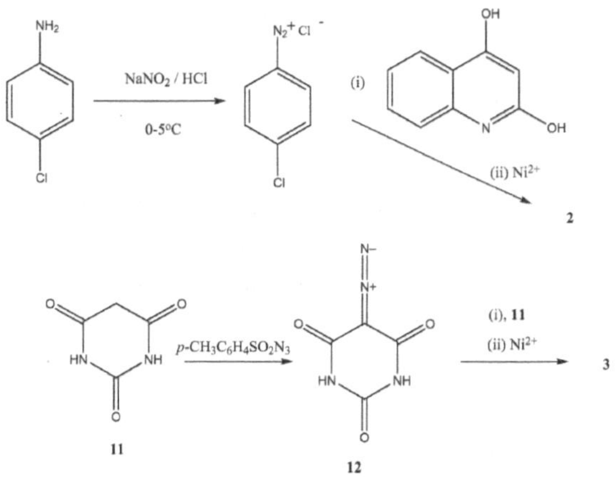

Figure 43.6: Synthesis of azo metal complex pigments (**2**) and (**3**).

43.4.2 Azomethine metal complex pigments

Azomethine metal complex pigments are synthesized from the reaction of *o*-hydroxyaldehydes with primary aromatic amines to form the azomethine molecule, followed by complexation with the transition metal ion. The syntheses of monoazomethine complex (**5**) and bisazomethine complex (**6**) from 2-hydroxy-1-naphthaldehyde (**13**) are illustrated in Figure 43.7.

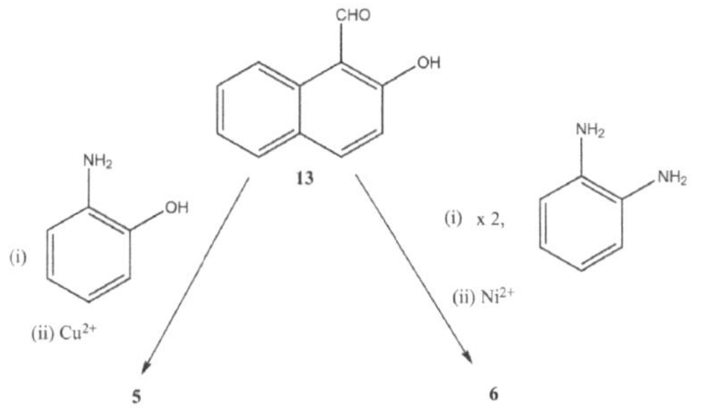

Figure 43.7: Syntheses of azomethine metal complex pigments (**5**) and (**6**).

43.4.3 Oxime metal complex pigments

CI Pigment Green 8 (**8**) is synthesized by complexing 1-nitroso-2-naphthol with an iron (II) salt. The "one-pot" synthesis of nickel dioxime complex pigments (**9a**) and (**9b**) from acetoacetanilides (**14**) initially involves nitrosation, using sodium nitrite and hydrochloric acid, followed by oxime formation with hydroxylamine and complexation with a nickel (II) salt, as shown in Figure 43.8.

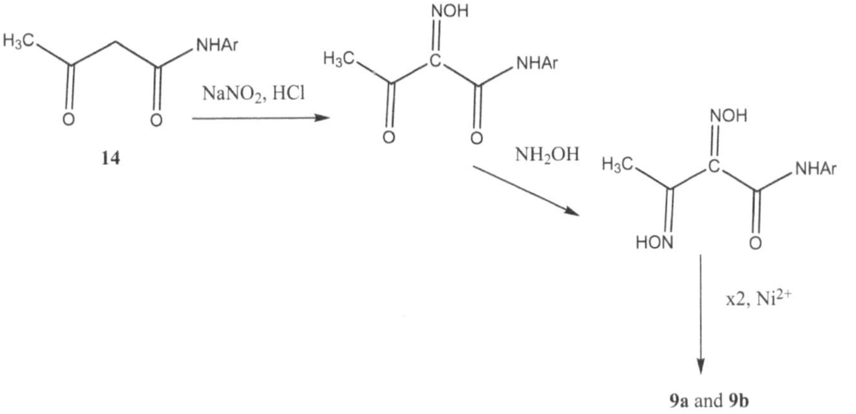

Figure 43.8: Synthesis of nickel dioxime complex pigments (**9**).

43.4.4 Isoindoline metal complex pigments

The synthesis of isoindoline pigments (**10a**) and (**10b**) involves condensation of iminoisoindolinone (**15**) with 2-cyanomethylbenzimidazole (**16**) or 2-aminobenzimidazole (**17**) (Figure 43.9), respectively, followed by complexation with a cobalt (II) salt.

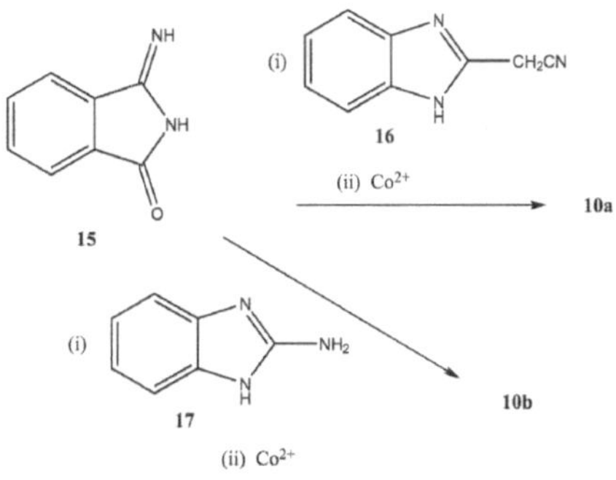

Figure 43.9: Synthesis of isoindoline metal complex pigments.

43.5 Applications

In general, with the obvious exception of the copper phthalocyanines, metal complex pigments tend to provide dull shades, especially in white reductions. On the other hand, several of the pigments provide brilliant effects when used in metallic and pearlescent paints and were often favored for automotive finishes. They can suffer from only moderate fastness to acids and alkalis, which can lead to de-metallization in storage and thus an unstable color, a feature that has limited their use in aqueous and nitrocellulose solvent-based systems. When incorporated in a paint film, however, most metal complex pigments meet current requirements for acid and alkali fastness.

43.5.1 CI Pigment Yellow 150 (3)

This pigment was first introduced by Bayer as Fanchon Fast Yellow Y. It is now offered by several manufacturers mainly, but not entirely, from Asia. It is a nickel azo complex with a dull greenish to mid shade yellow, used mainly for paints but with some limited application in plastics and inks. It offers very good fastness to light, but its fastness to weathering is less impressive, especially in paler shades. Its heat stability is sufficient for most paint systems up to 200°C when stoved for 30 minutes. It shows excellent fastness to solvents and, in an applied paint, it is stable to acids and alkalis. In plastics, it can be used in PVC, polyolefins, in which it is stable to 300°C, polystyrene, and PET. There have been new grades developed for the coloration of fibers, although specific recommendations vary among the individual manufacturers. In inks it is specially recommended for laminate printing.

43.5.2 CI Pigment Yellow 129 (5)

This pigment is an azomethine copper complex that was developed and introduced by Ciba Geigy as Irgazin Yellow 5GT at its manufacturing site in Paisley, Scotland. This pigment offers a very greenish shade and is used almost exclusively in industrial paints, in which it meets the highest demands for fastness to light and weather, as required for automotive original equipment finishes. Its high transparency leads to unique metallic effects in automotive paints, producing a very bright finish. However. it is less bright when reduced with white pigments. It shows heat stability up to 150°C but shows some bleeding when over-coated above 140°C.

43.5.3 CI Pigment Orange 68 (7)

This bisazomethine nickel complex pigment was first marketed by Sandoz as Sandorin Orange 5RLT. It is a dull, reddish orange pigment with excellent resistance to heat, and is recommended for use in in a broad range of plastics, even in nylon and other engineering polymers, subject to careful testing beforehand. In PVC it is fast to migration and has very good to excellent lightfastness. It is not recommended for cable sheathing. In other polymers it is stable to 300°C. Its lightfastness varies between very good and excellent according to the polymer used, except for ABS in which it is slightly lower. It does not cause dimensional instability in polyolefins and so can be used in injection moldings such as crates. It is not recommended for polyester and acrylonitrile fibers but may be used in polypropylene fibers, subject to the outcome of satisfactory preliminary tests.

43.5.4 CI Pigment Green 8 (8)

This product, also known as Pigment Green B, was the first synthetic metal complex organic pigment, and is still widely offered, although its use is at a much lower level than in former times. This iron complex pigment has a dull yellowish shade, intermediate between copper phthalocyanine greens and the inorganic chrome green but is available at a moderate cost. It is used in paints and plastics. In paints it exhibits good to very good fastness to light and good fastness to hydrocarbon solvents but is less fast to esters and ketones. It is not stable to acids, but is very stable to alkalis, making it suitable for the coloration of concrete. In plastics, it is stable to 220°C, allowing it to be used in low density polyethylene and polystyrene, but offers at best moderate fastness to light. It can still be used for the coloration of rubber for which it was originally developed.

The following pigments are either no longer available or have limited international availability. Except for the long-established CI Pigment Green 10 (**2**), they provide examples of developments accomplished mainly between 1970 and 1990.

43.5.5 CI Pigment Green 10 (2)

This azo nickel complex pigments offered a dull very yellowish green shade with low tinctorial strength. It was mainly used in industrial paints. Many grades were very transparent and so were used in metallic finishes. Its lightfastness was very good to excellent but inadequate for automotive finishes. The pigment lacked acid stability.

43.5.6 CI Pigment Yellow 117 (4)

This azomethine complex copper had a dull greenish yellow shade with relatively high tinctorial strength, not dissimilar to the structurally related CI Pigment Yellow 129 (**5**). It was used for automotive metallic finishes. Its fastness to solvents was generally good, although only moderate in ketones. Its lightfastness was very good in paint applications. It was stable to 180°C. It affected certain stabilizers used in PVC, which limited its use, but when suitable levels or alternative stabilizers were used, the pigment offered excellent lightfastness.

43.5.7 CI Pigment Orange 65 (6)

This pigment is a bisazomethine nickel complex that provided a dull reddish-orange shade. It was introduced to the market as Irgazin Orange 3GL aimed at the automotive original finish and refinish markets, where its high transparency was valued for metallic and pearlescent finishes. In this area it competed, apparently unsuccessfully, with the dibromanthanthrone red (CI Pigment Red 168).

43.5.8 CI Pigment Yellow 153 (9a)

Described as a reddish yellow shade, this dioxime nickel complex pigment was introduced by BASF mainly for the paint industry, although they no longer to offer it in some markets. It is rather dull in shade and has poor acid stability. Its fastness to aliphatic solvents is quite good but it performs much less well in aromatic solvents.

43.5.9 CI Pigment Orange 59 (9b)

This pigment is a dioxime nickel complex that is described as having a dull yellowish orange shade. It was introduced by BASF as Paliotol Orange L2370 for use in general industrial finishes. Its fastness to light and weather was rated as good and it was heat stable up to 180°C. It was not very stable to either acids or alkalis, a feature that limited its use.

43.5.10 CI Pigment Yellow 177 (10a)

This was an isoindoline cobalt complex pigment that provided a dull greenish yellow shade. The original pigment was introduced by Ciba, but there are no longer any products shown in the Colour Index. It was recommended in a narrow market for the spin dyeing of fibers, essentially polypropylene and nylon.

43.5.11 CI Pigment Yellow 179 (10b)

This pigment was another isoindoline cobalt complex, in this case with a reddish yellow shade. It was introduced by Ciba, as Irgazin Yellow 3R, mainly aimed at the high end industrial paint sector. It had very good fastness to light and weather, even in reduction, but not quite reaching the level of the best pigments in its shade area, such as Flavanthrone Yellow (CI Pigment Yellow 24). It was used mainly in metallic shades due to its high transparency.

43.5.12 CI Pigment Red 257

This unusual nickel complex pigment is described in the Colour Index as bluish red. It was introduced onto the market by Sandoz as Sandorin Red Violet 3RL aimed at high quality industrial paint markets, in which its good opacity was an advantage. It was incorporated into the Clariant/Hoechst range but is no longer included.

References

1. Cotton FA, Wilkinson G. Advanced inorganic chemistry, 4th ed, New York: John Wiley & Sons, ch 5 1980.
2. Lawrance GA. Introduction to coordination chemistry. New York: John Wiley & Sons, 2009.
3. Stallmann O. The use of metal complexes in organic dyes and pigments. J Chem Educ. 1960;37:220.

4. Hoffman O. Concerning the combination of nitrosonaphthols and nitrososulfonic acid acids with iron and cobalt. Chem Ber. 1885;18:46.
5. Kvalnes DE, Woodford HE. Metallised azo pigments. US 2,396,327. 12 Mar 1946.
6. Moser FH, Thomas AL. The Phthalocyanines vols I and II. Boca Raton, Florida: CRC Press Inc, 1983.
7. Leznoff CC, Lever AB. Phthalocyanines: properties and applications. Weinheim: VCH, 1989.
8. McKeown NB. Phthalocyanine materials: synthesis, structure and function. Cambridge: Cambridge University Press, 1998.
9. Zollinger H. Color chemistry: syntheses, properties and applications of organic dyes and pigments. 3rd ed. Weinheim: Wiley-VCH, 2003:245.
10. Hunger K, Schmidt MU. Industrial organic pigments, 4th ed, Weinheim: Wiley-VCH Verlag GmbH, Ch 2 2019.
11. Kaul BL. Advances in the science and technology of pigments. J Oil Col Chem Assoc. 1987;70:349–54.
12. Smith WE. J Oil Col Chem Assoc. 1985;68:170.
13. Brown DH, Oates G, Smith WE. The chromophore in some bisazomethine Schiff base complexes of nickel. Spectrochim Acta A: Mol Spectros. 1982;38:333–8.
14. Czajkowski W, Kraska J. Studies on polymorphism of Ni complexes of acetoacetanilide dioxime pigments during their preparation. Dyes Pigm. 1984;5:3–14.

Gerhard Pfaff, Michael R. Bartelt and Frank J. Maile

44 Metal effect pigments

Abstract: Metal effect pigments are luster pigments consisting of platelet-like metallic particles, mostly of aluminum (so-called silver bronzes), copper, copper/zinc (so-called gold bronzes), and zinc flakes. After parallel orientation in their application medium, they show a metal-like luster by reflection of light at the surface of the flat metal particles in one direction. Thus, the pigment particles act similar to small mirrors and lead when orientated parallel in the application system to a reflecting metal luster (metallic effect). Metal effect pigments are used in all relevant application systems such as coatings, paints, plastics, artist paints, cosmetics, printing inks, leather, construction materials, paper, glass, and ceramics. Specific composition, particle size distribution and surface quality determine the coloristic and application technical properties of the individual pigments. Many pigments are coated with a specific surface treatment to improve the quality concerning the stability and the compatibility with the application system. Metal effect pigments are offered in form of powders, pastes, pellets, suspensions or color concentrates. They are manufactured starting from a metal granulate (grit), which is ground under formation of flakes.

Keywords: aluminum pigments, metal effect pigments, metallic effect, metal flakes

44.1 Fundamentals and properties

Effect pigments are subdivided into metal effect pigments (metallic effect pigments) and special effect pigments including pearl luster pigments (pearlescent pigments, nacreous pigments) and interference pigments. There are own chapters for Effect Pigments and Special Effect Pigments, where a detailed description of these pigment categories can be found. All effect pigments consist of μm-sized thin platelets that show strong lustrous effects when oriented in parallel alignment in application systems [1]. The term "luster pigments" is also often used for effect pigments because almost all effect pigments provide lustrous effects in their applications. A decisive difference between metal effect pigments and special effect pigments is the transparent or semitransparent nature of the latter ones [2–6]. The optical principles for the interaction of visible light with different types of pigments is shown in Figure 44.1.

This article has previously been published in the journal Physical Sciences Reviews. Please cite as G.Pfaff, M. R. Bartelt, F. J. Maile, Metal Effect Pigments *Physical Sciences Reviews* [Online] 2021, 6. DOI: 10.1515/psr-2020-0182

https://doi.org/10.1515/9783110587104-044

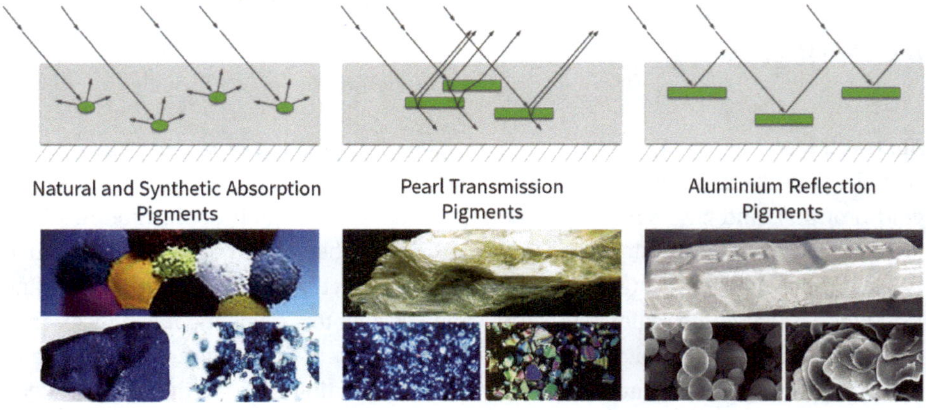

Figure 44.1: Optical principles for the interaction of light with different pigments.

Metal effect pigments consist of platelet-like metallic particles, so-called metal flakes. They are supplied in form of powders, pastes, pellets, suspensions or color concentrates [7]. Most of the metal effect pigments consist of aluminum (so-called silver bronzes), copper and copper/zinc alloys (brass but so-called gold bronzes).

There are also metallic pigments and powders besides the metal effect pigments, which are used for functional purposes in coatings for anticorrosion, IR reflectivity, heat resistance and electrical conductivity. They are based mainly on the metals zinc, stainless steel, and silver.

Aluminum pigments with their silver color are used in their pure form, but may also be combined with other pigments or with dyes. A typical combination is that of aluminum pigments with special effect pigments. The color of gold bronze pigments depends on the composition of the copper/zinc alloy and can range from copper (0 wt % Zn) over pale gold (approx. 10 wt % Zn), rich pale gold (approx. 20 wt % Zn) to rich gold (approx. 30 wt % Zn). The pigments provide the appearance of silver and gold with the aura of quality and prestige that these metals imply.

Oxidized gold bronze pigments are also available. They achieve deeper color shades by a controlled oxidizing process during manufacturing [7, 8].

The origin of metallic pigments goes back to the ancient art of gold beating. Early civilizations, for example the Egyptians, used metallic gold to treat it mechanically in order to form thin sheets, which were applied to overlay wood, bone, or other materials with the precious metal. This technique spread to other regions and the demand for thin gold sheets increased. It became necessary to produce thinner and thinner foils. In a step further, very thin gold foils were crushed to generate gold powders consisting of small platelets. Such gold powders were used for ornamental artwork, inks and cosmetics. The extremely high costs for gold were most probably the reason for the interest in alternative materials. Thus, gold bronze came into use, which was neither metallic gold nor bronze, but showed golden effects. Gold bronze as the oldest substitute for metallic gold is with its components copper and zinc, of course, not a bronze but a brass alloy. In later times, silver and tin were combined to form silver bronze powders. However, the discovery of aluminum smelting and the resulting prize reduction led to the replacement of silver/tin alloys by aluminum and thus to the development of the largest group of metal pigments.

The manufacture of gold bronze flake pigments started in Germany and can be dated back to the time around 1820. A breakthrough in the process was obviously the development of the stamping process by Bessemer in the middle of the nineteenth century [9]. Main parts of the equipment for the process were steel hammers, which fell on steel anvils, thus forming a desired metal into the flake form comparable with that of modern metallic pigments. From this time on, it was possible to replace the very expensive gold and silver bronze powders by cheaper metals. A further decisive progress was the development of the electrolytic processes for the aluminum production in the 1880s and 1890s. The now available metal was quickly introduced in the manufacture of metallic flakes. The aluminum powder used as the raw material for flakes came initially from the production of thin foils or from the waste of the aluminum production.

From about 1910 on, it was possible to produce aluminum powder and granules specifically by industrial-scale production. In addition, a grinding process was invented in which aluminum powders and granules could be deformed to flakes using large ball mills [10]. This process was not yet explosion-proof because the production of the aluminum flakes took place in a dry way. Important steps for the development and improvement of metal flakes, which were more and more used as effect pigments, were the introduction of the Hametag process (dry grinding) for gold bronze pigments and of the Hall process (wet grinding) for aluminum pigments [11, 12]. It was only possible with the introduction of mineral spirit as grinding medium by Hall to produce aluminum flakes safely and in large quantities. The processes were refined in the course of the years associated with an improvement of the pigment qualities and the production efficiency by a continuous milling process [13]. Typical stages of development for aluminum pigments were the conventional "cornflake type", the lenticular "silverdollar type" [14] and the "VMP type" (vacuum metallized pigment). The later one was prepared for the first time in the 1970s and is also called "PVD type" (physical vapor deposition) [15, 16].

Zinc flake pigments are produced for a long time with ball milling technology in hydrocarbon solvent but are of minor importance [17]. The main products are today leafing flakes for anticorrosive applications.

Silver and nickel flakes are known as well in the industry. They are used in industry only for technical application where electrically conductive pigments are necessary.

Stainless steel flakes are manufactured mainly due to the outstanding stability. The color is inferior to aluminum but the corrosion and abrasion resistance qualify the steel pigments for the use in master batch and polymer coatings [18].

The development and manufacture of colored aluminum pigments is based on the idea to expand the effects achievable with metallic pigments, especially to combine the advantages of metal effect pigments with color. Concepts used for this purpose include the formation or deposition of absorption and interference layers on the surface of aluminum flakes or the inclusion of colored pigment particles in a coated layer on the flakes consisting mostly of silica. Pigments based on metal oxide coated aluminum flakes take an intermediate position between metal effect pigments and special effect pigments. They offer strong reflectivity together with color effects from interference and absorption.

Pigments with strong angle-dependent color effects based on the Fabry-Perot structure, which have aluminum as a central layer, could be produce using the PVD technology. Diffractive pigments consisting of two dielectric layers and a central non-transparent aluminum layer can also be produced by the PVD process. A structured polymer film template is used to achieve the diffractive properties of these pigments [1]. Pigments with the Fabry-Perot structure as well as diffractive pigments are mostly counted as special effect pigments.

44.2 Optical properties

The metallic effect of metal effect pigments has its origin in the reflection of light at the plane surfaces of the flakes, which is overlaid by scattering of light at the edges and at irregularities on the metal surface. The metallic effect can be optimized by increasing the part of directed reflection at the surface of the platelets and simultaneous reduction of scattering at the edges and at irregularities.

The optical impression depends primarily on the following factors [19]:
- type of metal (aluminum, copper, copper/zinc alloy)
- wetting behavior of the pigment (leafing or non-leafing)
- surface smoothness of the pigment particles
- particle size and particle size distribution
- particle thickness and particle thickness distribution
- aspect ratio of the platelets (ratio of diameter to thickness)
- nature of the interface (air or application medium)

Metal effect pigments, with the exception of vacuum metalized pigments, are subjected to strong mechanical forces during their manufacture. These forces lead to surface defects. The absence of surface irregularities leads to a large degree of directed reflection.

The description of the visual impression of the metallic effect is not easy without having information on the orientation of the particles. Visual impression and measurement of the effect consist of the sum of characteristic single effects [19]:

- color shade
- brightness (whiteness)
- brilliance
- hiding power (tinting strength)
- optical roughness (sparkle effect)

The brightness describes how bright an effect pigment appears to an observer. The brilliance characterizes the reflectivity of a pigment and the "metallic" character. Pigments are brilliant if they reflect the incident light to a large degree and if only little undirected light scattering or absorption takes place. Metal effect pigments are in general the more brilliant the more perfect the pigment surfaces are or the narrower the particle size distribution is. Very small and irregularly shaped pigment particles lead to a shift of the color shade to grey. Coarse metal effect pigments with average particle sizes above 25 µm are known as sparkle types. The human eye may perceive in this case the mirror planes of single particles during incidence of light.

Three characteristics are of special importance with regard to the gloss of a metallic coating, especially an automotive coating [19]:
- flop effect (travel effect, two-tone-effect, brightness flop)
- distinctiveness of image (DOI)
- hue saturation

The flop effect describes the phenomenon that the physically experienced brightness at metallic effect coatings depends very much on the viewing angle. It varies from extremely bright at or close to the gloss angle to rather dark at inclined angles. The flop behavior of metallic coatings is determined in practice by using the brightness contrast of coated surfaces during observation under different viewing angles.

The DOI value is determined by a gloss measurement and describes the topcoat gloss close to the reflection angle. The determination is done either subjectively in a glow box or by measurement. The measured values are relative numbers from 0 to 100. As higher the DOI value as sharper appears the image of objects with high contrast by reflection at the coated surface. The DOI value of an effect coating is influenced by the particle size and the particle size distribution of the pigment used, the binder compatibility of the pigment, and the general spreading properties of the clear coat.

The hue saturation is defined as the property of a metal effect pigment to cover or influence a color shade generated by a colored pigment. There is a direct relation to the hiding power. Generally, the hue saturation is as higher as finer the metal effect pigment and as broader its particle size distribution is [19].

The particle orientation within the dry pigmented film is an important factor for the quality of the final coating. The orientation of the pigments depends to a large extent on the dispersion and wetting of the metal flakes, the formulation, the concentration, the method of application and on the smoothness of the substrate. The optimum situation is that all the flakes are oriented parallel in the coating and reflect the light in a parallel manner. The result in such a case is a maximum of brightness, brilliance and flop. Poor orientation results in an irregular reflection, are causing "salt and pepper effects", "cloudiness" and a poor metallic effect and flop. A disorientation of the flakes can be caused by too wet films (or too slow solvents) or by too dry films (or too fast solvents) [7].

An important factor for the quality of the final metallic effect is also the shrinkage during the solvent evaporation in the drying stage. During drying, the flakes find their final parallel orientation in the system. There is less or no shrinkage in high solid systems or UV-curing systems and even more extremely in powder coatings, which explains that the pigment orientation is poorer here than in low solid or medium high solid coatings [7].

Optical measurements for the characterization of the metallic effect of coatings containing metal effect pigments are typically carried out with goniophotometers or gonio-spectrophotometers in order to take the flop effect into account [19–22].

44.3 Production of metal effect pigments

44.3.1 Aluminum pigments

Aluminum pigments are manufactured starting from aluminum granulate (grit). This grit is produced from aluminum ingots, which are mostly melted in induction furnaces at temperatures above 700 °C, and thus above the melting point of aluminum. The molten aluminum is then sprayed under high air pressure through a nozzle. The atomized metal cools down immediately after this process step and is now available in passivated form as isometric powder. This granulate, commonly referred to as aluminum grit or atomized aluminum, is the raw material for the following grinding step. The grinding leads to effect pigments with platelet-shaped particles, so-called flakes [19].

Grinding aluminum grit under formation of flakes is necessary because the spherical grit is too coarse for the application in thin coating or printing films. In addition, the grit does not have the required pigment properties – due to the low aspect ratio brilliance and hiding power is very low.

The aluminum grid is now subjected to a grinding procedure in ball mills in the presence of mineral spirit and a lubricant (Hall process) [12]. C18 carboxylic acids, such as oleic acid or stearic acid, are usually added as lubricants. The grit is not only deformed to flakes during this process step but also crushed to smaller particles. The resulting highly reactive surfaces of the formed flakes are immediately oxidized and a hydrophilic aluminum oxide layer is formed. The lubricant is adsorbed from the oxidized surface of the aluminum flakes. The lubricant layer at the surface of the particles is important for the prevention of uncontrolled reactions (cold welding, formation of agglomerates). The process heat formed during grinding is dissipated via cooling systems. The concentration of the aluminum pigments in the mineral spirit during the production, especially during grinding and sieving, is typically between 10–30%.

After the grinding process, the pigment slurry is sieved and classified (agglomerates and coarse particles are separated) and pressed on a filter press to remove the excess solvent. The press cake, which consists of around 80% aluminum and 20% mineral spirit is blended with organic solvents to form an aluminum paste, typically containing 65% pigment and 35% solvent [19]. Depending on the desired product quality and the ball milling equipment the milling process could be either a continues-process or a batch-process [13]. Average size and size distribution as well as the shape of the aluminum particles in the paste depend strongly on the manufacturing conditions. Derived from this, a distinction of aluminum pigments produced by the Hall process is made between the conventional "cornflake type" and the more sophisticated, lenticular "silverdollar type" (Figure 44.2 and Figure 44.3). The later one shows the more perfect shape and a smoother surface. Consequently, aluminum pigments of the "silverdollar type" provide the more attractive metallic effects in coatings and other applications.

Figure 44.2: Aluminum Cornflake pigment (source: Carl Schlenk AG).

Figure 44.3: Aluminum lenticular Silverdollar pigment (source: Carl Schlenk AG).

A scheme showing the main production steps for flaky aluminum pigments is shown in Figure 44.4. A scanning electron micrograph of aluminum grit as it is used for the manufacture of aluminum platelets is shown in Figure 44.5.

There are applications, where aluminum powder is requested, e. g. in powder coatings, or where remaining mineral spirit would not be compatible with the application medium (e. g. waterborne coatings, waterborne inks, and master batches). In such cases the mineral spirit of the press cake is removed in vacuum driers and substituted by any kind of solvent, water, plasticizer, mineral oil or other liquids [7].

Aluminum pigments as well as other metal effect pigments are also supplied in form of pellets or granulates. These are dust-free and easy to handle preparations of the pigments with a certain percentage of resin. Pellets and granulates offer a broad variety of formulation possibilities to the end user besides other advantages. They are mainly used in printing inks and master batches [7]. Special types of aluminum pigments are coated with inorganic (mostly silica) or organic (polymer) materials to improve their chemical and thermal stability in waterborne coatings, powder coatings, or master batches. Such aluminum types are used in all relevant pigment applications but especially in automotive coatings and other outdoor uses.

The production of aluminum pigments prepared by vacuum metallization, respectively, physical vapor deposition leads to very thin flakes with a very unique mirror effect (Figure 44.6). They differ considerably from aluminum pigments of the

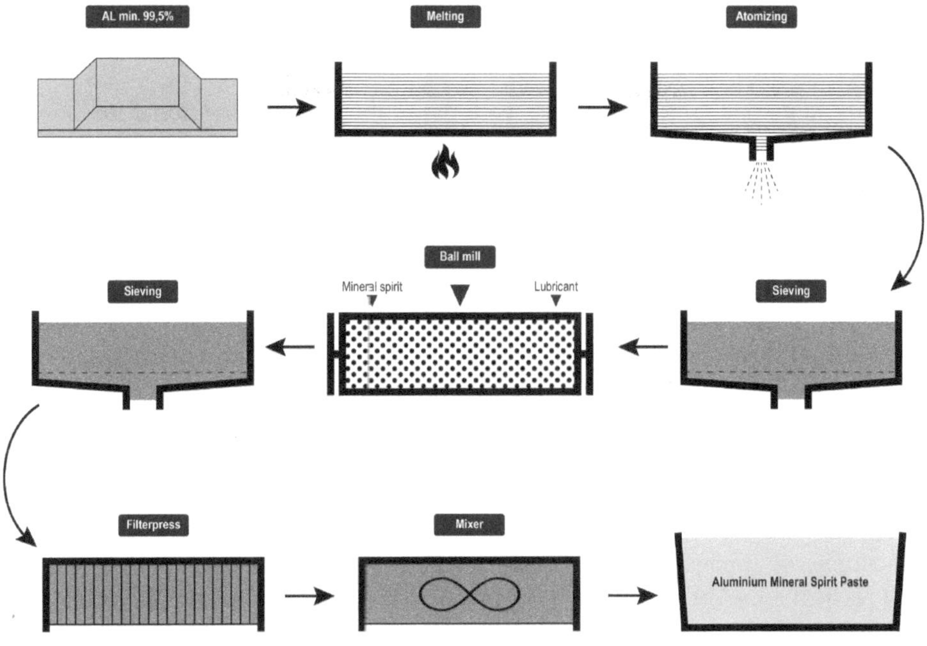

Figure 44.4: Production process for aluminum pigments.

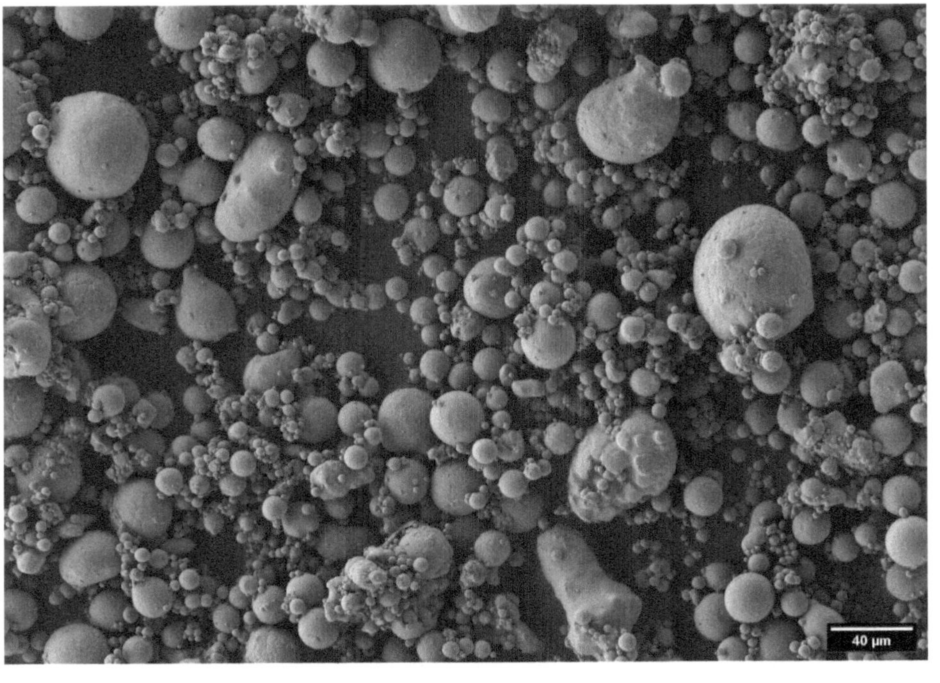

Figure 44.5: Atomized aluminum powder (source: Carl Schlenk AG).

"cornflake type" or the "silverdollar type." The production starts from highly purified aluminum, which is vaporized in high vacuum and deposited in form of a very thin film on a carrier material.

The actual manufacturing starts with the coating of a conventional foil (mostly based on polyethylene polypropylene) with a so-called release coat. Materials for the release coat are coatings or binders comparable with those used in printing applications, e. g., acrylates, cellulose systems, or vinyl resins. The foil coated with the release layer is now vacuum-deposited with an extremely thin aluminum layer. The vaporized aluminum is in the high vacuum directly deposited on the release coat. The evaporation step plays a key role for the pigment quality which is tried to be achieved at the end of the process. The decisive pigment properties, such as particle thickness and surface quality, are derived from the evaporation parameters.

In the further steps of the process, the foil coated with the release layer and the aluminum layer runs through a solvent bath (typically ketones, esters, and alcohols) where the release coat is dissolved and the aluminum layer is removed in shape of coarse irregular formed particles. The resulting dispersion consisting of the aluminum fragments, residues of the release coat, and solvents is concentrated and washed. The filter cake is dispersed now in the desired solvent and comminuted by means of a high-speed stirrer or by ultrasonic treatment to the final average particle size [19]. The newest VMP products have a narrow particle size distribution ranging from ~ 6 to 22 µm.

44.3.2 Gold bronze pigments

Gold bronze pigments are based on copper or copper/zinc alloys made from electrolytically manufactured copper and zinc. Copper and zinc are alloyed with the addition of some aluminum as reducing agent. The copper-zinc ratio determines the color of the alloy. The copper content is usually above 70%.

In a first production step, irregularly shaped, spattered metal grit is generated by atomization of a metal melt followed by cooling of the atomized material. Grinding of the metal grit to platelets takes place with steel balls in the presence of stearic acid as lubricant in a dry milling process. The latter serves primarily to prevent the build-up of cold welding of pigment particles, comparable with the situation during the production of aluminum pigments. A scanning electron micrograph of a gold bronze pigment with its irregularly shaped particles is shown in Figure 44.7.

Dry gold bronze pigments do not tend to dust explosions in the presence of atmospheric oxygen and can therefore be treated under appropriate precautions in the dry state in ball mills (Hametag process) [11]. The pigment powders formed are universally usable in nearly all solvents and binders. However, for the use in water-based system the pigment should be passivated. Without passivation, the zinc could react

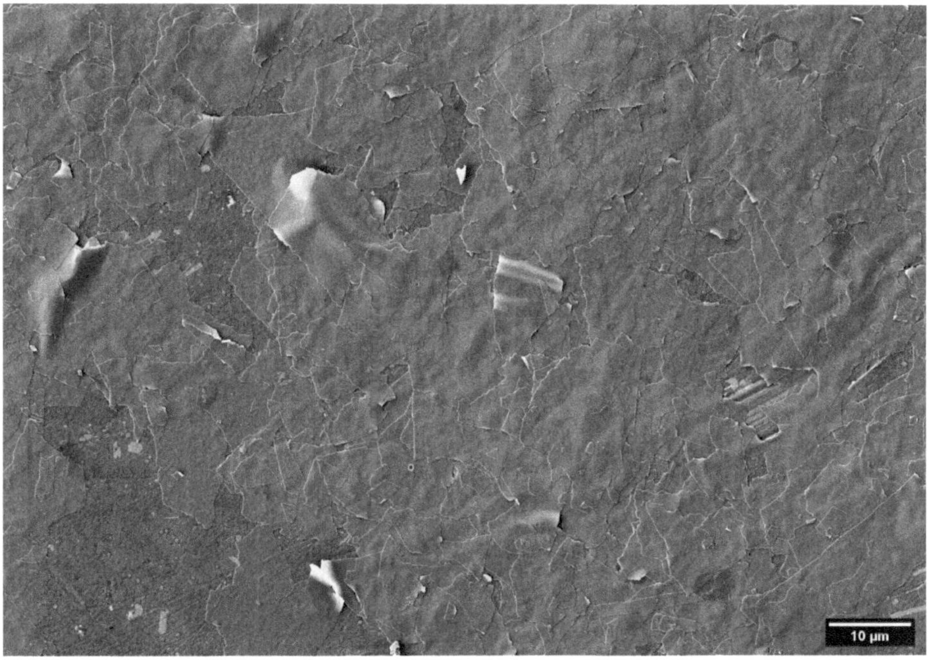

Figure 44.6: Vacuum Metallized Pigment (VMP, source: Carl Schlenk AG).

with water and a color shift into red and an agglomeration of binder components are possible. The pigment particles also have a platelet-like shape. Their density is, however, three times higher than for aluminum flakes.

The adjustment of particle size and particle size distribution is done by using several grinding steps and different grinding conditions. Design of the mill, grinding speed, size of the grinding balls, grinding time and the lubricant play an important role for the quality of the final pigment. Classification is done with cyclones. The pigment is divided into fractions in an air stream using gravitational and centrifugal forces. The properties of gold bronze pigments in coatings can be improved by a subsequent additional treatment of the surface with lubricants. Thus, the best possible metallic gloss for a pigmented coating can be achieved [19].

The range of copper and brass colored natural shades can be expanded by the formation of an oxide layer on the surface of the metal particles. Atmospheric oxygen reacts with the metal under defined conditions (temperature, time) to form thin oxide layers on the metal platelets. Interference and reflection in combination lead to interesting color nuances. Common color shades are lemon yellow, fine gold, and flame red.

Figure 44.7: Goldbronze flake pigment (source: Carl Schlenk AG).

44.3.3 Zinc pigments

The production of flaky zinc powders takes place analogous to the manufacture of aluminum flakes. Zinc grit is obtained by spraying of molten zinc metal under high air pressure through a nozzle. In the following step the grit is converted to flakes by wet grinding in mineral spirit or by dry grinding and deformation. Stearic acid is mostly used as lubricant for the production of zinc flakes [23].

44.3.4 Silver and Nickel pigments

Silver and Nickel flakes are of minor importance for the industry. They are not used for the optical appearance due to the superior properties of aluminum flakes. Silver flake pigments are bead milled in the present of water and with additional fatty acids as lubricants.

Nickel flakes are ball milled from atomized nickel powders with lubricants like stearic acid (leafing grades) or polyhydric alcohols for water compatible non-leafing grades [24].

44.3.5 Stainless steel pigments

Stainless steel flakes are manufactured from an atomized steel powder. A typical alloy composition (75% Fe/15% Cr/10% Ni) could be manufactured with milling condition similar to aluminum flakes. Due to the different deforming behavior of stainless steel the production of steel flakes requires higher milling energy [18].

44.3.6 Colored aluminum pigments

There are several technical concepts for colored aluminum pigments:
- controlled wet-chemical oxidation of aluminum flakes, formation of an aluminum oxide/hydroxide layer on the surface of the flakes; champagne effect based on interference [25]
- fixation of colored pigment particles on the surface of aluminum flakes by embedding in a coating matrix preferably consisting of silica, the silica layer is typically formed using a sol–gel process and has a thickness of about 200 nm; effects depend on the color and amount of the colored pigment and the quality of the aluminum flakes; color based on absorption [26–30]
- coating of iron oxide layers on aluminum flakes by fluidizing them at 450 °C in an atmosphere of nitrogen in a fluidized bed reactor; the reactants $Fe(CO)_5$ and O_2 are streamed in the reactor together with the agitated aluminum flakes; α-Fe_2O_3 is formed directly on the surface of the aluminum flakes:

$$2\,\mathrm{Fe(CO)_5} + 6.5\,\mathrm{O_2} + \mathrm{Al} \rightarrow \alpha\text{–}\mathrm{Fe_2O_3/Al} + 10\,\mathrm{CO_2} \tag{44.1}$$

- golden, orange, and red effect pigments with high color brilliance are possible; color based on interaction of absorption and interference; angle-dependent color effects are possible by introducing a layer with a low-refractive index (preferably silica) between the Al flake and the iron oxide layer (optically variable pigments, pigments with strong color travel), the pigments are often considered to be special effect pigments [1, 31, 32]
- coating of iron oxide layers on aluminum flakes by wet-chemical procedures starting with the deposition of a passivation layer (preferably silica) on the Al flakes followed by the precipitation of iron oxide hydroxide; the conversion to iron oxide takes place by thermal treatment; an example for the final pigment structure is $Fe_2O_3/SiO_2/Al/SiO_2/Fe_2O_3$; golden, orange, and red effect pigments with high color brilliance are possible; color based on interaction of absorption and interference; the pigments are often considered to be special effect pigments [33–35]
- PVD technology for the manufacture of pigments with Fabry-Perot structure, which show strong angle dependent color effects; the resulting pigments are based on a metal-dielectric multilayer structure, the thicknesses of the single

layers determine the color travel and the color intensity; the pigments are produced in a series of special coating installations (roll coaters) that are situated in vacuum chambers; a release layer is firstly deposited on a moving transfer foil before the PVD process starts; typically five layers are deposited successively in the roll coaters to generate the desired color effect of the final pigment: layer of a semitransparent absorber metal (e. g. chromium), dielectric layer (e. g. MgF_2), non-transparent metal layer (e. g. aluminum), dielectric layer (e. g. MgF_2), layer of a semitransparent absorber metal (e. g. chromium); a typical pigment arrangement after removal of the multilayer system from the belt by dissolving of the release layer consists therefore of $Cr/MgF_2/Al/MgF_2/Cr$, the resulting pieces of the deposited material are collected, broken down into particles and fractionated; the pigments are typically considered to be special effect pigments [1, 4, 36]

- PVD technology for diffractive pigments, which create special rainbow-like color effects by the bending of light at surfaces with a regularly repeating structure; production of the pigments by vacuum deposition of thin films onto a structured polymer film template; typical examples for the layer arrangement on the polymer film are $MgF_2/Al/MgF_2$ or $SiO_2/Al/SiO_2$; a release layer is used for the removal of the deposited layer structure; the resulting pieces of the deposited material are collected, broken down into particles and fractionated; typical particle thicknesses of less than 1 µm are achieved; the periodic repetition of the structure is around 1 µm with a depth of several nanometers; the pigments are typically considered to be special effect pigments [1, 37, 38].

44.4 Pigment properties and uses

The wetting behavior in different liquid systems is of special interest for the application of metal effect pigments. These are basically divided into leafing and non-leafing types (Figure 44.8).

Figure 44.8: Leafing behavior of metallic pigments.

Leafing pigments move towards the surface of a wet, not yet dried film (coating) and orient themselves parallel to the surface. They form thereby a dense metallic layer on top of the film with high reflectivity and significant barrier properties for the protection of the film. Leafing pigments can be used to create a distinct metallic chrome-like effect ("chrome effect") in decorative coatings and printing inks. They are also valuable for manifold applications in functional coatings e. g. for roof coatings, tank coatings, or anticorrosive coatings.

Non-leafing pigments show a favorable wetting behavior in most of the application systems. They orient themselves in the most cases equally distributed in the wet film. The dried films are rub-resistant and exhibit a beneficial adhesion behavior for following clear coats. Coatings pigmented with non-leafing metallic pigments can easily be tinted with colored pigments to achieve polychromatic metallic effects.

The particle shape of metal effect pigments depends primarily on the process parameters used for their production. The question whether the irregularly shaped "cornflake type" or the lenticular "silverdollar type" of an aluminum pigment is formed is already pre-decided with the selection of the grit quality and the milling condition. The thickness of the particles can vary from <0.1 to 1 µm, depending on the treatment. Particles of vacuum metallized pigments, however, are extremely thin with a thickness of about 0.03 µm. Their surface is very smooth and highly reflective.

Average particle size (diameter of the metal flakes) and particle size distribution depend also on the manufacturing conditions. The size of the particles may differ from a few micrometers up to 100 µm, depending on the optical effect required.

The flaky shape of the metal effect pigments should be considered during dispersion steps and application. High shear forces may injure the pigments and the desired effects and should therefore be avoided.

Main applications of metal effect pigments are paints and coatings, printing inks, graphic arts and plastics.

Automotive coatings (OEM and refinish coatings, interior and exterior vehicle parts), general industrial coatings, can coatings, coil coatings, powder coatings, anticorrosion coatings, roof coatings, marine paints, and conductive coatings are areas of application for metallic pigments. The selection of a specific pigment depends strongly on the final application. There is a broad variety of high-performance pigments for automotive coatings, waterborne coatings and powder coatings available besides the conventional leafing and non-leafing types.

Metal effect pigments are used in all common printing inks and in graphic arts. Typical areas of application are offset, gravure, flexographic, silk-screen and textile printing but also paper, textile and leather coating. Furthermore, also in new digital printing processes like ink-jet or electrostatic offset printing metal pigments are used. Pigments used in these application segments are supplied in form of powders, pastes, granulates, suspensions (VMP types), color concentrates, or ready-for-press printing inks. Particle sizes used depend on the printing process and range from 2 to 6 µm for offset printing, via 6 to 14 µm for flexographic printing, 6 to 16 µm for rotogravure printing, 10 to 48 µm for screen-printing up to 12 to 40 µm for textile printing. The selection of solvents in pigment pastes and suspensions depends on the printing technique as well. Solvents commonly used are mineral oils (offset inks), alcohols, esters and aromatic hydrocarbons (rotogravure and flexographic inks), water, alcohols and glycols (waterborne inks) and reactive thinners (UV inks) [7]. Important end products for printing inks containing metal effect pigments are packages for cigarettes, food and non-food articles, illustrations, labels, gift wrap papers, wallpapers and tissues.

Aluminum as well as gold-bronze pigments are applied in plastics mainly for the tinting of master batches which are used for injection molding, blow molding, extrusion, calendaring and other processing operations for thermoplastics. Thermosetting materials like polyester putties, polyester and epoxy floorings, and art objects are another application segment for metal effect pigments. The pigments are here preferably supplied in form of powders (only gold-bronze pigments), pastes, or pellets. There are specifically coated types of gold-bronze pigments which are in use for high-temperature processes (heat-resistant gold-bronze pigments for plastics). Special chemically resistant aluminum and gold-bronze pigments are used in PVC.

Increasing demands to reduce the amounts of organic solvents (VOCs) in coatings and printing inks have led to the development of waterborne and high-solid systems as well as of solvent-free powder coatings. Metal effect pigments are applied in all these application systems. Another way is the use of radiation-curing systems (UV radiation). Reactive thinners are used here instead of organic solvents. These thinners are incorporated in the dry film without the emission of solvents.

Aluminum reacts in aqueous systems, especially in alkaline or acidic media under formation of hydrogen gas. The use of conventional aluminum pigments in waterborne coatings would be a safety risk and would destroy the metallic effect. The way to overcome this problem is the use of stabilized aluminum pigments. High-performance aluminum pigments are equipped with an inhibiting treatment, which protects the aluminum against the attack by water. Stabilization can be achieved for example with layers of phosphor organic compounds, chrome treatments or by organic or inorganic encapsulation [39, 40].

Metal effect pigments in powder coatings can basically be applied using co-extrusion, dry blending or bonding. However, co-extrusion is not recommended for the application of metallic pigments because the high shear forces during extrusion could mechanically destroy the flaky particles and thereby reduce the metallic appearance. The advantage of dry blending is the gentle handling of the metal flakes and a maximum of metallic effect. However, there is the disadvantage that the metal platelets seem to be differently charged compared with the resin particles and therefore tend to separate in the overspray. In the bonding process the metallic flakes are thermally and mechanically bonded to the surface of the resin powder avoiding a separation in the overspray and thereby a problem free reuse of the material. There are also optical advantages in bonded metallic coatings due to the very homogeneous distribution of the metal effect pigments [7]. Applications of metallic effect pigments in powder coatings are found in steel furniture, tools, transportation, household equipment and architectural elements.

Uncoated aluminum and gold bronze pigments are used also in anhydrous cosmetics and personal care formulations. The recommended pH value for the formulations is between 6 and 8. Examples for the use of metallic pigments in cosmetic products are lipsticks, cosmetic pencils, eye shadows, mascara and nail polishes. Pigments for cosmetics are manufactured using the purest metals [41].

Zinc powders have been used since long time for corrosion protection. The common pigment type used for this application is zinc dust. Flake-shaped zinc pigments offer an alternative to zinc dust and can provide a comparable anticorrosive effect with significantly less zinc in the dry coating [42].

After long-term experience in production and application, aluminum, gold bronze and zinc pigments are regarded as nontoxic. However, the inhalation of copper vapor may cause the so-called "metal vapor fever." Copper-containing dust may have a similar effect if it is present in the ambient atmosphere in higher concentrations. Chronic poisoning initiated by aluminum, gold bronze and zinc pigments is not known [43].

Metal effect pigments in form of powders and preparations can be handled, stored and transported safely under specified conditions. The risk potential is different for the various pigments and dosage forms. Assessment and handling require therefore careful consideration of the safety data sheet of each individual product.

Hazards caused by aluminum and zinc pigments are flammability, formation of explosive dust-air mixtures and formation of flammable hydrogen when getting into contact with acids or bases (also with water over a longer period). A hazard of gold bronze pigments is their flammability under certain conditions. Furthermore, pure copper is classified as marine pollutant and falls under transport regulations for dangerous goods [44].

The properties of metal effect pigments are strongly affected by the specific surface area of the particles. Fine powders have a much higher tendency to react than coarser types because of the larger surface area.

Very fine aluminum powder can easily be whirled up and an air metal dust cloud is formed. When such a dust comes into contact with an ignition source, an explosion can happen. Raising of aluminum or zinc dust has to be avoided strictly. Precautions must be taken to avoid sources of ignition (sparks, hot surfaces, electrostatic discharges). Inert atmosphere (nitrogen) has to be used where it is appropriate [44].

Aluminum and zinc powders may become electrostatically charged by friction during production and application and sparks can arise. The processing equipment must therefore be grounded. This is done for containers and other equipment by using of grounding cables during handling and storage. The original metal containers have to be used only when transferring and filling aluminum or zinc pigments.

The use of pigment preparations reduces the risk of dust formation and explosion very efficiently. The type of solvent used for the preparation can cause new hazards, especially in case of irritating or flammable solvents [44].

Metal powder fires are extinguished with dry sand or special fire-fighting resources. A sufficient amount of dry sand should always be placed in close proximity to the working environment. Long-handled shovels must be available to form immediately a boundary around the fire and cover it. Whirling up the burning metal powder must strictly be avoided because explosion is possible already with a small dust cloud [44].

References

1. Pfaff G. In special effect pigments. Pfaff G, editor. 2nd ed. Hannover:Vincentz Verlag, 2008:16.
2. Franz K-D, Emmert R, Nitta K. Kontakte (Darmstadt). 1992;(2)3.
3. Pfaff G. Perlglanzpigmente Chem Unserer Zeit. 1997;31:6.
4. Pfaff G, Reynders P. Angle-dependent optical effects deriving from submicron structures of films and pigments. Chem Rev. 1999;99:1963.
5. Pfaff G. In industrial inorganic pigments. Buxbaum G, Pfaff G, editors. 3rd ed. Weinheim: Wiley-VCH Verlag, 2005:230.
6. Pfaff G. Inorganic pigments. Berlin/Boston: Walter de Gruyter GmbH, 2017:205.
7. Besold R. In industrial inorganic pigments. Buxbaum G, Pfaff G, editors. 3rd ed. Weinheim: Wiley-VCH Verlag, 2005:252.
8. Wheeler I. Metallic pigments in polymers. Shawbury: Rapra Technology Limited, 1999:7.
9. Hoenigsberg O. in Beiträge zur Geschichte der Technik und Industrie. vol. Band 2. Matschoss C, editor. Berlin: Springer, 1910:271.
10. Patent US. 964,024 (G. S. Montgomery, C. R. Hotchkiss). 1908.
11. Patent US. 1,832,868 (Hartstoff Metall AG). 1928.
12. Patent US. 2,002,891 (Metals Disintegration Co). 1931.
13. Patent GB. 1, 588,026 (Aluminium Company of America). 1976.
14. Patent EP. 0 651 777 B1 (Silberline). 1993.
15. Patent US. 3,949,139 (Avery Corporation). 1972.
16. Patent US. 4,321,087 (Revlon, Inc.). 1978.
17. Patent GB. 846,903 (W. Marx; Walter Marx & Co. kG). 1956.
18. Wissling P. In metallic effect pigments. 2nd ed. Hannover: Vincentz Verlag, 2013:77.
19. Wissling P, Kiehl A. In metallic effect pigments, Wissling P, editor. Hannover: Vincentz Verlag, 2006;12.
20. Rodrigues AB. Measurement of metallic and pearlescent colors. Die Farbe. 1990;37:65.
21. Sung LP, Nadal ME, McNight ME, Marx E, Laurenti B. Optical reflectance of metallic coatings: Effect of aluminum flake orientation. J Coat. Technol. 2002;74:55.
22. Cloppenburg H, Schmittmann D. Farbmessungen an Metalliclacken. Farbe + Lack. 1989;95:631.

23. Kiehl A, Kigle-Böckler, Wissling P. In metallic effect pigments. Wissling P, editor. 2nd ed. Hannover:Vincentz Verlag, 2013:21.
24. Wheeler I. Metallic pigments in polymers, Shawbury: Rapra Technology Limited, 1999:20.
25. Patent DE. 195 20 312 (Eckart). 1995.
26. Patent US. 5,037,475 (Showa). 1988.
27. Patent EP. 0 737 212 (Silberline). 1994.
28. Patent DE. 195 01 307 (Eckart). 1995.
29. Patent EP. 0 755 986 (Toyo Aluminium). 1995.
30. Patent EP. 0 810 370 (Toyo Aluminium). 1996.
31. Ostertag W, Mronga N, Hauser P. Eisenoxidbeschichtete Aluminiumpigmente. Farbe + Lack. 1987;93:973.
32. Ostertag W, Mronga N. Properties and applications of luster pigments. Macromol Chem Macromol Symp. 1995;100:163.
33. Patent EP. 1 412 433 (Merck). 2001.
34. Pfaff G, Fritsche K. A more intense colour for all systems. Eur Coat J. 2014;(11)34.
35. Patent WO. 2016/165832 (Schlenk). 2016.
36. Phillips RW, Bleikolm AF. Optical coatings for document security. Appl Opt. 1996;35:5529.
37. Argoitia A, Witzmann M. Society of vacuum coaters. 45th Annual technology conference proceeding. 2002:539.
38. Maile FJ, Pfaff G, Reynders P. Effect pigments – past, present and future. Progr Org Coat. 2005;54:150.
39. Besold B, Reisser W, Roth E. Aluminiumpigmente für wäßrige Beschichtungen – Widerspruch oder Wirklichkeit. Farbe + Lack. 1991;97:311.
40. Müller B, Holland A. Stabilization of aluminium pigments in water-borne systems by paint resins. Surf. Coat Int. 1997;80:321.
41. Graziano M, Kaupp G, Krueger P. In metallic effect pigments, Wissling P, editor. Hannover: Vincentz Verlag, 2006:214.
42. Fetz A, Wissling P. In metallic effect pigments, Wissling P, editor. Hannover: Vincentz Verlag, 2006:208.
43. Hartl-Gunselmann A, Hartmann T. In metallic effect pigments, Wissling P, editor. Hannover: Vincentz Verlag, 2006;223.
44. Lutz D. In metallic effect pigments, Wissling P, editor. Hannover: Vincentz Verlag, 2006;224.

Index

https://doi.org/10.1515/9783110587104-205